Lecture Notes in Mathematics 1819

Springer
Berlin
Heidelberg
New York
Hong Kong
London
Milan
Paris
Tokyo

D. Masser
Yu. V. Nesterenko
H.P. Schlickewei
W. M. Schmidt
M. Waldschmidt

Diophantine Approximation

Lectures given at the
C.I.M.E. Summer School
held in Cetraro, Italy,
June 28 – July 6, 2000

Editors: F. Amoroso
U. Zannier

Fondazione
C.I.M.E.

Springer

Authors and Editors

Francesco Amoroso
Laboratoire de Mathématiques
Nicolas Oresme, CNRS UMR 6139
Université de Caen, BP 5186
14032 Caen, France

e-mail: amoroso@math.unicaen.fr

Umberto Zannier
Instituto Universitario Architettura-D.C.A.
Santa Croce 191
300135 Venezia, Italy

e-mail: zannierqdimi.uniud.it

David Masser
Institute of Mathematics
Basel University
Rheinsprung 21
4051 Basel, Switzerland

e-mail: masser@math.unibas.ch

Yuri V. Nesterenko
Faculty of Mechanics
and Mathematics
Moscow State University
Vorob'evy Gory
119899 Moscow, Russia

e-mail: nest@trans.math.msu.su

Hans Peter Sclickewei
Department of Mathematics
Phillips University of Marburg
Hans-Meerwein-Str., Lahnberge
35032 Marburg, Germany

e-mail: hps@mathematik.uni-marburg.de

Wolfgang Schmidt
Department of Mathematics
University of Colorado
Boulder, CO 80309-0395, USA

e-mail: schmidt@euclid.colorado.edu

Michel Waldschmidt
Institut de Mathématiques
Université Paris VI
175 rue du Chevaleret
75013 Paris, France

e-mail: miw@math.jussieu.fr

Cataloging-in-Publication Data applied for
Bibliographic information published by Die Deutsche Bibliothek
Die Deutsche Bibliothek lists this publication in the Deutsche Nationalbibliografie; detailed bibliographic data is available in the Internet at http://dnb.ddb.de

Mathematics Subject Classification (2000): 11J68, 11J86, 11B37

ISSN 0075-8434
ISBN 3-540-40392-2 Springer-Verlag Berlin Heidelberg New York

Springer-Verlag Berlin Heidelberg New York a member of BertelsmannSpringer
Science + Business Media GmbH

http://www.springer.de

Typesetting: Camera-ready TEX output by the authors

SPIN: 10936825 41/3142/du - 543210 - Printed on acid-free paper

Preface

Diophantine Approximation is a branch of Number Theory having its origins in the problem of producing "best" rational approximations to given real numbers. Since the early work of Lagrange on Pell's equation and the pioneering work of Thue on the rational approximations to algebraic numbers of degree ≥ 3, it has been clear how, in addition to its own specific importance and interest, the theory can have fundamental applications to classical diophantine problems in Number Theory. During the whole 20th century, until very recent times, this fruitful interplay went much further, also involving Transcendental Number Theory and leading to the solution of several central conjectures on diophantine equations and class number, and to other important achievements. These developments naturally raised further intensive research, so at the moment the subject is a most lively one.

This motivated our proposal for a C.I.M.E. session, with the aim to make it available to a public wider than specialists an overview of the subject, with special emphasis on modern advances and techniques. Our project was kindly supported by the C.I.M.E. Committee and met with the interest of a large number of applicants; forty-two participants from several countries, both graduate students and senior mathematicians, intensively followed courses and seminars in a friendly and co-operative atmosphere.

The main part of the session was arranged in four six-hours courses by Professors D. Masser (Basel), H. P. Schlickewei (Marburg), W. M. Schmidt (Boulder) and M. Waldschmidt (Paris VI).

This volume contains expanded notes by the authors of the four courses, together with a paper by Professor Yu. V. Nesterenko (Moscow) – who was unable to accept our invitation to give an expected fifth course – concerning recent work by Matveev.

We shall now briefly illustrate the corresponding contents.

Masser's contribution concerns, roughly speaking, the modern theory of heights, starting with the most basic notions and then turning to the more sophisticated context of algebraic groups. This ample overview describes fundamental results and techniques in the subject, together with applications to transcendence problems. Masser also outlines the transcendence theory of elliptic logarithms and abelian functions (which he originally developed), and its important recent consequences toward outstanding diophantine problems on curves and abelian varieties.

Nesterenko's article is devoted to the proof of the nowadays best known lower bounds in Baker's theory of linear forms in logarithms of algebraic numbers. With the aim of stressing the new ideas introduced by Matveev, the author concentrates on a situation slightly simpler in detail than the most general one, but containing all the important features of the methods.

Schlickewei deals with the celebrated Subspace Theorem. This result, originally discovered by W. M. Schmidt, is a far-reaching extension of Roth's Theorem on the approximations of an algebraic number by rationals, also covered in the lectures. Schlickewei describes the most recent sharpenings (such as the "absolute version"), obtained mainly in joint work by himself and J.-H. Evertse. Finally, he presents here his very recent work on a version of the theorem for approximation by algebraic numbers of bounded degree (obtained jointly with H. Locher).

Schmidt's article concerns the diophantine theory of linear recurrences, whose famous prototype is the Fibonacci sequence. He gives a general survey of the most important problems, methods and results, involving also S-unit equations and intersections of varieties with finitely generated multiplicative groups. In particular, he also illustrates the general strategy underlying his recent solution of an outstanding conjecture in the field; namely, *the zero-multiplicity of a non-degenerate linear recurrence is bounded only in terms of the "length" of the recurrence.*

Waldschmidt's contribution is on transcendence and linear independence over $\overline{\mathbb{Q}}$ of logarithms of algebraic numbers. Starting with Lindemann's classical theorems on the exponential function, he proceeds with the sophisticated results by A. Baker, which yield fundamental applications to effective diophantine analysis. Waldschmidt describes several approaches to the technically complicated proofs, clarifying the main ideas underlying methods which may confound the non-expert. He also details certain modern devices to obtain the best numerical bounds for the involved quantities.

The topics presented in such fine lecture notes incorporate many of the most fundamental methods and applications of Diophantine Approximation, giving an extremely broad viewpoint, precious for both beginners and experts. Also, the style of exposition has little in common with other contributions to the topic and the volume substantially enriches the existing literature.

It is a pleasure for us to thank the authors for their difficult work in coordinating the respective contributions, for their efforts in explaining the subtle points in the simplest and most effective style, and for working out these beautiful papers. We also thank the participants, whose enthusiasm was fundamental for the success of the session.

Finally, the editors express their thanks to Carlo Viola for his valuable advice and help concerning both the organization of the session and the preparation of the present volume.

Francesco Amoroso Umberto Zannier

C.I.M.E.'s activity is supported by:

Ministero dell'Università e della Ricerca Scientifica e Tecnologica, COFIN '99;

Ministero degli Affari Esteri - Direzione Generale per la Promozione e la Cooperazione - Ufficio V;

Consiglio Nazionale delle Ricerche;

E.U. under the Training and Mobility of Researchers Programme;

UNESCO-ROSTE, Venice Office.

Contents

Heights, Transcendence, and Linear Independence on Commutative Group Varieties

David Masser

Mathematisches Institut, Universität Basel

1 First lecture. Introduction and basic techniques

Of course it is impossible for four lecturers to cover the whole of diophantine approximation and transcendence theory in 24 hours. So each one has to restrict himself to special aspects.

These notes expand slightly on my original lectures, and I am grateful to Sinnou David for his comments on an earlier manuscript.

Let us start with perhaps the most basic problems, analogous to the Goldbach conjecture in analytic number theory. In 1744 Euler proved that the number e is irrational, and shortly after in 1761 Lambert did the same for π. We still don't know if $e+\pi$ is irrational, and no-one expects a proof soon.

Much later in 1873 Hermite proved that e is transcendental; that is, the only polynomial $P(X)$ with coefficients in the field $\mathbb{Q}$ of rational numbers satisfying $P(e)=0$ is the zero polynomial. Shortly afterwards in 1882 Lindemann did the same for π. And a general 1934 result of Gelfond and Schneider implies the same for e^π.

It follows in particular that the value $\Gamma(1/2)=\sqrt{\pi}$ of the classical gamma function is also transcendental; for example if we have a non-trivial equation $P(\sqrt{\pi})=0$ then we can write $P(X)=XQ(X^2)-R(X^2)$ and it would follow that $\pi(Q(\pi))^2=(R(\pi))^2$ giving a non-trivial equation for π. More generally, if $\mathbb{C}$ denotes the field of all complex numbers, the subset

$$\overline{\mathbb{Q}}=\{\alpha \text{ in } \mathbb{C}\ ;\ \text{there is } P\neq 0 \text{ in } \mathbb{Q}[X] \text{ with } P(\alpha)=0\}$$

is known to be a field. So $\sqrt{\pi}$ in $\overline{\mathbb{Q}}$ would imply π in $\overline{\mathbb{Q}}$, a contradiction.

We also know that $\Gamma(1/3)$ is transcendental, although this is a relatively recent result of 1976 or so obtained by Chudnovsky. The proof is however different in several respects: for $\Gamma(1/2)$, π and e one uses in an essential way the exponential function e^z, whereas for $\Gamma(1/3)$ one uses the Weierstrass elliptic function $\wp(z)$ satisfying the differential equation $(\wp'(z))^2 = 4(\wp(z))^3 - 4$ (see Lecture 3). To this function is associated the elliptic curve E whose affine part is defined by $y^2 = 4x^3 - 4$; and so we have a commutative group variety or algebraic group. Actually we already had one with e^z; this function parametrizes the multiplicative group $\mathbb{G}_m$ whose complex points $\mathbb{G}_m(\mathbb{C})$ are the non-zero complex numbers $\mathbb{C}^*$.

In fact Chudnovsky uses also the function $\zeta(z)$ satisfying $\zeta'(z) = -\wp(z)$; this corresponds not to E but to a group extension G in the exact sequence

$$0 \to \mathbb{G}_a \to G \to E \to 0$$

with the additive group $\mathbb{G}_a$ (see Lecture 6). In fact we know rather more: the proof delivers the algebraic independence of $\Gamma(1/3)$ and π, which means that the only polynomial P in $\mathbb{Q}[X,Y]$ satisfying $P(\Gamma(1/3),\pi) = 0$ is $P = 0$. Taking P in $\mathbb{Q}[X]$ gives the transcendence of $\Gamma(1/3)$.

The topic of algebraic independence will however not be treated in these lectures. It was recently the subject of an instructional conference in Luminy; see Springer Lecture Notes 1752 "Introduction to algebraic independence theory".

Similarly by considering $y^2 = 4x^3 - 4x$ Chudnovsky proved the transcendence of $\Gamma(1/4)$.

More recently Nesterenko proved the algebraic independence of the three numbers π, e^π, $\Gamma(1/4)$, which was new even if $\Gamma(1/4)$ is omitted. The proof uses modular forms, for which there is no underlying group variety.

Going further, one hopes that in the next ten years the transcendence of $\Gamma(1/5)$, via the algebraic independence of π, $\Gamma(1/5)$ and $\Gamma(2/5)$, will be established using the curve $y^2 = 4x^5 - 4$. Right now we know only the algebraic independence of at least two of these numbers (see for example Chapter 3 of the recent Ph.D. Thesis of P. Grinspan). This curve has genus 2, and so we have to use the apparatus of Jacobians or more generally abelian varieties A or even extensions G satisfying

$$0 \to L \to G \to A \to 0$$

with $L = \mathbb{G}_a^r \times \mathbb{G}_m^s$ a linear group variety. And already such extensions include all commutative group varieties over $\overline{\mathbb{Q}}$.

It would be possible to start the present lecture course with a discussion of such general objects G; and possibly this suits the taste of several people in the audience. But I prefer to start with $\mathbb{G}_m$ and gradually work upwards; thus the present Lecture 1 as well as Lecture 2 will stay on the $\mathbb{G}_m$ level. Then Lectures 3 and 4 will go elliptic, Lecture 5 abelian, and finally Lecture 6 will treat aspects of the general case. At present, as might be expected, we can do a lot more for simpler group varieties, and one would lose many subtleties by going rightaway to the general case.

Back to basics. Once we know that a number like π is transcendental it is natural to ask for more quantitative results; for example how small can $P(\pi)$

be for nonzero $P(X)$ in $\mathbb{Q}[X]$, or nearly equivalently how small can $|\pi - \alpha|$ be for α in $\overline{\mathbb{Q}}$? Thus Mahler in 1953 used Hermite's methods to prove

$$|\pi - r/s| > s^{-42} \tag{1.1}$$

for all rational integers r, s in $\mathbb{Z}$ with $s \geq 2$. This looks like a curiosity, but in fact in the case $G = \mathbb{G}_m^n$ (corresponding to Baker's Theorem – see the lectures of Michel Waldschmidt) these quantitative results are as important, if not more so, as the purely qualitative ones. Thus lower bounds for linear forms in logarithms can be applied to diophantine equations, class number problems, and so forth.

The elliptic analogues for E^n also have applications of a different sort to diophantine equations, and also to isogeny problems for elliptic curves (see Lecture 4). And those for A^n can be applied not only to isogenies but also to answer some interesting polarization questions for abelian varieties (see Lecture 5).

It will be unavoidable to talk about heights. These started life as a mere tool in the proofs, but they have gradually acquired a life of their own and are today the subject of intensive research. The original methods of transcendence theory are employed in order to prove results about heights which can then be applied in other areas of number theory. In particular the theory of lower bounds for heights is very active at the moment, and we will see examples in Lectures 2, 3 and 5.

Again back to basics. Before we begin with group varieties, let us give an example in Mahler's Method, where there is no natural underlying group variety, only a kind of "2-action". This method seems also to have provided some of the inspiration for the precursors of Nesterenko's Theorem mentioned above.

In general the technical nature of most of the material makes it impossible to give complete proofs. But as an exception we will now prove the irrationality of the number

$$\eta = 2/3 + (2/3)^2 + (2/3)^4 + (2/3)^8 + \cdots = \sum_{m=0}^{\infty} (2/3)^{2^m}.$$

Consider the "auxiliary polynomial"

$$P(X, Y) = 2XY^2 + 4XY - 3Y^2 + X - Y. \tag{1.AP}$$

This may well be the first explicit specimen that the reader has ever seen; however it will not be the last. Define the numbers

$$\alpha_n = P((2/3)^{2^{n+1}}, \eta_n) \qquad (n = 0, 1, 2, \ldots)$$

with

$$\eta_n = (2/3)^{2^{n+1}} + (2/3)^{2^{n+2}} + \cdots = \text{"tail" of } \eta.$$

Clearly α_n is rather small; certainly $O((2/3)^{2^{n+1}})$ as $n \to \infty$. But P was chosen to make it even smaller. One finds after a short calculation that

$$\alpha_n = -2(2/3)^{6.2^{n+1}} + \text{ higher powers of } (2/3)^{2^{n+1}}$$

and so

$$|\alpha_n| \leq 3(2/3)^{6.2^{n+1}} < (1/10)^{2^{n+1}} \tag{1.UB}$$

if n is sufficiently large. Thus we gain an extra 6 in the exponent.

Now assume to the contrary that η is rational. It follows that

$$\alpha_n = P\left((2/3)^{2^{n+1}}, \eta - 2/3 - (2/3)^2 - \cdots - (2/3)^{2^n}\right)$$

is rational. We can estimate its denominator: if $\eta = r/s$ (r, s in $\mathbb{Z}$, $s \geq 1$) we find that $s_n = 3^{2^{n+1}}(s.3^{2^n})^2$ is one; that is, $s_n\alpha_n$ is in $\mathbb{Z}$. The "Fundamental Theorem of Transcendence" says that $|N| \geq 1$ if N is in $\mathbb{Z}$ with $N \neq 0$. It follows that

$$|\alpha_n| \geq 1/s_n = s^{-2}(1/9)^{2^{n+1}} \tag{1.LB}$$

provided

$$\alpha_n \neq 0. \tag{1.NV}$$

Now if n is large enough, (1.LB) contradicts (1.UB).

But why is $\alpha_n \neq 0$? This innocent question is here easy to answer, but it will become more and more of a nuisance until it almost takes over the subject. We will see the outcome in Lecture 6.

There are many ways of proving $\alpha_n \neq 0$. The fastest is analytic, and consists of checking that

$$\lim_{n\to\infty} \alpha_n/(2/3)^{6.2^{n+1}} = -2. \tag{1.2}$$

Thus $\alpha_n \neq 0$ for all n large enough. This is strong but the argument doesn't generalize too well, for example to several variables. An algebraic way is to observe that

$$\alpha_n = P(\xi_n, \eta_n), \qquad \xi_n = (2/3)^{2^{n+1}}.$$

Now the curve defined by $P = 0$ has infinitely many points, so there is no immediate contradiction from $\alpha_n = 0$. But

$$\alpha_{n+1} = Q(\xi_n, \eta_n), \qquad Q(X,Y) = P(X^2, Y - X).$$

The equations $P = Q = 0$ define an intersection of two curves and therefore probably a finite number of points. And indeed the resultant of P and Q (with respect to Y) is readily computed as

$$R(X) = 16X^{10} - 48X^9 + 36X^8 - 16X^7$$

(why are there so many zeros at $X = 0$?). Now if n is large enough then

$$\alpha_n \neq 0 \quad \text{or} \quad \alpha_{n+1} \neq 0$$

otherwise $R(\xi_n) = 0$, which is impossible. So (1.NV) holds for infinitely many n, and this weak assertion suffices for the above irrationality proof.

In some ways this proof is typical. The steps can be designated as follows (their order is not too important, and it is sometimes logically better to interchange the last two):

- (AP) - construction of auxiliary polynomial,
- (UB) - obtaining an upper bound,
- (LB) - obtaining a lower bound,
- (NV) - proving the non-vanishing.

One can replace 2/3 in the preceding example by any other rational ζ with $0 < |\zeta| < 1$, although the step (AP) can no longer be done explicitly, making the other steps correspondingly more difficult.

What lies behind (1.AP)? Of course $\eta = f(2/3)$ for the analytic function

$$f(z) = \sum_{m=0}^{\infty} z^{2^m}.$$

Now P is chosen such that the function

$$\varphi(z) = P(z, f(z))$$

has a zero of order 6 at $z = 0$; its Taylor expansion there starts with $-2z^6$. Further $\alpha_n = \varphi\big((2/3)^{2^{n+1}}\big)$ and so we see (1.2) more clearly.

For an explanation with interpolation determinants see Waldschmidt's lectures (section 2.1).

If we replace 2/3 with say 1999/2000 then $P(X,Y)$ has to have degree at most 22796 in each variable, with φ having a zero of order at least 519703208, and its coefficients are rational integers of size probably about $10^{10^{13}}$. So there is no hope of seeing the polynomial explicitly.

In general if P has degree at most L in each variable, then φ can have a zero of order at least $T = (L+1)^2 - 1$; and the proofs work because L^2 grows faster than L. Of course the functional equation $f(z^2) = f(z) - z$ also plays a crucial role.

We should also note the following. Even though we cannot write $P(X,Y)$ down, it must of course be non-zero. The algebraic independence over $\mathbb{Q}$ (and even over $\mathbb{C}$) of the functions z and $f(z)$ is easy to verify, and therefore $\varphi(z) = P(z, f(z))$ is not identically zero. So the above analytic proof of (1.NV) generalizes immediately.

A more serious difficulty is the step from irrationality to transcendence. It was Mahler in 1929 who proved that $f(\zeta)$ is transcendental whenever ζ is algebraic with $0 < |\zeta| < 1$. He used an *ad hoc* version of (LB); in particular it no longer suffices to consider the denominator alone. See the book [56] of Nishioka for a complete proof (pp. 1-5), as well as an excellent general account.

A substitute for the denominator can easily be found. Any α in $\overline{\mathbb{Q}}$ satisfies an equation $P(\alpha) = 0$ with $P(X)$ in $\mathbb{Q}[X]$, and one can assume that

(i) $P(X)$ is irreducible over $\mathbb{Q}$,
(ii) $P(X) = a_0X^d + \cdots + a_d$ with coprime $a_0, \ldots, a_d$ in $\mathbb{Z}$ and $a_0 > 0$.

Then $P(X)$ is unique, and its degree d is the degree $[\mathbb{Q}(\alpha) : \mathbb{Q}]$ of α; and if $d = 1$ this a_0 is the denominator of α. Unfortunately any inequality $|\alpha| \geq 1/F(a_0)$ is no longer true if $d > 1$, so we need more than just a_0. Formerly one took $\max\{|a_0|, \ldots, |a_d|\}$, but one gets much better functorial properties first by factorizing $P(X)$ over $\mathbb{C}$ (or $\overline{\mathbb{Q}}$) as

$$P(X) = a_0 \prod_{i=1}^{d} (X - \alpha_i)$$

for the conjugates $\alpha_1, \ldots, \alpha_d$; and then by defining

$$H(\alpha) = \left(a_0 \prod_{i=1}^{d} \max\{1, |\alpha_i|\} \right)^{1/d}. \tag{1.3}$$

The exponent $1/d$ is needed for properties such as

$$H(\alpha^m) = (H(\alpha))^{|m|}$$

for all m in $\mathbb{Z}$, which is not quite trivial to prove even for $m = 2$ (see Lecture 2).

Examples are $H(1999/2000) = 2000$, $H(26/65) = 65$ and more generally

$$H(r/s) = \max\{|r|, s\}$$

for coprime r, s in $\mathbb{Z}$ with $s \geq 1$. Or $H(\sqrt{2}) = \sqrt{2}$ and more generally $H(2^{1/d}) = 2^{1/d}$ for every $d \geq 1$ in $\mathbb{Z}$. And $H(1 - \sqrt{2}) = \sqrt{1 + \sqrt{2}}$, $H(\sqrt{-6}) = \sqrt{6}$, $H(1 + \sqrt{-6}) = \sqrt{7}$, $H(1 + \sqrt[3]{2}) = \sqrt[3]{3}$. Or $H(e^{\pi i/5}) = 1$ and more generally $H(\tau) = 1$ for every root of unity τ. And finally

$$H(1 - e^{\pi i/5}) = \sqrt{\frac{1}{2}(1 + \sqrt{5})} = H\left(\frac{1}{2}(1 + \sqrt{5})\right)$$

and, selected at random,

$$H(1/(\rho^2 - 5\rho + 50)) = \sqrt[3]{147200}$$

for $\rho^3 - 7\rho + 10 = 0$.

Now (LB) takes the form

$$|\alpha| \geq (H(\alpha))^{-d} \tag{1.4}$$

for any $\alpha \neq 0$ in $\overline{\mathbb{Q}}$. For the proof one notes that $H(\alpha^{-1}) = H(\alpha)$ is easy, and then

$$(H(\alpha^{-1}))^d \geq \max\{1, |\alpha^{-1}|\} \geq |\alpha|^{-1}.$$

As soon as one is accustomed to this height function, one may go ahead and prove the general result on $f(\zeta)$ referred to above. For example, to show that $f(2/3)$ is not an algebraic number of degree at most 1999 one uses an auxiliary polynomial $P(X,Y)$ of degree at most $L = 8119$ in each variable, with the parameter n tending to infinity. But to establish the transcendence of $f(2/3)$ this degree L must also be allowed to go to infinity, independently of n.

The algebraic independence over $\mathbb{Q}$ of z and $f(z)$ is now necessary not only for the proof, but also for the truth of Mahler's result. A vivid illustration of this was accidentally provided by Mahler himself. In [43] he used the functions z and

$$g(z) = \sum_{m=0}^{\infty} z^{2^m}/(1 - z^{2^{m+1}})$$

apparently to prove the transcendence of $\eta = g(\frac{1}{2}(1 - \sqrt{5}))$. In terms of the Fibonacci numbers $f_0 = 1$, $f_1 = 1$, $f_2 = 2$, ... of Wolfgang Schmidt's lectures (section 1), we find that

$$\sum_{n=0}^{\infty} 1/f_{2^n} = 2 - \eta$$

and is therefore also transcendental. After publication it was however pointed out that the sum is $\frac{1}{2}(7 - \sqrt{5})$! The explanation is that z and $g(z)$ are not algebraically independent and indeed $g(z) = z/(1 - z)$.

A similar phenomenon occurred in my thesis [44] (Chapter 3) when I was attempting to prove the linear independence of five numbers connected with an elliptic function in the case of complex multiplication. I failed; apart from incompetence the only imaginable explanation was the algebraic dependence of certain functions, and it turned out that this dependence did lead back to an unexpected (at least by me) linear relation between three of the five numbers.

Very recently Corvaja and Zannier have given another approach to the transcendence of numbers like $f(\zeta) = \sum_{m=0}^{\infty} \zeta^{2^m}$ which does not use functional equations. It is based on the Subspace Theorem with several valuations, and in view of the accompanying lectures of Hans Peter Schlickewei it seems appropriate to sketch the ideas, for simplicity in the context of irrationality.

We then require the p-adic valuations $|\ \ |_p$ on $\mathbb{Q}$ defined for each positive prime p by $|p|_p = 1/p$ and $|q|_p = 1$ for every integer q not divisible by p, together with multiplicativity $|xy|_p = |x|_p|y|_p$ and $|0|_p = 0$. We already have $|x|_\infty = |x|$ the standard archimedean valuation.

For example to prove the irrationality of $\eta = f(2/3)$ above, one could note that $\eta = r/s$ would differ from the rational number $\eta - \eta_n$ by the small quantity η_n. This by itself does not lead to a contradiction, even if we take into account the special denominator of $\eta - \eta_n$ by working also 3-adically. Instead one subtracts off an extra $(2/3)^{2^{n+1}}$ from η_n to get $\eta_{n+1} = O((2/3)^{2^{n+2}})$ which is even smaller, and one works 2-adically as well. In terms of the linear forms

$$\begin{array}{ll} M_\infty(X,Y)=X+Y, & N_\infty(X,Y)=Y, \\ M_2(X,Y)=X+Y, & N_2(X,Y)=Y, \\ M_3(X,Y)=X, & N_3(X,Y)=Y, \end{array}$$

evaluated at $(x_n, y_n) = (3^{2^n} r_n, -2^{2^{n+1}} s)$ with $r_n = 3^{2^n} s\,\eta_n$ in $\mathbb{Z}$ one finds

$$\begin{array}{ll} |M_\infty(x_n,y_n)|_\infty = 3^{2^{n+1}} s\,\eta_{n+1}, & |N_\infty(x_n,y_n)|_\infty = 2^{2^{n+1}} s, \\ |M_2(x_n,y_n)|_2 \le 1, & |N_2(x_n,y_n)|_2 \le 2^{-2^{n+1}}, \\ |M_3(x_n,y_n)|_3 \le 3^{-2^n}, & |N_3(x_n,y_n)|_3 \le 1, \end{array}$$

and that as $n \to \infty$ the product is $O(\theta^{2^n})$ with $\theta = 16/27$. Now the rational Subspace Theorem with $S = \{\infty, 2, 3\}$ (see Schlickewei's lectures, section 1 or his original papers [65], [66] especially Theorem 4.1 p. 395, [67] or also [70] Theorem 1D p. 177) easily supplies the required contradiction; all we need is $\theta < 1$ (and earlier results like Ridout's would have sufficed).

If we again replace 2/3 by $\zeta = 1999/2000$ then it is now the S-units $\zeta^{2^{n+1}}, \ldots, \zeta^{2^{n+13}}$ that should be subtracted off from the "almost S-unit" $f(\zeta) - \zeta - \cdots - \zeta^{2^n}$ with $S = \{\infty, 2, 5, 1999\}$, and the Subspace Theorem in 14 variables can be used. However an extra argument is needed to eliminate the exceptional subspaces of dimension 13.

See [18] for several other applications of these ideas.

One down, five to go...

2 Second lecture. More on heights

In the last lecture we saw how to define a height function H from the set $\overline{\mathbb{Q}}$ of all algebraic numbers to the real interval $[1, \infty)$, principally as a measuring device. But it has remarkable functorial properties, making it useful for a variety of problems, and many of these problems lead to the same fundamental question: how small can its values be?

We already noted in Lecture 1 that $H(\tau) = 1$ for all roots of unity τ. Kronecker's Theorem of 1857 says that $H(\alpha) > 1$ for all other $\alpha \neq 0$. But $H(2^{1/d}) = 2^{1/d}$ for all positive integers d, so $H(\alpha)$ can get arbitrarily close to 1. On the other hand $(H(2^{1/d}))^d = 2$ is bounded away from 1, and in 1933 Lehmer [42] asked if $(H(\alpha))^d$ is generally bounded away from 1 for all algebraic numbers $\alpha \neq 0$ of degree d that are not roots of unity. In fact Lehmer restricted himself to algebraic integers, because otherwise $a_0 \geq 2$ in (1.3) of Lecture 1 and so already $(H(\alpha))^d \geq 2$. The answer is still unknown, and Lehmer himself found the smallest value so far, which is $(H(\alpha_{10}))^{10} = 1.176\ldots$, with $P(\alpha_{10}) = 0$ for

$$P(X) = X^{10} + X^9 - X^7 - X^6 - X^5 - X^4 - X^3 + X + 1.$$

If indeed $(H(\alpha))^d \geq 1.176\ldots$ then it would follow that

$$h(\alpha) = \log H(\alpha) \geq cd^{-1}$$

for all algebraic α of degree d which are not roots of unity, with $c = .162\ldots$. Up to now this is not proved for any $c > 0$, and the best approximation is the 1976 result [27] of Dobrowolski, which we state in slightly weaker form as

$$h(\alpha) \geq cd^{-1}\left((\log d)/(\log\log d)\right)^{-3} \tag{2.1}$$

for some absolute $c > 0$ and every $d \geq 3$. The missing cases $d = 1, 2$ (and indeed any fixed value of d) can be handled separately, and we find without difficulty that in these cases

$$h(\alpha) \geq h(\alpha_2) = \frac{1}{2}\log\left(\frac{1+\sqrt{5}}{2}\right) = .240\ldots$$

with $\alpha_2^2 + \alpha_2 = 1$.

Later on we will sketch the proof of (2.1) and also mention some more recent developments. As a warm-up let us consider not α by itself but also $1-\alpha$; this line of thought was started by Szpiro, in order to make a connexion between some older works on "generalized cyclotomic polynomials" and some ideas of Bogomolov about the discreteness of the Néron-Tate metric on curves in their Jacobians. In 1992 Zhang [98] proved some general results on curves of which

$$h(\alpha) + h(1-\alpha) \geq c > 0 \tag{2.2}$$

is a very special case. So this lower bound is actually independent of d; and moreover we don't have to exclude roots of unity, only $\alpha = 0, 1$ and $e^{\pm\pi i/3}$. In 1993 Zagier [97] gave a more direct proof of some of these results and determined the sharp lower bound in (2.2); coincidentally it is

$$\frac{1}{2}\log\left(\frac{1+\sqrt{5}}{2}\right)$$

again, but this time with α as any primitive tenth root of unity. Zagier's proof was "archimedean" in nature. As our warm-up for Dobrowolski (even though it came nearly twenty years later), we will use the 1995 non-archimedean method of Bombieri and Zannier [13] to establish (2.2) with

$$c = (1/5)\log(5/3) = .102\ldots$$

Zagier worked with the original definition of $H(\alpha)$ in Lecture 1, which involves only the standard valuations on $\mathbb{C}$ and $\mathbb{Q}$. We need an equivalent definition in terms of the p-adic valuations on $\mathbb{Q}$. As usual we throw all these together by writing v for a typical element of the set $\{\infty, 2, 3, 5, \ldots\}$.

Now let K be any finite extension of $\mathbb{Q}$ containing α; for example the smallest one is $\mathbb{Q}(\alpha)$ itself. It is known that each $|\ |_v$ extends to K, but often in more than one way. If there are $J(v) = J_K(v)$ such extensions, we could denote them somewhat inelegantly by $|\ |_{v,j}$ $(1 \leq j \leq J(v))$. See [17], especially Chapter 9, for an excellent and leisurely account. One then defines temporarily

$$H_K(\alpha) = \prod_v \prod_j (\max\{1, |\alpha|_{v,j}\})^{D(v,j)}$$

where the $D(v,j)$ are certain exponents relating to topological completions; in fact

$$D(v,j) = [K_{v,j} : \mathbb{Q}_v]$$

where $\mathbb{Q}_v$ is the completion of $\mathbb{Q}$ with respect to $|\ |_v$ and $K_{v,j}$ that of K with respect to $|\ |_{v,j}$. All we need to know here is that

$$\sum_j D(v,j) = [K : \mathbb{Q}] = D \quad \text{(say)}. \tag{2.3}$$

It is then a fact, not so easy to prove from scratch, that

$$H_K(\alpha) = (H(\alpha))^D$$

(so that $H(\alpha^m) = (H(\alpha))^{|m|}$ is now clear for any $m \geq 0$ in $\mathbb{Z}$; but not for $m = -1$!). See for example Lang's book on diophantine geometry [38] (p. 54), which remains one of the best expositions of the general theory of heights, or also Waldschmidt's recent book [90] (p. 79).

For example if $v = \infty$ then $J(\infty) = J_1 + J_2$, where J_1 is the number of different field embeddings σ_j of K in $\mathbb{R}$ and J_2 is the number of different field embeddings σ_j of K in $\mathbb{C}$ distinct from these and distinct modulo complex conjugation. Each σ_j defines $|\alpha|_{\infty,j} = |\sigma_j \alpha|$ and $D(\infty, j) = 1$ or 2 accordingly. So (2.3) says $J_1 + 2J_2 = D$. Thus in $H_K(\alpha)$ we see all $\sigma_j \alpha$ with appropriate multiplicities, and if $K = \mathbb{Q}(\alpha)$ these are nothing more than the α_i in the original definition of $(H(\alpha))^d$. Similar things hold for each $v \neq \infty$.

Our formula $H(1/\alpha) = H(\alpha)$ is oddly enough no longer obvious except with the aid of the so-called Product Formula

$$\prod_v \prod_j |\alpha|_{v,j}^{D(v,j)} = 1 \tag{2.4}$$

for any $\alpha \neq 0$ in K. See for example [17] p. 190 or [38] pp. 18-21. In fact we could have used this to prove the lower bound (1.4) in Lecture 1 (where however α in $\overline{\mathbb{Q}}$ was already in $\mathbb{C}$); it implies that no single $|\alpha|_{v,j}$, or even a product of such terms, can be too small as a function of the height. Or we can apply it directly and so avoid estimating the height.

Now let us prove (2.2); this will be our second (and last) complete proof. Choose any number field K containing α, and consider the number

$$\beta = \alpha^5 + (1-\alpha)^5 - 1; \tag{2.AP}$$

this formula is so labelled because it involves the value of another "auxiliary polynomial" $P(X)$ at $X = \alpha$; this is then our second explicit specimen.

If $\beta = 0$ then α is determined, and in fact it must be one of the forbidden values 0, 1, $e^{\pm \pi i/3}$. So we can assume

$$\beta \neq 0 \tag{2.NV}$$

another statement about non-vanishing.

Now for upper bounds, using all valuations $|\ \ |_{v,j} = |\ \ |$ for brevity.

Suppose first $v \neq \infty$ and $v \neq 5$. Then

$$|\beta| \leq (\max\{1, |\alpha|\})^5 (\max\{1, |1-\alpha|\})^5 \tag{2.UB}$$

because of the ultrametric inequality $|x+y| \leq \max\{|x|, |y|\}$. But if $v = \infty$ we get (2.UB) with an extra 3 because of $|x+y+z| \leq 3 \cdot \max\{|x|, |y|, |z|\}$.

If $v = 5$ we note that

$$\beta = -5\alpha + 10\alpha^2 - 10\alpha^3 + 5\alpha^4$$

with all coefficients divisible by 5. So we can now obtain (2.UB) with an extra $|5|_{v,j} = 1/5$. This can be referred to as the "Frobenius idea".

Multiplying these upper bounds together, and using (2.3), we find that

$$1 = \prod_v \prod_j |\beta|_{v,j}^{D(v,j)} \leq (3/5)^D \left(H_K(\alpha)\right)^5 \left(H_K(1-\alpha)\right)^5, \tag{2.LB}$$

where we have used the Product Formula (2.4) as a substitute for the lower bound. Taking logarithms and dividing by D, we arrive at

$$h(\alpha) + h(1-\alpha) \geq (1/5)\log(5/3)$$

exactly as required.

Zhang actually proved a more general result about the group variety $\mathbb{G}_m^n$ with $\mathbb{G}_m^n(\mathbb{C}) = (\mathbb{C}^*)^n$. We can define an *ad hoc* height on $\mathbb{G}_m^n(\overline{\mathbb{Q}}) = \left(\overline{\mathbb{Q}}^*\right)^n$ by the formula

$$h(\pi) = h(\alpha_1) + \cdots + h(\alpha_n)$$

for $\pi = (\alpha_1, \ldots, \alpha_n)$ in $\left(\overline{\mathbb{Q}}^*\right)^n$. Zhang's original results concerned points π restricted to a curve C over $\overline{\mathbb{Q}}$, and they took the form

$$h(\pi) \geq c(C) > 0 \tag{2.5}$$

for all π in $C(\overline{\mathbb{Q}})$; but there are some exceptional situations.

Of course we should exclude torsion points π, because all their coordinates are roots of unity. A theorem of Liardet (see for example [38] Theorem 6.4 p. 203) implies that there are at most finitely many of these unless some component C^0 of C is parametrized by equations

$$x_1 = \tau_1 x^{a_1}, \ldots, x_n = \tau_n x^{a_n}$$

for roots of unity $\tau_1, \ldots, \tau_n$ and rational integers $a_1, \ldots, a_n$ not all zero. In that case taking x itself as a root of unity gives $h(\pi) = 0$ for infinitely many π in $C(\overline{\mathbb{Q}})$. Or taking $x = 2^{1/d}$ for $d \to \infty$ gives $h(\pi) > 0$ as near to 0 as we like. So we have to exclude such torsion points and such components. But C^0 is not only a subvariety but also a translate by the torsion point $\tau = (\tau_1, \ldots, \tau_n)$ of the connected group variety or algebraic subgroup H parametrized by the equations

$$x_1 = x^{a_1}, \ldots, x_n = x^{a_n}.$$

For example if C is absolutely irreducible then (2.5) holds provided $C \neq \tau H$ and π itself is not torsion. This is clearly true for C defined by $x_1 + x_2 = 1$ and so we recover the above result (2.2).

In 1995 Zhang [99] extended his results to varieties V in $\mathbb{G}_m^n$ of arbitrary dimension, also defined over $\overline{\mathbb{Q}}$. It suffices to remove from V all torsion translates $\tau H \subseteq V$, where now H is an arbitrary connected group subvariety. These latter are defined by a finite collection of monomial equations of the form

$$x_1^{b_1} \cdots x_n^{b_n} = 1 \tag{2.6}$$

for $b_1, \ldots, b_n$ in $\mathbb{Z}$. The notation V^* is often used for what remains of V after these removals. As in Liardet's Theorem, it was proved by Laurent [41] (Lemme 4 p. 308 – see also section 8 of Schmidt's lectures), or see [13] for a different proof, that at most finitely many removals suffice, so that V^* is a quasi-projective variety. Zhang showed that

$$h(\pi) \geq c(V) > 0 \tag{2.7}$$

for all π in $V^*(\overline{\mathbb{Q}})$. The paper [13] of Bombieri and Zannier contains also for this a different proof; the Frobenius idea amounts to selecting a polynomial P vanishing on V and applying the Product Formula to $\beta = P(\alpha_1^p, \ldots, \alpha_n^p)$ for $p = p(V)$ big enough. Now the possibility $\beta = 0$ provides another polynomial vanishing at π, and one can hope to use induction on the dimension of V.

For recent work on the finer structure of the constant $c(V)$ in (2.7) we refer the reader to the papers of Amoroso-David [3], Bombieri-Zannier [13], David-Philippon [25], Schlickewei-Wirsing [68], and Schmidt [71], [72].

Actually (2.7) itself is also a consequence of the 1997 Galois Equidistribution Theorem of Bilu [10], to the effect that if $h(\pi)$ is "small" then "most" of the conjugates of π are "near" the unit polycircle in $\mathbb{C}^n$, and are "uniformly distributed" around it.

So much at present for Zhang's (2.2); adding the term $h(1-\alpha)$ changes significantly the nature of the lower bounds. What about the result (2.1) of Dobrowolski?

We can try the same Frobenius idea, this time taking some polynomial $P(X)$ in $\mathbb{Z}[X]$ with $P(\alpha) = 0$ and noting that $(P(X))^p = P(X^p) + pQ(X)$ over $\mathbb{Z}[X]$. Thus the Product Formula on $\beta = P(\alpha^p)$ yields, as in (2.LB) above, the inequality

$$1 \leq (\Sigma/p)^D (H_K(\alpha))^{pL}$$

for any number field K of degree D containing α, if say P has degree at most L, and its length Σ is defined in the usual way as the sum of the absolute values of the coefficients. This leads to the lower bound

$$h(\alpha) \geq (pL)^{-1} \log(p/\Sigma).$$

To render the logarithmic term harmless we could assume from Bertrand's Postulate that $2\Sigma \leq p < 4\Sigma$, giving

$$h(\alpha) \geq (\log 2)/(4L\Sigma). \tag{2.8}$$

It is now a question of minimizing $L\Sigma$, essentially the degree of P multiplied by its length.

Certainly the natural choice of P as the minimal polynomial of α minimizes L as d. It looks easy to estimate Σ in terms of the zeros $\alpha_1, \ldots, \alpha_d$ of P, but the sheer numbers of terms pose problems; these numbers are binomial coefficients that are exponential in d, and this would lead to a similar bound in (2.8). If $dh(\alpha)$ is sufficiently small (which we could assume if necessary) then some results of Mignotte would enable us to improve the length estimates to something essentially exponential in $\sqrt{d}$, but this is not much better.

So we have to construct a less natural P, rather as in Lecture 1. To estimate its coefficients we need an effective version of (AP), which comes from the so-called Siegel Lemma. We discuss this further at the end of the present lecture. It gives for example the following. Assuming $(H(\alpha))^d < 2$ as we may, there is $P \neq 0$ in $\mathbb{Z}[X]$ with $L = 2d-1$ and length $\Sigma < 12d^{3/2}$. So just by allowing the degree to double, we get the length to fall dramatically. Now (2.8) leads to a lower bound for $h(\alpha)$ of order $d^{-5/2}$, which is at least no longer exponential.

Making L a big multiple of d enables us to reduce the power $5/2$ to $2 + \varepsilon$ for any fixed $\varepsilon > 0$. In fact one obtains a lower bound of order $1/(d^2 \log d)$ as Stewart did in [83] (the first application of transcendence techniques to such problems). But to get down to the desired $1 + \varepsilon$ in (2.1) we have to use "multiplicities" as in Lecture 1. If P has a zero of order at least T at α, we find that $\beta = P(\alpha^p)$ now involves a factor p^T. Localizing p as above leads to $h(\alpha) \geq (\log 2)/(2L(2\Sigma)^{1/T})$ so we gain in the exponent by $1/T$. When Σ is estimated by Siegel's Lemma, and the parameters L and T optimized, we arrive at a lower bound of the order $1/(d \log d)$.

But this is too good! And in fact we completely overlooked the difficulty (NV) of proving $\beta \neq 0$; now definitely non-trivial. If P had been the minimal polynomial then $P(\alpha^p) = 0$ implies that α^p is a conjugate of α, so $h(\alpha) = h(\alpha^p) = ph(\alpha)$ or $h(\alpha) = 0$ which was excluded right at the beginning. But we do not know our actual P so well. Dobrowolski's idea was to use not just one prime p but all in the interval provided by Bertrand's Postulate, counted by the Prime Number Theorem. One can prove that α^p has degree d for most of these, and the above trick with heights shows that the α^p are not conjugate. So the condition $P(\alpha^p) = 0$ means that P is divisible by many different polynomials of degree d. If the parameters are adjusted to rule this out, one ends up with the required result (2.1); the optimal value of T turns out to be of order $(\log d)/(\log\log d)$ and then L has order dT^2. For a complete proof see the original article [27], or, in terms of interpolation determinants, Schinzel's book [64] (section 4.2); and for both together with a comparison see Waldschmidt's book [90] (pp. 89-102).

Incidentally taking the much smaller degree $L = (1+\varepsilon)d$ and applying an analytic estimate of Erdős-Turán (Theorem 1 p. 100 of [28]) to P leads to Bilu's equidistribution result (at least with respect to angles in dimension 1), as Mignotte [50] (pp. 83, 84) observed. Again the minimal polynomial is of no use because of its dangerously large length.

More recently there have been several developments of this method. In 1999 Amoroso and David ([2]) extended Dobrowolski's Theorem to $\mathbb{G}_m^n$ in the following sense. There is a lower bound

$$h(\pi) \geq c(n)\delta^{-1}(\log 2\delta)^{-\kappa}$$

with $c(n) > 0$ and $\kappa = \kappa(n) = (n+1)((n+1)!)^n - n$, provided one removes all proper torsion translates $\tau H \neq \mathbb{G}_m^n$ from $\mathbb{G}_m^n$ (in other words, the coordinates $\alpha_1, \ldots, \alpha_n$ of π are multiplicatively independent). But here δ is not the natural degree $d = [\mathbb{Q}(\alpha_1, \ldots, \alpha_n) : \mathbb{Q}]$, otherwise the result would not be new. Rather $\delta = \delta(\pi)$ is the smallest degree of any $P \neq 0$ in $\mathbb{Z}[X_1, \ldots, X_n]$ vanishing at π. If $n = 1$ it is d after all; but if $n > 1$ it is usually smaller, for example $\delta(\alpha, 1-\alpha) = 1$, and in fact the inequality $\delta \leq nd^{1/n}$ holds. A valuable corollary in terms of d itself, not quite immediate, is that the product satisfies

$$h(\alpha_1) \ldots h(\alpha_n) \geq c'(n)d^{-1}(\log 2d)^{-\kappa'}$$

for $c'(n) > 0$ and $\kappa' = \kappa'(n)$, again if $\alpha_1, \ldots, \alpha_n$ are independent.

The proof follows the general lines laid down by Dobrowolski, mixed with elements of Baker's Method (see Lecture 4), but now (NV) causes much trouble. Even for $n = 2$ it is difficult to prove $P(\alpha_1^p, \alpha_2^p) \neq 0$, and it is necessary to introduce a second prime q and look at the quantities $P(\alpha_1^{pq}, \alpha_2^{pq})$. These are values of new polynomials $P^{[q]}(X_1, X_2) = P(X_1^q, X_2^q)$ at the points (α_1^p, α_2^p). In this way a whole batch of polynomials is produced, rather like $P(X,Y)$ and $Q(X,Y)$ as in Lecture 1. Now the geometric machinery of zero-estimates can be applied, to which topic we will return in Lecture 6.

Another generalization of Dobrowolski's result has been obtained by Amoroso and Zannier [5], and concerns the possibility of a large abelian extension k of $\mathbb{Q}$ contained in $\mathbb{Q}(\alpha)$. Then $d = [\mathbb{Q}(\alpha) : \mathbb{Q}]$ in (2.1) can be replaced by the relative degree $[\mathbb{Q}(\alpha) : k]$, at the expense of slightly increasing the power of the logarithm. As Amoroso remarked at Oberwolfach in April 2000, this result implies that if $n = 2$ and the curve C is defined over $\mathbb{Q}$ the "Zhang constant" $c(C)$ in (2.5) can be taken as $c(\varepsilon)(\deg C)^{-1-\varepsilon}$ for any $\varepsilon > 0$, where $\deg C$ is the degree of the closure of C in some projective embedding, say in $\mathbb{P}_2$ or better $(\mathbb{P}_1)^2$. In particular it does not depend on the coefficients defining C, as had been already observed by David and Philippon [25] (Théorème 1.5 p. 492) for arbitrary varieties V over $\mathbb{Q}$.

Also the special case $k = \mathbb{Q}(\alpha)$ of the Amoroso-Zannier Theorem says that the height $h(\alpha)$ of any $\alpha \neq 0$ in a field generated by roots of unity (which is as usual not itself a root of unity) is bounded below by a positive absolute constant c. This result, with $c = (\log 5)/12$, had already been proved in 2000 by Amoroso and Dvornicich [4]; and they also found an example with height $(\log 7)/12$.

We finish with a discussion of Siegel's Lemma. Suppose we have a system $\underline{\underline{A}}\,\underline{X} = 0$ with $\underline{X}$ the transpose $(x_1, \ldots, x_N)^t$ and a $M \times N$ matrix $\underline{\underline{A}}$ with entries in a number field K of degree D. In our application above we want to solve non-trivially for x_j in $\mathbb{Z}$, and this will certainly be possible if $N > DM$. Assume that $\underline{\underline{A}}$ has rank M. Then there is a solution with

$$0 < \max_j |x_j| \leq \left(\mathcal{H}(\underline{\underline{A}})\right)^{DM/(N-DM)},$$

where $\mathcal{H}(\underline{\underline{A}})$ is a somewhat sophisticated "Grassmann height" essentially due to Schmidt [69] in 1967 and Bombieri-Vaaler [12] in 1983. Theirs is actually the M-th power of ours.

For example we needed above $P(\alpha) = 0$ or

$$\alpha^L x_1 + \cdots + \alpha x_L + x_{L+1} = 0,$$

so $M = 1$, $N = L + 1$ and $D = d$. It turns out that

$$\mathcal{H}(\alpha^L, \ldots, \alpha, 1) \leq \sqrt{N}\, H(\alpha^L)$$

due to a difference in the norms. So choosing $L = 2d - 1$ gives

$$\max_j |x_j| \leq \sqrt{N}\, H(\alpha^L) = \sqrt{2d}\,(H(\alpha))^{2d-1} < 4\sqrt{2d}$$

since we assumed $(H(\alpha))^d < 2$. Thus the length $\Sigma < 4N\sqrt{2d} < 12d^{3/2}$ as claimed. And making the quotient L/d larger decreases the estimate, also in line with what we said.

How is the Siegel Lemma proved?

For $K = \mathbb{Q}$ the system defines a subspace Z in $\mathbb{R}^N$ of dimension $N - M$, and we need a small non-zero point of $\mathbb{Z}^N \cap Z$. This intersection is a lattice in Z, and its determinant is none other than $\left(\mathcal{H}(\underline{\underline{A}})\right)^M$. So Minkowski's First Theorem delivers a point with size of order $\left(\mathcal{H}(\underline{\underline{A}})\right)^{M/(N-M)}$.

For example, with the system

$$ax + by + cz = a'x + b'y + c'z = 0$$

with coefficients in $\mathbb{Z}$ one finds that $\left(\mathcal{H}(\underline{\underline{A}})\right)^2$ is the square root of the sum of the squares of the minors $bc' - b'c$, $ca' - c'a$, $ab' - a'b$ divided by their highest common factor. Apart from choice of norm, this is essentially

$$\prod_v \max\{|bc' - b'c|_v,\ |ca' - c'a|_v,\ |ab' - a'b|_v\}$$

taken over all v in our set $\{\infty, 2, 3, 5, \ldots\}$ above; and so we get a good hint on how to generalize to an arbitrary number field K.

For $K = K$ the system $\underline{\underline{A}}\,\underline{X} = 0$ is equivalent to some rational system $\underline{\underline{B}}\,\underline{X} = 0$ which can be solved above in terms of $\mathcal{H}(\underline{\underline{B}})$. It is also equivalent to a highly non-rational system $\underline{\underline{C}}\,\underline{X} = 0$ over $\overline{\mathbb{Q}}$ whose matrix $\underline{\underline{C}}$ is formed out of blocks $\sigma_1\underline{\underline{A}}, \ldots, \sigma_D\underline{\underline{A}}$ with the complex embeddings $\sigma_1, \ldots, \sigma_D$ of K. Now

$$\mathcal{H}(\underline{\underline{B}}) = \mathcal{H}(\underline{\underline{C}}) \leq \mathcal{H}(\underline{\underline{A}})$$

by nice functorial properties of the height. For further details see Theorem 12 with $k = \mathbb{Q}$ (p. 29) of the article [12] of Bombieri and Vaaler, or Lemma 9A (p. 33) of Schmidt's book [70].

It is sometimes convenient, as in Lecture 3, to solve the system over K. Now we need only $N > M$, and there is an algebraic integer solution $\underline{X} = (x_1, \ldots, x_N)^t \neq 0$ in K^N with

$$\|\underline{X}\| \leq \left\{(2/\pi)^{J_2}\sqrt{|\Delta_K|}\,\right\}^{1/D} \cdot \left(\mathcal{H}(\underline{\underline{A}})\right)^{M/(N-M)},$$

where J_2 is as above, Δ_K is the discriminant of K and

$$\|\underline{X}\| = \max_k \max\{1, |\sigma_k(x_1)|, \ldots, |\sigma_k(x_N)|\}.$$

Related estimates can be found in [12] (see for example Corollary 11 p. 28), but this particular one comes from a proof scheme implicitly to be found in Schmidt's paper [69] and worked out by Thunder (see also Chapter 1 of the recent Ph.D. Thesis of C. Liebendörfer). One embeds now in $\mathbb{R}^{DN}$ and calculates the lattice determinant.

Use of Minkowski's Second Theorem enables a basis of solutions to be found. And Vaaler's "cube-slicing" leads to the neat constants. But neither of these refinements will be needed in the remaining lectures; furthermore the crude estimate

$$\mathcal{H}(\underline{\underline{A}}) \leq C(M)\sqrt{N} \max_k \max_{ij} |\sigma_k(a_{ij})|$$

with some $C(M) \leq \sqrt{e}$ will usually be enough, which holds provided the entries a_{ij} of $\underline{\underline{A}}$ are algebraic integers.

Now $H(2)$ lectures down; $H(7 + \sqrt{65})$ to go.

3 Third lecture. Elliptic functions and elliptic curves

In Lecture 1 we proved that $f(2/3)$ is irrational with the function $f(z) = \sum_{m=0}^{\infty} z^{2^m}$, and we quoted Mahler's generalization that $f(\zeta)$ is transcendental for all algebraic $\zeta \neq 0$ in the disc of convergence. This function is not uninteresting. For example it has a natural boundary on the unit circle. And its behaviour near this circle may be very accurately determined using another strange property. Thus the recurrence sequence with term 2^m $(m = 0, 1, 2, \dots)$ may easily be extended backwards as in Schmidt's lectures (section 1). The corresponding sum over z^{2^m} is not very meaningful but for real z with $0 < z < 1$ the terms themselves are defined and tend to 1, and in fact the modified sum $g(z) = \sum_{m=-1}^{-\infty}(z^{2^m} - 1)$ converges. Now the function

$$C(z) = g(z) + f(z) + (\log(-\log z))/\log 2$$

has two remarkable properties; first $C(z) = C(z^2)$ and second $C(z)$ is extremely close to, but not quite equal to the constant $-.332747\dots$ for all z (the total variation is less than .000003).

Despite this sort of thing, the function does remain rather special, and there may be an understandable desire to see more natural transcendence results. The pioneering work of Hermite, Lindemann and Gelfond on exponential functions and Schneider on elliptic functions was unified by Schneider himself in 1949, and his result was simplified by Lang in 1962.

This result holds for functions $f_1, \dots, f_N$ meromorphic on $\mathbb{C}$. A technical (but necessary) assumption is that of "finite growth order"; in fact this will be automatic for all our examples, and since we don't need the concept later on, we skip the definition here. More significantly, there are "polynomial differential equations" which assert that all the derivatives $f_i' = df_i/dz$ $(1 \leq i \leq N)$ lie in the ring $K[f_1, \dots, f_N]$ for some number field K. Finally suppose at least two among $f_1, \dots, f_N$ are algebraically independent over K; we have already seen the importance of such conditions. Then the Schneider-Lang Theorem (in slightly simplified form) says that the set

$$\mathcal{S} = \{\zeta \text{ in } \mathbb{C} \; ; \; f_1(\zeta), \dots, f_N(\zeta) \text{ defined and in } K\}$$

is finite.

So nothing is asserted to be transcendental or even outside K; but by using extra structure we can often determine $\mathcal{S}$ exactly and deduce transcendence results. This extra structure almost always comes from an underlying group variety G. Thus to get the transcendence of $e = f_2(1) = 2.71828\dots$ with $f_2(z) = e^z$ we suppose to the contrary that e lies in a number field K. Then with $f_1(z) = z$ we find that the above set $\mathcal{S}$ contains $\mathbb{Z}$ and is therefore infinite. More generally one gets the transcendence of e^ζ for every algebraic $\zeta \neq 0$ (Theorem of Hermite-Lindemann); and now contrapositivity gives the transcendence of each non-zero value of $\log \alpha$ for every algebraic $\alpha \neq 0$. In particular $2\pi i = \log 1$ and so also π and $\sqrt{\pi} = \Gamma(1/2)$ are transcendental.

This application corresponds to the group variety $G = \mathbb{G}_a \times \mathbb{G}_m$, with $G(\mathbb{C}) = \mathbb{C} \times \mathbb{C}^*$, or alternatively to the functional equations

$$f_1(z+w) = f_1(z) + f_1(w), \quad f_2(z+w) = f_2(z)f_2(w),$$

which are usually referred to as addition theorems.

Or using the two exponential functions e^z and $e^{\beta z}$ for algebraic β, with a set S containing $\mathbb{Z} \cdot \log\alpha$ for $\log\alpha$ as above, we conclude the transcendence of $\alpha^\beta = e^{\beta \log \alpha}$; the famous Gelfond-Schneider Theorem of 1934. Only now one needs β to be irrational, not only for the truth of the Theorem but also for the algebraic independence of the functions. And the Theorem is equivalent to the $\overline{\mathbb{Q}}$-linear independence of numbers $\log\alpha_1$, $\log\alpha_2$ in case they are $\mathbb{Q}$-linearly independent; where now any values, including 0, of the logarithms of the non-zero algebraic numbers α_1, α_2 are permissible.

This application corresponds to the group variety $G = \mathbb{G}_m \times \mathbb{G}_m$.

We give a very short sketch of the proof of the Schneider-Lang Theorem. We are trying to prove that the set S is finite, but it is possible to derive a contradiction just from the assumption that S is sufficiently large. Suppose (without loss of generality) it is f_1 and f_2 which are algebraically independent. One constructs as in step (AP) a non-zero P in $K[X_1, X_2]$ such that

$$\varphi(z) = P(f_1(z), f_2(z))$$

has a zero of large order at least T at each ζ in the set S. If L is a bound for the degree of P, then T has to be chosen as a small multiple of L^2, as opposed to the maximal allowable $(L+1)^2 - 1$ used in Lecture 1. Here the Siegel Lemma is needed, and with coefficients and solutions in K, because the differential equations imply that each derivative of each monomial $f_1^{\ell_1} f_2^{\ell_2}$ is in $K[f_1, \ldots, f_N]$ and so its values on S lie in K. A crude version without the Grassmann height suffices.

The step (UB) requires new analytic tools, which are discussed in more detail by Waldschmidt (his lectures, section 1.2). Typically, suppose that φ is analytic on the disc defined by $|z| \leq R$, and we know the upper bound $|\varphi| \leq B$ on this disc. Then we can improve this bound if we also know that φ has a zero of order at least T at say $z = 0$. For then φ/z^T is also analytic on the disc, and the maximum modulus principle gives

$$|\varphi/z^T| \leq \sup_{|z| \leq R} |\varphi/z^T| = \sup_{|z|=R} |\varphi/z^T| \leq B/R^T.$$

Thus

$$|\varphi(z)| \leq (|z|/R)^T \cdot B$$

on the disc, improving the bound B if $|z| < R$ and $T > 0$.

For example, take $\varphi = 2zf^2 + 4zf - 3f^2 + z - f$ for f as in Lecture 1. With $R = 1/2$ we have $|f| \leq f(1/2) < 1$, so $|\varphi| \leq 2+4+3+1+1 = 11 = B$. And $T = 6$, so our numbers α_n there may be estimated by

$$|\alpha_n| = \left|\varphi((2/3)^{2^{n+1}})\right| \leq 11.2^6(2/3)^{6.2^{n+1}} \qquad (n = 0, 1, 2 \ldots) \qquad \text{(3.UB)}$$

which is not quite (1.UB) of Lecture 1 but suffices for the proof there.

Returning to our present φ, we see that high order zeros imply that φ is small on large discs, and then Cauchy's Integral Formula for the derivatives implies that these derivatives are similarly small. And if the set $\mathcal{S}$ of zeros is itself large, then so much the better. We find in fact that the values

$$\alpha_{t\zeta} = \varphi^{(t)}(\zeta) \qquad (0 \leq t < 2T, \ \zeta \text{ in } \mathcal{S})$$

are all small. Note here the increase in the order of differentiation from T to $2T$. Of course we are not allowed to increase the set $\mathcal{S}$ in a similar way, because it is given at the beginning.

Now (LB), in the form of lower bounds (1.4) involving heights as in Lecture 1, or alternatively the Product Formula (2.4) as in Lecture 2, leads to a contradiction provided some $\alpha = \alpha_{t\zeta}$ is non-zero.

Again it is this (NV) problem $\alpha \neq 0$ that presents trouble. It can be solved with resultants (provided we replace $2T$ by a large constant multiple of T), but this is unnecessarily complicated. For if all $\alpha = \alpha_{t\zeta}$ are zero then we have succeeded in obtaining more zeros than we started off with; and we can simply iterate the whole argument to yield zeros of orders $2T$, $4T$, $8T$, ... on the set $\mathcal{S}$. And nothing prevents us from getting infinite order; but even at a single point ζ this means $\varphi = 0$, which contradicts the algebraic independence of f_1 and f_2.

The contradiction completes our sketch of the proof of the Schneider-Lang Theorem. For complete accounts see Schneider's book [74] (Satz 13 p. 55), Lang's book [36] (Theorem 1 p. 21), Waldschmidt's book [86] (Théorème 3.3.1 p. 77), or Baker's book [6] (Theorem 6.1 p. 55).

In giving more examples of this theorem, the main constraints are the differential equations. Actually these often exist, as for example with Bessel functions like

$$J(z) = \sum_{m=0}^{\infty} z^m/(m!)^2$$

and $zJ'' + J' = J$; but despite Example 2 (p. 357) of [91] there is no longer an extra structure with which to calculate the set $\mathcal{S}$ precisely. One can deduce only rather weak results like the irrationality of at least one of $J(\zeta)$ and $J'(\zeta)$ for all but finitely many rational ζ. In fact $J(\zeta)$ is transcendental for all non-zero algebraic ζ, and even $J(\zeta)$ and $J'(\zeta)$ are algebraically independent (Siegel – see for example Chapter 5 of Schneider's book [74]). By very carefully estimating the size of $\mathcal{S}$, Bertrand [7] (p. 192) did succeed in 1977 to prove the transcendence of $J'(\zeta)/J(\zeta)$ for all such algebraic ζ, but this could be regarded as luck.

Another example of a differential equation involves the modular function

$$j(z) = q^{-1}\left(1 + 240\sum_{m=1}^{\infty} m^3 q^m (1-q^m)^{-1}\right)^3 \prod_{m=1}^{\infty}(1-q^m)^{-24}$$

($q = e^{2\pi i z}$) meromorphic on the upper half plane $\mathbb{H}$ of all complex numbers z with positive imaginary part. This satisfies

$$2j^2(j-1728)^2 j' j''' - 3j^2(j-1728)^2 j''^2 + (j^2 - 1968j + 2654208)j'^4 = 0.$$

However the restricted set $\mathbb{H}$ now makes additional difficulties. The elliptic analogue of the Gelfond-Schneider Theorem (see below) implies that if ζ in $\mathbb{H}$ is algebraic (except of degree 2) then $j(\zeta)$ is transcendental (see for example [87] p. 63). But despite Nesterenko's Theorem (see [9] and [55]) it is still unknown if $j(\zeta)$ and $j'(\zeta)$, let alone when taken together with $j''(\zeta)$, are algebraically independent.

The exceptions above by the way are genuine; thus if ζ in $\mathbb{H}$ has degree 2 then $j(\zeta)$ is not only algebraic but also integral over $\mathbb{Z}$, like its value

$$-7805727756261891959906304000 - 999421027517377348595712000\sqrt{61}$$

(calculated in the 1995 M. Sc. Thesis of A. Kessler) at $\zeta = \frac{1}{2}(-1+\sqrt{-427})$. Behind such things lies a whole family of functional equations (so many, in fact, that it is suprising how j manages to exist at all), of which the simplest are $j(z+1) = j(z)$, $j(-1/z) = j(z)$ and the third successive minimum is $F(j(z), j(2z)) = 0$ with

$$\begin{aligned} F(X,Y) = X^3 + Y^3 - X^2Y^2 + 1488XY(X+Y) - 162000(X^2+Y^2) \\ + 40773375XY + 8748000000(X+Y) - 157464000000000. \end{aligned}$$

Nevertheless as for $J(z)$ there is no addition theorem, and indeed in both examples above there are no underlying group varieties.

With elliptic functions there are – elliptic curves. Analytically one takes a lattice Ω in $\mathbb{C}$ and defines the Weierstrass function

$$\wp = \wp(z) = \wp(z, \Omega) = z^{-2} + \sum_{0 \neq \omega \in \Omega} \{(z-\omega)^{-2} - \omega^{-2}\}.$$

Then with the "invariants"

$$g_2 = 60 \sum_{0 \neq \omega \in \Omega} \omega^{-4}, \qquad g_3 = 140 \sum_{0 \neq \omega \in \Omega} \omega^{-6}$$

one has the famous differential equation

$$\wp'^2 = 4\wp^3 - g_2\wp - g_3. \tag{3.1}$$

This is not quite polynomial, so one has to adjoin $\wp'$ with $(\wp)' = \wp'$ and $(\wp')' = 6\wp^2 - \frac{1}{2}g_2$. Thus there is a chance of applying Schneider-Lang if g_2 and g_3 are algebraic numbers. Actually it is always possible to choose the lattice Ω so that this is the case, and in fact all values with $g_2^3 \neq 27g_3^2$ are permissible. See Silverman's books [80] and [81] for an excellent account of the general theory.

Now the Schneider-Lang Theorem will deliver information about complex numbers u, not in Ω, with $\wp(u)$ and $\wp'(u)$ in $\overline{\mathbb{Q}}$ (in fact we could here omit the derivative). These are the elliptic analogues of the numbers $\lambda = \log \alpha$ with e^λ in $\overline{\mathbb{Q}}$. So they could be called elliptic logarithms of algebraic numbers, or in the past the shorter but slightly misleading name "algebraic points" has also been used. In fact they almost form an additive group; the map $\exp = (\wp, \wp')$ takes $\mathbb{C} \setminus \Omega$ to the elliptic curve E_0 in affine space $\mathbb{A}^2$ defined by

$$y^2 = 4x^3 - g_2 x - g_3 \tag{3.2}$$

and the algebraic points go to the set $E_0(\overline{\mathbb{Q}})$ of points defined over $\overline{\mathbb{Q}}$. The group structure on $E_0(\overline{\mathbb{Q}})$ induced by that on $E_0(\mathbb{C})$ is the usual geometric "chord-and-tangent" law; given two points π_1, π_2 one draws the connecting line and reflects the third intersection point in the x-axis to obtain the sum. In terms of the coordinates it is given by rational maps. And in terms of the Weierstrass functions it corresponds to functional equations or addition theorems expressing $\wp(z+w)$ and $\wp'(z+w)$ as quotients of polynomials in $\wp(z)$, $\wp'(z)$, $\wp(w)$, $\wp'(w)$.

To get an exact group one adjoins the periods of Ω, which are then taken under exp to the single point at infinity in the projective completion E of E_0 in $\mathbb{P}_2$. Now exp becomes a surjective map from $\mathbb{C}$ to $E(\mathbb{C})$. Furthermore the double periodicity of the elliptic functions means that it induces a map from $\mathbb{C}/\Omega$ to $E(\mathbb{C})$, and this quotient map is injective.

Using such a group structure after the application of the Schneider-Lang Theorem, one deduces the transcendence of every non-zero ω in Ω, or indeed every non-zero algebraic point u (elliptic analogue of the Hermite-Lindemann Theorem, due to Schneider in 1937). This application corresponds to the group variety $G = \mathbb{G}_a \times E$. With $g_2 = 0$, $g_3 = 4$ an example is

$$\omega_3 = 2\int_1^\infty (4x^3 - 4)^{-1/2} dx = (\Gamma(1/3))^3/(2^{4/3}\pi). \tag{3.3}$$

So we deduce that $\Gamma(1/3)/\pi^{1/3}$ is transcendental. Or with $g_2 = 4$ and $g_3 = 0$ we get $\omega_4 = \Gamma(1/4)/(2\sqrt{2\pi})$ so the transcendence of $\Gamma(1/4)/\pi^{1/2}$. But we cannot prove the transcendence of $\Gamma(1/3)$ or $\Gamma(1/4)$ by themselves this way.

The elliptic analogue of the Gelfond-Schneider Theorem (also proved by Schneider in 1937) treats two algebraic points u_1, u_2. Assuming first that these are non-zero, and with

$$f_1(z) = \wp(z), \quad f_2(z) = \wp(\beta z), \quad f_3 = f_1', \quad f_4 = f_2'$$

and $\beta = u_2/u_1$ with the set S containing $\mathbb{Z}u_1$, we deduce that u_1 and u_2 are $\overline{\mathbb{Q}}$-linearly independent provided the functions f_1 and f_2 are algebraically independent.

Unlike the earlier situation with e^z and $e^{\beta z}$, this functional independence is not always guaranteed by the irrationality of β. In fact the set

$$\{\beta \text{ in } \mathbb{C}^* \text{ ; } \wp(z) \text{ and } \wp(\beta z) \text{ algebraically dependent}\}$$

together with $\beta = 0$, is a field $k = k(\Omega)$, and it is usually $\mathbb{Q}$ but it can be a complex quadratic extension. This latter case is called "complex multiplication" (CM), and in some sense it is rare; for example choosing g_2 and g_3 in $\mathbb{Z}$ with $1 \leq g_2, g_3 \leq X$ gives at least $X^2 - 3X$ possibilities, of which at most $39X$ have complex multiplication. In some finer sense, over the rationals there are essentially only 13 examples, and only 9 possibilities for the field k (Baker-Stark-Heegner; see Chapter 5 of [6]).

Now the elliptic analogue of the Gelfond-Schneider Theorem holds for any algebraic points u_1 and u_2 including 0, and it says that u_1, u_2 are $\overline{\mathbb{Q}}$-linearly independent if and only if they are k-linearly independent. This corresponds to the group variety $G = E \times E$.

There is a handful of other interesting applications of the Schneider-Lang Theorem. One can introduce a second elliptic curve $\widetilde{E}$ (see Lecture 4) and work with $G = E \times \widetilde{E}$. Or with group varieties G sitting in exact sequences

$$0 \to L \to G \to E \to 0$$

with $L = \mathbb{G}_a$ or $\mathbb{G}_m$ (as mentioned briefly in Lecture 1). Or finally $G = A$ an abelian variety of dimension 2 (see Lecture 5). These exhaust all possible types of commutative group variety of dimension 2 (see [87] pp. 61-73 or [74] pp. 57-64).

So much for transcendence. We next turn to heights and lower bounds of Dobrowolski type; a discussion of analogues of Zhang's results will be postponed to Lecture 5.

It is now natural to take the invariants g_2 and g_3 in a fixed number field K, and then a point $\pi = (\alpha, \beta)$ in $E_0(\overline{\mathbb{Q}})$ with degree

$$d = [K(\alpha, \beta) : \mathbb{Q}].$$

We can define the absolute logarithmic height $h(\pi) = h(\alpha)$ (the extra $h(\beta)$ is superfluous because α more or less determines β); and also $h(0) = 0$. But the vanishing $h(\pi) = 0$ doesn't quite fit together with the elliptic group law. This is because $h(m\pi)$ is not precisely related to $h(\pi)$ like $h(\alpha^m) = |m|\, h(\alpha)$. It is however approximately related to $m^2 h(\pi)$. Néron and Tate independently observed that the limit

$$\widehat{h}(\pi) = \lim_{m \to \infty} h(m\pi)/m^2 \geq 0$$

exists and can be used to recover a precise relation $\widehat{h}(m\pi) = m^2 \widehat{h}(\pi)$ for all m in $\mathbb{Z}$. It follows that if π is a torsion point then $\widehat{h}(\pi) = 0$. There is also a constant $C(E)$, depending only on E, such that

$$|\widehat{h}(\pi) - h(\pi)| \leq C(E) \tag{3.4}$$

for all π in $E(\overline{\mathbb{Q}})$, and it follows conversely that if $\widehat{h}(\pi) = 0$ then π is a torsion point.

In view of the limiting process the non-zero values of the Néron-Tate height $\widehat{h}(\pi)$ are arithmetically quite mysterious; for example with $\pi_0 = (7107, -602054)$ on the curve

$$y^2 + xy + y = x^3 + x^2 - 125615x + 61201397$$

(not in the standard Weierstrass form (3.2)) one has

$$\widehat{h}(\pi_0) = .00891432\ldots \tag{3.5}$$

and no-one knows if such values are algebraic or transcendental (unlike $h(\alpha) \neq 0$, which is always the logarithm of an algebraic number and therefore transcendental).

Anyway, it makes sense to ask for lower bounds for these values of $\widehat{h}(\pi)$. The inequality (3.4), although it will be useful in Lecture 4, is no help here if $h(\pi) \leq C(E)$.

Just as with $2^{1/d}$ one can make $\widehat{h}$ arbitrarily small. Thus if π_0 is as above and π satisfies $e\pi = \pi_0$ for a positive integer e, we have $\widehat{h}(\pi) = e^{-2}\widehat{h}(\pi_0)$. To estimate the degree d of π consider any conjugate π' of π. It too satisfies $e\pi' = \pi_0$ and so $\pi' - \pi$ is a torsion point whose order divides e. Because $(e^{-1}\Omega)/\Omega$ has e^2 elements, the exponential isomorphism shows that there are at most e^2 such torsion points, so at most e^2 conjugates π'. Thus $d \leq e^2$, and we conclude $\widehat{h}(\pi) \leq \widehat{h}(\pi_0)/d$ (this argument also works with $2^{1/d}$ and roots of unity).

The analogue of Lehmer's Question would be $\widehat{h}(\pi) \geq cd^{-1}$ for all non-torsion π in $E(\overline{\mathbb{Q}})$, but now $c = c(E) > 0$ ought to depend on the elliptic curve E; so for the time being we keep E fixed. The best known result remains my own

$$\widehat{h}(\pi) \geq c(E)d^{-3}(\log 2d)^{-2}$$

established way back in 1989 [46] (see Corollary 1 p. 249). The proof is worth a remark: it goes much as Schneider-Lang with $\varphi(z) = P(\wp(z), \wp(\beta z))$, but here $\beta = b$ is a suitable positive integer, actually of order $\sqrt{d}$. Now we must be careful; b is in k and so the two functions are not algebraically independent. However any relation must involve $\wp(z)$ to a degree at least b^2, and we are safe as long as the degree of the auxiliary polynomial $P(X_1, X_2)$ in X_1 stays less than b^2.

In the case of complex multiplication Laurent [40] in 1981 did succeed in obtaining an exact analogue of Dobrowolski's Theorem in the form

$$\widehat{h}(\pi) \geq c(E)d^{-1}\left((\log 3d)/(\log\log 3d)\right)^{-3}$$

for $c(E) > 0$. He uses again $P(\wp(z), \wp(bz))$, now mixed with Frobenius considerations as in Dobrowolski's proof, but with primes from the quadratic field $k = k(\Omega)$ — in $\mathbb{Q}$ there aren't enough for the job.

Finally, returning back to unrestricted E, we remark that these constants $c(E)$ have been intensively studied. Even over $K = \mathbb{Q}$ with $d = 1$ (so everything rational) their form is highly intriguing. Maybe $\widehat{h}(\pi_0)$ in (3.5) above is actually the smallest positive height occurring – at the time of writing no-one knows a smaller one. Lang even conjectures [37] (p. 92) that $\widehat{h}(\pi) \to \infty$ as "$E \to \infty$" in a precise sense. But the best result so far is David's 1997 lower bound [21] (see Corollaire 1.5 p.110)

$$\widehat{h}(\pi) \geq c(K)(h(E))^{-7/8}$$

(with $c(K) > 0$) for all non-torsion π in $E(K)$. Here $h(E)$ is any sort of crude but logarithmic height of the curve E such as $1 + h(g_2) + h(g_3)$. The proof uses p-adic analysis and Tate elliptic curves.

If we do want a lower bound that tends to infinity and not zero, we must accept 5 points on $E(K)$ that are independent with respect to the group law. David in [21] also proved (see Corollaire 1.6 p. 111) that at least one of these must have $\widehat{h}(\pi) \geq c(K)(1 + h(j))^{1/120}$ with $c(K) > 0$ and the j-invariant

$$j = j(E) = 1728 g_2^3/(g_2^3 - 27 g_3^2).$$

In this lower bound we cannot allow $h(E)$ in place of $h(j)$.

As mentioned above, we postpone our discussion of elliptic analogues of Zhang's results on $h(\alpha) + h(1 - \alpha)$ until Lecture 5.

At least e down; at least $\Gamma(1/3)$ to go.

4 Fourth lecture. Linear forms in elliptic logarithms

We saw in the previous lecture that if E is an elliptic curve defined over $\overline{\mathbb{Q}}$, then two algebraic points u_1 and u_2 are linearly independent over $\overline{\mathbb{Q}}$ if and only if they are linearly independent over the associated field k; here k is either $\mathbb{Q}$ or a complex quadratic extension of $\mathbb{Q}$. One of our goals in the present lecture is to extend this result to several algebraic points $u_1, \ldots, u_n$ and several elliptic curves $E_1, \ldots, E_n$ defined over $\overline{\mathbb{Q}}$. So now we have several elliptic functions $\wp_1, \ldots, \wp_n$.

The case $n = 2$ is not quite done; we had the restriction $E_1 = E_2$. But there is no difficulty in applying the Schneider-Lang Theorem without this restriction. It shows that if $u_1 \neq 0$ and $u_2 \neq 0$ are $\overline{\mathbb{Q}}$-linearly dependent then $\wp_1(z)$ and $\wp_2(\beta z)$ are algebraically dependent for some $\beta \neq 0$. We proceed to reformulate this analytic statement in a more geometric way as follows.

We have exponential maps from $\mathbb{C}$ to $\Pi_i = E_i(\mathbb{C})$ $(i = 1, 2)$ which are group homomorphisms. So there is a homomorphism exp from $\mathbb{C}^2$ to $G(\mathbb{C}) = \Pi_1 \times \Pi_2$ for the group variety $G = E_1 \times E_2$. Consider the subgroup $Z \subset \mathbb{C}^2$ defined by $z_2 = \beta z_1$. Then $\exp(Z)$ is a subgroup of $\Pi_1 \times \Pi_2$. So the Zariski closure Γ of $\exp(Z)$ is also a subgroup of the product $\Pi_1 \times \Pi_2$.

A purely group-theoretical description of subgroups of products now tells us that there are groups $\Delta_i \subseteq \Gamma_i \subseteq \Pi_i$ (with Δ_i normal in Γ_i – but everything here is commutative) for $i = 1, 2$, together with an isomorphism γ from Γ_1/Δ_1 to Γ_2/Δ_2, such that

$$\Gamma = \{(\pi_1, \pi_2) \text{ in } \Gamma_1 \times \Gamma_2;\ \pi_2 + \Delta_2 = \gamma(\pi_1 + \Delta_1)\}.$$

Now all our groups, except Z and $\exp(Z)$, are in fact group varieties, and in particular they have dimensions; for example $\dim \Pi_i = 1$ $(i = 1, 2)$. Since $\wp_1(z)$ and $\wp_2(\beta z)$ are algebraically dependent this means that $\dim \Gamma \leq 1$ as well; and it is easy to see that equality holds. It is equally easy to see that this forces $\dim \Delta_i = 0$ and $\dim \Gamma_i = 1$ $(i = 1, 2)$. This in turn means that Δ_i is finite and $\Gamma_i = \Pi_i = E_i(\mathbb{C})$ $(i = 1, 2)$. Such a conclusion reflects a general rigidity property of group subvarieties, another example of which we saw in Lecture 2 with the subgroups H of $\mathbb{G}_m^n$ defined by (2.6). We will see more examples later.

Anyway, we can now form the sequence

$$E_1(\mathbb{C}) = \Gamma_1 \to \Gamma_1/\Delta_1 \to \Gamma_2/\Delta_2 \to \Gamma_2 = E_2(\mathbb{C}) \tag{4.1}$$

where the three inner maps are respectively the canonical one, the isomorphism γ, and multiplication by the order of Δ_2. We end up with a non-zero group homomorphism from $E_1(\mathbb{C})$ to $E_2(\mathbb{C})$ which is also algebraic; that is, a rational map of varieties. Such a map is called an isogeny. Given arbitrary E_1 and E_2 it might not exist, but if it does then E_1 and E_2 are said to be isogenous. This relation is an equivalence relation between elliptic curves.

Thus in our situation we have shown that if non-zero algebraic points u_1 and u_2 are $\overline{\mathbb{Q}}$-linearly dependent then indeed E_1 and E_2 are isogenous. Actually this does not include the original result of Lecture 3 because the conclusion is empty if $E_1 = E_2$. So if $E_1 = E_2$ we have to add the original conclusion about k-linear dependence.

Now we can at least formulate the version for several points. If non-zero algebraic points $u_1, \ldots, u_n$ are $\overline{\mathbb{Q}}$-linearly dependent, then some E_i and E_j $(i \neq j)$ are isogenous; further if $E_1 = \cdots = E_n$ then $u_1, \ldots, u_n$ are k-linearly dependent. This is not yet the optimal version, but we will come to that in Lecture 5 in a more general context.

The proofs are another matter. Dealing even with $n = 3$ and ordinary logarithms $\log \alpha_1$, $\log \alpha_2$ and $\log \alpha_3$ requires working in $\mathbb{C}^2$ with Baker's Method. At present we have no exact analogue of the Schneider-Lang Theorem because of special problems with the analytic step (UB) of Lecture 3. Lang [36] (Theorem 1 p. 33) did prove a version for several variables which can sometimes be applied; for example it settles the $\overline{\mathbb{Q}}$-linear independence of n ordinary logarithms above (see for example Chapter 4 of [90]). But it is not suited to quantitative refinements like (1.1) in Lecture 1. Also Bombieri [11] obtained a profound generalization of Lang's version, but so far this has not yet found applications back to transcendence.

Because Baker's Method (and other variants) are discussed in detail in Waldschmidt's lectures (section 3 – see also Chapter 10 of his book [90]), we will give only a very brief sketch of what happens for $n = 3$, using the language of several complex variables.

Now we are starting with a relation

$$\beta_1 u_1 + \beta_2 u_2 = u_3$$

with algebraic β_1 and β_2. For the step (AP) we construct a non-zero auxiliary polynomial P in $\overline{\mathbb{Q}}[X_1, X_2, X_3]$ such that the function

$$\varphi(z_1, z_2) = P\big(\wp_1(z_1),\ \wp_2(z_2),\ \wp_3(\beta_1 z_1 + \beta_2 z_2)\big)$$

has zeros of large orders at least T on a large subset $\mathcal{S}$ of $\mathbb{Z}(u_1, u_2)$. This means of course that all the derivatives

$$(\partial/\partial z_1)^{t_1}(\partial/\partial z_2)^{t_2}\,\varphi \qquad (t_1 + t_2 < T)$$

vanish on $\mathcal{S}$. The construction is a standard application of the Siegel Lemma; for example at $(z_1, z_2) = (u_1, u_2)$ the three functions in φ take the algebraic values $\wp_1(u_1)$, $\wp_2(u_2)$, $\wp_3(u_3)$ respectively. Now if L is a bound for the degree of P then T^2 multiplied by the cardinality of $\mathcal{S}$ should be a small multiple of L^3, similar to the situation in Lecture 3 with the Schneider-Lang Theorem for $n = 2$.

For the step (UB) we can no longer increase the order T, for example to $2T$ as in Lecture 3. This is the special problem referred to just above. For example, the function $\varphi = (u_2 z_1 - u_1 z_2)^T$ has zeros of order T at every point of $\mathbb{Z}(u_1, u_2)$ and even the complex line $\mathbb{C}(u_1, u_2)$; but $(\partial/\partial z_1)^T \varphi = T!\, u_2^T$ is not small anywhere. Such examples are impossible with a single complex variable.

As explained by Waldschmidt (his lectures, section 3.1), Baker's analytic method leads instead to a decrease in the orders of vanishing, say from T to $T/2$; and this must therefore be compensated by enlarging the set $\mathcal{S}$ (which also was not possible in the Schneider-Lang situation). We will be a bit more precise in Lecture 6.

The step (LB) is standard, and it is the non-vanishing problem (NV) that causes the serious complications. Baker solved this problem for ordinary logarithms with ingenious *ad hoc* techniques involving huge determinants and Kummer Theory, but these don't generalize easily to commutative group varieties. It was Wüstholz who discovered in 1989 how to solve the problem in general – again see Lecture 6.

That is all we will say about the proof of the result on $u_1, \ldots, u_n$. Unfortunately it is not possible to find the full details in the literature except in the rather more general context of commutative group varieties; see Lectures 5 and 6. By considering $G = \mathbb{G}_a \times E_1 \times \cdots \times E_n$ in place of $G = E_1 \times \cdots \times E_n$ it is possible also to include the number 1; thus if $1, u_1, \ldots, u_n$ are $\overline{\mathbb{Q}}$-linearly dependent we get the same conclusion. One can also consult Wüstholz's original paper [93], which implies this conclusion for periods $u_1, \ldots, u_n$; as well as [94] for related work in the case $E_1 = \cdots = E_n$.

All these purely qualitative results come from Wüstholz's general Analytic Subgroup Theorem of 1989, and it is this that will be examined in Lectures 5 and 6. In line with Waldschmidt's course (section 4) we shall now discuss quantitative refinements, like Mahler's inequality $|\pi - r/s| > s^{-42}$ in (1.1).

In principle once one has a qualitative result of the form say $\Lambda \neq 0$, it is usually possible to refine it without too much trouble to $|\Lambda| \geq C$ with some explicit $C > 0$ depending on suitable parameters.

A simple example comes out of Lecture 1. If $\eta = f(2/3)$ is very close to a rational $\tilde{\eta} = r/s$, then the real numbers $\alpha_n = P(\xi_n, \eta_n)$ are very close to the rational numbers $\tilde{\alpha}_n = P(\xi_n, \tilde{\eta}_n)$ with

$$\tilde{\eta}_n = \tilde{\eta} - 2/3 - \cdots - (2/3)^{2^n} \qquad (n = 0, 1, 2, \ldots).$$

In fact it is easy to prove that $|\alpha_n - \tilde{\alpha}_n| \leq 25\,|\eta - \tilde{\eta}|$. Using the upper bound (3.UB) for $|\alpha_n|$ and a lower bound like (1.LB) for $|\tilde{\alpha}_n|$ if $\tilde{\alpha}_n \neq 0$, we find that

$$|\eta - \tilde{\eta}| \geq \frac{1}{25}\left(s^{-2}9^{-2^{n+1}} - 704\,(2/3)^{6.2^{n+1}}\right).$$

As usual we forgot the non-vanishing problem $\tilde{\alpha}_n \neq 0$. The analytic argument (1.2) is no use here because it works only for α_n. But, perhaps surprisingly, the algebraic argument continues to work for $\tilde{\alpha}_n$, since $\tilde{\eta}_{n+1} = \tilde{\eta}_n - \xi_n$ and so

$$\tilde{\alpha}_n = P(\xi_n, \tilde{\eta}_n), \quad \tilde{\alpha}_{n+1} = Q(\xi_n, \tilde{\eta}_n).$$

Thus if n is larger than some absolute constant, either $\tilde{\alpha}_n \neq 0$ or $\tilde{\alpha}_{n+1} \neq 0$, and we must accordingly allow for two possible optimal values of n. The details can sometimes become irritating but apart from this non-vanishing problem no new ideas are involved. One obtains in fact $|\eta - r/s| > c\,s^{-77}$ for some positive absolute constant c (see also the original papers of Galochkin [31] or Miller [51]).

But in general getting lower bounds of such quality for $|\Lambda|$ can be highly troublesome. Thus if $\omega \neq 0$ is a period of an elliptic function with algebraic invariants, it was not known until very recently if there exists a constant $\kappa = \kappa(\omega)$ such that

$$|\,\omega - r/s\,| > s^{-\kappa} \tag{4.2}$$

for all rational integers r, s with $s \geq 2$.

In 1995 David [20] collected together the works of several authors, added a vast amount of computation, and suceeded in obtaining a rather good estimate like Mahler's with all the constants explicitly calculated. This remains the only completely explicit estimate of its type, and so we take some time to describe it in detail.

Let $u_1, \ldots, u_n$ and $E_1, \ldots, E_n$ be as above, and define

$$\Lambda = \beta_0 + \beta_1 u_1 + \cdots + \beta_n u_n$$

for algebraic numbers $\beta_0, \beta_1, \ldots, \beta_n$ not all zero. To begin with, suppose that these β's lie in a number field K of degree D, with heights measured by a quantity H defined by

$$\log H = \max\{e, h(\beta_0), h(\beta_1), \ldots, h(\beta_n)\}.$$

In 1991 Hirata-Kohno [33] proved in the context of commutative group varieties a result that from an inequality

$$\log |\Lambda| < -C(u_1, \ldots, u_n, E_1, \ldots, E_n, D)(\log H)(\log \log H)^{n+1} \tag{4.3}$$

implies $\Lambda = 0$ and so the consequences that we have mentioned. Previously Philippon and Waldschmidt [62] had obtained a similar estimate involving $(\log H)^{n+1}$.

This does not quite imply the above result (4.2). If the rational number r/s is at all near to ω then $h(r/s)$ is at most of order $\log s$, and so with $\Lambda = \omega - r/s$ we get only $|\omega - r/s| > s^{-\kappa(\log\log 3s)^2}$ with $\kappa = \kappa(\omega)$. But it is already far more than good enough for many applications.

Quite a recent application is to find effectively and explicitly all integral points on a given elliptic curve E. Two ideas are involved. First suppose that $\pi = (x, y)$ satisfies $y^2 = 4x^3 - g_2 x - g_3$. We know that $y = \wp'(u)$. If $y > 0$ is large then it can only be because u is near to a pole or period of the associated lattice Ω. In fact it is easy to prove that the corresponding distance $\mathrm{dist}(u, \Omega)$ is of order at most $|\wp'(u)|^{-1/3} = y^{-1/3}$. More precisely $\mathrm{dist}(u, \Omega) \leq C_1 y^{-1/3}$ for some constant C_1 depending, like $C_2, \ldots, C_{10}$ below, only on E.

It is not so easy to prove, but it is true, that

$$\mathrm{dist}(u, \Omega) \leq (1.7666\ldots) y^{-1/3}$$

independently of E and π. Here the strange constant is $2^{1/3} \cdot 3^{-1/2} \cdot \omega_3$ for the special period (3.3) mentioned in Lecture 3, and it is best possible.

So we get

$$|\Lambda| < 2y^{-1/3} \tag{4.4}$$

for $\Lambda = u - \omega$ and some ω in Ω.

Now (assuming g_2 and g_3 are in $\mathbb{Q}$) Mordell's part of the Mordell-Weil Theorem says that the group $E(\mathbb{Q})$ of rational points is finitely generated. If the rank $m = 0$ the argument can easily be finished without using (4.3). Otherwise, if $m \geq 1$ let $\pi_1, \ldots, \pi_m$ be basis elements modulo the torsion group Δ. Thus $\pi \equiv b_1\pi_1 + \cdots + b_m\pi_m \pmod{\Delta}$ for certain integer coefficients. It follows that $u \equiv b_1 u_1 + \cdots + b_m u_m \pmod{\Omega/b}$ for the corresponding algebraic points $u_1, \ldots, u_m$ and the order b of Δ. Choosing basis elements ω_1 and ω_2 of Ω we end up with

$$\Lambda = b_1 u_1 + \cdots + b_m u_m + b_{m+1}\omega_1/b + b_{m+2}\omega_2/b$$

to which we can apply (4.3). Trivially $\Lambda \neq 0$ and so we get a lower bound for $|\Lambda|$. For the present purpose it is enough to replace the $(\log H)(\log\log H)^{n+1}$ by H, and we get

$$|\Lambda| \geq e^{-C_2 H}$$

where here $H \leq \max\{e^e, |b_1|, \ldots, |b_m|, |b_{m+1}|, |b_{m+2}|, |b|\}$. Comparing this with the upper bound (4.4) gives

$$y \leq e^{C_3 H} \tag{4.5}$$

which is not yet conclusive. But now the theory of heights provides good upper bounds for H. Thus the Néron-Tate height

$$\widehat{h}(\pi) = \widehat{h}(b_1\pi_1 + \cdots + b_m\pi_m)$$

is a positive definite quadratic form in the coefficients $b_1, \ldots, b_m$. So it is bounded below by $C_4^{-1}B^2$ (with $C_4 > 0$) for $B = \max\{|b_1|, \ldots, |b_m|\}$. But the relation (3.4) of Lecture 3 between the Néron-Tate height and the conventional height implies that $\widehat{h}(\pi) \leq h(\pi) + C_5$. If $\pi = (x, y)$ is an integral point in $\mathbb{Z}^2$, we have $h(\pi) = \log|x|$, and we conclude $\widehat{h}(\pi) \leq C_6 \log y$ assuming $y \geq 2$. This yields the estimate $B \leq C_7\sqrt{\log y}$. A similar estimate follows for H, and now (4.5) says

$$y \leq e^{C_8\sqrt{\log y}},$$

which implies $y \leq C_9$, $|x| \leq C_{10}$. All these constants can be explicitly calculated, and in this way (with the help of sophisticated computational devices) equations like

$$y^2 + xy = x^3 - 5818216808130x + 5401285759982786436$$

(not in Weierstrass form) have been solved. Here $m = 8$ (the largest rank so far handled), and according to Stroeker and Tzanakis [84] (p. 147) there are exactly 69 integral points.

Of course linear forms in ordinary non-elliptic logarithms are also applicable in principle, but not in practice; the bounds for x and y are exponential in the coefficients of the equation. On the other hand one does need in the above approach some tiny piece of luck to be able to calculate the Mordell-Weil group.

One needs also to calculate all the above constants $C_1, \ldots, C_{10}$ explicitly, as well as the crucial $C = C(u_1, \ldots, u_n, E_1, \ldots, E_n, D)$ in the inequality (4.3). It turns out to be C_4 which needs most of the computational trickery, but we confine ourselves here to a discussion of C, out of respect for David's efforts.

We return to the general situation with $E_1, \ldots, E_n$ in (4.3). First assume that these elliptic curves have invariants in the given number field K of degree $D \geq 2$, and that this field also contains the coefficients $\beta_0, \beta_1, \ldots, \beta_n$. We measure the curves using the quantities $h(E_1), \ldots, h(E_n)$ of Lecture 3 and the sum

$$h = h(E_1) + \cdots + h(E_n).$$

Assume also that each $\pi_i = (\wp_i(u_i), \wp_i'(u_i))$ lies in $E_i(K)$. Now this π_i determines u_i only modulo the period lattice, and to make life easier we will suppose that $|u_i|$ is chosen as small as possible (this is the elliptic analogue of normalizing the ordinary logarithm to its principal value). Define then

$$\log V_i = \max\{h, \widehat{h}(\pi_i)\} \qquad (1 \leq i \leq n),$$

and assume for definiteness that $V_1 \geq \cdots \geq V_n$. Then David's result for the constant C in (4.3) may be expressed as

$$C \leq c(n) D^{2n+2} (\log D)^{n+2} \cdot h^{n+1} (\log V_1) \cdots (\log V_n)(\log \log V_1)^{n+2},$$

with $c(n)$ depending only on n. This last quantity has not been studied so closely as in the case of ordinary logarithms (see Waldschmidt's lectures especially section 5.2, d)), and it comes out of David's work as $c^n n^{2n^2}$ for a small explicit absolute constant c.

As in the case of ordinary logarithms, there is also a factor E, which in our case corresponds to the possibility that some of the $|u_i|$ are exceptionally small, but we do not discuss this here (and of course we could not use this notation!).

Finally David also goes deeper into the consequences of $\Lambda = 0$. As we saw, there is then an isogeny between E_i and E_j for some $i \neq j$. Such an isogeny is always surjective but not usually injective. From the discussion at the beginning of this lecture, and especially the sequence (4.1), it can be seen to be "M to 1" for a positive integer M called the degree. In the context of that discussion this degree is closely related to the degree of the auxiliary polynomial $P(X_1, X_2)$ that connects $\wp_1(z)$ and $\wp_2(\beta z)$, and so it is not surprising that an estimate for this degree comes out of the proofs. We will see more details in the more general context of Lecture 6.

Anyway, David proves the upper bound

$$M \leq C' (\log \log H)^{n+1}$$

for a constant C' like the constant C in (4.3).

The good behaviour of this C' is crucial to certain applications. Namely, given an elliptic curve E over a number field K there are up to isomorphism only finitely many other elliptic curves $\widetilde{E}$ over K isogenous to E. Here an isomorphism means simply an isogeny of degree 1 (which can be expressed very simply in terms of the invariants). This result was first proved by Shafarevich in 1962 using reduction theory and finiteness results for S-unit equations, but the proof that we are about to sketch, due to Wüstholz and myself [47] in 1990, extends also to abelian varieties (see Lecture 5).

It suffices to bound the degree of some isogeny from $\tilde{E}$ to E. For there is then a similarly bounded isogeny in the opposite direction with kernel Δ, and $\tilde{E}$ is isomorphic to E/Δ. Accordingly let M_0 be the minimal such degree. We have the maps $\mathbb{C} \to \mathbb{C}/\Omega \to E(\mathbb{C})$ and similarly for $\tilde{E}$; and it turns out that the isogeny from $\tilde{E}(\mathbb{C})$ to $E(\mathbb{C})$ is induced by multiplication by some algebraic number $\beta \neq 0$. In particular β maps the lattice $\tilde{\Omega}$ of $\tilde{E}$ into Ω. Pick basis elements ω_1, ω_2 of Ω and some $\tilde{\omega} \neq 0$ of $\tilde{\Omega}$. Then there are rational integers b_1, b_2 such that

$$\beta\tilde{\omega} = b_1\omega_1 + b_2\omega_2$$

is a relation $\Lambda = 0$, with $E_1 = E_2 = E$ and $E_3 = \tilde{E}$. Admittedly E_1 and E_2 are trivially isogenous, but it can be concluded that E and $\tilde{E}$ themselves are isogenous, with a connecting isogeny of degree $M \leq C'(\log\log H)^4$. We have to be careful with $C' = C'(\omega_1, \omega_2, \tilde{\omega}, E, \tilde{E}, D)$ since it depends on the "unknown" $\tilde{E}$. However, $\tilde{E}$ is related to E through the original isogeny, and it can be shown for example that $h(\tilde{E}) \leq C_1'(E,D)\log M_0$. The $\log V_i$ are similarly bounded, as well as $\log H$, and so we end up with say

$$M \leq C_2'(E,D)(\log M_0)^{\kappa}$$

with κ absolute.

Now the minimality of M_0 implies $M_0 \leq M$, and it follows that $M_0 \leq C_3'(E,D)$. This suffices for the proof of Shafarevich's Theorem. The sharpest and most explicit result to date is Pellarin's

$$M_0 \leq CD^4(\log 2D)^2(h(E))^2$$

with C round about 10^{60}; this is obtained [57] using two linear forms instead of the single Λ, as in [47], and more careful analytic estimates.

In some recent work the annoying power of $\log\log H$ in (4.3) has been eliminated, at least for elliptic curves. Thus results like (4.2) are now proved. In the CM case this was achieved by Ably [1] using a clever analogue of the Feldman polynomials (see section 4.2 of Waldschmidt's lectures). In the general case David and Hirata-Kohno [23] proceed as follows. The alert reader will have noticed that we are always having to differentiate a large number of times. If z_0 is a fixed algebraic point then the derivatives $(d/dz)^t\wp$ at z_0 are algebraic numbers whose heights are growing like a factorial $t!$ for large t. This is harmless for the Schneider-Lang Theorem, but not for (4.2). In the proofs of results like the Thue-Siegel-Roth-Schmidt Theorems (see Schlickewei's lectures, section 5) or Stepanov's proof of the Weil estimates, such a growth can be reduced by using "divided derivatives" or "hyperderivatives" or "Hasse derivatives", which just means dividing out by the $t!$. This doesn't work for $\wp$. However, the "exponential" relation $x = \wp(z)$ can be locally inverted to a "logarithmic" relation $z = \mathcal{L}(x)$. And now the numbers $\alpha_t = (1/t!)(d/dx)^t\mathcal{L}$ at $x_0 = \wp(z_0)$ do have heights growing only like C^t. The reason is that the integral $\int_{\infty}^{x} dx/\sqrt{4x^3 - g_2x - g_3}$ becomes, after the substitution $x = \wp(z)$, just z; so it is none other than $\mathcal{L}$. The algebraic function in the integrand can be expanded about x_0 and its Taylor coefficients satisfy Eisenstein's Theorem on growth. The above α_t are the coefficients in the integral and so have nearly the same growth. More succinctly, $\mathcal{L}$ is a G-function.

Now $\dim(E^2 \times \widetilde{E}^2)$ lectures down; and the number left is the minimal degree of any isogeny between $y^2 = x(x+1)(x+4)$ and $y^2 = x(x-1)(x-9)$.

5 Fifth lecture. Abelian varieties

An elliptic curve defined in affine space by $y^2 = 4x^3 - g_2x - g_3$ embeds into projective space $\mathbb{P}_2$ and the chord-and-tangent group law makes it into a group variety. In Lecture 3 we came to this cubic equation through elliptic functions, but for the definition of abelian variety it is simpler to start with a projective variety in some $\mathbb{P}_N$ having a group law defined by rational maps. It turns out that the projectivity forces such a group law to be automatically commutative. For the transcendence proofs we do need the analogue of elliptic functions, which are then called abelian functions, but we skip this for the moment.

Thus an abelian variety over a field $\mathcal{K}$ may be defined as a connected projective variety over $\mathcal{K}$ together with a rational group law also defined over $\mathcal{K}$. The embedding dimension N is usually a lot bigger than the actual dimension, say n; some natural constructions like the Lefschetz Theorem give $N = 3^n$ or so, although in principle $N = 2n + 1$ is theoretically possible. For excellent modern accounts see the books of Mumford [54] or Lange-Birkenhake [39], as well as Milne's articles [52], [53].

So an elliptic curve $X_0X_2^2 = 4X_1^3 - g_2X_0^2X_1 - g_3X_0^3$ is an abelian variety, but the additive group $\mathbb{G}_a$ is not, because it is $\mathbb{P}_1$ with a point "∞" or $(X_0, X_1) = (0, 1)$ missing. Similarly $\mathbb{G}_m$ is $\mathbb{P}_1$ with two points ∞ and "0" or $(1, 0)$ missing.

We can generate more examples by taking products. The Segre mapping takes a product of projective spaces into a projective variety, and so the product of abelian varieties is another abelian variety. Thus $E \times \widetilde{E}$ is an abelian variety, but we are now in $\mathbb{P}_8$. In this way we can often state results about several abelian varieties in terms of a single one; we will see soon that the results of Lecture 4 concern the abelian variety $E_1 \times \cdots \times E_n$.

To get more examples of abelian varieties we must consider curves of higher genus and their Jacobians. In the following example it is possible to get some idea of the group law by generalizing the chord-and-tangent construction.

Let $F(x)$ be a squarefree polynomial of degree 5. Then the curve C_0 defined by $y^2 = F(x)$ has genus 2, and a first approximation to the Jacobian is the set J_0 of unordered pairs (p_1, p_2) with p_1, p_2 in C_0. We can add two such points (p_1, p_2) and (q_1, q_2), at least generically, as follows. First find a cubic curve $gy = G(x)$ passing through p_1, p_2, q_1, q_2, using the five free coefficients available. This curve will intersect C_0 in two other points, say $r_1^\dagger$ and $r_2^\dagger$.

Reflect these in the x-axis to get r_1 and r_2; then the sum $(p_1, p_2) + (q_1, q_2)$ is (r_1, r_2).

This sum does satisfy a generic associative law, and so we have an excellent chance for a group law. However it is not so clear that J_0 is a variety. As a set it is the quotient of the variety $C_0 \times C_0$ by an equivalence relation. Such quotients can cause trouble, and we will see this in a slightly different context later on and in Lecture 6; but here we can simply take bisymmetric functions like $x_1 + x_2$ or $(y_1 - y_2)/(x_1 - x_2)$ to get an affine embedding in $\mathbb{A}^4$.

To get an exact group as in Lecture 3 one has to take into account the point at infinity ∞, and adjoin to J_0 the origin (∞, ∞) as well as all pairs (p, ∞) with p on C_0. Unfortunately ∞ is singular; this can be resolved by "blowing-down" or more concretely identifying all the pairs $(p, p^\dagger)$, where $p^\dagger$ is the reflection of p, with (∞, ∞). The resulting object J is then an abelian variety of dimension 2, generally not related to a product of two elliptic curves. It has a fairly natural embedding in $\mathbb{P}_8$ (see for example p. 107 of Grant's paper [32] for the defining equations, and much more besides).

Now by the Abel-Jacobi Theorem the function

$$\mathcal{L}(x_1, x_2) = \left(\int_\infty^{x_1} dx/y + \int_\infty^{x_2} dx/y, \ \int_\infty^{x_1} x dx/y + \int_\infty^{x_2} x dx/y \right) \tag{5.1}$$

with $y = \sqrt{F(x)}$ can be interpreted as a local inverse of the exponential map for J, as at the end of Lecture 4. The exponential map itself is given in [32] (p. 101) in terms of what could be called the "Baker $\wp$-functions", which are special quotients of theta functions.

In the context of the curve C_0 we cannot resist giving a final example of an explicit auxiliary polynomial $P(X)$. It is

$$8\left(F^{(p-1)/2} + 1\right) F^2 + 4\Phi F F' + 2\Phi^2 F F'' - \Phi^2 F'^2$$

with $\Phi = \Phi(X) = X^p - X$ and the derivatives F', F''; and it can be used (following Stepanov) to prove that C_0 contains at least one point modulo every prime $p \geq 17$, provided F in $\mathbb{Z}[X]$ remains of degree 5 when reduced modulo p. To see this one notes (with some labour) that the falsity would lead to at least triple zeros at all integers modulo p. But the degree of P is too small!

Given an abelian variety A over $\mathcal{K}$, how does one define an abelian logarithm u, or an algebraic point if $\mathcal{K}$ is $\overline{\mathbb{Q}}$? It suffices for the moment to do this abstractly in terms of derivations. Suppose A is in $\mathbb{P}_N$ and the coordinate X_0 is non-zero at the origin 0 of A. Consider the set of derivations ∂ from the coordinate ring $\mathcal{K}[X_1/X_0, \ldots, X_N/X_0]$ to $\mathcal{K}$ satisfying $\partial(f+g) = \partial f + \partial g$ and $\partial(fg) = f(0)g + g(0)f$. This is the abstract tangent space (at the origin), usually denoted by $T(A)$, but we will use the notation $\mathcal{T}A$. It is a $\mathcal{K}$-vector space whose dimension is equal to the dimension n of A.

If $\mathcal{K} \subseteq \mathbb{C}$ then the theory of complex Lie groups delivers an exponential map exp from $\mathcal{T}A(\mathbb{C})$ to $A(\mathbb{C})$ which is surjective. When we pick a basis of the tangent space we get a map from $\mathbb{C}^n$ to $A(\mathbb{C})$ and then the X_i/X_0 become abelian functions on $\mathbb{C}^n$ just like $\wp$ and $\wp'$ for $n = 1$.

If $\mathcal{K} = \overline{\mathbb{Q}}$ then an algebraic point of A is just an element of $\exp^{-1} A(\overline{\mathbb{Q}})$. The set of such points is again a group.

We will need also the concept of a group subvariety B of A. This is a subvariety whose complex points themselves form a subgroup under the group law of A. There are plenty of subvarieties and plenty of subgroups, but as in earlier examples of group varieties the simultaneous property is subject to a great deal of rigidity. For example there are at most countably many, which implies in particular that if A is defined over $\overline{\mathbb{Q}}$ then so is B.

For example when A is an elliptic curve E then only $B = E$ or finite B are possible; and if we further assume that B is connected then the only finite B is $\{0\}$. Similarly if A is the product of two non-isogenous elliptic curves E_1 and E_2 then the only connected group subvarieties are $B = E_1 \times E_2$, $E_1 \times \{0\}$, $\{0\} \times E_2$ or $\{0\} \times \{0\}$; the proof is essentially the argument near the beginning of Lecture 4. Similar things hold for $A = E_1 \times \cdots \times E_n$ if the factors are mutually non-isogenous.

But if $A = E \times E$ then there are "diagonal" examples like the set B of all (π, π) as π runs through E. More generally if $A = E^n$ then every "linear equation" with integer coefficients

$$b_1\pi_1 + \cdots + b_n\pi_n = 0 \tag{5.2}$$

for $(\pi_1, \ldots, \pi_n)$ in A defines a group subvariety B, and if E has no complex multiplication then every connected such B is defined by a finite collection of such equations, almost (but not quite) as in Lecture 2 with $x_1^{b_1} \cdots x_n^{b_n} = 1$. But if E has complex multiplication then one has to allow coefficients in the associated field k or more precisely in a suitable order.

Returning to arbitrary A, if B is connected then it too is an abelian variety ("abelian subvariety") and so has a tangent space $\mathcal{T}B$. This can naturally be considered as a subspace of $\mathcal{T}A$. For example with B in $A = E^n$ defined by the above linear equation then $\mathcal{T}B$ is defined by the equation

$$b_1 z_1 + \cdots + b_n z_n = 0 \tag{5.3}$$

in $\mathcal{T}E$, where $\mathcal{T}A$ is now identified with $(\mathcal{T}E)^n$.

The first transcendence theorems for abelian varieties were proved in 1941 by Schneider [73]. Most of the subsequent advances regarding linear independence are included in Wüstholz's 1989 Analytic Subgroup Theorem [96] about commutative group varieties. This will come in Lecture 6, but at the moment we state the special case for abelian varieties.

Let A be an abelian variety over $\overline{\mathbb{Q}}$ and let u be an algebraic point. The assertion is that the smallest $\overline{\mathbb{Q}}$-vector subspace Z_u of $\mathcal{T}A$ containing u has the form $\mathcal{T}B_u$ for some (unique) connected group subvariety B_u of A.

The strength of this assertion lies in the relating of the subspace Z_u to a group subvariety B_u, which as we have seen is subject to considerable rigidity. We will see in Lecture 6 how these group subvarieties arise in the proof.

Let us deduce from this assertion our results of Lecture 4 on $u_1, \ldots, u_n$ and $E_1, \ldots, E_n$. Suppose that $u_1 \neq 0, \ldots, u_n \neq 0$ are $\overline{\mathbb{Q}}$-linearly dependent. Then with $A = E_1 \times \cdots \times E_n$ and $u = (u_1, \ldots, u_n)$ the space Z_u is not the whole of $\mathcal{T}A$. It follows that B_u is not the whole of A.

Assume first that no two E_i, E_j $(i \neq j)$ are isogenous. Then B_u must be a product $\Delta_1 \times \cdots \times \Delta_n$ with each factor either $\{0\}$ or the full elliptic curve, and without loss of generality we can suppose $\Delta_1 = \{0\}$. But this implies $u_1 = 0$ a contradiction.

Secondly assume that $E_1 = \cdots = E_n = E$ say, for simplicity without complex multiplication. Then there is some non-zero $(b_1, \ldots, b_n)$ in $\mathbb{Z}^n$ such that (5.2) holds on B_u. But then (5.3) holds on $\mathcal{T}B_u = Z_u$, which means that $u_1, \ldots, u_n$ are linearly dependent over $\mathbb{Q}$. A similar proof works in the case of complex multiplication.

We also get new results very easily. For example, with A as the Jacobian of $y^2 = 4x^5 - 4$ one obtains as in (3.3) or (5.1) the transcendence of the number $\pi\Gamma(1/5)/(\Gamma(2/5))^3$; this partly explains the remark about $\Gamma(1/5)$ made in the introduction to Lecture 1. More generally it can be determined whether several given beta-values $B(a, b) = \Gamma(a)\Gamma(b)/\Gamma(a+b)$ for rational a and b are $\overline{\mathbb{Q}}$-linearly dependent or not (see Satz 4 p. 7 of the article [92] of Wolfart and Wüstholz). Further by considering $G = \mathbb{G}_a \times A$ the number 1 can be adjoined and Schneider's 1941 result on the transcendence of $B(a, b)$ (and so for example of $(\Gamma(1/5))^2/\Gamma(2/5)$ or $\Gamma(1/5)(\Gamma(2/5))^2/\pi$) thereby recovered; but here G is no longer an abelian variety and thus we would have to wait until Lecture 6 for this application.

We postpone a discussion of the proof of Wüstholz's result also until Lecture 6. Instead we want to talk about some quantitative refinements analogous to "$\Lambda = 0$" in Lecture 4. In this case the zero estimates deliver an upper bound for the degree of B_u as a projective variety. Such a quantity behaves rather like the quantity M of Lecture 4, and the upper bounds can be used for a similar purpose to estimate isogenies.

What do these upper bounds look like? One may expect as in Lecture 4 some sort of term $h(A)$ measuring the coefficients in the defining polynomials for A, and some very general 1997 results of Waldschmidt [89] confirm this expectation. In the isogeny problem A is fixed but the second abelian variety $\widetilde{A}$ is not, and we apply the Analytic Subspace Theorem to the product $\widetilde{A} \times A^{2n}$. So if M_0 is the minimal degree of some isogeny, the quantitative version can be expected to yield a new isogeny whose degree is of order some power of $h(A) + h(\widetilde{A}) + \log M_0$. The minimality argument then provides an upper bound for M_0 in terms of $h(A)$ and $h(\widetilde{A})$. However the second abelian variety $\widetilde{A}$ is not fixed, and it is much more difficult to estimate $h(A)$; in fact even its definition depends on a projective embedding which might not be known in practice. For example $\widetilde{A}$ might be a product of two fixed elliptic curves divided by a large finite subgroup, and then we are back to our previous difficulties of controlling quotients.

It is here that the use of the Faltings height $h^{\mathrm{F}}(A)$ is crucial. It is much more difficult to define (see for example [29], [78] and [79]), but it has excellent functorial properties, and in our situation these imply

$$h^{\mathrm{F}}(\widetilde{A}) \leq h^{\mathrm{F}}(A) + n \cdot \log M_0;$$

this solves our present problem with $h(\widetilde{A})$. But the price is that the quantitative Analytic Subgroup Theorem must involve the Faltings height and not the more primitive height corresponding to coefficients in equations.

With our original definition of abelian variety it is not possible to give a meaning to the Faltings height, except if $A = E$ is an elliptic curve, when $h^{\mathrm{F}}(E)$ is essentially $h(j(E))$ as at the end of Lecture 3. An alternative more abstract definition, also more in line with our approach to elliptic curves, consists of a lattice Ω, this time in $\mathbb{C}^n$, together with a polarization, which can be regarded as a positive definite Hermitian form $R(z,w)$ on $\mathbb{C}^n$ whose imaginary part takes rational integer values on $\Omega \times \Omega$. On the other hand with this abstract definition it is no longer so clear when the abelian variety is defined over $\overline{\mathbb{Q}}$. If $n = 1$ the presence of R is automatic – take $R(z,w)$ as the product of z with the complex conjugate of w divided by the area of a fundamental parallelogram of Ω – but if $n > 1$ its existence is a considerable restriction on Ω. This form, called a Riemann form, has a certain degree $\deg R \geq 1$ in $\mathbb{Z}$. If $\deg R = 1$, as is the case for elliptic curves, then it is possible to regard $h^{\mathrm{F}}(A)$ and $h(A)$ as equivalent, but if $\deg R > 1$ this is no longer possible; and $\deg R$ is liable to be large for the above quotient example of $\widetilde{A}$.

The two apparently different definitions of abelian variety can be reconciled by noting that $\mathbb{C}^n$ is $\mathcal{T}A$ after choosing a basis, and then Ω is $\exp^{-1}(0)$. Now exp is an analytic isomorphism from $\mathbb{C}^n/\Omega$ to $A(\mathbb{C})$. And the form R provides a natural class of projective embeddings of A, which helps to determine in the abstract definition if A is defined over $\overline{\mathbb{Q}}$ or not.

Let us return to the Analytic Subgroup Theorem and the estimates for the degree $\deg_R B_u$ of B_u now chosen relative to these embeddings. In the literature so far these have been proved with the Faltings height only when u is a period. In the original definition of abelian variety this means that $\exp u$ is the zero of A, and in the abstract definition this means that u is in Ω. The estimate involves the degree D of a number field of definition of A, as well as the distance function $\mathrm{dist}(u,v) = \sqrt{R(u-v, u-v)}$ coming from the Riemann form. It takes the shape

$$\deg_R B_u \leq C(n)\big(\max\{\mathrm{dist}(u,0),\, \deg R,\, h^{\mathrm{F}}(A),\, D\}\big)^{\kappa} \tag{5.4}$$

with $\kappa = \kappa(n) = 4^n(n+2)!$, due to Wüstholz and myself [48] in 1993. We did not prove that $C(n)$ is effective but this point has since been settled using the approach of Bost in his 1995 Séminaire Bourbaki talk [16].

Also in [16], and the references therein, can be found more details about applications to isogenies and finiteness theorems for abelian varieties; as well as to Galois representations for elliptic curves E. It is in these latter applications that one needs not only the abelian variety $E \times E$ but also its quotient by a large finite subgroup.

These estimates for $\deg_R B_u$ also point to an interesting possibility of bounding $\deg R$ itself in (5.4) in terms of $h^{\mathrm{F}}(A)$ and D. One knows in principle that such upper bounds exist, because up to isomorphism there are at most finitely many abelian varieties A defined over a field of degree D with $h^{\mathrm{F}}(A) \leq X$; but this argument is not by itself effective. There are grounds for believing that R always exists with degree controlled by some power of $\max\{h^{\mathrm{F}}(A), D\}$ analogously to (5.4), but so far this has been proved only in special cases like $n \leq 7$ or if A is simple (which means that the only group subvarieties $B \neq A$ are finite) with squarefree dimension n. The proof [49] involves a detailed study of involutions on division algebras which we cannot yet extend to a certain case in dimension $n = 8$.

This completes our discussion of transcendence and diophantine approximation for abelian varieties. We next turn to lower bounds for heights.

We now assume that A in projective space is defined over a fixed number field K, and consider a point π on $A(\overline{\mathbb{Q}})$. This has an absolute logarithmic height $h(\pi)$ as for the elliptic case in Lecture 3; only now one cannot restrict to a single coordinate and one must define it using all the coordinates as in Schlickewei's lectures (section 1) or Schmidt's course (section 8). There is also a Néron-Tate height $\widehat{h}(\pi)$ obtained with the same limiting procedure as in Lecture 3. If π is not a torsion point then $\widehat{h}(\pi) > 0$.

Take now such a π whose affine coordinates generate over K a field $K(\pi)$ of degree d over $\mathbb{Q}$. The analogue of Lehmer's Question involves the inequality $\widehat{h}(\pi) \geq cd^{-\nu}$ for some $\nu = \nu(n) > 0$ and some $c = c(A) > 0$. But now one must be a little more careful about the optimal value of ν. If A is simple then one expects $\nu = 1/n$, but if A is the product $E_1 \times E_2$ of two elliptic curves then degenerate points such as $\pi = (\pi_1, 0)$ pull $\widehat{h}(\pi)$ down to order at most d^{-1} as in Lecture 3.

In general all that is known is an old 1986 result of mine holding with any $\nu > 2n + 1$, which for $n = 1$ reduces to my even older (1989!) result [46] for elliptic curves. A proof was sketched in a letter to Bertrand, but as far as I am aware the full details have never been published; however they could be extracted without difficulty from David's 1993 article [19].

Assuming an extra condition of complex multiplication David and Hindry [22] have recently extended Laurent's Theorem from elliptic curves to abelian varieties. In terms of the description of A as $\mathbb{C}^n/\Omega$ this condition when A is simple means that the ring of matrices $\mathcal{M}$ with $\mathcal{M}\Omega \subseteq \Omega$ after tensoring with $\mathbb{Q}$ over $\mathbb{Z}$ becomes a field k of degree $2n$. They then prove that

$$\widehat{h}(\pi) \geq c(A)d^{-1/n}\big((\log 3d)/(\log\log 3d)\big)^{-\lambda}$$

with $c(A) > 0$ and $\lambda = \lambda(n) = (2n(n+1)!)^{n+2}$. The proof is long and intricate, involving in particular zero estimates in the context of isogenies which we will briefly mention in Lecture 6.

Their results are stated also for non-simple A with an additional assumption on π to exclude the degenerate points like $(\pi_1, 0)$ above. It suffices that π is of infinite order modulo every proper group subvariety $B \neq A$ of A. For example this for $A = E^n$ and a CM elliptic curve E implies the elliptic analogue of the Amoroso-David Theorem of Lecture 2 in the form

$$\widehat{h}(\pi_1)\cdots\widehat{h}(\pi_n) \geq c(n,E)d^{-1}\left((\log 3d)/(\log\log 3d)\right)^{-\lambda'}$$

for $c(n,E) > 0$, $\lambda' = \lambda'(n)$ and any k-linearly independent $\pi_1, \ldots, \pi_n$ in $E(\overline{\mathbb{Q}})$ whose coordinates generate over K a field of degree d.

Let us return back to unrestricted A. As for the elliptic case in Lecture 3, some of the constants $c(A)$ in the above lower bounds have been further studied. The conjecture of Lang was generalized to abelian varieties in 1981 by Silverman [77] (p. 396), and some partial results were obtained in 1993 by David [19]. But for arbitrary A only negative exponents are known. To be precise we must take into account the Riemann form R and now write $\widehat{h}_R$ for the Néron-Tate height associated with the natural class of projective embeddings. Then my old 1988 result in [45] (Corollary 1 p. 110), stated in terms of something like $h(A)$, implies in terms of the Faltings height that

$$\widehat{h}_R(\pi) \geq c(K)\left(\max\{1,\, h^{\mathrm{F}}(A),\, \log\deg R\}\right)^{-(2n+1)}$$

(with $c(K) > 0$) for all non-torsion π in $A(K)$.

If we do want a lower bound that tends to infinity and not to zero, then we must be patient and wait for the abelian analogues of David's work [21].

Finally we discuss lower bounds of Zhang type, which we omitted in Lectures 3 and 4 (Sections 3, 4).

Let A be an abelian variety defined over $\overline{\mathbb{Q}}$, and let V be a subvariety also defined over $\overline{\mathbb{Q}}$, not necessarily a group subvariety. Now we have to remove from V all torsion translates $\tau + B$ in V of all connected group subvarieties B, proper or not, of A; let V^* be what remains. Again this is Zariski open in V. It was Zhang in 1998 who proved (via Galois Equidistribution shortly after the work of Ullmo [85] on curves) in [101] that

$$\widehat{h}(\pi) \geq c(V) > 0 \tag{5.5}$$

for all π in $V^*(\overline{\mathbb{Q}})$. The CM case had been settled earlier in 1996 by Bombieri and Zannier [14] (Theorem 2 p. 782) using Frobenius arguments as in Lecture 2.

To give effective or explicit forms for the constant $c(V)$ one must work harder; great advances have been made recently by David and Philippon [24]. In particular their step (UB) involves delicate volume computations of an apparently new kind. But even for curves V in a product $E \times E$ of elliptic curves their step (AP) with auxiliary polynomials is very elaborate; they use not quite a function $P(\wp(z), \wp(bz))$ as in Lecture 3, but rather

$$\varphi(z) = P\big(\wp(z), \wp(2z), \ldots, \wp(2^\ell z)\big)$$

where the integer ℓ is also large depending on V; almost a "2-action" as in Mahler's Method. Again one has to be careful: the various elliptic constituents in $\varphi(z)$ are not only algebraically dependent but any two of them are, and now Hilbert counting functions are needed to ensure that $\varphi(z)$ is not identically zero. However David points out to me that the simpler $P(\wp(z), \wp(bz))$ could also have been used.

In fact most of these lower bound results are formulated for a Néron-Tate height $\widehat{h}(V)$ of V itself, which was introduced by Philippon [59] in 1991 (see also [60], [61]). If V is a single point π then $\widehat{h}(V) = \widehat{h}(\pi)$. If V has dimension $m \geq 1$ then $\widehat{h}(V)$ is rather more complicated, and we omit the definition. At any rate the link with (5.5) can be described as follows. For any quasi-projective variety U defined over $\overline{\mathbb{Q}}$ denote by $\mu(U)$ the infimum of $\widehat{h}(\pi)$ as π runs over all points of $U(\overline{\mathbb{Q}})$. Thus we are interested in $\mu(V^*)$. Define $\mu^{\text{ess}}(V)$ as the supremum of $\mu(V \setminus W)$ as W runs over all subvarieties of V. Then this is closely related to $\widehat{h}(V)$, and in 1995 Zhang [100] proved that

$$\mu^{\text{ess}}(V) \leq \widehat{h}(V)/(\deg V) \leq (m+1)\mu^{\text{ess}}(V).$$

So lower bounds for $\widehat{h}(V)$ lead immediately to lower bounds for $\mu^{\text{ess}}(V)$; however V^* is a $V \setminus W$ as above and therefore $\mu^{\text{ess}}(V) \geq \mu(V^*)$ goes in the wrong direction. So lower bounds for $\mu(V^*)$ are not so immediate. For further details see [24].

Similar considerations apply to the multiplicative situation of Lecture 2; see [25] and [3].

Now $\deg\{(\pi_1, \pi_2) \text{ in } E^2\,;\ \pi_2 = 2\pi_1\}$ down; and the number left is the minimal degree of any polarization of $J^2/\{(\pi,\pi)\,;\ 5\pi = 0\}$.

6 Sixth Lecture. Commutative group varieties

We take as in Lecture 5, a connected group variety G in some $\mathbb{P}_N$, but now we drop the requirement that it is a projective variety. In fact it may be assumed without loss of generality to be quasi-projective of the form $V \setminus W$ for projective V and W. But then commutativity is no longer automatic, and if we don't want to let in examples like GL_2 then we have explicitly to demand that the group is commutative.

So $\mathbb{G}_a$ and $\mathbb{G}_m$ $(= GL_1)$ are allowed back in, as well as products like $L = \mathbb{G}_a^r \times \mathbb{G}_m^s$ or $L \times A$ for an abelian variety A. These are not quite all examples; but every commutative group variety over $\mathbb{C}$ or $\overline{\mathbb{Q}}$ sits in an exact sequence

$$0 \to L \to G \to A \to 0 \qquad (6.1)$$

which splits if $G = L \times A$ but usually not.

The simplest example of non-splitting is with $L = \mathbb{G}_a$ and $A = E$ an elliptic curve; this was referred to in Lectures 1 and 3. Then G can be embedded in $\mathbb{P}_4$ as $V \setminus W$; here V is the surface defined by

$$X_0X_2^2 = 4X_1^3 - g_2X_0^2X_1 - g_3X_0^3, \qquad X_0X_4 - X_2X_3 = 2X_1^2,$$

and W is the line defined by $X_0 = X_1 = X_2 = 0$. The two inner maps in (6.1), call them "in" from $\mathbb{G}_a \subset \mathbb{P}_1$ to G and "pr" from G to $E \subset \mathbb{P}_2$, are given by

$$\mathrm{in}(X_2, X_4) = (0, 0, X_2, 0, X_4) \qquad (X_2 \neq 0)$$

and of course

$$\mathrm{pr}(X_0, X_1, X_2, X_3, X_4) = (X_0, X_1, X_2).$$

These equations are perhaps more amusing than illuminating, and for example I don't know how to express the group law without using the exponential map (see below).

The case $L = \mathbb{G}_m$, $A = E$ can be similarly embedded in $\mathbb{P}_4$.

The case $r = 0$ in (6.1) with $L = \mathbb{G}_m^s$ is called a semiabelian variety.

There seem to be no convenient modern expositions of the theory of commutative group varieties, although most of the standard texts on linear algebraic groups (see for example [15], [34], [82]) contain some material on algebraic groups or group varieties in general. One could also consult Serre's book [75] and the references therein, or parts of [87] especially the Appendix [76] also written by Serre.

As for $G = A$ there is the tangent space $\mathcal{T}G$ at the origin and the exponential map exp from $\mathcal{T}G(\mathbb{C})$ to $G(\mathbb{C})$. In the above example $L = \mathbb{G}_a$, $A = E$, we can identify $\mathcal{T}G(\mathbb{C})$ with $\mathbb{C}^2$ and then

$$\exp(z, y) = \big(1,\, \wp(z),\, \wp'(z),\, y + \zeta(z),\, y\wp'(z) + \zeta(z)\wp'(z) + 2(\wp(z))^2\big)$$

at least for z not in the period lattice, contains the Weierstrass zeta-function $\zeta(z)$. The group law involves the addition theorem expressing $\zeta(z+w)-\zeta(z)-\zeta(w)$ in terms of $\wp(z)$, $\wp'(z)$, $\wp(w)$, $\wp'(w)$.

Similarly the case $L = \mathbb{G}_m$, $A = E$ contains the Weierstrass sigma-function $\sigma(z)$. For more details on these examples see [87] (pp. 64-68), and see the article [30] of Faltings and Wüstholz for the general case (6.1).

We can also define group subvarieties H in G and the corresponding tangent spaces $\mathcal{T}H$. Only now examples like $y = \pi x$ ($\pi = 3.14159\ldots$) in $\mathbb{G}_a^2$ show that H need no longer be defined over $\overline{\mathbb{Q}}$ if G is. So the rigidity is slightly relaxed; and furthermore if the exact sequence (6.1) is not split then it might not be so easy to describe the group subvarieties so explicitly as we have done up to now (see however p. 8 of Bertrand's article [8]).

The first transcendence theorems on group varieties as such were given by Lang [35] in 1962. As with abelian varieties, most of the subsequent advances regarding linear independence are included in Wüstholz's Analytic Subgroup Theorem [96]. This can be stated almost exactly as in Lecture 5.

Suppose that G is defined over $\overline{\mathbb{Q}}$ and u is in $\exp^{-1} G(\overline{\mathbb{Q}})$. Then the smallest $\overline{\mathbb{Q}}$-vector subspace Z_u of $\mathcal{T}G$ containing u has the form $\mathcal{T}H_u$ for some (unique) connected group subvariety H_u of G itself defined over $\overline{\mathbb{Q}}$.

When G has dimension 2 this result includes all the known applications of the Schneider-Lang Theorem mentioned in Lecture 3.

Or we saw that all group subvarieties of $G = \mathbb{G}_m^n$ are defined by relations $x_1^{b_1} \cdots x_n^{b_n} = 1$, and we thereby recover Baker's Theorem on $\log \alpha_1, \ldots, \log \alpha_n$. Similarly if we take $G = \mathbb{G}_a \times \mathbb{G}_m^n$ then we get the same Theorem on $1, \log \alpha_1, \ldots, \log \alpha_n$. The abelian case $G = A$ is exactly as in Lecture 5.

We will try to give some idea of the proof according to the basic pattern of Lecture 1, assuming for simplicity that Z_u has dimension $n-1$, where $n = \dim G$. This is the most frequently occurring situation for applications.

For the step (AP) we construct an auxiliary polynomial $P(X_0, \ldots, X_N)$, now homogeneous, that does not vanish identically on G but has a zero of large order at least T "along Z_u" at many points of $\mathbb{Z}u$. What does "along Z_u" mean? We have maps

$$\mathbb{Z}u \to Z_u \to \mathcal{T}G(\mathbb{C}) \to G(\mathbb{C}) \to \mathbb{C},$$

where the first two are natural inclusions, the third is exp, and the last comes from P after making a choice of affine coordinates. Choose any basis elements $\partial_1, \ldots, \partial_{n-1}$ of Z_u; then a meaning can be given to $\mathcal{D} = \partial_1^{t_1} \cdots \partial_{n-1}^{t_{n-1}}$ applied to $\varphi(z) = P(\exp z)$ at any point z of $\mathcal{T}G(\mathbb{C})$. In fact the group law can be used to extend the derivations from the coordinate field to itself thus allowing composition. More concretely the group law amounts essentially to addition theorems involving $\exp(z+w)$ and these imply partial differential equations (generalizing (3.1) of Lecture 3, and those needed in the Schneider-Lang Theorem). For example if ∂ is in $\mathcal{T}G$ then $\partial\varphi(z)$ can be defined as $(d/dy)\varphi(z + y\partial)$ at $y = 0$.

The conditions

$$\mathcal{D}\varphi(z) = 0 \qquad (|\mathcal{D}| = t_1 + \cdots + t_{n-1} < T)$$

are now independent of the basis.

We choose $z = su$ for rational integers s subject to some extra condition $0 \leq s < S$. The number of linear equations for Siegel's Lemma is of order $T^{n-1}S$. Of course for this Lemma we must choose the basis over $\overline{\mathbb{Q}}$ or more precisely over a suitable number field. If L is the degree of P, the elementary theory of Hilbert functions shows that the number of free coefficients in P is of order L^n. Thus we take

$$T^{n-1}S = \varepsilon(n)L^n$$

for a small positive quantity $\varepsilon(n)$ depending only on n.

The next step (UB) for upper bounds is carried out with Baker's analytic method, after making a final choice of basis elements $\partial_1, \ldots, \partial_{n-1}$ to get well-defined meromorphic functions on $\mathbb{C}^{n-1}$. It delivers $\widetilde{T} = T/2$ and $\widetilde{S} = CS$ with a big constant C, so that the $\mathcal{D}\varphi(su)$ $(|\mathcal{D}| < \widetilde{T},\ 0 \le s < \widetilde{S})$ are small.

The step (LB) for lower bounds then enables us to assume that these values are zero; that is

$$\mathcal{D}\varphi(su) = 0 \qquad (|\mathcal{D}| < \widetilde{T},\ 0 \le s < \widetilde{S}), \tag{6.2}$$

and this is an advance on the previous situation because we now have

$$\widetilde{T}^{n-1}\widetilde{S} = \mathcal{E}(n)L^n \tag{6.3}$$

for some large $\mathcal{E}(n)$ (capital epsilon!) depending only on n.

The last step (NV), to find a non-vanishing value $\mathcal{D}\varphi(su)$, consists of "zero estimates" for G. We will see some details later, in particular how they deliver a connected group subvariety $H \neq 0$ defined over $\overline{\mathbb{Q}}$. If we are lucky then $\mathcal{T}H = Z_u$ and we are finished; but all we really know at this point is the inclusion $\mathcal{T}H \subseteq Z_u$.

We can finish anyway using induction on the dimension n of G. The quotient $\overline{G} = G/H$ turns out automatically to be a group variety; it has smaller dimension, with tangent space $\mathcal{T}\overline{G} = \mathcal{T}G/\mathcal{T}H$. And $Z_{\overline{u}} = Z_u/\mathcal{T}H$ is the smallest $\overline{\mathbb{Q}}$-subspace containing the coset $\overline{u} = u + \mathcal{T}H$. So the inductive result for $\overline{G}$ gives $Z_{\overline{u}} = \mathcal{T}H_{\overline{u}}$ for some connected group subvariety $H_{\overline{u}}$ in $\overline{G} = G/H$. As in ordinary group theory $H_{\overline{u}} = H_u/H$ for some connected group subvariety H_u in G; and we conclude $Z_u = \mathcal{T}H_u$ as desired.

This completes our sketch of the proof. For full details one can consult the original article [96] of Wüstholz, which uses his original zero-estimate [95], the first of its kind properly to treat multiplicities in more than one variable. Or there is a different proof in Waldschmidt's article [88] (see Corollary 3.1 p. 380).

What do the quantitative versions look like? We have already discussed the case "$\Lambda = 0$" for abelian varieties. A more general result estimating the degree $\deg H_u$ of H_u in terms of quantities like $h(G)$ was established by Waldschmidt [89] (see Théorème 2.1. p. 67) in 1997. A Faltings height $h^{\mathrm{F}}(G)$ has yet to be invented.

What about $\Lambda \neq 0$? And what is Λ anyway? We reformulate the Analytic Subgroup Theorem as follows. Take an algebraic point u as above and any $\overline{\mathbb{Q}}$-subspace Z in $\mathcal{T}G$ with the following property. For every connected group subvariety H of G defined over $\overline{\mathbb{Q}}$ with $\mathcal{T}H \subseteq Z$, the point u does not lie in $\mathcal{T}H$. In practice this is not too hard to check if we have a good description of all possible H. When it is indeed checked, the Analytic Subgroup Theorem enables us to conclude that u does not lie in Z. Thus $\Lambda \neq 0$ for some distance function $\Lambda = \mathrm{dist}(u, Z)$, and so we can ask for lower bounds for this distance in terms of u and Z (and G).

To prove these bounds a new idea is required; even supposing that we have projective embeddings of G and H, it may be difficult to derive a projective embedding of $\overline{G} = G/H$ in the inductive argument above. The quotient makes its usual trouble. It was Philippon and Waldschmidt [62] in 1988 who saw how to use the zero-estimate itself to overcome this difficulty, by interpreting the resulting information in terms of Hilbert functions to construct a new auxiliary polynomial. In 1991 Hirata-Kohno [33] was able to improve their lower bound to one of the shape

$$\log \operatorname{dist}(u, Z) \geq -C(?)(\log H_\nabla)(\log\log H_\nabla)^{n+1}. \tag{6.4}$$

But now the quantities H_∇ and $C(?)$ need some extra explanation.

As soon as we fix a basis ∇ for $\mathcal{T}G$ then the subspace Z can be described by linear equations over $\overline{\mathbb{Q}}$, and so its (non-logarithmic) height H_∇ can be defined as in Lecture 2, either by simply using the coefficients or with Grassmannians.

In $C(?)$ one needs the basis ∇ and of course u and G but also some choice of distance function on $\mathcal{T}G$. For abelian varieties we saw that this comes naturally with a polarization (or embedding), but in general there seems to be no natural choice. Finally one needs the degree D of some number field containing everything under consideration. Then Waldschmidt [89] (also in Théorème 2.1 p. 67) has given estimates of the form

$$C(?) \leq C(n, \operatorname{dist})\big(\max\{\operatorname{dist}(u,0),\, h(G),\, h(\exp u),\, D\}\big)^{\nu}$$

for an exponent $\nu = \nu(n)$ of order n.

In principle such a result for abelian varieties should suffice for applications to integral points on curves of arbitrary genus as described in Lecture 4 for elliptic curves, provided one can find a basis for the rational Mordell-Weil group. Despite the considerable recent progress in explicit computations of this sort for genus $g = 2$ (see again [32] and the references therein), it will probably be a long time before anyone carries out such a programme in practice.

It is not yet proved that the term $\log\log H_\nabla$ in (6.4) can be eliminated, but it should be soon, especially for abelian varieties.

Now we will discuss the above step (NV) of the qualitative proof of the Analytic Subgroup Theorem in more detail. Our eventual aim is to show how the equations (6.2) lead to the connected group subvariety $H \neq 0$ with $\mathcal{T}H \subseteq Z_u$. The fact that H is defined over $\overline{\mathbb{Q}}$ will be clear without further comment. So we can henceforth forget about $\overline{\mathbb{Q}}$ and think of everything as defined just over $\mathbb{C}$. But first we discuss the zero-estimates themselves.

The counting behind the relation (6.3) suggests that these equations (6.2), when considered as linear equations in the coefficients of the auxiliary polynomial P, form an over-determined system. If so, they would force the identical vanishing of P on G and thereby provide us with a contradiction. And indeed for u that is "generic" in some sense it really is so. Here "non-generic" means that a connected group subvariety H of G exists with $H \neq G$ and

$$\widetilde{T}^e \cdot \widetilde{S}_H \cdot (\deg H)L^m \leq C(n)(\deg G)L^n. \tag{6.5}$$

Here

$$m = \dim H, \qquad e = n - 1 - \dim_{\mathbb{C}}(Z_u \cap \mathcal{T}H),$$

for the algebraic and vector space dimensions respectively, and $\widetilde{S}_H$ is the cardinality of the set of $s\pi$ $(0 \le s < \widetilde{S})$ taken modulo H, with $\pi = \exp u$. Actually we could omit the term $\deg H \ge 1$, as it is needed only for the case $\Lambda = 0$; and we could also omit the term $\deg G$, which for our present purposes is constant. This zero-estimate, telling us that (6.2) implies (6.5) for any $\widetilde{T}$, $\widetilde{S}$ and P, was proved in 1986 by Philippon [58] in a rather more general form.

But we wanted the conditions $H \ne 0$ and $\mathcal{T}H \subseteq Z_u$, not $H \ne G$. How do we get these?

First, if $H = 0$ then $m = 0$, $e = n - 1$ and $\widetilde{S}_H \ge \widetilde{S}$; this last at least assuming that π is a non-torsion point. So (6.5) becomes, using for brevity the symbol $\ll$ for constants depending only on G (and n), just $\widetilde{T}^{n-1}\widetilde{S} \ll L^n$. Now we can rule this out by (6.3) if $\mathcal{E}(n)$ is sufficiently large. So $H \ne 0$ as required.

Secondly, if $\mathcal{T}H \not\subseteq Z_u$ then $Z_u \cap \mathcal{T}H$ is not $\mathcal{T}H$, and on recalling that $\dim Z_u = n - 1$ we see that $\dim(Z_u \cap \mathcal{T}H) = m - 1$. So $e = n - m$. Now (6.5) implies $\widetilde{T}^{n-m} \ll L^{n-m}$. As $m < n$ we get $\widetilde{T} \ll L$. In practice $\widetilde{S}$ should certainly be large, but not that large and one can ensure from (6.3) that $\widetilde{T} \gg L^{\tau}$ for some $\tau > 1$. So this possibility can also be ruled out, and $\mathcal{T}H \subseteq Z_u$ also as required.

To get $\widetilde{S}_H \ge \widetilde{S}$ just above we had assumed that $\pi = \exp u$ was non-torsion. But the torsion case needs a word; for example the case $\pi = 0$ of periods was used in the application to isogenies in Lecture 5. Here one can argue either as in Kummer theory with division values $\exp(u/q)$, or with a more subtle use of derivatives.

And finally, let us try to explain what sort of arguments lie behind Philippon's zero-estimate (6.5). Similar estimates were also referred to by Schlickewei (Roth's Lemma and a version of the Faltings Product Theorem in his lectures, section 5) and by Waldschmidt (results of Pólya in section 1.4 of his lectures, Philippon in section 3.2 c, Tijdeman in section 3.3, and others) so we feel obliged to give some idea of what is going on. Here we are especially interested in explaining the appearance of the "obstruction subgroup" H. We retain our assumption above that π is not a torsion point of G.

To avoid complications with large dimension let us restrict ourselves to the case $n = 2$. We listed all the types of two-dimensional group variety in Lecture 3, but we want to sketch here a proof independent of the type. In fact for $n = 2$ we are in the situation of Schneider-Lang and we saw that there was no need for zero-estimates, but nevertheless they are often indispensable in obtaining quantitative refinements.

Now $\mathcal{T}G(\mathbb{C})$ can be identified with $\mathbb{C}^2$, and we suppose that Z_u is defined by an equation $y = \beta z$. So $u = (\zeta, \beta\zeta)$ for some complex number ζ. We can take $\partial_1 = d/dz$, so that our conditions (6.2) and (6.3) look like

$$(d/dz)^t P(\exp(z, \beta z)) = 0 \qquad (0 \le t < \widetilde{T},\ z = s\zeta,\ 0 \le s < \widetilde{S})$$

$$\widetilde{T}\widetilde{S} = \mathcal{E}L^2.$$

We could attack such equations with resultants as mentioned in Lecture 1, but these are not the optimal tools in $\mathbb{P}_N$ and to see more clearly the obstruction subgroup it seems better to use the language of intersections and Bézout's Theorem. The vanishing of P defines a curve V in G passing through the points $s\pi = \exp(s\zeta,\ s\beta\zeta)$ $(0 \le s < \widetilde{S})$, and the vanishing of the derivatives as well means that P passes through these points with multiplicity at least $\widetilde{T}$ in some suitable sense.

Consider the translate $V_1 = V - \pi$. This curve passes through the points $s\pi - \pi$, and if we ignore one of these we get the points $s\pi$ $(0 \le s < \widetilde{S} - 1)$. Similarly for $V_2 = V - 2\pi$; and so on. If we go all the way up to $\widetilde{S}$ then we lose all interesting information, but if we stop halfway then we see that the curves $V_i = V - i\pi$ $(0 \le i < \widetilde{S}/2)$ pass through the points $s\pi$ $(0 \le s \le \widetilde{S}/2)$. And translating doesn't affect the multiplicity, which remains at least $\widetilde{T}$.

As well as translating we can "differentiate" V, also along Z_u, to get a curve V'. This is less easy to see geometrically but the differential equations for the exponential map imply that if ∂ is in $\mathcal{T}G$ and $\mathcal{X}$ is a monomial in the affine coordinates $X_1/X_0, \ldots, X_N/X_0$ then $\partial\mathcal{X}$ is a rational function on G. In fact under reasonable asssumptions $\partial\mathcal{X}$ is in the affine coordinate ring, and this in turn implies that $(d/dz)P(\exp(z, \beta z))$ has the form $Q(\exp(z, \beta z))$ for some homogeneous polynomial Q. So the derived V' is defined by the vanishing of Q. But it vanishes now only to multiplicity at least $\widetilde{T} - 1$.

And similarly with $V'', \ldots$, going for the same reasons only up to halfway. Finally we can translate and differentiate simultaneously to get curves $V_i^{(j)}$ $(0 \le j < \widetilde{T}/2,\ 0 \le i < \widetilde{S}/2)$, and these each have the property that they pass through the points $s\pi$ $(0 \le s \le \widetilde{S}/2)$ with multiplicities bigger than $\widetilde{T}/2$.

Consider now the intersection W of all these $V_i^{(j)}$, which has the same property. In particular W is not empty. There are two possibilities for the dimension of W.

(a) <u>$\dim W = 0$</u>. This case is somehow generic; as we observed in Lecture 1, it is to be expected that several curves intersect in a finite set. Since π is not torsion, W consists of more than $\widetilde{S}/2$ points; but the multiplicities mean that these are to be considered as $\widetilde{T}\widetilde{S}/4$ points. On the other hand, any primitive version of Bézout's Theorem gives the upper bound $\ll L^2$. It follows that $\widetilde{T}\widetilde{S} \ll L^2$, which implies (6.5) with the obstruction subgroup $H = 0$.

(b) <u>$\dim W = 1$</u>. This is non-generic, and therefore can be expected to provide unusual information. Now some irreducible component W^0 of W goes through the origin 0. Its stabilizer

$$H = \{g \text{ in } G;\ g + W^0 = W^0\}$$

is a group subvariety of G with $H \subseteq W^0$. So $\dim H \le 1$.

The case $\dim H = 0$ is handled much as in (a) above, so we suppose right away that $\dim H = 1$. It follows that $H = W^0$ is connected.

Let us pretend at first that there are no derivatives involved (as if $\widetilde{T} = 1$). Now

$$H = W^0 \subseteq W = \bigcap_i V_i \subseteq V_i$$

so every translate $H_{-i} = H + i\pi$ $(0 \leq i < \widetilde{S}/2)$ lies in V. How many different such H_{-i} are there? Well, a coincidence $H_{-i} = H_{-i'}$ would mean

$$(i\pi - i'\pi) + H = H$$

and so $i\pi - i'\pi$ lies in H. Thus we have to count the multiples of π modulo H, and this leads essentially to the quantity $\widetilde{S}_H$ defined above (we are going up to $\widetilde{S}/2$ not $\widetilde{S}$ but that makes no essential difference). Each H_{-i} has degree $\gg \deg H$ (something might be lost in the translation!), and so the total degree of all the different H_{-i} is $\gg \widetilde{S}_H(\deg H)$. This lies in V with degree $\deg V \ll L$. We conclude $\widetilde{S}_H(\deg H) \ll L$ which is (6.5) with $m = 1$ (and $\widetilde{T} = \widetilde{T}^e = 1$).

If we do take derivatives into account then we get inclusions like $H \subseteq V^{(j)}$ $(0 \leq j < \widetilde{T}/2)$. These are analogous to saying that X divides $(d/dX)^j P(X)$ $(0 \leq j < \widetilde{T}/2)$ in the simple polynomial ring $\mathbb{C}[X]$. It follows from the latter that X^ℓ divides $P(X)$ for some integer $\ell \geq \widetilde{T}/2$. So the multiplicities might again be expected to push the above inequalities up to $\widetilde{T}\widetilde{S}_H(\deg H) \ll L$ as in (6.5).

But where has the missing exponent e of $\widetilde{T}$ gone? Here we approach the heart of the general zero-estimate, and a proper explanation requires Wüstholz's Lemma [95] (p. 488) for which there is no space here. Roughly speaking the above analogy with $\mathbb{C}[X]$ was a little too simple. The inclusions $H \subseteq V^{(j)}$ $(0 \leq j < \widetilde{T}/2)$ can be shown to imply either that the length ℓ associated with H as a component of V satisfies $\ell \geq \widetilde{T}/2$; or that H is identical with its derived curve H'. In the former case the multiplicities do have the desired effect, and we are finished. In the latter case H must vanish along Z_u to infinite order, which is possible only if $Z_u = \mathcal{T}H$. This means $e = 0$ and so the apparently missing e was never there anyway. Such is typical of the general situation.

This completes our sketch of some of the ideas behind (6.5), at least for two-dimensional G. The proof for general G is similar but there are rather more possibilities for the various dimensions. For full details one can consult the original article [58] of Philippon as well as Roy's exposition [63], and there is a more geometric version of the proof due to Denis [26].

The scheme of the zero-estimates used by Amoroso and David (Lecture 2) is also similar. In two dimensions, from a curve V defined by the vanishing of $P(X_1, X_2)$ one gets $V^{[q]}$ defined by the vanishing of $Q(X_1, X_2) = P(X_1^q, X_2^q)$ and one looks at intersections. This corresponds to multiplication by q on $\mathbb{G}_m^2$; which however can be performed on any group variety (commutative or not), and in this way one arrives at the zero estimates used by David and Hindry (Lecture 5).

Coming full circle, we note that it is possible to do similar things in Mahler's Method (Lecture 1) with the polynomials $P(X,Y)$ and $Q(X,Y) = P(X^2, Y - X)$. Only here there is no underlying group variety.

[At this point the speaker thanked Francesco Amoroso and Umberto Zannier for their excellent organization of the conference in such an elegant hotel with its charming gardens and magnificent views over land and sea; and, exchanging his lecturing glasses for sunglasses, strolled off to enjoy a holiday in the same surroundings.]

References

1. M. Ably. Formes linéaires de logarithmes de points algébriques sur une courbe elliptique de type CM; Annales Inst. Fourier. 50 (1998), 1-33.
2. F. Amoroso and S. David. Le problème de Lehmer en dimension supérieure; J. reine angew. Math. 513 (1999), 145-179.
3. F. Amoroso and S. David. Minoration de la hauteur normalisée des hypersurfaces; Acta Arithmetica 92 (2000), 339-366.
4. F. Amoroso and R. Dvornicich. Lower bound for the height in abelian extensions; J. Number Theory 80 (2000), 260-272.
5. F. Amoroso and U. Zannier. A relative Dobrowolski's lower bound over abelian extensions; Ann. Scuola Norm. Sup. Pisa 29 (2000), 711-727.
6. A. Baker. Transcendental number theory; Cambridge 1975.
7. D. Bertrand. A transcendence criterion for meromorphic functions, Transcendence theory: advances and applications (eds. A. Baker and D.W. Masser); Academic Press 1977 (pp. 187-193).
8. D. Bertrand. Endomorphismes de groupes algébriques; applications arithmétiques, Approximations diophantiennes et nombres transcendants (eds. D. Bertrand and M. Waldschmidt); Progress in Math. 31, Birkhäuser 1983 (pp. 1-45).
9. D. Bertrand. $\Theta(\tau, z)$ and transcendence, Introduction to algebraic independence theory (eds. Yu.V. Nesterenko and P. Philippon); Lecture Notes in Math. 1752, Springer 2001 (pp. 1-11).
10. Yu. Bilu. Limit distribution of small points on algebraic tori; Duke Math. J. 89 (1997), 465-476.
11. E. Bombieri. Algebraic values of meromorphic maps; Inventiones Math. 10 (1970), 267-287.
12. E. Bombieri and J. Vaaler. On Siegel's Lemma; Inventiones Math. 73 (1983), 11-32.
13. E. Bombieri and U. Zannier. Algebraic points on subvarieties of $\mathbb{G}_m^n$; Int. Math. Research Notices 7 (1995), 333-347.
14. E. Bombieri and U. Zannier. Heights of algebraic points on subvarieties of abelian varieties; Ann. Scuola Norm. Sup. Pisa 23 (1996), 779-792.

15. A. Borel. Linear algebraic groups; Benjamin 1969.
16. J.-B. Bost. Périodes et isogénies des variétés abéliennes sur les corps de nombres; Astérisque 237 (1996), 115-161; and corrected version dated June 1998.
17. J.W.S. Cassels. Local fields; London Math. Soc. Student Texts 3, Cambridge 1986.
18. P. Corvaja and U. Zannier. Some new applications of the Subspace Theorem; Compositio Math. 131 (2002), 319-340.
19. S. David. Minorations de hauteurs sur les variétés abéliennes; Bull. Soc. Math. France 121 (1993), 509-544.
20. S. David. Minorations de formes linéaires de logarithmes elliptiques; Mém. Soc. Math. France 62 (1995), 1-143.
21. S. David. Points de petite hauteur sur les courbes elliptiques; J. Number Theory 64 (1997), 104-129.
22. S. David and M. Hindry. Minoration de la hauteur de Néron-Tate sur les variétés abéliennes de type C.M.; J. reine angew. Math. 529 (2000), 1-74.
23. S. David and N. Hirata-Kohno. Recent progress on linear forms in elliptic logarithms; to appear in A Panorama of Number Theory or The View from Baker's Garden (ed. G. Wüstholz), Cambridge 2002.
24. S. David and P. Philippon. Minorations des hauteurs normalisées des sous-variétés de variétés abéliennes; Contemporary Math. 210 (1998), 333-364.
25. S. David and P. Philippon. Minorations des hauteurs normalisées des sous-variétés des tores; Ann. Scuola Norm. Sup. Pisa 28 (1999), 489-543. Errata, ibid. 29 (2000), 729-731.
26. L. Denis. Lemmes de multiplicités et intersection; Commentarii Math. Helv. 70 (1995), 235-247.
27. E. Dobrowolski. On a question of Lehmer and the number of irreducible factors of a polynomial; Acta Arithmetica 34 (1979), 391-401.
28. P. Erdős and P. Turán. On the distribution of roots of polynomials; Annals of Math. 51 (1950), 105-119.
29. G. Faltings. Finiteness theorems for abelian varieties over number fields, in Arithmetic geometry (eds. G. Cornell and J.H. Silverman); Springer 1986 (pp. 9-27).
30. G. Faltings and G. Wüstholz. Einbettungen kommutativer algebraischer Gruppen und einige ihrer Eigenschaften; J. reine angew. Math. 354 (1984), 175-205.
31. A.I. Galochkin. Transcendence measures of values of functions satisfying certain functional equations; Math. Notes 27 (1980), 83-88.
32. D. Grant. Formal groups in genus two; J. reine angew. Math. 411 (1990), 96-121.
33. N. Hirata-Kohno. Formes linéaires de logarithmes de points algébriques sur les groupes algébriques; Inventiones Math. 104 (1991), 401-433.
34. J.E. Humphries. Linear algebraic groups; Graduate Texts in Math. 21, Springer 1975.
35. S. Lang. Transcendental points on group varieties; Topology 1 (1962), 313-318.
36. S. Lang. Introduction to transcendental numbers; Addison-Wesley 1966.
37. S. Lang. Elliptic curves: diophantine analysis; Grundlehren math. Wiss. 231, Springer 1978.
38. S. Lang. Fundamentals of diophantine geometry; Springer 1983.
39. H. Lange and Ch. Birkenhake. Complex abelian varieties; Grundlehren math. Wiss. 302, Springer 1992.

40. M. Laurent. Minoration de la hauteur de Néron-Tate, Séminaire de théorie des nombres de Paris 1981-82 (ed. M.J. Bertin); Progress in Math. 38, Birkhäuser 1983 (pp. 137-152).
41. M. Laurent. Equations diophantiennes exponentielles; Inventiones Math. 78 (1984), 299-327.
42. D.H. Lehmer. Factorization of certain cyclotomic functions; Annals of Math. 34 (1933), 461-479.
43. K. Mahler. On the transcendency of the solutions of a special class of functional equations; Bull. Australian Math. Soc. 13 (1975), 389-410, and Corrigendum, ibid. 14 (1976), 477-478.
44. D.W. Masser. Elliptic functions and transcendence; Lecture Notes in Math. 437, Springer 1975.
45. D.W. Masser. Small values of heights on families of abelian varieties, Diophantine approximation and transcendence theory (ed. G. Wüstholz); Lecture Notes in Math. 1290, Springer 1988, (pp. 109-148).
46. D.W. Masser. Counting points of small height on elliptic curves; Bull. Soc. Math. France 117 (1989), 247-265.
47. D.W. Masser and G. Wüstholz. Estimating isogenies on elliptic curves; Inventiones Math. 100 (1990), 1-24.
48. D.W. Masser and G. Wüstholz. Periods and minimal abelian subvarietes; Annals of Math. 137 (1993), 407-453.
49. D.W. Masser and G. Wüstholz. Polarization estimates for abelian varieties; Preprint 2000 (40 pages).
50. M. Mignotte. Sur un théorème de M. Langevin; Acta Arithmetica 54 (1989), 81-86.
51. W. Miller. Transcendence measures by a method of Mahler; J. Australian Math. Soc. A 32 (1982), 68-78.
52. J.S. Milne. Abelian varieties; Arithmetic geometry (eds. G. Cornell and J.H. Silverman), Springer 1986 (pp. 103-150).
53. J.S. Milne. Jacobian varieties; Arithmetic geometry (eds. G. Cornell and J.H. Silverman), Springer 1986 (pp. 167-212).
54. D. Mumford. Abelian varieties; Oxford 1974.
55. Yu.V. Nesterenko. Algebraic independence for values of Ramanujan functions, Introduction to algebraic independence theory (eds. Yu.V. Nesterenko and P. Philippon); Lecture Notes in Math. 1752, Springer 2001 (pp. 27-46).
56. K. Nishioka. Mahler functions and transcendence; Lecture Notes in Math. 1631, Springer 1996.
57. F. Pellarin. Sur une majoration explicite pour un degré d'isogénie liant deux courbes elliptiques; Acta Arithmetica 100 (2001), 203-243.
58. P. Philippon. Lemmes de zéros dans les groupes algébriques commutatifs; Bull. Soc. Math. France 114 (1986), 355-383 and ibid. 115 (1987), 397-398.
59. P. Philippon. Sur les hauteurs alternatives I; Math. Annalen 289 (1991), 255-283.
60. P. Philippon. Sur les hauteurs alternatives II; Annales Inst. Fourier 44 (1994), 1043-1065.
61. P. Philippon. Sur les hauteurs alternatives III; J. Math. Pures Appl. 74 (1995), 345-365.
62. P. Philippon and M. Waldschmidt. Formes linéaires de logarithmes sur les groupes algébriques commutatifs; Illinois J. Math. 32 (1988), 281-314.

63. D. Roy. Zero estimates on commutative algebraic groups, Introduction to algebraic independence theory (eds. Yu.V. Nesterenko and P. Philippon); Lecture Notes in Math. 1752, Springer 2001 (pp. 167-186).
64. A. Schinzel. Polynomials with special regard to reducibility; Encyclopaedia of mathematics and its applications 77, Cambridge 2000.
65. H.P. Schlickewei. Linearformen mit algebraischen Koeffizienten; Manuscripta Math. 18 (1976), 147-185.
66. H.P. Schlickewei. On products of special linear forms with algebraic coefficients; Acta Arithmetica 31 (1976), 389-398.
67. H.P. Schlickewei. The p-adic Thue-Siegel-Roth-Schmidt theorem; Arch. Math. 29 (1977), 267-270.
68. H.P. Schlickewei and E. Wirsing. Lower bounds for the heights of solutions of linear equations; Inventiones Math. 129 (1997), 1-10.
69. W.M. Schmidt. On heights of algebraic subspaces and diophantine approximations; Annals of Math. 85 (1967), 430-472.
70. W.M. Schmidt. Diophantine approximations and diophantine equations; Lecture Notes in Math. 1467, Springer 1991.
71. W.M. Schmidt. Heights of points on subvarieties of $\mathbb{G}_m^n$; Number Theory 93-94 (ed. S. David), London Math. Soc. Lecture Notes 235, Cambridge 1996 (pp. 157-187).
72. W.M. Schmidt. Heights of algebraic points lying on curves or hypersurfaces; Proc. Amer. Math. Soc. 124 (1996), 3003-3013.
73. Th. Schneider. Zur Theorie der Abelschen Funktionen und Integrale; J. reine angew. Math. 183 (1941), 110-128.
74. Th. Schneider. Einführung in die transzendenten Zahlen; Grundlehren math. Wiss. 81, Springer 1957.
75. J.-P. Serre. Groupes algébriques et corps de classes; Hermann 1959.
76. J.-P. Serre. Quelques propriétés des groupes algébriques commutatifs; Astérisque 69-70 (1987), 191-202.
77. J.H. Silverman. Lower bounds for height functions; Duke Math. J. 51 (1984), 395-403.
78. J.H. Silverman. The theory of height functions, Arithmetic geometry (eds. G. Cornell and J.H. Silverman); Springer 1986 (pp. 151-166).
79. J.H. Silverman. Heights and elliptic curves, Arithmetic geometry (eds. G. Cornell and J.H. Silverman); Springer 1986 (pp. 253-265).
80. J.H. Silverman. The arithmetic of elliptic curves; Graduate Texts in Math. 106, Springer 1986.
81. J.H. Silverman. Advanced topics in the arithmetic of elliptic curves; Graduate Texts in Math. 151, Springer 1994.
82. T.A. Springer. Linear algebraic groups; Progress in Math. 9, Birkhäuser 1998.
83. C.L. Stewart. Algebraic integers whose conjugates lie near the unit circle; Bull. Soc. Math. France 196 (1978), 169-176.
84. R.J. Stroeker and N. Tzanakis. On the elliptic logarithms method for elliptic diophantine equations: reflections and an improvement; Experimental Math. 8 (1999), 135-149.
85. E. Ullmo. Positivité et discrétion des points algébriques des courbes; Annals of Math. 147 (1998), 167-179.
86. M. Waldschmidt. Nombres transcendants; Lect. Notes in Math. 402, Springer 1974.

87. M. Waldschmidt. Nombres transcendants et groupes algébriques; Astérisque 69-70 (1987), 1-162.
88. M. Waldschmidt. On the transcendence methods of Gelfond and Schneider in several variables, new advances in transcendence theory (ed. A. Baker); Cambridge 1988 (pp. 375-398).
89. M. Waldschmidt. Approximation diophantienne dans les groupes algébriques commutatifs (I): Une version effective du théorème du sous-groupe algébrique; J. reine angew. Math. 493 (1997), 61-113.
90. M. Waldschmidt. Diophantine approximation on linear algebraic groups; Grundlehren math. Wiss. 326, Springer 2000.
91. E.T. Whittaker and G.N. Watson. A course in modern analysis; Cambridge 1965.
92. J. Wolfart and G. Wüstholz. Überlagerungsradius gewisser algebraischer Kurven; Math. Annalen 273 (1985), 1-15.
93. G. Wüstholz. Zum Periodenproblem; Inventiones Math. 78 (1984). 381-391.
94. G. Wüstholz. Transzendenzeigenschaften von Perioden elliptischer Integrale; J. reine angew. Math. 354 (1984), 164-174.
95. G. Wüstholz. Multiplicity estimates on group varieties; Annals of Math. 129 (1989), 471-500.
96. G. Wüstholz. Algebraische Punkte auf analytischen Untergruppen algebraischer Gruppen; Annals of Math. 129 (1989), 501-517.
97. D. Zagier. Algebraic numbers close to both 0 and 1; Math. Comp. 61 (1993), 485-491.
98. S. Zhang. Positive line bundles on arithmetic surfaces; Annals of Math. 136 (1992), 569-587.
99. S. Zhang. Positive line bundles on arithmetic varieties; J. Amer. Math. Soc. 8 (1995), 187-221.
100. S. Zhang. Small points and adelic metrics; J. Algebraic Geom. 4 (1995), 281-300.
101. S. Zhang. Equidistribution of small points on abelian varieties; Annals of Math. 147 (1998), 159-165.

Linear Forms in Logarithms of Rational Numbers

Yuri Nesterenko

Faculty of Mechanics and Mathematics, Moscow State University

1 Introduction

The history of the theory of linear forms in logarithms is well known. We shall briefly sketch only some of the moments connected with new technical progress and important for our article. This theory was originated by pioneer works of A.O. Gelfond (see, for example, [5, 6]); with the help of the ideas which arose in connection with the solution of 7-th Hilbert problem (construction of auxiliary functions, extrapolation of zeros and small values), the bounds for the homogeneous linear forms in two logarithms were proved. In the middle of the sixties A. Baker, [1], using auxiliary functions in several complex variables, for the first time obtained bounds for linear forms in any number of logarithms, both in the homogeneous and non-homogeneous cases. All further development of the theory is connected with improvements of these bounds. So N.I. Feldman introduced the so-called binomial polynomials in the construction of the auxiliary function, and due to this the dependence of the bounds on the coefficients of the linear forms was improved. The Kummer theory was used by A. Baker and H. Stark, [2], to improve the dependence of the estimates on the parameters related to the algebraic numbers appearing as arguments of the logarithms. The further improvements of this dependence are connected with the introduction of bounds for the number of zeros of polynomials on algebraic groups in works of G. Wüstholz, [14], P. Philippon and M. Waldschmidt, [11], A. Baker and G. Wüstholz, [3].

The best up today general bounds for homogeneous rational linear forms in logarithms were proved very recently by E.M. Matveev, [10]. The proof of these results is rather technical, abundant in computational details and is difficult for a first reading. The aim of our article is to follow the proof from [10] but in some simpler situation and to focus one's attention on the main new ideas introduced by Matveev.

This research was partially supported by INTAS-RFBR grant No 97-1904.

First of all we assume that all the arguments of the logarithms are rational. This excludes the degree of the field from the estimate. Secondly, we simplify the definition of parameters and constants but preserve the exponential dependence of the constants in the number of logarithms. Our bound is slightly worse than Matveev's one. Actually, our aim was not to prove the best one. Moreover it seems that any application requires an individual estimate of linear forms that is the best in some specific parameters. We preferred simplicity of the formulas and inequalities to the power of the final result. Nevertheless, the dependence of the parameters in the number of logarithms is rather delicate, and in the proof we shall meet difficult enough inequalities.

The proof is rather close to the proof by A. Baker and G. Wüstholz from [3]. But we shall use zero estimate by P. Philippon, as stated in the book of M. Waldschmidt [13]. Our exposition of the proof can be arranged in traditional steps. Section 2 contains the formulation of the main result of our article and the inductive assumption. In section 3 we construct an auxiliary function. Section 4 is devoted to extrapolation of zeros of the auxiliary function and the last Section 5 contains zero bounds and the end of induction.

2 Main result and induction assumption

Let α be an algebraic number and let $\mathbf{K}$ be an algebraic number field of finite degree $[\mathbf{K}:\mathbb{Q}]$ containing α. The absolute logarithmic height of α is defined by the equality

$$h(\alpha) = \frac{1}{[\mathbf{K}:\mathbb{Q}]} \sum_{\sigma\in\mathcal{M}_\mathbf{K}} d_\sigma \max(0, \log|\alpha|_\sigma),$$

where $|\ |_\sigma$ runs over all normalised absolute values (archimedean and non-archimedean) of the field $\mathbf{K}$, see [8, Chapter 3] or [13, Chapter 3]. The number d_σ is the local degree of $|\ |_\sigma$. This height does not depend on the choice of the field $\mathbf{K}$. Due to the product formula

$$\prod_{\sigma\in\mathcal{M}_\mathbf{K}} |\alpha|_\sigma^{d_\sigma} = 1,$$

the following representation holds

$$h(\alpha) = \frac{1}{2[\mathbf{K}:\mathbb{Q}]} \sum_{\sigma\in\mathcal{M}_\mathbf{K}} d_\sigma \big|\log|\alpha|_\sigma\big|. \tag{2.1}$$

If $\alpha \neq 0$ then $h(\alpha^{-1}) = h(\alpha)$ and $h(\alpha^x) = |x|h(\alpha)$ for any rational number x. For any set of algebraic numbers $\alpha_1, \dots, \alpha_m$ the inequality

$$h(\alpha_1 \cdots \alpha_m) \leq h(\alpha_1) + \cdots + h(\alpha_m)$$

holds. If $\alpha = \frac{c}{d}$ with coprime integers c, d, then

$$h(\alpha) = \log \max(|c|, |d|).$$

Our main result is

Theorem 2.1. *Let $\alpha_1, \dots, \alpha_n$ be positive rational numbers such that the real values of logarithms $\log \alpha_1, \dots, \log \alpha_n$ are linearly independent over $\mathbb{Q}$. Then for any set of integers $b_1, \dots, b_n$, $B = \max |b_j| > 0$, the following inequality holds:*

$$\log |b_1 \log \alpha_1 + \dots + b_n \log \alpha_n| \geq -2.9(2e)^{2n+6}(n+2)^{9/2} h(\alpha_1) \cdots h(\alpha_n) \log(eB). \quad (2.2)$$

Now we introduce some parameters convenient for the proof. Denote

$$\Lambda = b_1 \log \alpha_1 + \dots + b_n \log \alpha_n,$$

Let $A_1, \dots, A_n$ be real numbers such that

$$A_j \geq \max\{h(\alpha_j), |\log \alpha_j|, 1\}, \quad A = \max A_j, \quad 1 \leq j \leq n. \quad (2.3)$$

Denote for brevity

$$\Omega = A_1 \cdots A_n, \qquad C(n) = 2\,(2e)^{2n+6}(n+2)^{9/2}, \qquad \lambda = 2e^2.$$

Theorem 2.1 is a consequence of

Theorem 2.2. *Let $\alpha_1, \dots, \alpha_n$ be positive rational numbers such that the real values of logarithms $\log \alpha_1, \dots, \log \alpha_n$ are linearly independent over $\mathbb{Q}$. Then for any set of integers $b_1, \dots, b_n$ and real numbers $A_1, \dots, A_n, B$ satisfying* (2.3) *and $B \geq \max |b_j| > 0$, the following inequality holds*

$$\log |\Lambda| \geq -C(n)\Omega \log(eB). \quad (2.4)$$

Proof of Theorem 2.1. Since $h(\alpha) \geq |\log \alpha|$ for any positive $\alpha \in \mathbb{Q}$ and $h(\alpha) > 1$ if α is distinct from $1, 2, 2^{-1}$, one can choose

$$A_j = \begin{cases} h(\alpha_j), & \text{if } \alpha_j \neq 2,\, 2^{-1}, \\ 1, & \text{if } \alpha_j = 2,\, 2^{-1}. \end{cases}$$

and $B = \max |b_j|$. Theorem 2.1 follows from (2.4) with this choice of parameters, since $2 < 2.9 \log 2$.

Lemma 2.3 (Liouville inequality). *The inequality*

$$|e^{\Lambda} - 1| \geq e^{-nAB}$$

holds.

Proof. Let d_j be the least positive integer such that $d_j\alpha_j \in \mathbb{Z}$, if $b_j > 0$, and $d_j\alpha_j^{-1} \in \mathbb{Z}$ if $b_j < 0$. Since $\log d_j \le h(\alpha_j) \le A$ and $(d_1 \cdots d_n)^B \left(e^{\Lambda} - 1\right)$ is a nonzero integer, we derive the sought inequality.

The following inequality allows to connect the bounds of Theorem 2.1 and of Lemma 2.3

$$|e^z - 1| \le \sum_{k=1}^{\infty} \frac{|z|^k}{k!} \le \frac{|z|}{1-|z|} \le 2|z|, \quad \text{if} \quad z \in \mathbb{C},\ |z| \le \frac{1}{2}. \tag{2.5}$$

Corollary 2.4. *Theorem 2.2 is true in case $n = 1$.*

Proof. If the assertion is wrong then

$$e^{-2A_1 \log(eB)} > |\Lambda| = |b_1 \log \alpha_1| \ge |\log \alpha_1|.$$

On the other hand, applying (2.5) for $z = \log \alpha_1$ and using Lemma 2.3 we derive

$$|\log \alpha_1| \ge \frac{1}{2}\, |\alpha_1 - 1| \ge \frac{1}{2} e^{-A_1} \ge e^{-2A_1}.$$

These upper and lower bounds for $|\log \alpha_1|$ are contradictory, and this proves the Corollary.

Corollary 2.5. *If the parameters $B, A_1, \ldots, A_n$ satisfy*

$$(n+1)AB \le C(n)\Omega \log(eB),$$

then Theorem 2.2 holds.

Proof. Applying inequality (2.5) and Lemma 2.3 we derive

$$|\Lambda| \ge \frac{1}{2}\, |e^{\Lambda} - 1| \ge e^{-(n+1)AB}.$$

Theorem 2.2 will be proved by induction on n. We assume now that $n \ge 2$ and that the inequality corresponding to (2.4) is true for every linear form containing less than n logarithms.

One can suppose that

$$(n+1)AB > C(n)\Omega \log(eB). \tag{2.6}$$

Since the function $\dfrac{x}{1+\log x}$ increases for $x \ge 1$, the last inequality (2.6) implies that

$$B \ge C(n). \tag{2.7}$$

Besides, since $A_j \ge 1$, one can assume without loss of generality that $b_k \ne 0$ for $k = 1, \ldots, n$, and that $A_1 \le \cdots \le A_n$.

Let us define a norm $||\mathbf{x}||_A$ on the space $\mathbb{R}^n$ by the formula

$$||\mathbf{x}||_A = A_1|x_1| + \cdots + A_n|x_n|.$$

For every sublattice $\Phi \subset \mathbb{Z}^n$ of dimension $\nu = \dim \Phi \geq 1$, we will denote by $< \Phi >$ the linear subspace of $\mathbb{R}^n$ generated over $\mathbb{R}$ by the vectors of Φ and by $\mathfrak{K}$ the set of points $\mathbf{x}$ in $< \Phi >$ satisfying the inequality $||\mathbf{x}||_A \leq 1$. Let $V(\Phi)$ be the volume of the fundamental parallelepiped of Φ in $< \Phi >$ and let $V(\mathfrak{K})$ be the ν-dimensional volume of the set $\mathfrak{K}$.

Proposition 2.6. *Let $n \geq 2$ and assume that for every linear form containing less than n logarithms the inequality corresponding to* (2.4) *is true. Assume that there exists a sublattice $\Phi \subset \mathbb{Z}^n$ of dimension $\nu \in \{1, \ldots, n-1\}$ such that*

$$\mathbf{b} = (b_1, \ldots, b_n) \in < \Phi >,$$

and

$$V(\Phi) \leq 2^{-1-\nu}(2e)^{2n-2\nu}V(\mathfrak{K}) \cdot \Omega. \tag{2.8}$$

Then the inequality (2.4) *holds.*

For the proof of this Proposition we need Minkowski's Theorem on Successive Minima. Let Φ be a full lattice and $\mathfrak{K}$ be a compact convex set in $\mathbb{R}^\nu$, containing 0 and symmetric about 0. Let us denote by $V(\Phi)$ and $V(\mathfrak{K})$ the volume of the fundamental parallelepiped of Φ and the volume of $\mathfrak{K}$. Let $\lambda_j = \lambda_j(\mathfrak{K})$ be the minimum of those numbers $\lambda > 0$ such that $\lambda\mathfrak{K}$ contains j linearly independent points from Φ. We refer to these numbers as *successive minima*.

Theorem (Minkowski's Theorem on Successive Minima). *Suppose that $\mathfrak{K}$ is a compact convex set in $\mathbb{R}^\nu$, containing* 0 *and symmetric about* 0. *Let Φ be a full lattice in $\mathbb{R}^\nu$. Then*

$$\frac{2^\nu}{\nu!}V(\Phi) \leq \lambda_1 \cdots \lambda_\nu V(\mathfrak{K}) \leq 2^\nu V(\Phi).$$

Proof. See [4, Ch.VIII, Theorem V].

Proof (of Proposition 2.6). Let us denote by $\mathfrak{K}$ the set of points $\mathbf{x}$ in $< \Phi >$ satisfying the inequality $||\mathbf{x}||_A \leq 1$ and let $\lambda_1, \ldots, \lambda_\nu$ be the successive minima of Φ with respect to $\mathfrak{K}$. For $i = 1, \ldots, \nu$ denote $\mathbf{z}_i = (z_{i1}, \ldots, z_{in}) \in \Phi$ linearly independent points of Φ such that $||\mathbf{z}_i||_A = \lambda_i$.

Define rational numbers $\theta_1, \ldots, \theta_\nu$ by the equalities

$$\log\theta_i = z_{i,1}\log\alpha_1 + \ldots + z_{in}\log\alpha_n, \quad 1 \leq i \leq \nu.$$

Since $\mathbf{b} \in < \Phi >$ one can find coprime integers $m_0, m_1, \ldots, m_\nu$ such that

$$m_0\mathbf{b} = m_1\mathbf{z}_1 + \ldots + m_\nu\mathbf{z}_\nu. \tag{2.9}$$

The numbers $m_0, m_1, \dots, m_\nu$ are uniquely defined (up to a sign). The equality (2.9) implies that

$$m_0 \Lambda = m_1 \log \theta_1 + \dots + m_\nu \log \theta_\nu. \qquad (2.10)$$

We need an upper bound for $|m_i|$. In what follows we can assume without loss of generality that $\Delta = \det(z_{ij})_{\substack{1 \le i \le \nu \\ 1 \le j \le \nu}} \neq 0$, and that the number $A = \max A_i$ is contained in $\{A_1, \dots, A_\nu\}$. By Cramer's rule applied to (2.9) one can express the rational numbers m_j/m_0 as ratios of determinants with Δ in denominator. The determinant Δ_j in the numerator can be computed by replacing the coordinates of the point $\mathbf{z}_j$ with the corresponding coordinates of the point $\mathbf{b}$.

By Hadamard inequality we derive

$$|m_0| \cdot A_1 \cdots A_\nu \le |\Delta| \cdot A_1 \cdots A_\nu \le ||\mathbf{z}_1|| \cdots ||\mathbf{z}_\nu|| \qquad (2.11)$$

and

$$|m_j| \cdot A_1 \cdots A_\nu \le |\Delta_j| \cdot A_1 \cdots A_\nu \le \nu B \cdot \frac{||\mathbf{z}_1||_A \cdots ||\mathbf{z}_\nu||_A}{||\mathbf{z}_j||_A}. \qquad (2.12)$$

According to Minkowski's theorem and (2.8)

$$\Omega' = ||\mathbf{z}_1||_A \cdots ||\mathbf{z}_\nu||_A \le 2^\nu \frac{V(\Phi)}{V(\mathfrak{K})} \le 2^{-1}(2e)^{2n-2\nu}\Omega. \qquad (2.13)$$

Let us set

$$\Lambda' = m_1 \log \theta_1 + \dots + m_\nu \log \theta_\nu,$$
$$B' = \max(|m_1|, \dots, |m_\nu|), \quad A'_j = ||\mathbf{z}_j||_A, \quad j = 1, \dots, \nu.$$

According to the definition of A_j we find

$$|\log \theta_i| \le \sum_{k=1}^{n} |z_{ik}| \cdot |\log \alpha_k| \le \sum_{k=1}^{n} A_k |z_{ik}| = A'_j,$$
$$h(\theta_i) \le \sum_{k=1}^{n} |z_{ik}| \cdot h(\alpha_k) \le \sum_{k=1}^{n} A_k |z_{ik}| = A'_j.$$

Since $z_{jk} \in \mathbb{Z}$ and $A'_j \ge 1$ then

$$A'_j \ge \max(|\log \theta_j|, h(\theta_j), 1), \quad j = 1, \dots, \nu.$$

Inequalities (2.12) and (2.6) imply

$$B' \le \nu B \frac{||\mathbf{z}_1||_A \cdots ||\mathbf{z}_\nu||_A}{A_1 \cdots A_\nu} \le \nu 2^{-1}(2e)^{2n-2\nu} B \frac{\Omega}{A} \le B^2.$$

Similarly, from (2.11) and (2.6) we obtain

$$|m_0| \leq B.$$

Now one can apply the induction assumption to the linear form Λ'. From (2.10) and (2.13) it follows that

$$\begin{aligned}\log|\Lambda| = \log|\Lambda'| - \log|m_0| &\geq -C(\nu)\Omega' \log(eB') - \log B \\ &\geq -2(2e)^{2n+6}(\nu+2)^{9/2}\Omega \log(eB) - \log B \\ &\geq -C(n)\Omega \log(eB).\end{aligned}$$

These inequalities prove the Proposition.

We will use the following strategy for the proof of Theorem 2.2. Assume that the inequality (2.4) is wrong for a linear form Λ, namely

$$\log|\Lambda| < -C(n)\Omega \log(eB). \tag{2.14}$$

In sections 2–5 we shall construct a sublattice $\Phi \subset \mathbb{Z}^n$ satisfying the conditions of Proposition 2.6. A contradiction of this Proposition with (2.14) will finish the proof of Theorem 2.2.

According to what stated above we will use the inequality (2.6) and the inequalities $A_1 \leq \ldots \leq A_n = A$. Besides, without loss of generality we assume that the vector $(b_1, \ldots, b_n)$ is primitive.

3 Construction of auxiliary function

This section contains four subsections where some technical aspects of the construction are discussed. These are binomial polynomials, Siegel's lemma with weights, topics from geometry of numbers and an upper bound for the index of a special lattice. The fifth subsection contains in essence the construction of the auxiliary function.

3.1 Binomial polynomial

For ℓ a non-negative integer and z a complex number, let us define

$$\Delta(z,\ell) = \frac{z(z+1)\cdots(z+\ell-1)}{\ell!}.$$

For any pair of integers $L \geq 0$, $H \geq 1$, we introduce the following binomial polynomials $\Delta(z,L,H)$. Let $\Delta(z,0,H) = 1$ and

$$\Delta(z,L,H) = \Delta(z,H)^q \cdot \Delta(z,r), \tag{3.1}$$

where

$$L = qH + r, \quad 1 \leq r \leq H.$$

Note that $\deg_z \Delta(z, L, H) = L$. If s is a non-negative integer, we write $\Delta^{(s)}(z, L, H)$ for the derivative $(d/dz)^s \Delta(z, L, H)$.

The first idea of eliminating the factorials from the derivatives of auxiliary functions using such polynomials was introduced (in the case $H = L$) by Feldman in 1960. He used this tool to improve the estimates of the measure of transcendence of π and of the logarithms of algebraic numbers. Later in 1968, binomial polynomials were one of his key tools in order to obtain the best possible dependence of the estimate for linear forms in logarithms in terms of the heights of the coefficients. The introduction of this kind of polynomials in the case $r = H$ and $H < L$ into the transcendence theory is due to A. Baker, who in 1972, improved in this way the dependence of lower bounds for linear forms in logarithms in terms of the heights of the α_i. These polynomials (3.1) were used by Matveev in 1993, [9].

Proposition 3.1. *The following assertions hold:*

1) *Let $H \geq 1$, $m \geq 1$ be integers and $d(m, H)$ denote the least common denominator of the numbers*

$$\frac{1}{s!}\Delta^{(s)}(x, \ell, H), \qquad x \in \mathbb{Z}, \qquad \ell \geq 0, \qquad 0 \leq s \leq m.$$

Then

$$d(m, H) \leq \exp(1.04Hm). \tag{3.2}$$

2) *Let $H \geq 1$ be integer and z, c be complex numbers, $c \neq 0$. For any triple of nonnegative integers s, t, ℓ, $s + t \leq \ell$ define $\delta(\ell, s, t) = \delta(\ell, s, t, c, H, z)$ by the equality*

$$\left|\frac{1}{s!t!}c^t \Delta^{(s+t)}(cz, \ell, H)\right| = \delta(\ell, s, t) \cdot \left(e\left(1 + |c| \cdot \frac{|z| + 1}{H}\right)\right)^{\ell}.$$

Then

$$\sum_{\ell=0}^{\infty} \sum_{s+t\leq\ell} \delta(\ell, s, t) \leq 2e^{H/e}.$$

Proof. For each positive integer k and real number a, we denote by d_k the least common multiple of $1, 2, \ldots, k$ and by $[a]$ the integer part of a.

Let p be a prime number and let $b_1 \leq b_2 \leq \ldots \leq b_L$ be integers. For any integer $k > 0$ we denote by r_k the number of b_i which are multiples of p^k. Then

$$\operatorname{ord}_p(b_1 \cdots b_L) = r_1 + r_2 + \cdots.$$

If we delete in any way s numbers from $b_1, \ldots, b_L$ and if $b_{j_1}, \ldots, b_{j_{L-s}}$ denote the remaining $L - s$ numbers, we obtain

$$\operatorname{ord}_p(b_{j_1} \cdots b_{j_{L-s}}) \geq \sum_{k\geq1} \max(r_k - s, 0).$$

We define now the numbers $b_1, \ldots, b_L$ as the factors of type $x+t$ in the product

$$\left(x(x+1)\cdots(x+H-1)\right)^q \cdot x(x+1)\cdots(x+r-1).$$

In this case

$$r_k \geq q\left[\frac{H}{p^k}\right] + \left[\frac{r}{p^k}\right], \quad k \geq 1.$$

Now, from the identity

$$\Delta^{(s)}(z,L,H) = u! \cdot \Delta(z,L,H) \cdot \sum (z+j_1)^{-1} \cdots (z+j_s)^{-1}, \tag{3.3}$$

where the summation is taken over all sets $\{j_1, \ldots, j_s\}$ such that the polynomial $(z+j_1)\cdots(z+j_s)$ divides $\Delta(z,L,H)$, we see that

$$\begin{aligned}
&\operatorname{ord}_p\left(d_H^m \cdot \Delta^{(s)}(x,L,H)\right) \\
&\geq m \cdot \left[\frac{\log H}{\log p}\right] - \sum_{p^k \leq H}\left(q\left[\frac{H}{p^k}\right] + \left[\frac{r}{p^k}\right]\right) + \sum_{k\geq 1} \max\left(q\left[\frac{H}{p^k}\right] + \left[\frac{r}{p^k}\right] - s, 0\right) \\
&\geq \sum_{p^k \leq H} \max\left(m-s, m - q\left[\frac{H}{p^k}\right] - \left[\frac{r}{p^k}\right]\right) \geq 0.
\end{aligned}$$

This proves the assertion $d_H^m \cdot \Delta^{(s)}(x,L,H) \in \mathbb{Z}$, $0 \leq s \leq m$, $x \in \mathbb{Z}$.

The estimate (3.2) follows from the well known inequality $\log d_k \leq \frac{107}{103} \cdot k \leq 1.04k$.

For the proof of the second assertion we note that the polynomial $\Delta(z,\ell,H)$ is dominated by the polynomial $(z+H-1)^\ell (H!)^{-1}(r!)^{-1}$ and this yields the inequalities

$$\left|\Delta^{(s)}(z,\ell,H)\right| \leq \frac{\ell!}{(\ell-s)!}(|z|+H-1)^{\ell-s}(H!)^{-q}(r!)^{-1}$$

and

$$\begin{aligned}
&\sum_{s+t\leq \ell}\left|\frac{c^t}{t!s!}\Delta^{(s+t)}(cz,\ell,H)\right| \\
&\qquad \leq \sum_{s+t\leq \ell} \frac{\ell!}{t!s!(\ell-s-t)!}(|cz|+H-1)^{\ell-s-t}\frac{|c|^t}{(H!)^q r!} \\
&\qquad = \frac{(|cz|+H+|c|)^\ell}{(H!)^q r!} \\
&\qquad = \frac{(H/e)^\ell}{(H!)^q r!}\left(e(1+|c| \cdot \frac{|z|+1}{H})\right)^\ell,
\end{aligned}$$

where q, r are defined by the conditions $\ell = Hq + r$, $1 \leq r \leq H$. Hence we derive

$$\sum_{s+t\leq \ell} \delta(\ell,s,t) \leq \frac{(H/e)^{\ell}}{(H!)^{u}v!}, \quad \ell \geq 0, \quad \ell = Hu+v, \quad 0 \leq v < H,$$

and

$$\sum_{\ell=0}^{\infty}\sum_{s+t\leq \ell} \delta(\ell,s,t) \leq \sum_{\ell=0}^{\infty} \frac{(H/e)^{\ell}}{(H!)^{u}v!} = \sum_{u=0}^{\infty}\left(\frac{1}{H!}\cdot\left(\frac{H}{e}\right)^{H}\right)^{u} \cdot \sum_{v=0}^{H-1}\frac{1}{v!}\cdot\left(\frac{H}{e}\right)^{v}$$
$$\leq \sum_{u=0}^{\infty} e^{-u}e^{H/e} \leq 2e^{H/e}.$$

The inequality $H! \geq H^{H}e^{-H+1}$ was used here.

3.2 Siegel's lemma with weights

Let J be an integer $J \geq 2$ and Q be a set of real numbers $q_j > 0$, $1 \leq j \leq J$, (weights). We will use the notation Q' for the set $\{q_j^{-1}\}$.

For every vector $\mathbf{x} = (x_1, \ldots, x_J) \in \mathbb{Q}^J$ denote

$$\|\mathbf{x}\|_Q = \left(\sum_{j=1}^{J} |x_j|^2 q_j^2\right)^{1/2} \prod_{p} \max_{1\leq j\leq J} |x_j|_p,$$

where $|x|_p$ is the p-adic absolute value. It is easy to see that for any three vectors $\mathbf{a}, \mathbf{b} \in \mathbb{Q}^J$ and $\mathbf{c} = (c_1, \ldots, c_J) \in \mathbb{Q}^J$ the following inequality holds

$$\|\mathbf{a}\|_Q \leq \|\mathbf{b}\|_Q \cdot \prod_{\sigma} \max_{1\leq j\leq J} |c_j|_{\sigma}, \tag{3.4}$$

where σ runs over all valuations of the field of rational numbers $\mathbb{Q}$.

Besides we shall use the ordinary absolute value

$$|\mathbf{x}| = \left(\sum_{k=1}^{J} x_k^2\right)^{1/2}.$$

Proposition 3.2. *Let* $\mathbf{a}_i = (a_{i,1}, \ldots, a_{i,J}) \in \mathbb{Q}^J$, $1 \leq i \leq I$ *and* $J > I$. *Then the system*

$$\sum_{j=1}^{J} a_{i,j}x_j = 0, \quad 1 \leq i \leq I, \tag{3.5}$$

has a nontrivial solution $\mathbf{x} \in \mathbb{Z}^J$ *satisfying the inequality*

$$\|\mathbf{x}\|_{Q'} \leq J^{1/2}\left(q^{-J}\prod_{i=1}^{I}\max(q, \|\mathbf{a}_i\|_Q)\right)^{\frac{1}{J-I}},$$

where

$$q = \left(\prod_{j=1}^{J} q_j \right)^{1/J}.$$

For the proof of the Proposition we need the following lemma.

Lemma 3.3. *Let $\phi : \mathbb{R}^J \longrightarrow \mathbb{R}^I$, $I < J$, be a linear mapping of euclidean spaces, having in some orthogonal coordinate systems in $\mathbb{R}^J$ and $\mathbb{R}^I$ the matrix* $\mathbf{A}$ *of rank* $\operatorname{rank} \mathbf{A} = I$.

1) *If $\Pi \subset \mathbb{R}^J$ is an arbitrary solid and Π' is the orthogonal projection of Π on $(\operatorname{Ker} \phi)^{\perp}$, then $\phi(\Pi) = \phi(\Pi')$ and*

$$V_I(\phi(\Pi)) = V_I(\Pi') \cdot (\det(\mathbf{A} \cdot \mathbf{A}^t))^{1/2}.$$

2) *If $\mathfrak{N}$ is a full lattice in $\mathbb{R}^J$ and $\phi(\mathfrak{N})$ is a full lattice in $\mathbb{R}^I$, then there exists $\mathbf{x} = (x_1, \ldots, x_J) \in \mathfrak{N} \cap \operatorname{Ker} \phi$, $\mathbf{x} \neq 0$, such that*

$$|\mathbf{x}| \leq 2 \left(\frac{V_J(\mathfrak{N}) \cdot (\det AA^t)^{1/2}}{V_I(\phi(\mathfrak{N})) \cdot \omega_{J-I}} \right)^{\frac{1}{J-I}},$$

where $\omega_k = \dfrac{\pi^{k/2}}{\Gamma(k/2+1)}$ *is the volume of the unit ball in* $\mathbb{R}^k$.

In the last equality $V_I(\,\cdot\,)$ is the I-dimensional volume and $\mathbf{A}^t$ is the transpose matrix.

Proof. The equality $\phi(\Pi) = \phi(\Pi')$ is evident. Let $\mathbf{A}$ be the matrix of ϕ in the orthogonal bases $\mathbf{f}_1, \ldots, \mathbf{f}_J$ in $\mathbb{R}^J$ and $\mathbf{u}_1, \ldots, \mathbf{u}_I$ in $\mathbb{R}^I$. Let $\mathfrak{L} = \operatorname{Ker} \phi$ and $\mathbf{e}_1, \ldots, \mathbf{e}_J$ be an orthogonal basis of $\mathbb{R}^J$ such that $\mathfrak{L}^{\perp} = (\mathbf{e}_1, \ldots, \mathbf{e}_I)$. The matrix of ϕ in the basis $\mathbf{e}_1, \ldots, \mathbf{e}_J$ is the $I \times J$ matrix $(\mathbf{B}|0)$ where $\mathbf{B}$ is a square matrix of size I, $\det \mathbf{B} \neq 0$. Since the contraction of ϕ on $\mathfrak{L}^{\perp}$ has the matrix $\mathbf{B}$ in bases $\mathbf{e}_1, \ldots, \mathbf{e}_I$ and $\mathbf{u}_1, \ldots, \mathbf{u}_I$, the equality $\phi(\Pi) = \phi(\Pi')$ implies

$$V_I(\phi(\Pi)) = V_I(\phi(\Pi')) = V_I(\Pi') \cdot |\det \mathbf{B}|. \tag{3.6}$$

If $\mathbf{e}_k = \mathbf{C}\mathbf{f}_k$ with an orthogonal $J \times J$ matrix $\mathbf{C}$ then $(\mathbf{B}|0) = \mathbf{AC}$ and we derive the equality

$$\mathbf{BB}^t = (\mathbf{B}|0|)(\mathbf{B}|0)^t = \mathbf{AC}(\mathbf{AC})^t = \mathbf{ACC}^t\mathbf{A}^t = \mathbf{AA}^t.$$

Hence $(\det \mathbf{B})^2 = \det(\mathbf{AA}^t)$ and together with (3.6) this proves the first assertion of the Lemma.

For the proof of the second one denote $\mathfrak{L} = \operatorname{Ker} \phi \subset \mathbb{R}^J$ and $\mathsf{B}_0 = \mathsf{B} \cap \mathfrak{L}$, where B is the J-dimensional ball of radius 1 with the centre at the origin.

The set $\mathfrak{N}_0 = \mathfrak{N} \cap \mathfrak{L}$ is a lattice in $\mathfrak{L}$. Define $d = \operatorname{rank} \mathfrak{N}_0$. Since $\dim \mathfrak{L} = J - I$, then $d \leq J - I$. The rank of $\phi(\mathfrak{N})$ equals I. Denote by $\phi(\mathbf{e}_1), \ldots, \phi(\mathbf{e}_I)$

a basis of $\phi(\mathfrak{N})$. Every vector from $\mathfrak{N}$ can be expressed as a linear combination of vectors $\mathbf{e}_1, \dots, \mathbf{e}_I \in \mathfrak{N}$ and vectors from $\mathfrak{N}_0$. This is why $I + d \geq J$ and therefore $d = J - I$.

Denote by λ_1 the first minimum of $\mathfrak{N}_0$ with respect to the usual norm $|\mathbf{x}|$. By Minkowski's theorem

$$\lambda_1^{J-I} \leq 2^{J-I} \cdot \frac{V_{J-I}(\mathfrak{N}_0)}{V_{J-I}(\mathsf{B}_0)}. \tag{3.7}$$

Denote by $\mathbf{e}_1, \dots, \mathbf{e}_J$ a basis of $\mathfrak{N}$ such that $\mathbf{e}_{I+1}, \dots, \mathbf{e}_J$ is a basis of $\mathfrak{N}_0$. Then the parallelepiped Π spanned by the vectors $\mathbf{e}_1, \dots, \mathbf{e}_J$ is a fundamental parallelepiped of $\mathfrak{N}$ and the parallelepiped Π_0 spanned by the vectors $\mathbf{e}_{I+1}, \dots, \mathbf{e}_J$ is a fundamental parallelepiped of $\mathfrak{N}_0$. Denote by Π' the orthogonal projection of Π on $\mathfrak{L}^{\perp}$. Hence

$$V_J(\mathfrak{N}) = V_J(\Pi).$$

and

$$V_{J-I}(\mathfrak{N}_0) = V_{J-I}(\Pi_0) = \frac{V_J(\mathfrak{N})}{V_I(\Pi')}.$$

By the first assertion of the Lemma one can write

$$V_I(\phi(\Pi)) = V_I(\Pi') \cdot (\det AA^t)^{1/2}$$

and therefore

$$V_{J-I}(\mathfrak{N}_0) = \frac{V_J(\mathfrak{N})}{V_I(\phi(\Pi))} \cdot (\det AA^t)^{1/2}. \tag{3.8}$$

Now with the help of (3.7), (3.8) and $V_I(\phi(\Pi)) = V_I(\phi(\mathfrak{N}))$ we derive the second assertion.

Proof of Proposition 3.2. Since $\|z \cdot \mathbf{x}\|_Q = \|\mathbf{x}\|_Q$ for any $z \in \mathbb{Q}$ and $\mathbf{x} \in \mathbb{Q}^J$ one can assume without loss of generality that $\mathbf{a}_i$ are vectors with integer and coprime coordinates. Then

$$\|\mathbf{a}_i\|_Q = \left(\sum_{j=1}^{J} |a_{i,j}|^2 q_j^2 \right)^{1/2}.$$

Denote by r the rank of the system (3.5).

Consider at first the case $r = 0$. Without loss of generality we may assume that $q_1 = \max_{1 \leq j \leq J} q_j$. Setting $x_1 = 1$, $x_2 = \ldots = x_J = 0$ we have

$$\|\mathbf{x}\|_{Q'} = q_1^{-1} \leq q^{-1} \leq \left(q^{-J} \prod_{i=1}^{I} \max(q, \|\mathbf{a}_i\|_Q) \right)^{\frac{1}{J-I}}.$$

Now suppose that $1 \le r = I \le J - 1$. Denote by $\mathfrak{N}$ the full lattice in $\mathbb{R}^J$ consisting of vectors $\mathbf{z} = (z_1, \dots, z_J) \in \mathbb{R}^J$ satisfying $q_j z_j \in \mathbb{Z}$ for $1 \le j \le J$. Then

$$V_J(\mathfrak{N}) = (q_1 \cdots q_J)^{-1}. \tag{3.9}$$

Define the map ϕ from $\mathbb{R}^J$ onto $\mathbb{R}^I$ by the formula

$$\phi(\mathbf{x}) = \underline{y} = (y_1, \dots, y_I), \quad \text{where} \quad y_i = \sum_{j=1}^{J} a_{i,j} q_j x_j, \quad 1 \le i \le I.$$

Then $\phi(\mathfrak{N})$ is a full lattice in $\mathbb{R}^I$, and since $\phi(\mathfrak{N}) \subset \mathbb{Z}^I$ the inequality holds:

$$V_I(\phi(\mathfrak{N})) \ge 1.$$

According to Lemma 3.3 there exists a non-zero vector $\mathbf{z} \in \mathfrak{N}$ such that

$$|\mathbf{z}| \le 2 \left(\frac{(q_1 \cdots q_J)^{-1}}{\omega_{J-I}} (\det A_Q A_Q^t)^{1/2} \right)^{\frac{1}{J-I}}, \tag{3.10}$$

where A_Q is the $I \times J$ matrix with elements $a_{i,j} q_j$. Since $\left(\det A_Q A_Q^t\right)^{1/2}$ is the volume of the parallelepiped in $\mathbb{R}^I$ generated by the vectors

$$(a_{i,1} q_1, \dots, a_{i,J} q_J), \qquad 1 \le i \le I,$$

by Hadamard inequality we derive

$$\left(\det A_Q A_Q^t\right)^{1/2} \le \|\mathbf{a}_1\|_Q \cdots \|\mathbf{a}_I\|_Q,$$

where $\mathbf{a}_i = (a_{i,1}, \dots, a_{i,J})$. By this inequality and (3.10) we find

$$|\mathbf{z}| \le 2 \left(q^{-J} \|\mathbf{a}_1\|_Q \cdots \|\mathbf{a}_I\|_Q \omega_{J-I}^{-1}\right)^{\frac{1}{J-I}}.$$

It is easy to check that

$$\rho_n = \omega_n^{-\frac{1}{n}} = \frac{(\Gamma(n/2+1))^{1/n}}{\pi^{1/2}}, \quad n \ge 1,$$

is an increasing sequence and $2\rho_{n-1} \le \sqrt{n}$, $n \ge 2$. Hence

$$2\rho_{J-I} \le 2\rho_{J-1} \le 2\sqrt{J}.$$

The vector

$$\mathbf{x} = (x_1, \dots, x_J) \in \mathbb{Z}^J, \quad x_j = z_j q_j^{-1}, \quad j = 1, \dots, J,$$

satisfies the assertion of Proposition 3.2.

The last case is $1 \le r < I$. Denote

$$L_k = \max(q, \|\mathbf{a}_k\|), \qquad k = 1, \dots, I.$$

Without loss of generality one can assume that the vectors $\mathbf{a}_1, \dots, \mathbf{a}_r$ are linearly independent. Applying the previous case to the system

$$\sum_{j=1}^{J} a_{i,j} x_j = 0, \quad 1 \le i \le r,$$

one can find a non-zero vector $\mathbf{x} \in \mathbb{Z}^J$ such that

$$\|\mathbf{x}\|_Q \le \sqrt{J}\left(q^{-J} L_1 \cdots L_r\right)^{\frac{1}{J-r}}.$$

The inequality

$$\left(q^{-J} L_1 \cdots L_r\right)^{\frac{1}{J-r}} \le \left(q^{-J} L_1 \cdots L_I\right)^{\frac{1}{J-I}}$$

can be easily checked and this concludes the proof of Proposition 3.2.

3.3 Some topics from the geometry of numbers

The usual way to construct an auxiliary function $F(z)$ is to choose a polynomial

$$P \in \mathbb{Z}[z, x_1, \dots, x_n], \quad \deg_{x_j} P \le \frac{L}{A_j}, \quad j = 1, \dots, n, \quad \deg_z P \le L_0, \quad (3.11)$$

where L, L_0 are parameters and to define $F(z) = P(z, \alpha_1^z, \dots, \alpha_n^z)$. Conditions (3.11) imply that P is a sum of monomials $x_1^{k_1} \cdots x_n^{k_n}$ whose vectors of exponents $(k_1, \dots, k_n)$ belong to the parallelepiped $0 \le k_j \le \frac{L}{A_j}$, $j = 1, \dots, n$. Instead of this parallelepiped Matveev has proposed to use in the construction a set $\mathfrak{W}_0$ of points $\mathbf{w} = (w_1, \dots, w_n) \in \mathbb{R}^n$, satisfying

$$|w_j| \le \frac{L}{2A_j}, \quad j = 1, \dots, n, \qquad \left|\sum_{j=1}^{n} w_j \log \alpha_j\right| \le \eta^{-1} L, \qquad (3.12)$$

where $\eta > 1$ is a parameter depending only on n.

The main result of this subsection is the following

Proposition 3.4. *1. The volume of the set $\mathfrak{W}_0$ satisfies the inequality*

$$V_n(\mathfrak{W}_0) \ge \frac{2}{n\eta} \cdot \frac{L^n}{\Omega}.$$

2. If $\mathfrak{N} \subset \mathbb{R}^n$ is a full lattice with determinant $V(\mathfrak{N})$ and $m = \left[\frac{V_n(\mathfrak{W}_0)}{V(\mathfrak{N})}\right]$, then there exist $m+1$ points $\mathbf{x}_0, \dots, \mathbf{x}_m \in \mathfrak{W}_0$ such that

$$\mathbf{x}_i - \mathbf{x}_0 \in \mathfrak{N}. \qquad (3.13)$$

For the proof of this proposition we need two classical results of geometry of numbers. If $\mathfrak{A}$, $\mathfrak{B}$ are subsets of $\mathbb{R}^n$ then the sum $\mathfrak{A} + \mathfrak{B}$ of these subsets is a set consisting of all points $\mathbf{a} + \mathbf{b}$, $\mathbf{a} \in \mathfrak{A}$, $\mathbf{b} \in \mathfrak{B}$. For any positive λ the set $\lambda\mathfrak{A}$ consists of all points $\lambda\mathbf{a}$, $\mathbf{a} \in \mathfrak{A}$.

Theorem (Brunn-Minkowski). *For nonempty compact, convex subsets $\mathfrak{K}_0, \mathfrak{K}_1 \subset \mathbb{R}^n$ and any $\lambda \in \mathbb{R}$, $0 \le \lambda \le 1$ the following inequality holds*

$$V_n\left((1-\lambda)\mathfrak{K}_0 + \lambda\mathfrak{K}_1\right)^{1/n} \ge (1-\lambda)V_n(\mathfrak{K}_0)^{1/n} + \lambda V_n(\mathfrak{K}_1)^{1/n}.$$

Proof. For a proof, see [12, Chapter 6].

For nonempty compact convex sets $\mathfrak{K}, \mathfrak{K} \subset \mathbb{R}^n$ and any $w \in \mathbb{R}$ we will signify

$$\mathfrak{K}_w = \left\{\mathbf{v} \in \mathbb{R}^{n-1} \quad \text{such that} \quad (\mathbf{v}, w) \in \mathfrak{K}\right\}.$$

Corollary 3.5. *If $\mathfrak{K} \subset \mathbb{R}^n$ is a compact symmetric with respect to the origin convex set and $\lambda \in \mathbb{R}$, $0 \le \lambda \le 1$, then*

$$V_{n-1}(K_{\lambda w}) \ge V_{n-1}(K_w).$$

Proof of Corollary 3.5. First we note that if $w_1, w_2, \lambda \in \mathbb{R}$, $0 \le \lambda \le 1$ then

$$\lambda\mathfrak{K}_{w_1} + (1-\lambda)\mathfrak{K}_{w_2} \subset \mathfrak{K}_{\lambda w_1 + (1-\lambda)w_2}.$$

By Brunn–Minkowski theorem we have

$$\begin{aligned} V_{n-1}&(\mathfrak{K}_{\lambda w_1 + (1-\lambda)w_2})^{\frac{1}{n-1}} \\ &\ge V_{n-1}(\lambda\mathfrak{K}_{w_1} + (1-\lambda)\mathfrak{K}_{w_2})^{\frac{1}{n-1}} \\ &\ge \lambda V_{n-1}(\mathfrak{K}_{w_1})^{\frac{1}{n-1}} + (1-\lambda)V_{n-1}(\mathfrak{K}_{w_2})^{\frac{1}{n-1}}. \end{aligned} \tag{3.14}$$

In particular, in case $w_1 = w$, $w_2 = -w$ and $\lambda = \frac{1}{2}$, since $V_{n-1}(\mathfrak{K}_{-w}) = V_{n-1}(\mathfrak{K}_w)$ we find

$$V_{n-1}(\mathfrak{K}_0) \ge V_{n-1}(\mathfrak{K}_w);$$

in other words the central section of a symmetric convex set has maximal volume among all parallel sections.

Now applying (3.14) to $w_1 = w$, $w_2 = 0$ we derive

$$V_{n-1}(\mathfrak{K}_{\lambda w})^{\frac{1}{n-1}} \ge \lambda V_{n-1}(\mathfrak{K}_w)^{\frac{1}{n-1}} + (1-\lambda)V_{n-1}(\mathfrak{K}_0)^{\frac{1}{n-1}} \ge V_{n-1}(\mathfrak{K}_w)^{\frac{1}{n-1}}.$$

For any two points $\mathbf{v} = (v_1, \ldots, v_n)$ and $\mathbf{w} = (w_1, \ldots, w_n)$ in $\mathbb{R}^n$ we shall write for brevity

$$\mathbf{v} \cdot \mathbf{w} = v_1 w_1 + \ldots + v_n w_n.$$

Lemma 3.6. *Let $\mathfrak{W} \subset \mathbb{R}^n$ be a symmetric convex set, $\mathbf{v} \in \mathbb{R}^n$, $\mathbf{v} \neq 0$, and $0 \leq \eta \leq 1$. Denote*

$$\theta = \sup_{\mathbf{w} \in \mathfrak{W}} \{|\mathbf{v} \cdot \mathbf{w}|\} \tag{3.15}$$

and let $\mathfrak{E}$ be the set of all points $\mathbf{w} \in \mathbb{R}^n$ satisfying $|\mathbf{v} \cdot \mathbf{w}| \leq \theta$. Then

$$V_n(\mathfrak{W} \cap \eta\mathfrak{E}) \geq \eta V_n(\mathfrak{W}).$$

Proof. Without loss of generality one can assume that $\mathbf{v} = (0, \dots, 0, v_n)$. Otherwise we apply a suitable orthogonal transformation of $\mathbb{R}^n$.

If $\mathfrak{E}'$ is the orthogonal projection of $\mathfrak{E}$ on $\mathbf{v}\mathbb{R}$ then

$$V_n(\mathfrak{W} \cap \eta\mathfrak{E}) = \int_{\eta\mathfrak{E}'} V_{n-1}(\mathfrak{W}_w)dw = \eta \int_{\mathfrak{E}'} V_{n-1}(\mathfrak{W}_{\eta x})dx.$$

For the last equality we changed variable $w = \eta x$. According to Corollary 3.5 we have $V_{n-1}(\mathfrak{W}_{\eta x}) \geq V_{n-1}(\mathfrak{W}_x)$. Therefore

$$V_n(\mathfrak{W} \cap \eta\mathfrak{E}) \geq \eta \int_{\mathfrak{E}'} V_{n-1}(\mathfrak{E}_x)dx = \eta V_n(\mathfrak{W}).$$

Proof of Proposition 3.4. Denote

$$\mathbf{v} = (\log \alpha_1, \dots, \log \alpha_n)$$

and let $\mathfrak{W}$ be the set of points $\mathbf{w} = (w_1, \dots, w_n) \in \mathbb{R}^n$, satisfying

$$|w_j| \leq \frac{L}{2A_j}, \quad j = 1, \dots, n.$$

This set is compact, convex and symmetric about 0. For every point $\mathbf{w}$ of $\mathfrak{W}$ we have

$$|\mathbf{v} \cdot \mathbf{w}| = \left| \sum_{j=1}^{n} w_j \log \alpha_j \right| \leq \sum_{j=1}^{n} |w_j| A_j \leq \frac{nL}{2}.$$

This implies that the parameter θ for this set $\mathfrak{W}$ defined in (3.15) satisfies $\theta \leq \frac{nL}{2}$. Put $\eta_1 = \frac{2}{n\eta}$. Then $\eta_1\theta \leq \eta^{-1}L$ and we see that the set $\mathfrak{E}$ from Lemma 3.6 satisfies $\mathfrak{W} \cap \eta_1\mathfrak{E} \subset \mathfrak{W}_0$. By Lemma 3.6 we derive

$$V_n(\mathfrak{W}_0) \geq V_n(\mathfrak{W} \cap \eta_1\mathfrak{E}) \geq \eta_1 V_n(\mathfrak{W}) = \frac{2}{n\eta} \cdot \frac{L^n}{\Omega}.$$

The second assertion of Proposition 3.4 is a consequence of the following Blichfeldt's theorem applied to the set $\mathfrak{A} = \mathfrak{W}_0$.

Theorem (Blichfeldt). *Let m be a positive integer, $\mathfrak{N}$ a full lattice in $\mathbb{R}^n$ with the volume of fundamental parallelepiped $V(\mathfrak{N})$, and $\mathfrak{A}$ a point set of volume $V_n(\mathfrak{A})$, possibly $V_n(\mathfrak{A}) = \infty$. Suppose that either*

$$V_n(\mathfrak{A}) > mV(\mathfrak{N}),$$

or

$$V_n(\mathfrak{A}) = mV(\mathfrak{N})$$

and $\mathfrak{A}$ is compact. Then there exist $m+1$ distinct points $\mathbf{x_0}, \dots, \mathbf{x_m}$ of $\mathfrak{A}$ such that $\mathbf{x_i} - \mathbf{x_j}$ are all in $\mathfrak{N}$.

Proof. For a proof, see [4, Chapter III].

3.4 Upper bound for an index

Define $\mathfrak{N}$ as the lattice consisting of all $\lambda = (\lambda_1, \dots, \lambda_n) \in \mathbb{Q}^n$ such that

$$\alpha_1^{\lambda_1} \cdots \alpha_n^{\lambda_n} = \exp(\lambda_1 \log \alpha_1 + \cdots + \lambda_n \log \alpha_n) \in \mathbb{Q}.$$

Since $\mathbb{Z}^n \subset \mathfrak{N}$, $\mathfrak{N}$ is a full lattice in $\mathbb{R}^n$. Denote by $N = [\mathfrak{N} : \mathbb{Z}^n] \in \mathbb{Z}$ the index of $\mathbb{Z}^n$ in $\mathfrak{N}$. Then the volume $V_n(\mathfrak{N})$ of the fundamental parallelepiped of $\mathfrak{N}$ equals $\frac{1}{N}$ and $N\mathfrak{N} \subset \mathbb{Z}^n$.

The parameter N plays an important role in Matveev's proof of the bound for linear forms in logarithms. Here we estimate N in terms of n and Ω.

Proposition 3.7. *For any $n \geq 2$ the index N satisfies the inequality*

$$N \leq 2n(2e)^n \Omega. \tag{3.16}$$

To prove the proposition we denote

$$L = \frac{1}{2}, \qquad \eta = 2e^n,$$

and we let $\mathfrak{W}_0$ be the set defined by inequalities (3.12). According to Proposition 3.4 there exist $m+1$ points $\mathbf{x}_0, \dots, \mathbf{x}_m \in \mathfrak{W}_0$ satisfying (3.13) and

$$m + 1 > \frac{2}{n\eta} \cdot \frac{NL^n}{\Omega} = n^{-1}(2e)^{-n}\frac{N}{\Omega}. \tag{3.17}$$

It remains to prove that $m \leq 1$.

Denote $\mathfrak{L} = \{\mathbf{x}_0, \dots, \mathbf{x}_m\}$ and assume that $m \geq 2$. Fix a point $\mathbf{v} = (v_1, \dots, v_n) \in \mathfrak{L}$; then $\mathbf{l} - \mathbf{v} \in \mathfrak{N}$ for any $\mathbf{l} = (\ell_1, \dots, \ell_n) \in \mathfrak{L}$. Define

$$\beta_{\mathbf{l}} = \alpha_1^{\ell_1 - v_1} \cdots \alpha_n^{\ell_n - v_n} \in \mathbb{Q}, \qquad \mathbf{l} \in \mathfrak{L}$$

and

$$\Delta = \det\left(\beta_{\mathbf{l}}^{n-T}\right)_{\mathbf{l} \in \mathfrak{L}, 0 \leq n \leq m},$$

where $T = \frac{m}{2}$. By this definition of T we have

$$\sum_{n=0}^{m}(n-T)=0, \qquad \sum_{n=0}^{m}|n-T| \le \frac{m+1}{4}. \tag{3.18}$$

Since $\alpha_1, \dots, \alpha_n$ are multiplicatively independent, the numbers $\beta_{\mathfrak{l}}$ are distinct and $\Delta \ne 0$.

We observe that for any prime number p

$$|\log|\beta_{\mathfrak{l}}|_p| \le \sum_{j=1}^{n}(|\ell_j|+|v_j|)\,|\log|\alpha_j|_p| \le \sum_{j=1}^{n}\frac{L}{A_j}\,|\log|\alpha_j|_p| = S_p,$$

and

$$\sum_p S_p \le \sum_{j=1}^{n}\frac{L}{A_j}\sum_p|\log|\alpha_j|_p| \le \sum_{j=1}^{n}\frac{h(\alpha_j)}{A_j} \le n. \tag{3.19}$$

Hence

$$|\Delta|_p \le \max_{\tau}\prod|\beta_{\mathfrak{l}_{\tau(n)}}|^{n-T} \le e^{S_p\sum_{n=0}^{m}|n-T|} \le e^{S_p\frac{(m+1)^2}{4}}. \tag{3.20}$$

Due to the equality (3.18) one can write

$$\Delta = \det\left(\alpha_1^{\ell_1}\cdots\alpha_n^{\ell_n}\right)_{\mathfrak{l}\in\mathfrak{L},\,0\le n\le m}.$$

To find an upper bound for $|\Delta|$, we apply the following

Lemma 3.8. *Let r and R be two real numbers with $0 < r \le R$. Let also $\phi_0(z), \dots, \phi_m(z)$ be functions of one complex variable, which are analytic in the disc $|z| \le R$ of $\mathbb{C}$, and let $\zeta_0, \dots, \zeta_m$ belong to the disc $|z| \le r$. Then the absolute value of the determinant*

$$\Delta = \det\left(\phi_n(\zeta_\ell)\right)_{0\le n,\,\ell\le m}$$

is bounded from above by

$$|\Delta| \le \left(\frac{R}{r}\right)^{-m(m+1)/2}(m+1)!\prod_{n=0}^{m}|\phi_n(z)|_R.$$

Proof. See [13, Lemma 2.5].

In this lemma $|\phi(z)|_R = \max\{|\phi(z)|;\ |z| \le R\}$.

Apply this lemma to the set of functions $\phi_n(z) = e^{(n-T)z}$ and to the set of points

$$\zeta_{\mathfrak{l}} = \log\left(\alpha_1^{\ell_1}\cdots\alpha_n^{\ell_n}\right), \qquad \mathfrak{l}\in\mathfrak{L}.$$

Since by definition of the set $\mathfrak{L}$,

$$|\zeta_1| = |\ell_1 \log \alpha_1 + \ldots + \ell_n \log \alpha_n| \le \eta^{-1} L = \frac{1}{4} e^{-n}$$

one can apply Lemma 3.8 with $r = \frac{1}{4}e^{-n}$ and $R = \frac{2m}{m+1}$. Accordingly we derive

$$|\Delta| \le \left(\frac{8m}{m+1} e^n\right)^{-m(m+1)/2} (m+1)! \cdot e^{\sum_{n=0}^{m} |n-T| R}.$$

This inequality implies, in view of (3.18) and $(m+1)! \le e^{m(m+1)/2}$,

$$|\Delta| \le \left(\frac{8m}{m+1} e^n\right)^{-m(m+1)/2} e^{m(m+1)}. \tag{3.21}$$

It follows from the product formula and (3.20), (3.19) (3.21) that

$$1 = |\Delta| \prod_p |\Delta|_p \le e^{n(m+1)^2/4} \left(\frac{8m}{m+1} e^n\right)^{-m(m+1)/2} e^{m(m+1)},$$

or

$$e^{n\frac{m-1}{2m}} \le \frac{e^2}{8} \frac{m+1}{m} \le \frac{3e^2}{16} = 1,38\ldots.$$

But for $n \ge 2$ and $m \ge 2$ we have

$$n \frac{m-1}{2m} \ge \frac{m-1}{m} \ge \frac{1}{2}.$$

Since $e^{1/2} = 1,64\ldots$ we derive a contradiction. This proves $m \le 1$ and accomplishes the proof of Proposition 3.7.

3.5 Construction

Denote

$$L = 0.8\lambda^{n+3}(n+1)^{5/2} \frac{\log(eB)}{\log(eBN)} \Omega, \qquad \eta = e^{n+3}$$

Due to (3.16), the inequalities hold:

$$L \ge \frac{0.8\lambda^{n+3}(n+1)^{5/2}}{\log(n(2e)^{n+1})} \ge 10^6.$$

Define the set $\mathfrak{W}$ with the help of the inequalities (3.12). By Proposition 3.4 there exist $m+1$ points $\mathbf{x}_0, \ldots, \mathbf{x}_m \in \mathfrak{W}$ satisfying (3.13) and

$$m + 1 > \frac{2}{n\eta} \cdot N \cdot \frac{L^n}{\Omega}.$$

Denote by $\mathfrak{L}$ any subset of $\{\mathbf{x}_0, \ldots, \mathbf{x}_m\}$ such that

$$|\mathfrak{L}| = \left[\frac{2}{n\eta} \cdot N \cdot \frac{L^n}{\Omega}\right] + 1.$$

Here $|\mathfrak{L}|$ denotes the number of elements in $\mathfrak{L}$. Due to the inequality

$$N/(\log(eN))^n \geq e^{n-1}n^{-n}$$

we have $|\mathfrak{L}| \geq 10^{11}$.

We will construct an auxiliary function $\mathcal{F}(Y_0, \dots, Y_n)$ in the form

$$\mathcal{F}(Y_0, \dots, Y_n) = \sum_{\ell_0=0}^{L_0} \sum_{\mathbf{l}\in\mathfrak{L}} p(\ell_0, \mathbf{l}) \Delta(Y_0, \ell_0, H) Y_1^{\ell_1} \cdots Y_n^{\ell_n}, \tag{3.22}$$

where

$$L_0 = \left[0.9\lambda^{n+3}(n+1)^{7/2} \frac{\log(eB)}{\log(e^{n+4.4}N)} \Omega\right], \quad H = [0.96 \log(eBN)], \tag{3.23}$$

$Y_0, \dots, Y_n$ are variables, the vector $\mathbf{l} = (\ell_1, \dots, \ell_n)$ runs over the set $\mathfrak{L}$, and $p(\ell_0, \mathbf{l})$ are integers. Note that due to (3.16) and (2.7) we have $L_0 \geq 10^7$.

The set of coefficients $p(\ell_0, \mathbf{l})$ is a point in $J = (L_0 + 1) \cdot |\mathfrak{L}|$ dimensional space $\mathbb{R}^J$. Points of this space have coordinates $x(\ell, \mathbf{l})$. Define weights

$$q(\ell_0, \mathbf{l}) = (16^{n+8.4}N)^{L_0 - \ell_0};$$

also, denote $Q = \{q(\ell_0, \mathbf{l})\}$ and define the corresponding norm $\| \, \|_Q$ on $\mathbb{R}^J$ as before Proposition 3.2.

Define differential operators

$$\partial_0 = \frac{\partial}{\partial Y_0}, \quad \partial_k = b_n Y_k \frac{\partial}{\partial Y_k} - b_k Y_n \frac{\partial}{\partial Y_n}, \quad 1 \leq k \leq n-1.$$

With this definition the following equalities hold

$$\partial_0^s \cdots \partial_{n-1}^{m_{n-1}} (Y_1^{\ell_1} \cdots Y_n^{\ell_n}) = \mathfrak{X}_1(\mathbf{l})^{m_1} \cdots \mathfrak{X}_{n-1}(\mathbf{l})^{m_{n-1}} Y_1^{\ell_1} \cdots Y_n^{\ell_n},$$

where $\mathfrak{X}_k(\mathbf{x}) = b_n x_k - b_k x_n, \quad 1 \leq k \leq n-1$.

For any integer x and $\mathbf{m} = (m_0, \dots, m_{n-1}) \in \mathbb{Z}^n$, $m_j \geq 0$, we will signify

$$F(x, \mathbf{m}) = \partial_0^{m_0} \cdots \partial_{n-1}^{m_{n-1}} \mathcal{F}(Y_0, \dots, Y_n) \left|_{(2^S x, \alpha_1^x, \dots, \alpha_n^x)}\right., \tag{3.24}$$

where

$$S = n + 24 + [\log_2 N].$$

The main result of this subsection is the following

Proposition 3.9. *There exists a non-zero function* (3.22) *with*

$$\mathbf{p} = \{p(\ell_0, \mathbf{l})\} \in \mathbb{Z}^J$$

and coprime $p(\ell_0, \mathbf{l})$, *satisfying inequalities*

$$\|\mathbf{p}\|_{Q'} \le e^{(n+1)\Omega \log(eB)}$$

and such that

$$F(x, \mathbf{m}) = 0, \quad |x| \le X, \quad m_0 + \cdots + m_{n-1} \le \widehat{M} = \left(1 + \frac{1}{(n+2)^3}\right) M, \tag{3.25}$$

with

$$X = \lceil (9.9n + 17.8) \log(eBN) \rceil,$$

and

$$M = (2n+2)L = 1.6\lambda^{n+3}(n+1)^{7/2} \frac{\log(eB)}{\log(eBN)} \Omega.$$

We shall use from now on the notation $\lceil x \rceil$ for the unique integer satisfying $x \le \lceil x \rceil < x + 1$.

As usual we apply Siegel's lemma to solve the system of linear equations (3.25) in the $p(\ell_0, \mathbf{l})$. An upper bound for J is necessary for application of Proposition 3.2. From the inequalities

$$L \le 0.8\lambda^{n+3}(n+1)^{5/2} \frac{\log(eB)\Omega}{\log(eC(n))}, \qquad 1 \le N \le 2n(2e)^n \Omega$$

and

$$L_0 \ge 10^7, \qquad |\mathfrak{L}| \ge 10^{11}$$

we obtain

$$\begin{aligned} J &\le \frac{2\lambda^{n+3}(n+1)^{7/2}}{n\eta} \cdot \frac{N \log(eB)}{\log(e^{n+4.4}N)} L^n \\ &\le \frac{4\lambda^{n+3}(n+1)^{7/2}(1.6\lambda^{n+3}(n+1)^{5/2})^n}{e\eta(n+4.4)(\log(eC(n)))^n} (e\Omega \log(eB))^{n+1} \\ &\le (eC(n))^{n+1} (e\Omega \log(eB))^{n+1}. \end{aligned}$$

For every $x \ge 1$ the inequality $x \le e^{x/e}$ holds. That's why with $x = e\Omega \log(eB)$ we have

$$J \le e^{(2n+2)\Omega \log(eB)}. \tag{3.26}$$

Now we transform the system (3.25) to a form having better bounds of coefficients.

For any real number ξ we signify by $\ll \xi \gg$ the distance from ξ to the nearest integer number. Fix a point $\mathbf{v} = (v_1, \ldots, v_n) \in \mathfrak{L}$, then $\mathbf{l} - \mathbf{v} \in \mathfrak{N}$ for any $\mathbf{l} \in \mathfrak{L}$ and define

$$w_j = N^{-1} \ll N v_j \gg, \qquad \mathbf{w} = (w_1, \ldots, w_n).$$

Then

$$|w_j| \le \frac{1}{2N}, \qquad j = 1, \dots, n. \tag{3.27}$$

Let U be an integer. Assume that for any vector $\mathbf{m} = (m_0, \dots, m_{n-1}) \in \mathbb{Z}^n$, $m_j \ge 0$, $m_0 + \dots + m_{n-1} \le U$ one has a polynomial

$$f_{\mathbf{m}}(x_0, \dots, x_{n-1}) \in \mathbb{R}[x_0, \dots, x_{n-1}], \qquad \deg_{x_j} f_{\mathbf{m}} = m_j.$$

By induction in lexicographic ordering it is easy to prove that this set of polynomials forms a basis for the space of all polynomials of total degree not greater than U.

Define

$$f_{\mathbf{m}}(x_0, \dots, x_{n-1}) = x_0^{m_0} \prod_{k=1}^{n-1} \Delta(N(x_k - \mathfrak{X}_k(\mathbf{w})), m_k, m_k). \tag{3.28}$$

Then

$$f_{\mathbf{m}}(\partial_0, \dots, \partial_{n-1}) \mathcal{F}(Y_0, \dots, Y_n)$$
$$= \sum_{\ell_0=0}^{L_0} \sum_{\mathbf{l} \in \mathfrak{L}} p(\ell_0, \mathbf{l}) \Delta^{(m_0)}(Y_0, \ell_0, H) \prod_{k=1}^{n-1} \Delta\left(N\mathfrak{X}_k(\mathbf{l} - \mathbf{w}), m_k, m_k\right) Y_1^{\ell_1} \cdots Y_n^{\ell_n} \tag{3.29}$$

and the system (3.25) is equivalent to the system

$$\sum_{\ell_0}^{L_0} \sum_{\mathbf{l} \in \mathcal{L}} p(\ell_0, \mathbf{l}) a(\ell_0, \mathbf{l}, x, \overline{m}) = 0, \tag{3.30}$$

$$|x| \le X, \qquad m_0 + \ldots + m_{n-1} \le \widehat{M},$$

where

$$a(\ell_0, \mathbf{l}, x, \mathbf{m}) = d_0(m_0) \frac{1}{m_0!} \Delta^{(m_0)}(2^S x, \ell_0, H) =$$
$$\times \left(\prod_{k=1}^{n-1} \Delta(N\mathfrak{X}_k(\mathbf{l} - \mathbf{w}), m_k, m_k) \right) \alpha_1^{\lambda_1 x} \cdots \alpha_n^{\lambda_n x},$$

and $\lambda_j = \ell_j - v_j$, $j = 1, \dots, n$.

Now we should estimate the coefficients $a(\ell_0, \mathbf{l}, x, \mathbf{m})$ of the system (3.30). For further use we perform the necessary estimations in a little more general situation. Let us denote by $\mathbf{v}_0$ a proper point in the set $\mathfrak{L}$ and for any integer s, $0 \le s \le S$, define

$$\mathbf{v}_s = 2^{-s} \mathbf{v}_0 = (v_{s,1}, \dots, v_{s,n}) \in \mathbb{R}^n, \qquad \mathfrak{L}_s = 2^{-s} \mathfrak{L} \cap \{\mathbf{v}_s + \mathfrak{N}\} \tag{3.31}$$

and $\mathbf{w}_s = (w_{s,1}, \dots, w_{s,n})$ where

$$w_{s,j} = N^{-1}2^{-s} \ll N v_{0,j} \gg, \qquad j = 1, \dots, n. \tag{3.32}$$

Then $\mathbf{v}_s \in \mathfrak{L}_s$, and

$$|w_{s,j}| \le 2^{-s-1}N^{-1}, \qquad j = 1, \dots, n. \tag{3.33}$$

Next, define for any complex ζ

$$b_s(\ell_0, \mathbf{l}, \zeta, \mathbf{m}) = d(m_0, H)\frac{1}{m_0!}\Delta^{(m_0)}(2^{S-s}\zeta, \ell_0, H) \times \prod_{k=1}^{n-1} \Delta\left(2^s N \mathfrak{X}_k(\mathbf{l} - \overline{w}_s), m_k, m_k\right). \tag{3.34}$$

Lemma 3.10. *Let*

$$0 \le s \le S, \quad m_0 + \ldots + m_{n-1} \le \widehat{M}$$

and

$$\zeta \in \mathbb{C}, \quad |\, \zeta \,| \le R_s = 2^{s+n}e^{n+2.3}(9.9n + 19.8)\log(eBN).$$

Then the following inequality holds

$$\begin{aligned}\Sigma &= \sum_{\ell_0=0}^{L_0} \sum_{\mathbf{l} \in \mathfrak{L}_s} |b_s(\ell_0, \mathbf{l}, \zeta, \mathbf{m})| \cdot (16^{n+8.4}N)^{L_0-\ell_0} \\ &\le \exp(6.1\lambda^{n+3}(n+1)^{7/2}\Omega\log(eB)).\end{aligned}$$

Proof. Set

$$E = \max_{\mathbf{l} \in \mathfrak{L}_s}\left| d(m_0, H)\prod_{k=1}^{n-1} \Delta\left(2^s N \mathfrak{X}_k(\mathbf{l} - \mathbf{w}_s), m_k, m_k\right)\right|.$$

Then

$$\Sigma \le E \cdot |\mathfrak{L}_s| \sum_{\ell_0=0}^{L_0} \frac{1}{m_0!}\left|\Delta^{(m_0)}(2^{S-s}\zeta, \ell_0, H)\right|(16^{n+8.4}N)^{L_0-\ell_0}. \tag{3.35}$$

To prove an upper bound for E we denote

$$y_k = 2^s N \mathfrak{X}_k(\mathbf{l} - \mathbf{w}_s) = 2^s N(b_n(\ell_k - w_{sk}) - b_k(\ell_n - w_{s,n})).$$

Since $|b_j| \le B$, $|\ell_j| \le 2^{-s}\frac{L}{2A_j}$ and $|w_{s,j}| \le 2^{-s-1}N^{-1}$, we find

$$|y_k| \le \frac{BNL}{2}\left(\frac{1}{A_k} + \frac{1}{A_n} + \frac{2}{NL}\right) \le BNL + B. \tag{3.36}$$

Thereby for $U = m_1 + \ldots + m_{n-1}$ we derive

$$E \le e^{1.04Hm_0} \prod_{k=1}^{n-1} \frac{(BNL + B + m_k)^{m_k}}{m_k!}$$
$$\le e^{1.04Hm_0} \frac{1}{U!}((n-1)(BNL+B)+U)^U.$$

By definition of parameters we find $e^{1.04H} \le eBN$,

$$\frac{(n-1)(BNL+B)+U}{eBN} \le M\left(\frac{n-1}{e(2n+2)} + \frac{n-1}{eM} + \frac{3}{2eB}\right) \le M\frac{n}{e(2n+2)}$$

and

$$\begin{aligned} E &\le (eBN)^{m_0} \frac{1}{U!}\left(eBNM\frac{n}{e(2n+2)}\right)^U \\ &\le (eBN)^{\widehat{M}} \frac{1}{U!}\left(M\frac{n}{e(2n+2)}\right)^U \\ &\le (eBN)^{M\left(1+(n+2)^{-3}+\frac{n}{e(2n+2)\log eBN}\right)} \\ &\le \exp(1.7\lambda^{n+3}(n+1)^{7/2}\Omega\log(eB)) \end{aligned} \tag{3.37}$$

Let us go back to the inequality (3.35). With the notations of Proposition 3.1 one can write

$$\begin{aligned} &\sum_{\ell_0=0}^{L_0} \frac{1}{m_0!}\left|\Delta^{(m_0)}(2^{S-s}\zeta, \ell_0, H)\right| (16^{n+8.4}N)^{L_0-\ell_0} \\ &\quad = \sum_{\ell_0=0}^{L_0} \delta(\ell_0, m_0, 0)\left(e\left(1+2^{S-s}\frac{|\zeta|+1}{H}\right)\right)^{\ell_0} (16^{n+8.4}N)^{L_0-\ell_0} \end{aligned} \tag{3.38}$$

By definition of parameters and conditions of Lemma 3.10 one has inequalities

$$2^{S-s}\frac{|\zeta|}{H} \le 2^{S-s}\cdot 2^{s+n}e^{n+2.4}(9.9n+19.8) \le 2^{2n+24}e^{n+2.4}(9.9n+19.8)N,$$

$$1 + 2^{S-s}\frac{1}{H} \le 2^{S+1} \le 2^{n+25}N$$

and

$$e\left(1+2^{S-s}\frac{|\zeta|+1}{H}\right) \le 16^{n+8.4}N. \tag{3.39}$$

Therefore

$$\sum_{\ell_0=0}^{L_0} \delta(\ell_0, m_0, 0)\left(e\left(1+2^{S-s}\frac{|\zeta|+1}{H}\right)\right)^{\ell_0}(16^{n+8.4}N)^{L_0-\ell_0}$$
$$\leq (16^{n+8.4}N)^{L_0}\sum_{\ell_0=0}^{L_0}\delta(\ell_0, m_0, 0).$$

The last sum can be bounded by Proposition 3.1. Since $|\mathfrak{L}_s| \leq |\mathfrak{L}| \leq J$ and (3.35), (3.38) this implies

$$\Sigma \leq E \cdot J \cdot (16^{n+8.4}N)^{L_0} 2e^{H/e}$$

From (3.37), (3.26) and from the inequalities

$$\begin{aligned} L_0 \log(16^{n+8.4}N) &\leq 0.9\lambda^{n+3}(n+1)^{7/2}\frac{\log(eB)}{\log(e^{n+4.4}N)}\Omega\log(16^{n+8.4}N) \\ &\leq 0.9\lambda^{n+3}(n+1)^{7/2}\frac{\log(16^{n+8.4})}{n+4.4}\Omega\log(eB) \\ &\leq 4.1\lambda^{n+3}(n+1)^{7/2}\Omega\log(eB), \end{aligned} \tag{3.40}$$

and

$$\begin{aligned} \frac{H}{e} + \log 2 &\leq \frac{1}{2}H \leq 0,48\log(2n(2e)^n eB\Omega) \leq \log(eB\Omega) \\ &\leq \Omega\log(eB) \end{aligned} \tag{3.41}$$

we derive the assertion of Lemma 3.10.

It is easy to see that for any rational number x there exists an algebraic number field $\mathbf{K}$ such that for any vector $\mathbf{l} \in \mathfrak{L}_s$ and $\overline{\lambda} = \mathbf{l} - \mathbf{v}_s = (\lambda_1, \dots, \lambda_n) \in \mathfrak{N}$, the number $\alpha_1^{\lambda_1 x} \cdots \alpha_n^{\lambda_n x}$ belongs to the field $\mathbf{K}$.

Lemma 3.11. *For any rational number x the following inequality holds*

$$\frac{1}{[\mathbf{K}:\mathbb{Q}]}\sum_{\sigma\in\mathcal{M}_{\mathbf{K}}} d_\sigma \log\max_{\mathbf{l}\in\mathfrak{L}_s}\left|\alpha_1^{\lambda_1 x}\cdots\alpha_n^{\lambda_n x}\right|_\sigma \leq 2^{-s}nL|x|. \tag{3.42}$$

Proof. It is well known that the left hand side of (3.42) does not depend on the choice of the field $\mathbf{K}$. Therefore without loss of generality one can assume that $\alpha_j^{\lambda_j x} \in \mathbf{K}$, $1 \leq j \leq n$, for any $\overline{\lambda} \in \mathfrak{N}$.

If $|\ |_\sigma$ is an absolute value of the field $\mathbf{K}$ then

$$\begin{aligned} \log\left|\alpha_1^{\lambda_1 x}\cdots\alpha_n^{\lambda_n x}\right|_\sigma &= \sum_{j=1}^n \log|\alpha_j^{\lambda_j x}|_\sigma = \sum_{j=1}^n \ell_j x\log|\alpha_j|_\sigma - \sum_{j=1}^n v_{s,j}x\log|\alpha_j|_\sigma \\ &\leq \sum_{j=1}^n |\ell_j|\cdot|x|\cdot\left|\log|\alpha_j|_\sigma\right| = -\sum_{j=1}^n v_{s,j}x\log|\alpha_j|_\sigma \\ &\leq 2^{-s}|x|\frac{L}{2}\sum_{j=1}^n\frac{1}{A_j}\cdot\left|\log|\alpha_j|_\sigma\right| - \sum_{j=1}^n v_{s,j}x\log|\alpha_j|_\sigma. \end{aligned}$$

After multiplication of these inequalities by local degrees d_σ and summation over $\sigma \in \mathcal{M}_{\mathbf{K}}$ we derive, in view of (2.1), inequalities $h(\alpha_j) \le A_j$, and by the product formula the inequality (3.42).

Let us go back to the proof of Proposition 3.9. For any $\mathbf{l} \in \mathfrak{L}$ and $\overline{\lambda} = \mathbf{l} - \mathbf{v} = (\lambda_1, \dots, \lambda_n) \in \mathfrak{N}$, the number $c(\ell_0, \mathbf{l}, x, \mathbf{m}) = \alpha_1^{\lambda_1 x} \cdots \alpha_n^{\lambda_n x}$ is rational for integer x. Besides $N(\mathbf{l} - \mathbf{v}) \in \mathbb{Z}^n$ and by (3.32)

$$N(\mathbf{l} - \mathbf{w}) = N(\mathbf{l} - \mathbf{v}) + (N\mathbf{v} - N\mathbf{w}) \in \mathbb{Z}^n.$$

This implies $b(\ell_0, \mathbf{l}, x, \mathbf{m}) = b_0(\ell_0, \mathbf{l}, x, \mathbf{m}) \in \mathbb{Z}$. Consequently the vector $\mathbf{a}(x, \mathbf{m})$ with coordinates $a(\ell_0, \mathbf{l}, x, \mathbf{m})$, $0 \le \ell_0 \le L_0$, $\mathbf{l} \in \mathfrak{L}_s$ belongs to the space $\mathbb{Q}^J$ and one can define the norm $\|\mathbf{a}(x, \mathbf{m})\|_Q$ for it. Let

$$\mathbf{b}(x, \mathbf{m}), \mathbf{c}(x, \mathbf{m}) \in \mathbb{Q}^J$$

be vectors with coordinates $b(\ell_0, \mathbf{l}, x, \mathbf{m})$, $c(\ell_0, \mathbf{l}, x, \mathbf{m})$. Then

$$a(\ell_0, \mathbf{l}, x, \mathbf{m}) = b(\ell_0, \mathbf{l}, x, \mathbf{m}) \cdot c(\ell_0, \mathbf{l}, x, \mathbf{m}).$$

Due to the inequality (3.4) and Lemma 3.11 we derive

$$\|\mathbf{a}(x, \mathbf{m})\|_Q \le e^{nL|x|} \cdot \|\mathbf{b}(x, \mathbf{m})\|_Q. \tag{3.43}$$

Since $b(\ell_0, \mathbf{l}, x, \mathbf{m}) \in \mathbb{Z}$ we have $\|\mathbf{b}(\ell_0, \mathbf{l})\|_Q \le \Sigma$ and by Lemma 3.10 with $s = 0$ and $\mathbf{v}_0 = \mathbf{v}$ we find

$$\begin{aligned} \| \mathbf{a}(x, \mathbf{m}) \|_Q &\le \exp\left(nL \mid x \mid +6.1\lambda^{n+3}(n+1)^{7/2}\Omega \log(eB)\right) \\ &\le \exp\left(8.75\lambda^{n+3}(n+1)^{9/2}\Omega \log(eB)\right). \end{aligned} \tag{3.44}$$

The parameter q from Proposition 3.2 in our situation satisfies

$$\log q = \frac{1}{L_0 + 1} \sum_{\ell_0 = 0}^{L_0} (L_0 - \ell_0) \log(16^{n+8.4} N) = \frac{L_0}{2} \log(16^{n+8.4} N).$$

Hence by (3.40) we derive

$$\log q \le 2.1\lambda^{n+3}(n+1)^{7/2}\Omega \log(eB). \tag{3.45}$$

The inequalities (3.44) and (3.45) imply

$$p = \max(q, \|\mathbf{a}(x, \mathbf{m})\|_Q) \le \exp(8.75\lambda^{n+3}(n+1)^{9/2}\Omega \log(eB)). \tag{3.46}$$

On the other hand

$$\begin{aligned} \log q &\ge 0.45(1 - 10^{-7})\lambda^{n+3}(n+1)^{7/2} \frac{\log(16^{n+8.4}N)}{\log(e^{n+4.4}N)} \Omega \log eB \\ &\ge 0.45(1 - 10^{-7}) \frac{\log(16^{n+8.4} 2n(2e)^n)}{\log(e^{n+4.4} 2n(2e)^n)} \lambda^{n+3}(n+1)^{7/2}\Omega \log eB. \end{aligned} \tag{3.47}$$

Since $X \geq 1$ and

$$M \geq 1.6\lambda^{n+3}(n+1)^{7/2}\frac{\Omega}{\log(eN)} \geq 1.6\lambda^{n+3}(n+1)^{7/2}\frac{\Omega}{\log(n(2e)^{n+1}\Omega)}$$
$$\geq 1.6\lambda^{n+3}(n+1)^{7/2}\frac{1}{\log(n(2e)^{n+1})} \geq 1001n^2,$$

and by the definitions of parameters, we have the following bounds for the number of variables J and for the number of equations I:

$$I \leq (2X+1)\frac{(\widehat{M}+n)^n}{n!} \leq \frac{1.001}{n!}(2X+1)\widehat{M}^n$$
$$\leq 1.001(19.8n+35.8)\frac{(2n+2)^n}{n!}\left(1+\frac{1}{(n+2)^3}\right)^n \log(eBN)L^n,$$

$$J = (L_0+1)|\mathfrak{L}| > \frac{1.8\lambda^{n+3}(n+1)^{7/2}}{n\eta} \cdot \frac{N\log(eB)}{\log(e^{n+4.4}N)} \cdot L^n.$$

This implies that

$$\frac{I}{J} \leq \frac{1.001n(19.8n+35.8)(n+4.4)(2n+2)^n e^{n+3}(1+(n+2)^{-3})^n}{1.8 \cdot n!\lambda^{n+3}(n+1)^{7/2}}. \quad (3.48)$$

By (3.47), (3.46) we derive

$$\frac{I\log p}{J\log q} \leq 11 \cdot \frac{(19.8n+35.8)(n+4.4)(2n+2)^n e^{n+3}\log\left(e^{n+4.4}2n(2e)^n\right)}{(n-1)!\lambda^{n+3}(n+1)^{5/2}\log\left(16^{n+8.4}2n(2e)^n\right)}$$
$$\cdot\left(1+\frac{1}{(n+2)^3}\right)^n \leq 1,$$

and according to Proposition 3.2 there exists a non-zero vector $\mathbf{p} = \{p(\ell_0, \mathbf{l})\} \in \mathbb{Z}^J$ satisfying the inequality

$$||\mathbf{p}||_{Q'} \leq J^{1/2}$$

and the equations (3.30). Due to (3.26) this completes the proof of Proposition 3.9.

4 Extrapolation of zeros

For a proper vector $\mathbf{v}_0 \in \mathfrak{L}$ we defined in subsection 3.5 a sequence of vectors $\mathbf{v}_s$ and sets $\mathfrak{L}_s$, $0 \leq s \leq S$. Now we specify this vector $\mathbf{v}_0$ and both sequences $\mathbf{v}_s$, $\mathfrak{L}_s$ accordingly. In Proposition 3.9 we constructed a non-zero vector $\mathbf{p} = \{p(\ell_0, \mathbf{l})\} \in \mathbb{Z}^J$ of coefficients for the auxiliary function (3.22). Denote by $\mathbf{v}_0$ a vector such that there exists a non-zero coefficient $p(\ell_0, \mathbf{v}_0)$. Define correspondent vectors $\mathbf{v}_s$ and sets $\mathfrak{L}_s$.

The following functions play a central role in the extrapolation procedure:

$$\mathcal{F}_s(Y_0,\dots,Y_n) = \sum_{\ell_0=0}^{L_0} \sum_{\mathbf{l}\in\mathfrak{L}_s} p(\ell_0, 2^s\mathbf{l})\Delta(Y_0,\ell_0,H)Y_1^{\ell_1}\cdots Y_n^{\ell_n}. \tag{4.1}$$

Since $2^s\mathfrak{L}_s \subset \mathfrak{L}$ these functions are properly defined.

According to (3.24), we denote for any $\zeta \in \mathbb{C}$

$$\begin{aligned} F_s(\zeta,\mathbf{m}) &= \partial_0^{m_0}\cdots\partial_{n-1}^{m_{n-1}}\mathcal{F}_s(Y_0,\dots,Y_n)\,|_{(2^{S-s}\zeta,\alpha_1^\zeta,\dots,\alpha_n^\zeta)} \\ &= \sum_{\ell_0=0}^{L_0}\sum_{\mathbf{l}\in\mathfrak{L}_s}\Big(p(\ell_0,2^s\mathbf{l})\Delta^{(m_0)}(2^{S-s}\zeta,\ell_0,H) \\ &\qquad \times \mathfrak{X}_1(\mathbf{l})^{m_1}\cdots\mathfrak{X}_{n-1}(\mathbf{l})^{m_{n-1}}\alpha_1^{\ell_1\zeta}\cdots\alpha_n^{\ell_n\zeta}\Big). \end{aligned} \tag{4.2}$$

For any integers $s \geq 0$, $\nu \geq 0$ we now introduce two new parameters which are important for the extrapolation procedure

$$\begin{aligned} T_s &= \left[\frac{2^{-s-1}}{n+1}M\right] = [2^{-s}L], \\ X_s &= \left[\frac{XM}{(4n+4)(T_s+1)}\right] = \left[\frac{XL}{2(T_s+1)}\right]. \end{aligned} \tag{4.3}$$

We also introduce the following sets of nodes

$$\begin{aligned} \mathcal{X}_{0,0} &= \{x\in\mathbb{Z},\quad |x|\leq X_0\}, \\ \mathcal{X}_{s,0} &= \{x\in\mathbb{Z},\quad |x|\leq 2X_s-1,\ \ x\equiv 1 \pmod 2\}, \\ \mathcal{X}_{s,\nu} &= \{x\in\mathbb{Z},\quad |x|\leq 2^\nu X_s\}, \qquad \nu\geq 1. \end{aligned}$$

The main result of this section is

Proposition 4.1. *Let s,ν be integers satisfying*

$$0\leq\nu\leq n, \qquad 0\leq s\leq S.$$

Then the equalities hold

$$F_s(x,\mathbf{m}) = 0 \tag{4.4}$$

for all integers x, $m_0\geq 0,\dots,m_{n-1}\geq 0$,

$$\begin{aligned} &x\in\mathcal{X}_{s,\nu}, \\ &m_0+\cdots+m_{n-1}\leq M_{s,\nu} = M\left((n+2)^{-3}+2^{-s-1}\left(2-\frac{\nu}{n+1}\right)\right). \end{aligned} \tag{4.5}$$

This proposition shall be proved by induction on lexicographic ordering of pairs (s, ν). In the case $s = 0$, $\nu = 0$ the equalities (4.4) follow from Proposition 3.9, due to $X_0 \leq X$.

We use two different kinds of inductive steps in the proof. The first one, $(s, \nu) \to (s, \nu + 1)$, $\nu < n$, is based on classical Baker extrapolation scheme and is contained in the subsection 4.2. The second one, $(s, n) \to (s + 1, 0)$, uses Kummer descent. And this is the reason of the abnormal choice of the sets $\mathcal{X}_{s,0}$, $s \geq 1$. This step of induction is discussed in the subsection 4.3.

In the proof of Proposition 3.9 we changed the system of equalities (3.25) to (3.30). The new one had better estimates of coefficients. The same trick is useful in the proof of Proposition 4.1. Instead of (4.4) we shall prove the equivalent system of equalities

$$\Phi_s(x, \mathbf{m}) = 0, \tag{4.6}$$

for the same set of $(x, \mathbf{m})$. Here

$$\Phi_s(\zeta, \mathbf{m}) = \sum_{\ell_0=0}^{L_0} \sum_{\mathbf{l} \in \mathfrak{L}_s} p(\ell_0, 2^s \mathbf{l}) b_s(\ell_0, \mathbf{l}, \zeta, \mathbf{m}) \alpha_1^{\lambda_1 \zeta} \cdots \alpha_n^{\lambda_n \zeta}, \tag{4.7}$$

and $\overline{\lambda} = (\lambda_1, \dots, \lambda_n) = \mathbf{l} - \mathbf{v}_s$.

4.1 Interpolation formula

In this subsection we recall the interpolation formula, which follows from Cauchy's residue theorem and was first applied in the theory by Gelfond. Only a special case of this formula will be used, that is the case with equal multiplicities.

Consider a finite set $\mathcal{X} \subset \mathbb{C}$ (of the nodes of interpolation) and a positive integer T (the number $T + 1$ is the multiplicity of nodes). Put

$$Q(\zeta) = \prod_{x \in \mathcal{X}} (\zeta - x)^{T+1}.$$

Let $f(\zeta)$ be an analytic function in the circle $|\zeta| \leq R$ then $f(z) = f_1(z) + f_2(z)$, where

$$f_1(z) = \frac{1}{2\pi i} \int_{\Gamma(0,R)} \frac{Q(z)}{Q(\zeta)} \cdot \frac{f(\zeta) d\zeta}{\zeta - z}, \tag{4.8}$$

$$f_2(z) = \sum_{x \in \mathcal{X}} \sum_{\tau=0}^{T} \frac{f^{(\tau)}(x)}{2\pi i \tau!} \int_{\Gamma(x,r)} \frac{Q(z)}{Q(\zeta)} \cdot \frac{(\zeta - x)^\tau d\zeta}{\zeta - z}. \tag{4.9}$$

Here $\Gamma(x, r)$ is the circle of radius r centered in x oriented clockwise; it does not contain the point z and other points from the set $\mathcal{X}$; $\Gamma(0, R)$ is the circle of radius R centered in 0 oriented anti-clockwise and containing z and all points of $\mathcal{X}$.

Lemma 4.2. *Let $\mathcal{X} \subseteq \mathbb{R}$ be a finite arithmetic progression with the difference $\delta > 0$, placed symmetrically with respect to 0 and let ξ be a maximal point in $\mathcal{X}$. Assume that $z \in \mathbb{R}$, $z \notin \mathcal{X}$, and denote*

$$r = \frac{1}{2}\min(\delta, \min\{|z - x|, x \in \mathcal{X}\}), \qquad z_0 = \max(\xi + \delta, |z|).$$

Then

1) for each $\zeta \in \mathbb{C}$, with $|\zeta| > |z|$ we have

$$\left|\frac{Q(z)}{Q(\zeta)}\right| \leq \left|\frac{z_0}{\zeta}\right|^{|\mathcal{X}|(T+1)}. \tag{4.10}$$

2) for each $\zeta \in \Gamma(x, r)$, $x \in \mathcal{X}$, we have

$$\left|\frac{Q(z)}{Q(\zeta)}\right| \leq \left(\frac{3z_0}{r|\mathcal{X}|}\right)^{|\mathcal{X}|(T+1)}. \tag{4.11}$$

Here $|\mathcal{X}|$ is the number of elements in the set $\mathcal{X}$.

Proof. Set

$$q(\zeta) = \prod_{x \in \mathcal{X}} (\zeta - x).$$

Then $Q(z) = q(z)^{T+1}$. Since $|q(-z)| = |q(z)|$, one can assume $z > 0$. With assumption we have

$$\left|\frac{q(z+\delta)}{q(z)}\right| = \left|\frac{z + \xi + \delta}{z - \xi}\right| > 1,$$

or $|q(z + \delta)| > |q(z)|$. Due to this inequality one can suppose without loss of generality $z > \xi$.

Now (4.10) follows from the inequality

$$\left|\frac{z^2 - x^2}{\zeta^2 - x^2}\right| \leq \frac{z^2 - x^2}{|\zeta|^2 - x^2} \leq \frac{z^2}{|\zeta|^2} \leq \frac{z_0^2}{|\zeta|^2}$$

valid for any $x \in \mathcal{X}$.

To obtain (4.11) notice that the assumption $z > \xi$ implies $z^2 - x^2 \leq z^2$ for any $x \in \mathcal{X}$ and $|q(z)| \leq |z_0|^{|\mathcal{X}|}$. It remains only to verify the inequality

$$|q(\zeta)| \geq \left(\frac{r|\mathcal{X}|}{3}\right)^{|\mathcal{X}|} \tag{4.12}$$

Let x_0 be the point of the set $\mathcal{X}$ nearest to the origin. For any $k \in \mathbb{Z}$ define $x_k = x_0 + k\delta$. To prove (4.12) we state a more general inequality. Namely we check that for any pair of integers $U \geq V$ the following inequality holds

$$\prod_{k=V}^{U} |\zeta - x_k| \geq r^{U-V+1}(U-V)!. \tag{4.13}$$

We prove (4.13) by induction on $U - V$. Since

$$|\zeta - x_V| \geq |\zeta - x| = r,$$

the inequality (4.13) is valid in the case $U = V$. It is impossible to have both inequalities

$$|\zeta - x_V| < \frac{\delta}{2}(U - V + 1), \qquad |\zeta - x_{U+1}| < \frac{\delta}{2}(U - V + 1).$$

If $|\zeta - x_{U+1}| \leq |\zeta - x_V|$, then

$$\prod_{k=V}^{U+1} |\zeta - x_k| = |\zeta - x_V| \cdot \prod_{k=V+1}^{U+1} |\zeta - x_k| \geq \frac{\delta}{2}(U - V + 1) \cdot r^{U-V+1}(U-V)!.$$

This proves (4.13) in general case. The consequence of this inequality is a lower bound

$$|q(\zeta)| \geq r^{|\mathcal{X}|}(|\mathcal{X}| - 1)! \geq \left(\frac{r|\mathcal{X}|}{3}\right)^{|\mathcal{X}|}.$$

4.2 Extrapolation of zeros in $\mathbb{Q}$

Assume that $0 \leq s \leq S$, $0 \leq \nu < n$ and that the equality (4.4) holds for any set $(x, \mathbf{m})$, satisfying (4.5). To prove the induction step we choose an integer z and $\overline{\mu} = (\mu_0, \ldots, \mu_{n-1}) \in \mathbb{Z}^n$, $\mu_j \geq 0$, with

$$z \in \mathcal{X}_{s,\nu+1}, \quad \mu_0 + \cdots + \mu_{n-1} \leq M_{s,\nu+1}. \tag{4.14}$$

At the end of this subsection we shall state that $F_s(z, \overline{\mu}) = 0$. Notice that one can assume $z \notin \mathcal{X}_{s,\nu}$.

Since the linear form Λ is assumed to be small, as usual one can exclude α_n from analytical considerations. Assume that $\zeta \in \mathbb{C}$, $|\zeta| \leq R_s$ (see Lemma 3.10) and $\overline{\lambda} = \mathbf{1} - \mathbf{v}_s$. Then

$$\left|\alpha_1^{\lambda_1\zeta} \cdots \alpha_n^{\lambda_n\zeta} - \alpha_1^{\frac{x_1(\overline{\lambda})}{b_n}\zeta} \cdots \alpha_{n-1}^{\frac{x_{n-1}(\overline{\lambda})}{b_n}\zeta}\right| = \left|\alpha_1^{\lambda_1\zeta} \cdots \alpha_n^{\lambda_n\zeta}\right| \cdot \left|1 - e^{-\Lambda\frac{\lambda_n\zeta}{b_n}}\right|,$$

and due to the inequality (2.5) we have

$$\begin{aligned}\left|1 - e^{-\Lambda\frac{\lambda_n\zeta}{b_n}}\right| &\leq 2\left|\Lambda\frac{\lambda_n\zeta}{b_n}\right| \leq |\Lambda| 2^{1-s}\frac{L|\zeta|}{b_n A_n} \\ &\leq |\Lambda| \cdot 1.6\lambda^{n+3}(n+1)^{5/2}2^n e^{n+2.3}(9.9n + 19.8)\Omega\log(eB) \\ &\leq |\Lambda| e^{14n\Omega\log(eB)}.\end{aligned}$$

The inequality

$$\left|\Lambda \frac{\lambda_n \zeta}{b_n}\right| \le \frac{1}{2} \tag{4.15}$$

is important for an application of (2.5). It is valid since the left hand side of (4.15) can be bounded by

$$0.8\lambda^{n+3}(n+1)^{5/2}2^n e^{n+2.3}(9.9n+19.8)\Omega\log(eB)|\Lambda| \le e^{(14n-C(n))\Omega\log(eB)} \le (eB)^{-\Omega} \le e^{-1}.$$

Finally, applying (2.14) we find

$$\left|\alpha_1^{\lambda_1\zeta}\cdots\alpha_n^{\lambda_n\zeta} - \alpha_1^{\frac{x_1(\overline{\lambda})}{b_n}\zeta}\cdots\alpha_{n-1}^{\frac{x_{n-1}(\overline{\lambda})}{b_n}\zeta}\right| \le \left|\alpha_1^{\lambda_1\zeta}\cdots\alpha_n^{\lambda_n\zeta}\right|\cdot e^{-0.5C(n)\Omega\log(eB)}. \tag{4.16}$$

In the sequel the notation

$$\varepsilon(\zeta) = \max_{\mathfrak{l}\in\mathfrak{L}_s}\left|\alpha_1^{\lambda_1\zeta}\cdots\alpha_n^{\lambda_n\zeta}\right|\cdot e^{-0.5C(n)\Omega\log(eB)}$$

will be used. Furthermore we shall employ in the extrapolation procedure an entire function of a single complex variable ζ defined like in (3.29) by the equality

$$f(\zeta) = f_s(\zeta,\overline{\mu}) = \sum_{\ell_0=0}^{L_0}\sum_{\mathfrak{l}\in\mathfrak{L}_s} p(\ell_0, 2^s\mathfrak{l})b_s(\ell_0,\mathfrak{l},\zeta,\overline{\mu})\alpha_1^{\frac{x_1(\overline{\lambda})}{b_n}\zeta}\cdots\alpha_{n-1}^{\frac{x_{n-1}(\overline{\lambda})}{b_n}\zeta}. \tag{4.17}$$

For the definition of $b_s(\ell_0,\mathfrak{l},\zeta,\overline{\mu})$ see (3.34). The following Lemma will be used in the subsection 4.3 in the case $\nu = n$. So we prove it in a more general situation then one needs here.

Lemma 4.3. *Let (s,ν) be a pair of integers satisfying the assertion of Proposition 4.1 and let $\mu_0,\ldots,\mu_{n-1}$ satisfy (4.14). If t and x are integers such that*

$$0 \le t \le T_s, \qquad x \in \mathcal{X}_{s,\nu}, \tag{4.18}$$

then

$$\left|\frac{1}{t!}f^{(t)}(x)\right| \le \varepsilon(x)\cdot e^{6.1\lambda^{n+3}(n+2)^{7/2}\Omega\log(eB)}. \tag{4.19}$$

Proof. Due to the identity

$$\alpha_1^{\frac{\mathfrak{X}_1(\bar{\lambda})}{b_n}\zeta}\cdots\alpha_{n-1}^{\frac{\mathfrak{X}_{n-1}(\bar{\lambda})}{b_n}\zeta}=\alpha_1^{-\frac{\mathfrak{X}_1(\mathbf{v}_s)}{b_n}\zeta}\cdots\alpha_{n-1}^{-\frac{\mathfrak{X}_{n-1}(\mathbf{v}_s)}{b_n}\zeta}\cdot\alpha_1^{\frac{\mathfrak{X}_1(\mathbf{1})}{b_n}\zeta}\cdots\alpha_{n-1}^{\frac{\mathfrak{X}_{n-1}(\mathbf{1})}{b_n}\zeta}$$

the derivatives of the function $f(\zeta)$ can be presented in the form

$$\begin{aligned}\frac{1}{t!}f^{(t)}(\zeta)=\frac{d(\mu_0,H)}{\mu_0!}\sum_{t_0+\cdots+t_n=t}\Bigg(&b_n^{-t_1-\cdots-t_n}\cdot\frac{2^{t_0(S-s)}}{t_0!}\prod_{k=1}^{n-1}\frac{(\log\alpha_k)^{t_k}}{t_k!}\\&\times\frac{1}{t_n!}(-\mathfrak{X}_1(\mathbf{v}_s)\log\alpha_1-\cdots-\mathfrak{X}_{n-1}(\mathbf{v}_s)\log\alpha_{n-1})^{t_n}\cdot\Lambda(\mathbf{t})\Bigg)\end{aligned}\tag{4.20}$$

where

$$\begin{aligned}\Lambda(\mathbf{t})=\sum_{\ell_0=0}^{L_0}\sum_{\mathbf{l}\in\mathfrak{L}_s}\Bigg(&p(\ell_0,2^s\mathbf{l})\Delta^{(\mu_0+t_0)}(2^{S-s}\zeta,\ell_0,H)\\&\times\prod_{k=1}^{n-1}\left(\Delta(2^sN\mathfrak{X}_k(\mathbf{l}-\mathbf{w}_s),\mu_k,\mu_k)\mathfrak{X}_k(\mathbf{l})^{t_k}\alpha_k^{\frac{\mathfrak{X}_k(\bar{\lambda})}{b_n}\zeta}\right)\Bigg).\end{aligned}$$

Let $t_0,\dots,t_{n-1}$ be a set of non-negative integers such that

$$t_0+\ldots+t_{n-1}\le t.$$

According to (4.18), (4.14), (4.4) and the inequalities $M_{s,\nu+1}+T_s\le M_{s,\nu}$ if $\nu<n$, and $M_{s+1,0}+T_s\le M_{s,n}$, we derive

$$\begin{aligned}\sum_{\ell_0=0}^{L_0}\sum_{\mathbf{l}\in\mathfrak{L}_s}\Bigg(&p(\ell_0,2^s\mathbf{l})\Delta^{(\mu_0+t_0)}(2^{S-s}x,\ell_0,H)\\&\times\prod_{k=1}^{n-1}\left(\Delta(2^sN\mathfrak{X}_k(\mathbf{l}-\mathbf{w}_s),\mu_k,\mu_k)\mathfrak{X}_k(\mathbf{l})^{t_k}\alpha_k^{\lambda_k x}\right)\Bigg)=0.\end{aligned}\tag{4.21}$$

Define for brevity

$$x_k=\mathfrak{X}_k(\mathbf{l})\log\alpha_k,\quad y_k=2^sN\mathfrak{X}_k(\mathbf{l}-\mathbf{w}_s),\quad k=1,\dots,n-1,$$

$$x_n=\mathfrak{X}_1(\mathbf{v}_s)\log\alpha_1+\cdots+\mathfrak{X}_{n-1}(\mathbf{v}_s)\log\alpha_{n-1}.$$

and

$$\begin{aligned}\Sigma(\ell_0,\mathbf{l})=\frac{d(\mu_0,H)}{\mu_0!}\sum_{t_0+\cdots+t_n=t}\Bigg(&|b_n|^{-t_1-\cdots-t_n}\cdot\prod_{k=1}^{n}\frac{|x_k|^{t_k}}{t_k!}\\&\times\prod_{k=1}^{n-1}|\Delta(y_k,\mu_k,\mu_k)|\cdot\frac{2^{t_0(S-s)}}{t_0!}|\Delta^{(\mu_0+t_0)}(2^{S-s}x,\ell_0,H)|\Bigg).\end{aligned}$$

In this notation (4.20), (4.16) and (4.21) imply

$$\left|\frac{1}{t!}f^{(t)}(x)\right| \le \varepsilon(x)\sum_{\ell_0=0}^{L_0}\sum_{\mathbf{l}\in\mathfrak{L}_s}|p(\ell_0,2^s\mathbf{l})|\cdot\Sigma(\ell_0,\mathbf{l}). \tag{4.22}$$

To find an upper bound for $\Sigma(\ell_0,\mathbf{l})$ we first note that $|y_k| \le 2NBL$ (see (3.36)),

$$|x_k| \le B(|\ell_k|+|\ell_n|)|\log\alpha_k| \le BA_k\left(\frac{L}{2A_k}+\frac{L}{2A_n}\right) \le BL,$$

for $k=1,\dots,n-1$, and $|x_n| \le nBL$. Hence

$$\sum_{k=1}^{n}|x_k| + \sum_{k=1}^{n-1}|y_k| \le 4nNBL.$$

Besides, for $\mu = \mu_1+\dots+\mu_{n-1}$ we have

$$\mu_0+\mu+t \le M_{s,\nu} \le \widehat{M} \le 1.75\lambda^{n+3}(n+1)^{7/2}\frac{\log(eB)}{\log(eBN)}\Omega. \tag{4.23}$$

Since

$$\begin{aligned}
&\sum_{t_1+\dots+t_n=t-t_0}|b_n|^{-t_1-\dots-t_n}\prod_{k=1}^{n}\frac{|x_k|^{t_k}}{t_k!}\cdot\prod_{k=1}^{n-1}|\Delta(y_k,\mu_k,\mu_k)| \\
&\quad\le \sum_{t_1+\dots+t_n=t-t_0}\prod_{k=1}^{n}\frac{|x_k|^{t_k}}{t_k!}\prod_{k=1}^{n-1}\frac{(|y_k|+\mu_k)^{\mu_k}}{\mu_k!} \\
&\quad\le \frac{(x_1+\dots+x_n+y_1+\dots+y_{n-1}+\mu_1+\dots+\mu_{n-1})^{t-t_0+\mu}}{(t-t_0+\mu)!} \\
&\quad\le \frac{(4nNBL+\mu)^{t-t_0+\mu}}{(t-t_0+\mu)!},
\end{aligned}$$

and

$$\begin{aligned}
\frac{4nNBL+\mu}{eBN} &\le \frac{4n}{e}L+\frac{\widehat{M}}{eB} \\
&\le 1.2\lambda^{n+3}(n+1)^{7/2}\frac{\log(eB)}{\log(eBN)}\Omega \\
&\le 0.4\lambda^{n+3}(n+1)^{5/2}\Omega\log(eB)
\end{aligned}$$

$$d(\mu_0,H) \le (eBN)^{\mu_0},$$

we derive

$$\Sigma(\ell_0, \mathbf{l}) \le (eBN)^{\mu_0+t+\mu} e^{0.4\lambda^{n+3}(n+1)^{5/2}\Omega \log(eB)}$$
$$\times \sum_{t_0=0}^{\ell_0} \frac{2^{t_0(S-s)}}{\mu_0! t_0!} |\Delta^{(\mu_0+t_0)}(2^{S-s}x, \ell_0, H)|.$$

Due to Proposition 3.1, (3.39), (3.41) and (4.23) we have

$$\Sigma(\ell_0, \mathbf{l}) \le \exp\left(1.9\lambda^{n+3}(n+1)^{7/2}\Omega \log(eB)\right)\left(16^{n+8.4}N\right)^{\ell_0}.$$

Finally by (4.22), Proposition 3.9, (3.40) and (3.26) we find (4.19).

To find an upper bound for $f(z)$ we apply the interpolation formula with parameters

$$T = T_s, \qquad R = 2^{s+\nu} e^{n+2.3}(9.9n + 19.8)\log(eBN),$$
$$\delta = 1, \text{ if } \nu \ge 1, \quad \delta = 2, \text{ if } \nu = 0, \qquad r = \frac{1}{2},$$

and the set $\mathcal{X} = \mathcal{X}_{s,\nu}$. The degree of corresponding polynomial $Q(\zeta)$ equals to $|\mathcal{X}_{s,\nu}|(T_s + 1)$.

Notice that by definition for any $\nu \ge 0$ and $s \ge 0$ the following inequalities hold

$$2^{\nu+1}X_s \le |\mathcal{X}_{s,\nu}| \le 2^{\nu+1}X_s + 1, \qquad X_s \le 2^{s-1}X. \tag{4.24}$$

Besides, for all integers $s \ge 0, \nu \ge 0$, we have

$$\begin{aligned}|\mathcal{X}_{s,\nu}|(T_s + 1) &\ge 2^{\nu+1}X_s(T_s+1) \ge \frac{2^{\nu-1}XM}{(n+1)(T_s+1)} \cdot (T_s+1)\\ &= \frac{2^{\nu-1}}{n+1}XM,\end{aligned} \tag{4.25}$$

and

$$\begin{aligned}|\mathcal{X}_{s,\nu}|(T_s+1) &\le (2^{\nu+1}X_s + 1)(T_s+1)\\ &\le \left(2^{\nu+1}\left(\frac{XM}{(4n+4)(T_s+1)} + 1\right) + 1\right)(T_s+1)\\ &\le \frac{2^{\nu-1}}{n+1}XM\left(1 + (6n+6)\frac{T_s+1}{XM}\right).\end{aligned}$$

Due to the inequalities

$$T_s \le L, \quad L \ge 10^6, \quad X \ge (9.9n + 17.8)\log(eBN) \ge (9.9n+17.8)\log(eC(n))$$

we derive

$$|\mathcal{X}_{s,\nu}|(T_s+1) \le 2^{\nu} \cdot 8\lambda^{n+3}(n+2)^{7/2}\Omega \log(eB). \tag{4.26}$$

Thus by (4.25) and (4.26) we have a lower and upper bounds for the degree of the polynomial $Q(\zeta)$.

Let $\zeta \in \mathbb{C}$, $|\zeta| = R$ and vector $\overline{\lambda} = \mathbf{l} - \mathbf{v}_s$, $\mathbf{l} \in \mathfrak{L}_s$. By (4.15), by definition of parameters and definition of the set $\mathfrak{L}_s$ we have

$$\begin{aligned}\log\left|\alpha_1^{\frac{x_1(\overline{\lambda})}{b_n}\zeta} \cdots \alpha_{n-1}^{\frac{x_{n-1}(\overline{\lambda})}{b_n}\zeta}\right| &\le |\zeta|\left|\frac{\lambda_n}{b_n} \cdot \Lambda - \sum_{k=1}^{n} \lambda_k \log \alpha_k\right| \\ &\le \frac{1}{2} + |\zeta| \cdot 2^{1-s}\eta^{-1}L \\ &\le 2^{\nu}\lambda^{n+3}(n+1)^{5/2}(7.9n + 15.8)\Omega \log(eB). \end{aligned} \tag{4.27}$$

Now apply the definition of $f(\zeta)$, see (4.17), Lemma 3.10 and Proposition 3.9 to find an upper bound

$$\begin{aligned}|f(\zeta)| &\le e^{2^{\nu}\lambda^{n+3}(n+1)^{5/2}(7.9n+15.8)\Omega\log(eB)} \cdot \sum_{\ell_0=0}^{L_0} \sum_{\mathbf{l}\in\mathfrak{L}_s} |p(\ell_0, 2^s\mathbf{l}) \cdot b_s(\ell_0, \mathbf{l}, \zeta, \overline{\mu})| \\ &\le e^{(2^{\nu}\lambda^{n+3}(n+1)^{5/2}(7.9n+15.8)+n+1)\Omega\log(eB)} \cdot \Sigma \\ &\le e^{2^{\nu}\lambda^{n+3}(n+1)^{5/2}(14n+22)\Omega\log(eB)}. \end{aligned} \tag{4.28}$$

Now we use the inequality (4.10). In our case $z_0 \le 2^{\nu+1}X_s + 2$. That's why

$$\left|\frac{Q(z)}{Q(\zeta)}\right| \le \left(\frac{2^{s+\nu}X + 2}{R}\right)^{|\mathcal{X}|(T+1)} \le e^{-(n+2.3)|\mathcal{X}|(T+1)}.$$

Applying these bounds, the inequality $2|z| \le R$ and (4.25) to (4.8) we derive

$$|f_1(z)| \le e^{-7.92\cdot 2^{\nu}\lambda^{n+3}(n+1)^{9/2}\Omega\log(eB)}. \tag{4.29}$$

To estimate $f_2(z)$ first note that in (4.11), due to (4.24), we have

$$\frac{z_0}{|\mathcal{X}|} \le \frac{2^{\nu+1}X_s + 2}{2^{\nu+1}X_s} \le 2$$

and

$$\left|\frac{Q(z)}{Q(\zeta)}\right| \le 12^{|\mathcal{X}|(T+1)} \le 12^{2^{\nu}\cdot 8\lambda^{n+3}(n+2)^{7/2}\Omega\log(eB)}.$$

By (4.9) and Lemma 4.3 we derive

$$\begin{aligned}|f_2(z)| &\le |\mathcal{X}|(T+1)\max_{x\in\mathcal{X}} \varepsilon(x) \cdot e^{2^{\nu}\cdot 25.98\lambda^{n+3}(n+2)^{7/2}\Omega\log(eB)} \\ &\le \max_{x\in\mathcal{X}} \varepsilon(x) \cdot e^{2^{\nu}\cdot 26\lambda^{n+3}(n+2)^{7/2}\Omega\log(eB)} \\ &\le \max_{x\in\mathcal{X}} \varepsilon(x) \cdot e^{0.21C(n)\Omega\log(eB)}. \end{aligned} \tag{4.30}$$

To estimate $|\varepsilon(x)|$ notice that by (4.24) we have

$$|x| \le 2^{s+\nu}(9.9n + 17.9)\log(eBN)$$

and by definition of the set $\mathfrak{L}_s$

$$\log\left|\alpha_1^{\lambda_1 x}\cdots\alpha_n^{\lambda_n x}\right| \le |x|\left|\sum_{k=1}^{n}\lambda_k\log\alpha_k\right| \le |x|\cdot 2\eta^{-1}2^{-s}L \le 0.01C(n)\Omega\log(eB).$$

Hence

$$\log|\varepsilon(x)| \le -0.49C(n)\Omega\log(eB). \tag{4.31}$$

Combining this bound with (4.30) we derive

$$|f_2(z)| \le e^{-0.28C(n)\Omega\log(eB)}.$$

Finally, since $7.92\cdot 2^{\nu}\lambda^{n+3}(n+1)^{9/2} \le 0.28C(n)$, we have with (4.29) the inequality

$$|f(z)| \le |f_1(z)| + |f_2(z)| \le 2e^{-7.92\cdot 2^{\nu}\lambda^{n+3}(n+1)^{9/2}\Omega\log(eB)}. \tag{4.32}$$

Due to (4.17), (4.7), (4.16) we find

$$|\Phi_s(z,\overline{\mu}) - f(z)| \le \varepsilon(z)\sum_{\ell_0=0}^{L_0}\sum_{\mathbf{l}\in\mathfrak{L}_s}|p(\ell_0, 2^s\mathbf{l})\cdot b_s(\ell_0,\mathbf{l},z,\overline{\mu})| \tag{4.33}$$

Like in (4.31) it is easy to check that

$$\log|\varepsilon(z)| \le -0.49C(n)\Omega\log(eB).$$

and due to Lemma 3.10 and Proposition 3.9, like in (4.28) we derive

$$|\Phi_s(z,\overline{\mu}) - f(z)| \le e^{-0.4C(n)\Omega\log(eB)}.$$

Hence

$$\begin{aligned}|\Phi_s(z,\overline{\mu})| &\le |f(z)| + |\Phi_s(z,\overline{\mu}) - f(z)| \\ &\le 3e^{-7.92\cdot 2^{\nu}\lambda^{n+3}(n+1)^{9/2}\Omega\log(eB)}.\end{aligned} \tag{4.34}$$

Recall that the number $\Phi_s(z,\overline{\mu})$ is rational. Since $p(\ell_0, 2^s\mathbf{l})$, $b(\ell_0,\mathbf{l},z,\overline{\mu})$ are integers we have for any prime p the bound

$$|\Phi_s(z,\overline{\mu})|_p \le \max_{\mathbf{l}\in\mathfrak{L}_s}\left|\alpha_1^{\lambda_1 z}\cdots\alpha_n^{\lambda_n z}\right|_p.$$

Assume that $\Phi_s(z,\overline{\mu}) \ne 0$. Then by the product formula (in this case all local degrees are equal to 1) and Lemma 3.11 applied to the field $\mathbf{K} = \mathbb{Q}$ we derive

$$\begin{aligned}
1 &= \prod_v |\Phi_s(z,\overline{\mu})|_v \\
&= |\Phi_s(z,\overline{\mu})| \prod_p |\Phi_s(z,\overline{\mu})|_p \\
&\le |\Phi_s(z,\overline{\mu})| \cdot \prod_p \max_{\mathbf{l}\in\mathfrak{L}_s} \left|\alpha_1^{\lambda_1 z}\cdots\alpha_n^{\lambda_n z}\right|_p \\
&\le |\Phi_s(z,\overline{\mu})| \cdot \prod_v \max_{\mathbf{l}\in\mathfrak{L}_s} \left|\alpha_1^{\lambda_1 z}\cdots\alpha_n^{\lambda_n z}\right|_v \\
&\le |\Phi_s(z,\overline{\mu})| \cdot e^{2^{-s}nL|z|} \\
&\le 3e^{-7.92\cdot 2^{\nu}\lambda^{n+3}(n+1)^{5/2}\Omega\log(eB)} \\
&< 1.
\end{aligned}$$

Here we have used the inequality

$$\max_{\mathbf{l}\in\mathfrak{L}_s} \left|\alpha_1^{\lambda_1 z}\cdots\alpha_n^{\lambda_n z}\right| \ge 1, \tag{4.35}$$

following from the fact that $\mathbf{v}_s \in \mathfrak{L}_s$.

This contradiction proves that $\Phi_s(z,\overline{\mu}) = 0$ and completes the inductive step $(s,\nu) \to (s,\nu+1)$, $\nu < n$.

4.3 Extrapolation with Kummer descent

In this subsection we assume that s is an integer $0 \le s < S$ and for all integers x, $m_0 \ge 0, \dots, m_{n-1} \ge 0$, with

$$x \in \mathcal{X}_{s,n}, \quad m_0 + \cdots + m_{n-1} \le M_{s,n}$$

the equalities hold:

$$F_s(x,\mathbf{m}) = 0.$$

To prove the induction step $(s,n) \to (s+1,0)$ we choose an odd number $2y+1$ and a vector $\overline{\mu} = (\mu_0, \dots, \mu_{n-1}) \in \mathbb{Z}^n$, $\mu_j \ge 0$, with

$$2y+1 \in \mathcal{X}_{s+1,0}, \quad \mu_0 + \cdots + \mu_{n-1} \le M_{s+1,0}. \tag{4.36}$$

We shall first prove that $F_s(z,\overline{\mu}) = 0$, where $z = y + \frac{1}{2}$. According to Lemma 4.3 for the function $f(\zeta)$ defined in (4.17) and any pair of integers t, x satisfying

$$0 \le t \le T_s, \qquad x \in \mathcal{X}_{s,n}, \tag{4.37}$$

the inequality holds:

$$\left|\frac{1}{t!}f^{(t)}(x)\right| \le \varepsilon(x)\cdot e^{6.1\lambda^{n+3}(n+2)^{7/2}\Omega\log(eB)}. \tag{4.38}$$

In the same way as in subsection 4.2 we shall find an upper bound for $f(z) = f(y + \frac{1}{2})$.

Apply the interpolation formula with parameters

$$T = T_s, \quad R = 2^{s+n} e^{n+2.3}(9.9n + 19.8)\log(eBN), \quad \delta = 1, \quad r = \frac{1}{4},$$

and the set $\mathcal{X} = \mathcal{X}_{s,n}$. For $|\zeta| = R$ the inequality (4.27) is valid and the function $f(\zeta)$ satisfies (4.28).

The inequality $\frac{T_s}{2} \le \frac{2^{-s-2}M}{n+1}$ implies $\frac{T_s-1}{2} \le T_{s+1}$ and $T_s+1 \le 2(T_{s+1}+1)$. Therefore

$$X_{s+1} \le \left\lceil \frac{2XM}{(4n+4)(T_s+1)} \right\rceil \le \frac{2XM}{(4n+4)(T_s+1)} + 1 \le 2X_s + 1$$

and in this case

$$z_0 \le \max(2^n X_s + 1, X_{s+1}) = 2^n X_s + 1 \le 2^{n+s-1} X + 1.$$

For $|\zeta| = R$ we have

$$\frac{z_0}{R} \le \frac{1}{2e^{n+2.3}}$$

and

$$\left|\frac{Q(z)}{Q(\zeta)}\right| \le \left(2e^{n+2.3}\right)^{-|\mathcal{X}|(T+1)}.$$

Applying these bounds and (4.25) to (4.8) we derive

$$|f_1(z)| \le e^{-7.92 \cdot 2^n \lambda^{n+3}(n+1)^{7/2}(n+2)\Omega \log(eB)}. \tag{4.39}$$

To estimate $f_2(z)$ first note that in (4.11) due to $|\mathcal{X}_{s,n}| = 2^n X_s + 1$ we have

$$\left|\frac{Q(z)}{Q(\zeta)}\right| \le 12^{|\mathcal{X}|(T+1)} \le 12^{2^{n+3}\lambda^{n+3}(n+2)^{7/2}\Omega \log(eB)}.$$

By (4.9) and Lemma 4.3 we derive

$$\begin{aligned} |f_2(z)| &\le |\mathcal{X}|(T+1)\max_{x\in\mathcal{X}} \varepsilon(x) \cdot e^{26 \cdot 2^n \lambda^{n+3}(n+2)^{7/2}\Omega \log(eB)} \\ &\le \max_{x\in\mathcal{X}} \varepsilon(x) \cdot e^{0.41C(n)\Omega \log(eB)}. \end{aligned} \tag{4.40}$$

and in the same way as in subsection 4.2 we find

$$|\Phi_s(z, \overline{\mu})| \le 3e^{-7.92 \cdot 2^n \lambda^{n+3}(n+1)^{7/2}(n+2)\Omega \log(eB)}. \tag{4.41}$$

Denote by $\mathbf{K}$ the field generated over $\mathbb{Q}$ by the numbers $\sqrt{\alpha_j}$,

$$\mathbf{K} = \mathbb{Q}(\sqrt{\alpha_1}, \ldots, \sqrt{\alpha_n}).$$

Then $[\mathbf{K}:\mathbb{Q}] \le 2^n$ and

$$\Phi_s(z,\overline{\mu}) = \sum_{\ell_0=0}^{L_0} \sum_{\mathbf{l}\in\mathfrak{L}_s} p(\ell_0, 2^s\mathbf{l}) b_s\left(\ell_0, \mathbf{l}, y+\frac{1}{2}, \overline{\mu}\right) \alpha_1^{\lambda_1(y+1/2)} \cdots \alpha_n^{\lambda_n(y+1/2)} \in K, \tag{4.42}$$

where

$$b_s\left(\ell_0, \mathbf{l}, y+\frac{1}{2}, \overline{\mu}\right) = \frac{d(\mu_0,H)}{\mu_0!}\Delta^{(\mu_0)}\left(2^{S-s}(y+\frac{1}{2}), \ell_0, H\right)\cdot \prod_{k=1}^{n-1} \Delta\left(2^s N \mathfrak{X}_k(\mathbf{l}-\overline{w}_s), \mu_k, \mu_k\right) \in \mathbb{Z}.$$

For any non-archimedean absolute value $|\ |_v$ on the field $\mathbf{K}$ we have the upper bound

$$|\Phi_s(z,\overline{\mu})|_v \le \max_{\mathbf{l}\in\mathfrak{L}_s} \left|\alpha_1^{\lambda_1 z}\cdots\alpha_n^{\lambda_n z}\right|_v . \tag{4.43}$$

If $|\ |_v$ is an archimedean absolute value, then

$$|\Phi_s(z,\overline{\mu})|_v \le \max_{\mathbf{l}\in\mathfrak{L}_s} \left|\alpha_1^{\lambda_1 z}\cdots\alpha_n^{\lambda_n z}\right|_v \cdot \sum_{\ell_0=0}^{L_0}\sum_{\mathbf{l}\in\mathfrak{L}_s}\left|p(\ell_0,2^s\mathbf{l})b_s\left(\ell_0,\mathbf{l},y+\frac{1}{2},\overline{\mu}\right)\right|.$$

In the same way as in (4.28) we derive

$$|\Phi_s(z,\overline{\mu})|_v \le \max_{\mathbf{l}\in\mathfrak{L}_s} \left|\alpha_1^{\lambda_1 z}\cdots\alpha_n^{\lambda_n z}\right|_v \cdot e^{6.2\lambda^{n+3}(n+1)^{7/2}\Omega\log(eB)}. \tag{4.44}$$

Assume that $\Phi_s(z,\overline{\mu}) \ne 0$. Then by the product formula and (4.41), (4.43), (4.44) we find

$$\begin{aligned} 1 &= \prod_v |\Phi_s(z,\overline{\mu})|_v^{d_v} \\ &\le \prod_v \max_{\mathbf{l}\in\mathfrak{L}_s} \left|\alpha_1^{\lambda_1 z}\cdots\alpha_n^{\lambda_n z}\right|_v^{d_v} \cdot 3e^{-2^n\lambda^{n+3}(n+1)^{7/2}(7.92n+9)\Omega\log(eB)}. \end{aligned} \tag{4.45}$$

Here we used the inequality (4.35).

Now by Lemma 3.11 we have

$$\begin{aligned} \log \prod_v \max_{\mathbf{l}\in\mathfrak{L}_s} \left|\alpha_1^{\lambda_1 z}\cdots\alpha_n^{\lambda_n z}\right|_v &\le 2^{-s}nL|z|[\mathbf{K}:\mathbb{Q}] \\ &\le 2^{n-s}nLX_{s+1} \\ &\le 2^n nLX \\ &\le 2^n\lambda^{n+3}(n+1)^{5/2}(7.92n^2+16n)\Omega\log(eB). \end{aligned}$$

But this inequality contradicts (4.45). This contradiction proves that

$$\Phi_s\left(y+\frac{1}{2},\overline{\mu}\right)=0.$$

The system of equalities $\Phi_s(y+\frac{1}{2},\overline{\mu})=0$ for each pair $(2y+1,\ \overline{\mu})$ satisfying (4.36) is equivalent to the system

$$F_s\left(\frac{x}{2},\overline{\mu}\right)=0,\quad x\in\mathcal{X}_{s+1,0},\quad \mu_0+\cdots+\mu_{n-1}\leq M_{s+1,0}. \tag{4.46}$$

The Kummer descent in the proof of Proposition 4.1 exploits the following lemma (for the more general Baker–Stark lemma based on Kummer theory see [2]).

Lemma 4.4. *Let $\beta,\ \beta_1,\ldots,\beta_m,\ m\geq 0$, be rational numbers and $\sqrt{\beta}\in\mathbb{Q}(\sqrt{\beta_1},\ldots,\sqrt{\beta_m})$. Then there exist $i_k\in\{0;1\}$, $k=1,\ldots,m$, such that*

$$\beta=\beta_1^{i_1}\cdots\beta_m^{i_m}\gamma^2,\qquad \gamma\in\mathbb{Q}.$$

Proof. We prove this lemma by induction on m. For $m=0$ the assertion is certainly true. Assume now $m\geq 1$ and that the assertion is true for any β and any set of β_i containing less than m elements. Denote $\mathbf{K}=\mathbb{Q}(\sqrt{\beta_1},\ldots,\sqrt{\beta_{m-1}})$. If $\sqrt{\beta}\in\mathbf{K}$ we derive the assertion by induction assumption. So we suppose $\sqrt{\beta}\notin\mathbf{K}$, and this implies $\sqrt{\beta_m}\notin\mathbf{K}$. Then

$$\sqrt{\beta}=\gamma_0+\gamma_1\sqrt{\beta_m},\quad \gamma_0,\gamma_1\in\mathbf{K},\quad \gamma_1\neq 0. \tag{4.47}$$

By this equality the number $\sqrt{\beta}$ is a root of the polynomial

$$z^2-2\gamma_0 z+\gamma_0^2-\beta_m\gamma_1^2\in\mathbf{K}[z].$$

Since $\beta\in\mathbb{Q}$ and $\sqrt{\beta}\notin\mathbf{K}$, this fact implies $\gamma_0=0$ and by (4.47) we have $\sqrt{\beta/\beta_m}=\gamma_1\in\mathbf{K}$. Now we apply the induction assumption to the number β/β_m and the set $\beta_1,\ldots,\beta_{m-1}$.

Corollary 4.5. *Let $\overline{u}_i=(u_{i,1},\ldots,u_{i,n})\in\mathbb{Q}^n$, $1\leq i\leq n$, be a basis of the lattice $\mathfrak{N}$ and define rational numbers $\theta_1,\ldots,\theta_n$ by the equalities*

$$\log\theta_i=\sum_{j=1}^{n}u_{i,j}\log\alpha_j.$$

Then

$$\deg\mathbb{Q}\left(\sqrt{\theta_1},\ldots,\sqrt{\theta_n}\right)=2^n. \tag{4.48}$$

Proof. Denote $\mathbf{K}_j = \mathbb{Q}\left(\sqrt{\theta_1}, \dots, \sqrt{\theta_j}\right)$, $1 \le j \le n$ and $\mathbf{K}_0 = \mathbb{Q}$. The equalities $[\mathbf{K}_{j+1} : \mathbf{K}_j] = 2$, $0 \le j < n$ ensure (4.47). Assume that for an index j we have $[\mathbf{K}_{j+1} : \mathbf{K}_j] = 1$. This implies $\sqrt{\theta_{j+1}} \in \mathbf{K}_j$. By Lemma 4.4 one can find $\gamma \in \mathbb{Q}$ and integers $k_1, \dots, k_j \in \{0, 1\}$ such that

$$\theta_{j+1} = \theta_1^{k_1} \cdots \theta_j^{k_j} \gamma^2. \tag{4.49}$$

Hence

$$\gamma = \alpha_1^{v_1} \cdots \alpha_n^{v_n}, \qquad v_j \in \mathbb{Q},$$

and $\overline{v} = (v_1, \dots, v_n) \in \mathfrak{N}$. Since $\overline{u}_1, \dots, \overline{u}_n$ form a basis of $\mathfrak{N}$ then

$$\overline{v} = m_1\overline{u}_1 + \cdots + m_n\overline{u}_n, \qquad m_i \in \mathbb{Z}.$$

Now we can write (4.49) in the form

$$\overline{u}_{j+1} = \sum_{i=1}^{j} k_j\overline{u}_i + 2\sum_{i=1}^{n} m_i\overline{u}_i.$$

Since the vectors $\overline{u}_i$ are linearly independent over $\mathbb{Q}$, this gives a contradiction.

Let $\mathbf{U}$ be the matrix with elements $u_{i,j}$. For any vector $\overline{\lambda} = (\lambda_1, \dots, \lambda_n) \in \mathfrak{N}$ there exists a vector $\overline{\kappa} = (\kappa_1, \dots, \kappa_n) \in \mathbb{Z}^n$ such that $\overline{\lambda} = \overline{\kappa}\mathbf{U}$ and

$$\alpha_1^{\lambda_1} \cdots \alpha_n^{\lambda_n} = \theta_1^{\kappa_1} \cdots \theta_n^{\kappa_n}.$$

For brevity we shall denote expressions like in the last equality by $\overline{\alpha}^{\overline{\lambda}}$ and $\overline{\theta}^{\overline{\kappa}}$.

Let $x = 2y + 1$ and $\overline{\mu}$ satisfy conditions (4.36). Due to (4.46) and the definition of numbers $F_s(x, \overline{\mu})$ (see (4.2)) we derive

$$\begin{aligned} 0 &= \sum_{\ell_0=0}^{L_0} \sum_{\mathbf{l}\in\mathfrak{L}_s} \Big(p(\ell_0, 2^s\mathbf{l}) \Delta^{(\mu_0)}(2^{S-s-1}x, \ell_0, H) \\ &\qquad \times \mathfrak{X}_1(\mathbf{l})^{\mu_1} \cdots \mathfrak{X}_{n-1}(\mathbf{l})^{\mu_{n-1}} \alpha_1^{\lambda_1 x/2} \cdots \alpha_n^{\lambda_n x/2} \Big) \\ &= \sum_{\ell_0=0}^{L_0} \sum_{\mathbf{l}\in\mathfrak{L}_s} \Big(p(\ell_0, 2^s\mathbf{l}) \Delta^{(\mu_0)}(2^{S-s-1}x, \ell_0, H) \\ &\qquad \times \mathfrak{X}_1(\mathbf{l})^{\mu_1} \cdots \mathfrak{X}_{n-1}(\mathbf{l})^{\mu_{n-1}} \theta_1^{\kappa_1 x/2} \cdots \theta_n^{\kappa_n x} \Big) \end{aligned} \tag{4.50}$$

Every vector $\overline{\kappa}$ can be expressed in the form $\overline{\kappa} = 2\overline{\tau} + \overline{\delta}$, where $\overline{\tau} \in \mathbb{Z}^n$ and $\overline{\delta} \in \{0; 1\}^n$. Hence

$$\overline{\kappa} \cdot \frac{x}{2} = \overline{\tau} \cdot x + \overline{\delta} \cdot y + \frac{\overline{\delta}}{2},$$

and due to (4.50) we obtain

$$0 = \sum_{\overline{\delta}\in\{0;1\}^n} \left[\sum_{\ell_0=0}^{L_0} \sum_{\substack{\mathbf{l}\in\mathfrak{L}_s \\ \overline{\mu}\equiv\overline{\delta}(mod 2)}} p(\ell_0, 2^s\mathbf{l}) \Delta^{(\mu_0)}(2^{S-s-1}x, \ell_0, H) \right.$$

$$\left. \mathfrak{X}_1(\mathbf{l})^{\mu_1} \cdots \mathfrak{X}_{n-1}(\mathbf{l})^{\mu_{n-1}} \overline{\theta}^{x\overline{\tau}+y\overline{\delta}} \right] \overline{\theta}^{\overline{\delta}/2}.$$

Since all numbers in square brackets are rational numbers and since, according to Corollary 4.5, the numbers $\overline{\theta}^{\overline{\delta}}$, $\overline{\delta} \in \{0;1\}^n$ are linearly independent over $\mathbb{Q}$, we derive

$$\sum_{\ell_0=0}^{L_0} \sum_{\substack{\mathbf{l}\in\mathfrak{L}_s \\ \overline{\kappa}\equiv\overline{\delta}(mod 2)}} p(\ell_0, 2^s\mathbf{l}) \Delta^{(\mu_0)}(2^{S-s-1}x, \ell_0, H) \mathfrak{X}_1(\mathbf{l})^{\mu_1} \cdots \mathfrak{X}_{n-1}(\mathbf{l})^{\mu_{n-1}} \overline{\theta}^{x\overline{\tau}+y\overline{\delta}}$$

$$= 0,$$

for $\overline{\delta} \in \{0;1\}^n$. In particular, for $\overline{\delta} = (0, \dots, 0)$ we have

$$\sum_{\ell_0=0}^{L_0} \sum_{\substack{\mathbf{l}\in\mathfrak{L}_s \\ \overline{\kappa}\equiv 0(mod 2)}} p(\ell_0, 2^s\mathbf{l}) \Delta^{(\mu_0)}(2^{S-s-1}x, \ell_0, H) \mathfrak{X}_1(\mathbf{l})^{\mu_1} \cdots \mathfrak{X}_{n-1}(\mathbf{l})^{\mu_{n-1}} \overline{\theta}^{x\overline{\tau}}$$

$$= 0. \quad (4.51)$$

It is easy to see that a vector $\mathbf{l}$ satisfies conditions

$$\mathbf{l} \in \mathfrak{L}_s, \qquad \mathbf{l} - \overline{v}_s \in \mathfrak{N}$$

if and only if

$$\overline{\rho} = \frac{1}{2}\mathbf{l} \in \mathfrak{L}_{s+1}.$$

This implies that the equality (4.51) can be written in the form

$$\sum_{\ell_0=0}^{L_0} \sum_{\overline{\rho}\in\mathfrak{L}_{s+1}} \Big(p(\ell_0, 2^{s+1}\overline{\rho}) \Delta^{(\mu_0)}(2^{S-s-1}x, \ell_0, H)$$

$$\times \mathfrak{X}_1(\overline{\rho})^{\mu_1} \cdots \mathfrak{X}_{n-1}(\overline{\rho})^{\mu_{n-1}} \overline{\alpha}^{x(\overline{\rho}-\mathbf{v}_{s+1})} \Big) = 0,$$

or in equivalent form as $F_{s+1}(x, \overline{\mu}) = 0$. This completes the inductive step $(s, n) \to (s+1, 0)$ and the proof of Proposition 4.1.

5 Zero estimates and the end of the proof of Theorem 2.1

The last step of the extrapolation procedure ($s = S$, $\nu = n$ in Proposition 4.1) leads to equalities

$$F_S(x,\mathbf{m}) = 0, \quad x \in \mathcal{X}_{S,n}, \quad m_j \geq 0, \quad m_0 + \cdots + m_{n-1} \leq \frac{M}{(n+2)^3}. \tag{5.1}$$

Due to the definition of $\mathcal{X}_{S,n}$ the system (5.1) implies

$$\partial_0^{m_0} \cdots \partial_{n-1}^{m_{n-1}} \mathcal{G}(Y_0, \ldots, Y_n)|_{(x,\alpha_1^x,\ldots,\alpha_n^x)} = 0, \tag{5.2}$$

$$x \in \mathbb{Z}, \quad |x| \leq 2^n X_S, \quad m_0 + \cdots + m_{n-1} \leq \frac{M}{(n+2)^3},$$

where

$$\mathcal{G}(Y_0, \ldots, Y_n) = \sum_{\ell_0=0}^{L_0} \sum_{\mathbf{l} \in \mathfrak{L}_S} p(\ell_0, 2^S \mathbf{l}) Y_0^{\ell_0} Y_1^{\ell_1} \cdots Y_n^{\ell_n}, \qquad \mathcal{G} \not\equiv 0.$$

For any $\mathbf{l} \in \mathfrak{L}_S$ the vector $\mathbf{l} - \overline{v}_S$ belongs to $\mathfrak{N}$, hence $N(\mathbf{l} - \overline{v}_S) \in \mathbb{Z}^n$ and

$$P(Y_0, \ldots, Y_n) = Y_1^{-Nv_{S,1}} \cdots Y_n^{-Nv_{S,n}} \mathcal{G}(Y_0, Y_1^N, \ldots, Y_n^N) \tag{5.3}$$

is a polynomial in the variables $Y_0, Y_1, \ldots, Y_n, Y_1^{-1}, \ldots, Y_n^{-1}$.

By (5.2) the following equalities hold

$$\partial_0^{m_0} \cdots \partial_{n-1}^{m_{n-1}} P(Y_0, \ldots, Y_n)|_{(x,\xi_1^x,\ldots,\xi_n^x)} = 0, \tag{5.4}$$

$$x \in \mathbb{Z}, \quad |x| \leq 2^n X_S, \quad m_0 + \cdots + m_{n-1} \leq \frac{M}{(n+2)^3},$$

where $\xi_j = \alpha_j^{1/N}$.

The set $\mathfrak{L}_S$ is contained in $2^{-S}\mathfrak{L}$. By definition of the set $\mathfrak{L}$ and by the inequalities (3.12) we derive the following bounds for degrees of the polynomial P

$$\deg_{Y_0} P \leq L_0, \quad \deg_{Y_j} P \leq 2^{-S} \frac{NL}{A_j}, \quad \deg_{Y_j^{-1}} P \leq 2^{-S} \frac{NL}{A_j}. \tag{5.5}$$

Assume that $2^{-S} NL < 1$. According to inequalities (5.5) this assumption implies that P is a polynomial in single variable, $P \in \mathbb{C}[Y_0]$. By (4.25) the number of zeros of this polynomial can be bounded from below by

$$|\mathcal{X}_{S,n}|(T_S + 1) \geq 0.8 \cdot 2^n \lambda^{n+3} (n+1)^{5/2} (9.9n + 17.8) \Omega \log(eB).$$

Therefore

$$0.8 \cdot 2^n \lambda^{n+3} (n+1)^{5/2} (9.9n + 17.8) \Omega \log(eB). \leq \deg P \leq L_0.$$

But this inequality contradicts the definition of the parameter L_0. This contradiction implies

$$NL \geq 2^S. \tag{5.6}$$

5.1 Zero estimates on linear algebraic groups

The polynomial P defined in (5.3) has many multiple zeros at points

$$(x, \xi_1^x \cdots \xi_n^x).$$

The set of these points has a structure and this allows to find an additional information about the numbers α_j. General results of this kind were published in a series of articles by D. Masser, G. Wüstholz, P. Philippon. We allude here to the book [13] containing an exposition of the subject (D.Roy), essentially due to P.Philippon. We adapt the notations to our situation, simpler than in [13].

We denote by G the group $\mathbb{C} \times (\mathbb{C}^*)^n$ with its group law written additively and the ring

$$\mathbb{C}[G] = \mathbb{C}[Y_0, Y_1, \ldots, Y_n, Y_1^{-1}, \ldots, Y_n^{-1}].$$

Given nonnegative integers $D_0, D_1, \ldots, D_n$, we say that an element P of $\mathbb{C}[G]$ is of multidegree $(D_0, D_1, \ldots, D_n)$ if its degree in Y_0 is $\leq D_0$ and if, for $j = 1, \ldots, n$, its degree in Y_j and its degree in Y_j^{-1} are $\leq D_j$.

For each $\mathbf{w} = (\eta_0, \eta_1, \ldots, \eta_n) \in \mathbb{C}^{n+1}$ we introduce the derivation $\mathfrak{D}_{\mathbf{w}}$ of $\mathbb{C}[G]$ by putting

$$\mathfrak{D}_{\mathbf{w}} = \eta_0 \frac{\partial}{\partial Y_0} + \eta_1 Y_1 \frac{\partial}{\partial Y_1} + \eta_n Y_n \frac{\partial}{\partial Y_n}.$$

Given a point $g \in G$, a vector subspace $\mathfrak{W}$ of $\mathbb{C}^{n+1}$ and an integer $T \geq 0$, we say that an element P of $\mathbb{C}[G]$ vanishes to order $> T$ at g with respect to $\mathfrak{W}$ if

$$\mathfrak{D}_{\mathbf{w}_1} \cdots \mathfrak{D}_{\mathbf{w}_s} P(g) = 0$$

for any integer s with $0 \leq s \leq T$ and any $\mathbf{w}_1, \ldots, \mathbf{w}_s \in \mathfrak{W}$, this condition being interpreted as $P(g) = 0$ when $s = 0$.

Any algebraic subgroup G^* of G can be written in the form $G^* = \mathfrak{V} \times T_\Phi$ where $\mathfrak{V}$ is 0 or $\mathbb{C}$, Φ is a subgroup of $\mathbb{Z}^n$ and

$$T_\Phi = \left\{ (y_1, \ldots, y_n) \in (\mathbb{C}^*)^n;\ y_1^{\phi_1} \cdots y_n^{\phi_n} = 1 \text{ for all } (\phi_1, \ldots, \phi_n) \in \Phi \right\}.$$

The dimension of G^* is $d^* = d_0 + n - r$, where $d_0 = \dim \mathfrak{V}$, $r = \operatorname{rank}(\Phi)$ and G^* is connected if and only if Φ is a direct factor of $\mathbb{Z}^n$. We also attach to G^* the polynomial

$$\mathcal{H}(G^*; D_0, D_1, \ldots, D_n) = d^*! 2^{n-r} D_0^{d_0} \sum |\det M_{i_1, \ldots, i_r}| D_{j_1} \cdots D_{j_{n-r}}. \quad (5.7)$$

In this expression, the sum extends to all partitions of $\{1, 2, \ldots, n\}$ into disjoint subsets $\{i_1, \ldots, i_r\}$ and $\{j_1, \ldots, j_{n-r}\}$ with $i_1 < \ldots < i_r$ and $j_1 < \cdots < j_{n-r}$, and the symbol $M_{i_1, \ldots, i_r}$ denotes the $r \times r$ matrix formed by the columns of indices $i_1, \dots, i_r$ of a fixed $r \times n$ matrix M whose rows generate Φ. We also make the convention that the empty matrix has determinant 1. In particular, for the group G, we have

$$\mathcal{H}(G; D_0, D_1, \dots, D_n) = (n+1)! 2^n D_0 D_1 \cdots D_n.$$

Given an algebraic subgroup $G^* = \mathfrak{V} \times T_\Phi$ of G, we define the tangent space of G^* at the neutral element e to be the subspace $T_e(G^*) = \mathfrak{V} \times \mathcal{L} \subset \mathbb{C}^{n+1}$, where $\mathcal{L}$ denotes the subspace of $\mathbb{C}^n$ consisting of the common zeros of the linear forms $\phi_1 Y_1 + \cdots + \phi_n Y_n$ with $(\phi_1, \dots, \phi_n) \in \Phi$. In particular $T_e(G) = \mathbb{C}^{n+1}$.

Finally for any subset Σ of G containing the neutral element e of G and any positive integer d we will use the notation

$$\Sigma[d] = \{\sigma_1 + \cdots + \sigma_d; \quad (\sigma_1, \dots, \sigma_d) \in \Sigma^d\}.$$

Proposition 5.1. *Let Σ be a finite subset of G containing e, and let $\mathfrak{W}$ be a vector subspace of $\mathbb{C}^{n+1}$. Assume that for given integers $D_0, D_1, \dots, D_n \geq 0$ and $S_0 \geq 0$ there exists a nonzero element P of $\mathbb{C}[G]$ of multidegree $\leq (D_0, D_1, \dots, D_n)$ which vanishes to order $> (n+1)S_0$ with respect to $\mathfrak{W}$ at each point of $\Sigma[n+1]$. Then there exists a connected algebraic subgroup G^* of G of dimension $\leq n$ such that if we set*

$$\ell_0 = \dim_{\mathbb{C}} \left(\frac{\mathfrak{W} + T_e(G^*)}{T_e(G^*)} \right),$$

then

$$\binom{S_0 + \ell_0}{\ell_0} \cdot \mathrm{Card}\left(\frac{\Sigma + G^*}{G^*} \right) \cdot \mathcal{H}(G^*; D_0, D_1, \dots, D_n) \leq \mathcal{H}(G; D_0, D_1, \dots, D_n).$$

Proof. See [13, Theorem 8.1]

5.2 Construction of the sublattice Φ from Proposition 2.6

In this subsection we apply Proposition 5.1 to the polynomial P from (5.3). Due to (5.5) one can choose the following parameters

$$D_0 = L_0 + 1, \qquad D_j = \left[2^{-S} \frac{NL}{A_j} \right] + 1, \quad j = 1, \dots, n$$

Define Σ as the set of points

$$\Sigma = \left\{ (x, \xi_1^x, \dots, \xi_n^x) \in G; \qquad x \in \mathbb{Z}, \quad |x| \leq \frac{2^n}{n+1} X_S \right\}$$

and let $\mathfrak{W}$ be the linear subspace in $\mathbb{C}^{n+1}$ generated by the vectors $\mathbf{w} = (\eta_0, \eta_1, \dots, \eta_n) \in \mathbb{C}^{n+1}$ satisfying the equality $b_1 \eta_1 + \cdots + b_n \eta_n = 0$. Thanks to (5.4) we derive that the polynomial P satisfies the conditions of Proposition 5.1 with

$$S_0 = \left[\frac{M}{(n+2)^4}\right]. \tag{5.8}$$

We will use in our case notations introduced in the subsection 5.1. In particular for the subgroup G^* that exists according to Proposition 5.1, we have the representation $G^* = \mathfrak{V} \times T_\Phi$. This subgroup has some special properties and we start by clarifying them. Finally we will prove that the lattice Φ corresponding to G^* (see the subsection 5.1) satisfies conditions of Proposition 2.6, and this will complete the proof of Theorem 2.1.

Lemma 5.2. *The following equality holds:*

$$\Sigma \cap G^* = \{e\}.$$

Proof. Assume that for some integer x, satisfying $0 \le |x| \le \frac{2^n}{n+1} X_S$ the point $(x, \xi_1^x, \dots, \xi_n^x)$ belongs to G^*. This of course implies that $\mathfrak{V} = \mathbb{C}$. The group G^* is distinct from G, that's why $r = \operatorname{rank}(\Phi)$ is positive.

For every point $\overline{\phi} = (\phi_1, \dots, \phi_n) \in \Phi$ the equality

$$\xi_1^{x\phi_1} \cdots \xi_n^{x\phi_n} = 1$$

holds, and

$$x\phi_1 \log \alpha_1 + \dots + x\phi_n \log \alpha_n = 2\pi i N \phi_0 \tag{5.9}$$

with some integer ϕ_0. Since $\phi_1, \dots, \phi_n$ are integers and since by conditions of Theorem 2.1 the logarithms $\log \alpha_1, \dots, \log \alpha_n$ are linearly independent over $\mathbb{Q}$, the equality (5.9) implies $r = 1$ and consequently $\dim T_e(G^*) = n$.

Assume that $\mathfrak{W} \subset T_e(G^*)$. Since $\dim \mathfrak{W} = n$ this implies that $\mathfrak{W}$ and $T_e(G^*)$ coincide. The group G^* is connected. Hence one can assume that the vector $\overline{\phi}$ is primitive. But the vector $(b_1, \dots, b_n)$ is primitive too. That's why these two vectors differ from each other only by a sign, and we derive from (5.9) the inequality

$$2\pi N \le |x| \cdot |\Lambda| \le \frac{2^n}{n+1} X_S \cdot |\Lambda| < \frac{2^n}{n+1} 2^{S-1} X \exp(-C(n)\Omega \log(eB))$$

By definition of the parameters X, S and by Proposition 3.7 one can write

$$\begin{aligned} 2\pi &\le \frac{1}{n+1} 2^{2n+23} (9.9n + 18.8) \log(eBN) \exp(-C(n)\Omega \log(eB)) \\ &\le \frac{1}{n+1} 2^{2n+23} (9.9n + 18.8) \log(n(2e)^{n+1}) \exp(-C(n)) < 1. \end{aligned} \tag{5.10}$$

This contradiction implies $\mathfrak{W} \not\subset T_e(G^*)$ and $\mathfrak{W} + T_e(G^*) = \mathbb{C}^{n+1}$. Accordingly, the parameter ℓ_0 in Proposition 5.1 equals 1.

Since in our case $d_0 = 1$, $d^* = n$, $\ell_0 = 1$ and

$$\binom{\Sigma + G^*}{G^*} \geq 1,$$

the inequality of Proposition 5.1 implies

$$(S_0 + 1) \cdot n! 2^{n-1} \cdot D_0 \sum_{j=1}^{n} |\phi_j| \frac{D_1 \cdots D_n}{D_j} \leq (n+1)! 2^n \cdot D_0 D_1 \cdots D_n.$$

At least one of the ϕ_j is distinct from 0. That's why

$$(S_0 + 1) \frac{D_1 \cdots D_n}{D_j} \leq (2n+2) \cdot D_1 \cdots D_n$$

for some j, and due to the definition of S_0 we derive

$$M < (2n+2)(n+2)^4 \cdot D_j. \tag{5.11}$$

Since by definition of the parameters and Proposition 3.7

$$\begin{aligned} M &\geq 1.6\lambda^{n+3}(n+1)^{7/2} \frac{\Omega}{\log(eN)} \\ &\geq 1.6\lambda^{n+3}(n+1)^{7/2} \frac{1}{\log(n(2e)^{n+1}} \\ &> (2n+2)(n+2)^4, \end{aligned}$$

we have $D_j \geq 2$ and

$$D_j \leq 2^{1-S} \frac{NL}{A_j}.$$

The inequality (5.11) and the definition of parameters imply

$$M < (2n+2)(n+2)^4 \cdot 2^{1-S} \frac{NL}{A_j} \leq (2n+2)(n+2)^4 \cdot 2^{-n-22} L = (n+2)^4 \cdot 2^{-n-22} M.$$

This contradiction proves Lemma 5.2.

Now we simplify the inequality from Proposition 5.1. As a consequence of Lemma 5.2 one can write

$$\binom{\Sigma + G^*}{G^*} = \operatorname{Card} \Sigma \geq \frac{2^n}{n+1} X_S.$$

We prove now the bounds

$$X_S \geq 2^{n+22} X, \qquad \operatorname{Card} \Sigma \geq \frac{2^{2n+22}}{n+1} X. \tag{5.12}$$

The second inequality is a consequence of the first one. If $T_S = 0$, then by definitions of X_S, S and the inequality (5.6) we have

$$X_S \geq \frac{XM}{4n+4} = \frac{1}{2}XL \geq \frac{1}{2}2^S N^{-1} X \geq 2^{n+22}X.$$

If $T_S \geq 1$, then by definition of T_S one can write $T_S + 1 \leq \frac{2^{-S}}{n+1}M$ and

$$X_S \geq \frac{XM}{(4n+4)(T_S+1)} \geq 2^{S-2}X \geq 2^{n+22}X.$$

The inequalities (5.12) are proved in both cases.

Due to (5.8) the following inequalities hold

$$\binom{S_0+\ell_0}{\ell_0} \geq \frac{(S_0+1)^{\ell_0}}{\ell_0!} \geq \frac{1}{\ell_0!}\left(\frac{M}{(n+2)^4}\right)^{\ell_0},$$

and the inequality from Proposition 5.1 implies

$$\frac{1}{\ell_0!}\left(\frac{M}{(n+2)^4}\right)^{\ell_0} \cdot \frac{2^n}{n+1} X_S \cdot \mathcal{H}(G^*; D_0, D_1, \dots, D_n) \leq (n+1)!2^n D_0 D_1 \cdots D_n. \quad (5.13)$$

At least one of the coefficients $|\det M_{i_1,\dots,i_r}|$ of the polynomial (5.7) is a non-zero integer. That's why the inequality (5.13) implies

$$\begin{aligned} M^{\ell_0} X_S D_0^{d_0} &\leq (d^*!)^{-1}(n+1)!\ell_0!(n+1)(n+2)^{4\ell_0}2^{r-n} D_0 D_{i_1} \cdots D_{i_r} \\ &\leq (d^*!)^{-1}(n+1)!\ell_0!(n+1)(n+2)^{4\ell_0}2^{r-n} D_0 D_1 \cdots D_r. \end{aligned}$$

Finally by definition of parameters we have

$$D_0 \geq 0.9\lambda^{n+3}(n+1)^{7/2}\frac{\log eB}{\log(e^{n+4.4}N)}\Omega \geq \frac{9}{16}M$$

and

$$M^{\ell_0+d_0} X_S \leq \frac{16}{9}(d^*!)^{-1}(n+1)!\ell_0!(n+1)(n+2)^{4\ell_0}2^{r-n} D_0 D_1 \cdots D_r. \quad (5.14)$$

Lemma 5.3. *The tangent space $T_e(G^*)$ is a subspace of $\mathfrak{W}$, and the vector $\overline{b}$ belongs to the linear space generated by the lattice Φ over the field $\mathbb{Q}$.*

Proof. Assume that $T_e(G^*) \not\subset \mathfrak{W}$. Then

$$\dim(\mathfrak{W} \cap T_e(G^*)) = \dim(T_e(G^*)) - 1 = d_0 + n - r - 1$$

and

$$\ell_0 = \dim(\mathfrak{W}+T_e(G^*)) - \dim T_e(G^*) = \dim \mathfrak{W} - \dim(\mathfrak{W}\cap T_e(G^*)) = r+1-d_0.$$

We will derive a contradiction from (5.14).

Since in our case $\ell_0 + d_0 = r + 1$ and (5.12), the left hand side of (5.14) can be estimated as follows

$$X_S M^{r+1} \geq 2^{n+22} X M^{r+1} \geq 2^{n+22}(9.9n + 17.8)(2n + 2)^{r+1} \log(eBN) \cdot L^{r+1}. \quad (5.15)$$

To find an upper bound for the right hand side of (5.14) we need upper bounds for D_j. First note that

$$D_0 = L_0 + 1 > 10^7.$$

and consequently

$$D_0 \leq 2L_0 \leq \frac{9}{4}(n+1)\frac{\log(eBN)}{\log(e^{n+4.4}N)}L. \quad (5.16)$$

According to (5.6) we have the following upper bound for D_j, $1 \leq j \leq n$,

$$D_j \leq 1 + 2^{-S}NL \leq 2^{1-S}NL \leq 2^{-n-22}L. \quad (5.17)$$

Now we deduce from (5.14)

$$2^{n+22}(9.9n + 17.8)(2n + 2)^{r+1} \log(eBN) \cdot L^{r+1}$$
$$\leq (2n+2)^2(n+2)^{4r+4}2^{r-n}\frac{(n+1)!(r+1)!}{(n-r)!} \cdot \frac{\log(eBN)}{\log(e^{n+4.4}N)} \cdot 2^{-r(n+22)}L^{r+1},$$

and

$$9.9n + 17.8 \leq \frac{(n+1)!(r+1)!}{(n-r)!(n+4.4)}(2n+2)^2 2^{-1-n}\left(\frac{(n+2)^4}{n+1}\right)^{r+1} \cdot 2^{-(r+1)(n+22)}.$$

It is easy to see that the right hand side of last inequality is a decreasing function in r for $r \geq 0$. That's why

$$(9.9n + 17.8)(n + 4.4) \leq (2n + 2)^2(n + 2)^4 2^{-2n-23}.$$

This is impossible for $n \geq 2$ and the contradiction proves Lemma 5.3.

Herein after one needs an upper bound for the product $D_1 \cdots D_n$. According to our assumption the sequence A_j is increasing. That's why

$$D_1 \geq D_2 \geq \cdots \geq D_n.$$

If $D_n = 1$, then by (5.17) we have

$$D_1 \cdots D_n \leq 2^{-(n+22)(n-1)}L^{n-1}.$$

But in the case $D_n \geq 2$ one can use the inequality $D_j \leq 2^{-n-22}\frac{L}{A_j}$ and

$$D_1 \cdots D_n \leq 2^{-n(n+22)}\frac{L^n}{\Omega} \leq 2^{-n(n+22)}L^{n-1}0.8\lambda^{n+3}(n+1)^{5/2}.$$

This implies in both cases

$$D_1 \cdots D_n \leq 1.6\lambda^{n+3}(n+1)^{5/2}2^{-n(n+22)}L^{n-1}. \tag{5.18}$$

Due to Lemma 5.3 we have $\bar{b} \in \Phi$ and

$$r = \operatorname{rank}\Phi \geq 1.$$

Lemma 5.4. *The rank of Φ satisfies*

$$r = \operatorname{rank}\Phi < n.$$

Proof. Assume $r = n$. Since G^* is a connected algebraic subgroup of G, then $\Phi = \mathbb{Z}^n$. Besides $\dim T_e(G^*) = d_0$.

According to Lemma 5.3 we have $T_e(G^*) \subset \mathfrak{W}$. This implies

$$\ell_0 = \dim \mathfrak{W} - \dim T_e(G^*) = n - d_0.$$

Hence the inequality (5.14) can be written in the form

$$M^n X_S \leq \frac{16}{9}((n+1)!)^2(n+2)^{4n}D_0D_1\cdots D_n.$$

By the inequalities

$$X_S \geq 2^{n+22}X \geq (9.9n+17.8)2^{n+22}\log(eBN)$$

(5.16) and (5.18) we derive

$$\begin{aligned}
&2^{n+22}(n+4.4)(9.9n+17.8)(2n+2)^nL^n \\
&\quad \leq 4((n+1)!)^2(n+2)^{4n}\cdot(n+1)LD_1\cdots D_n \\
&\quad \leq 6.4((n+1)!)^2(n+2)^{4n}\cdot\lambda^{n+3}(n+1)^{7/2}2^{-n(n+22)}L^n
\end{aligned}$$

and this is a contradiction, that proves Lemma 5.4.

Now we are ready to prove that the lattice Φ satisfies the inequality (2.8). Let

$$\bar{a}_i = (a_{i,1}, \ldots, a_{i,n}), \qquad i = 1, \ldots, r,$$

be a basis of Φ, and let $< \Phi >$ be a linear subspace of $\mathbb{R}^n$ generated by vectors from Φ. Then the set

$$\{y_1\bar{a}_1 + \cdots + y_r\bar{a}_r \mid \; 0 \leq y_j \leq 1 \quad j = 1, \ldots, r\}$$

is the fundamental parallelepiped of Φ, and $V(\Phi)$ is the r-dimensional volume of this parallelepiped. As in Proposition 2.6 denote

$$\mathfrak{K} = \{\overline{x} = (x_1, \dots, x_n) \in <\Phi> \mid \quad A_1|x_1| + \dots + A_n|x_n| \le 1\}$$

and let $V(\mathfrak{K})$ be the r-dimensional volume of $\mathfrak{K}$.

Denote by $\mathfrak{K}_1$ the set

$$\mathfrak{K}_1 = \{\overline{x} = (x_1, \dots, x_n) \in <\Phi> \mid \quad A_1^2 x_1^2 + \dots + A_n^2 x_n^2 \le n^{-1}\}$$

and let $V(\mathfrak{K}_1)$ be the r-dimensional volume of $\mathfrak{K}_1$. For any point $\overline{x} \in \mathfrak{K}_1$ we have

$$A_1|x_1| + \dots + A_n|x_n| \le (A_1^2 x_1^2 + \dots + A_n^2 x_n^2)^{1/2} \cdot \sqrt{n} \le 1.$$

Therefore $\mathfrak{K}_1 \subset \mathfrak{K}$ and $V(\mathfrak{K}_1) \le V(\mathfrak{K})$.

The linear mapping $f : \mathbb{R}^r \longrightarrow \mathbb{R}^n$ defined by

$$x_j = a_{1,j} y_1 + \dots + a_{r,j} y_r, \qquad j = 1, \dots n,$$

moves the set

$$\mathfrak{C}_1 = \{\overline{y} = (y_1, \dots, y_r) \mid \quad \sum_{j=1}^{n} A_j^2 (a_{1,j} y_1 + \dots + a_{r,j} y_r)^2 \le n^{-1}\}$$

onto $\mathfrak{K}_1$. That's why

$$V(\mathfrak{K}_1) = V(\Phi) \cdot V(\mathfrak{C}_1). \tag{5.19}$$

Denote

$$\overline{b}_i = (A_1 a_{i,1}, \dots, A_n a_{i,n}), \qquad i = 1, \dots, r,$$

let $\Psi \subset \mathbb{Z}^n$ be the lattice spanned by the vectors $\overline{b}_1, \dots, \overline{b}_r$, let $V(\Psi)$ be the r-dimensional volume of the fundamental parallelepiped of Ψ and let $<\Psi>$ be the subspace in $\mathbb{R}^n$ generated by vectors from Ψ. Denote

$$\mathfrak{K}_2 = \{\overline{x} = (x_1, \dots, x_n) \in <\Psi> \mid \quad x_1^2 + \dots + x_n^2 \le n^{-1}\}.$$

Similarly to (5.19) we derive

$$V(\mathfrak{K}_2) = V(\Psi) \cdot V(\mathfrak{C}_1). \tag{5.20}$$

The set $\mathfrak{K}_2$ is a r-dimensional ball in $\mathbb{R}^n$, that's why

$$V(\mathfrak{K}_2) = \frac{\pi^{r/2}}{\Gamma(1 + r/2)} n^{-r/2}.$$

Combining these relations we find

$$\frac{V(\Phi)}{V(\mathfrak{K})} \le \frac{V(\Phi)}{V(\mathfrak{K}_1)} = \frac{V(\Psi)}{V(\mathfrak{K}_2)} = \frac{\Gamma(1 + r/2)}{\pi^{r/2}} n^{r/2} \cdot V(\Psi). \tag{5.21}$$

To find an upper bound for $V(\Psi)$ we will use the notations introduced in the subsection 5.1

$$\begin{aligned} V(\Psi) &= \left(\sum (|\det M_{i_1,\dots,i_r}| \cdot A_{i_1} \cdots A_{i_r})^2\right)^{1/2} \\ &\le \sum |\det M_{i_1,\dots,i_r}| \cdot A_{i_1} \cdots A_{i_r} \\ &= \Omega L^{r-n} \sum |\det M_{i_1,\dots,i_r}| \cdot \frac{L}{A_{j_1}} \cdots \frac{L}{A_{j_{n-r}}}. \end{aligned}$$

Since

$$\frac{L}{A_j} \le \frac{2^S}{N} D_j \le 2^{n+24} D_j$$

we derive

$$V(\Psi) \le 2^{(n-r)(n+24)} \Omega L^{r-n} \sum |\det M_{i_1,\dots,i_r}| \cdot D_{j_1} \cdots D_{j_{n-r}}.$$

Since $\mathrm{Card}(\frac{\Sigma+G^*}{G^*}) = \mathrm{Card}\,\Sigma$ (Lemma 5.2) and

$$\ell_0 = \dim \mathfrak{W} - \dim T_e(G^*) = r - d_0$$

(Lemma 5.3), we have by Proposition 5.1

$$V(\Psi) \le \frac{2^{(n+23)(n-r)}}{(n-r+d_0)!}(n+1)!2^n(r-d_0)!\Omega L^{r-n} \frac{D_0^{1-d_0} D_1 \cdots D_n}{\mathrm{Card}\,\Sigma (S_0+1)^{r-d_0}}.$$

Due to the inequalities

$$S_0 + 1 \le \frac{1.01}{(n+1)^4} M, \qquad D_0 \ge 0.9\lambda^{n+3}(n+1)^{7/2} \frac{\log(eB)}{\log(e^n N)} \Omega \ge \frac{9}{16} M$$

and consequently $D_0 > S_0 + 1$, (5.16), (5.18) and (5.12), we find

$$\begin{aligned} V(\Psi) &\le \frac{2^{(n+23)(n-r)}}{(n-r+d_0)!}(n+1)!2^n(r-d_0)!\Omega L^{r-n} \frac{D_0 D_1 \cdots D_n}{\mathrm{Card}\,\Sigma \cdot (S_0+1)^r} \\ &\le 3.6 \frac{2^{(n+23)(n-r)}}{(n-r+d_0)!}(n+1)!(r-d_0)!2^n \lambda^{n+3}(n+1)^{9/2} 2^{-n(n+22)} \\ &\quad \times (n+4.4)^{-1} \left(\frac{(n+2)^4}{2n+2}\right)^r 2^{-2n-22} \Omega \frac{\log(eBN)}{X} \\ &\le 3.6(n+4.4)^{-1}(9.9n+17.8)^{-1} \cdot 2^{-(r+1)(n+23)+n+1} \\ &\quad \times (n+1)! \frac{r!}{(n-r)!} \lambda^{n+3} (n+1)^{9/2} \left(\frac{(n+2)^4}{2n+2}\right)^r \cdot \Omega \end{aligned}$$

This inequality and (5.21) prove that

$$\frac{V(\Phi)}{V(\mathfrak{K})} \le 0.5(2\lambda)^n (4\lambda)^{-r} \Omega,$$

i.e. the lattice Φ satisfies the inequality from Proposition 2.6.

This finishes the proof of Theorem 2.2.

Note that the last inequality is a consequence of

$$14.4 \cdot \frac{\Gamma(1+r/2)(n+2)^{4r}}{\pi^{r/2}(2n+2)^r} \cdot n^{r/2} \cdot 2^{-(r+1)(n+23)} \cdot \frac{(n+1)!r!}{(n-r)!} \cdot \lambda^3(n+1)^{9/2}(4\lambda)^r \\ \leq (n+4.4)(9.9n+17.8). \quad (5.22)$$

The left-hand side of (5.22) is a decreasing function in r for $1 \leq r < n$. Therefore (5.22) follows from the inequality

$$14.4(n+2)^4 n^{3/2} 2^{-2n-46} \cdot \lambda^4 (n+1)^{9/2} \leq (n+4.4)(9.9n+17.8),$$

that is correct for any $n \geq 2$.

References

1. Baker A., Linear forms in the logarithms of algebraic numbers I, II, III, Mathematika, 1966, v.13, 204-216; 1967, v.14, 102-107, 220-228; 1968, v.15, 204-216.
2. Baker A., Stark H.M., On a fundamental inequality in number theory, Ann. Math., 1971, v.94, 190-199.
3. Baker A., Wüstholz G., Logarithmic forms and group varieties, J. reine angew. Math., 1993, v.442, 19-62.
4. Cassels J.W.S., An Introduction to the geometry of numbers, Springer-Verlag, 1959.
5. Gelfond A.O., On the approximation by algebraic numbers of the ratio of the logarithms of two algebraic numbers, Izvestia Acad. Sci. SSSR, 1939, v.3, no 5-6, 509-518.
6. Gelfond A.O., On the algebraic independence of transcendental numbers of certain classes, Uspechi Mat. Nauk SSSR, 1949, v.4, no 5, 14-48.
7. Gelfond A.O., Feldman N.I., On lower bounds for linear forms in three logarithms of algebraic numbers, Vestnik MGU, 1949, no 5, 13-16.
8. Lang S., Fundamentals of diophantine geometry, Springer, 1983.
9. Matveev E.M., On the arithmetic properties of the values of generalised binomial coefficients; Mat. Zam., 54 (1993), 76–81; Math. Notes, 54 (1993), 1031–1036.
10. Matveev E.M., An explicit lower bound for a homogeneous rational linear form in logarithms of algebraic numbers I, II, Izvestia: Mathematics, 1998, v.62, no 4, 723-772; 2000, v.64, no 6, 125-180.
11. Philippon P., Waldschmidt M. Lower bounds for linear forms in logarithms. New Advances in Transcendence Theory, ed. A.Baker. Cambridge Univ. Press. 1988, 280-312.
12. Schneider R., Convex bodies: The Brunn–Minkowski Theory, Cambridge Univ. Press.
13. Waldschmidt M., Diophantine approximation on linear algebraic groups, Springer, 2000.
14. Wüstholz G., A new approach to Baker's theorem on linear forms in logarithms. I, II in: Diophantine problems and transcendence theory, Lecture Notes Math., 1987, v.1290, 189-202, 203-211; III in: New Advances in Transcendence Theory, ed. A.Baker. Cambridge Univ. Press. 1988, 399-410.

Approximation of Algebraic Numbers

Hans Peter Schlickewei

Fachbereich Mathematik, Philipps-Universität Marburg

The following notes are an enriched exposition of the material which I presented during the C.I.M.E. summer school in Cetraro. My main goal is to illustrate the ideas behind the proofs of recent results generalizing the Subspace Theorem in diophantine approximation. I have tried to keep a balance between avoiding certain technical details (which would have made the notes by far too long and which rather would have hidden the basic ideas) and giving a presentation which is accessible for nonspecialists without simply waving hands.

Ringrazio Francesco Amoroso, Umberto Zannier e il C.I.M.E. per l'organizzazione della scuola estiva e per la cordiale ospitalità nello splendido ambiente di Cetraro.

1 Results

Roth's famous result on rational approximations of algebraic numbers says the following:

Theorem 1.1. (Roth, 1955, [6]). *Suppose $\delta > 0$. Let α be an irrational algebraic number. Then the inequality*

$$\left|\alpha - \frac{x}{y}\right| < y^{-2-\delta} \tag{1.1}$$

has only finitely many solutions in integers x, y with $y > 0$.

In homogeneous form (1.1) becomes

$$|y||y\alpha - x| < y^{-\delta}.$$

Essentially this is the same as

$$|y||y\alpha - x| < \max\{|x|, |y|\}^{-\delta}. \tag{1.2}$$

Notice that in (1.2) we have on the left hand side the product of the absolute values of the two linear forms

$$L_1(x, y) = x - \alpha y, \quad L_2(x, y) = y.$$

These forms are linearly independent and have algebraic coefficients.
This remark illustrates why the following theorem, due to W. M. Schmidt, is the generalization of Roth's Theorem 1.1 to n dimensions.

Theorem 1.2 (Subspace Theorem). (W. M. Schmidt, 1972, [9]). *Suppose $\delta > 0$. Let $n \geq 2$ and let $L_1(\boldsymbol{X}), \ldots, L_n(\boldsymbol{X})$ be linearly independent linear forms with algebraic coefficients in $\boldsymbol{X} = (X_1, \ldots, X_n)$. Consider the inequality*

$$|L_1(\boldsymbol{x}) \ldots L_n(\boldsymbol{x})| < |\boldsymbol{x}|^{-\delta} \quad \text{in } \boldsymbol{x} \in \mathbb{Z}^n, \tag{1.3}$$

where $|\boldsymbol{x}| = (x_1^2 + \ldots + x_n^2)^{1/2}$. Then there exist finitely many proper linear subspaces $T_1, \ldots, T_t$ of $\mathbb{Q}^n$ with the following property:
The set of solutions $\boldsymbol{x}$ of (1.3) *is contained in the union $T_1 \cup \ldots \cup T_t$.*

The principal problems that have been studied in the literature in the context of theorems 1.1 and 1.2 are as follows:

I Give an *algorithm to determine all solutions* in theorem 1.1, or *all subspaces* in theorem 1.2.

II Give *upper bounds for the number of solutions* in theorem 1.1, or *for the number of subspaces* in theorem 1.2.

III Prove generalizations of theorems 1.1 and 1.2 when the approximating elements are *arbitrary algebraic numbers,* also study *arbitrary absolute values.*

As is well known, problem I up to now is almost inaccessible.
In these notes we will discuss some of the recent developments made with respect to problems II and III.
Before we formulate the results we introduce some notation.
Given a number field K we write $\mathfrak{M}(K)$ for the set of its places. For $v \in \mathfrak{M}(K)$, $| \ |_v$ denotes the associated absolute value, normalized such that for $x \in \mathbb{Q}$ we have $|x|_v = |x|$, the standard absolute value, if $v \mid \infty$ and such that $|p|_v = p^{-1}$ if $v \mid p$, where p is a rational prime. Given $x \in K$, $v \in \mathfrak{M}(K)$ with $v \mid p$ where $p \in \mathfrak{M}(\mathbb{Q})$ we define the *normalized absolute value* $\| \ \|_v$ on K by

$$\|x\|_v = |x|_v^{d(v|p)}. \tag{1.4}$$

Here

$$d(v|p) = [K_v : \mathbb{Q}_p]/[K : \mathbb{Q}], \tag{1.5}$$

K_v and $\mathbb{Q}_p$ denoting the completions at v or at p respectively.
The absolute value $\|\ \|_v$ of K has a unique extension $\|\ \|'_v$, say, to $\overline{K}_v$, the algebraic closure of the completion K_v. We extend $\|\ \|_v$ to $\overline{\mathbb{Q}}$ as follows: we fix an embedding $\tau_v : \overline{\mathbb{Q}} \to \overline{K}_v$ over K. Then our extension from K to $\overline{\mathbb{Q}}$ of $\|\ \|_v$ is given by

$$\|x\|_v = \|\tau_v(x)\|'_v \quad \text{for } x \in \overline{\mathbb{Q}}. \tag{1.6}$$

We point out that for any finite extension E of K there exists a place $w \in \mathfrak{M}(E)$ with $w \mid v$ such that the absolute value $\|\ \|_w$ on E (normalized with respect to E in the sense of (1.4), (1.5)) is related with the nonnormalized absolute value $\|\ \|_v$ on E defined in (1.6) by

$$\|x\|_w = \|x\|_v^{d(w|v)} \quad \text{for } x \in E, \tag{1.7}$$

where

$$d(w|v) = [E_w : K_v]/[E : K]. \tag{1.8}$$

Given $\boldsymbol{x} = (x_1, \ldots, x_n) \in \overline{\mathbb{Q}}^n$, let E be a number field with $\boldsymbol{x} \in E^n$. For $w \in \mathfrak{M}(E)$, we introduce the w-adic norm of $\boldsymbol{x}$ by

$$\|\boldsymbol{x}\|_w = \begin{cases} (|x_1|_w^2 + \ldots + |x_n|_w^2)^{d(w|\infty)/2} & \text{for } w \in \mathfrak{M}_\infty(E) \\ \max\{\|x_1\|_w, \ldots, \|x_n\|_w\} & \text{for } w \in \mathfrak{M}_0(E). \end{cases} \tag{1.9}$$

Here $\mathfrak{M}_\infty$ and $\mathfrak{M}_0$ stand for the set of archimedean or nonarchimedean places respectively. The (absolute) height $H(\boldsymbol{x})$ then is defined by

$$H(\boldsymbol{x}) = \prod_{w \in \mathfrak{M}(E)} \|\boldsymbol{x}\|_w. \tag{1.10}$$

Notice that $H(\boldsymbol{x})$ is independent of the particular number field E such that $\boldsymbol{x} \in E^n$.
Given a polynomial P with coefficients in $\overline{\mathbb{Q}}$, the w-adic norm $\|P\|_w$ as well as the height $H(P)$ will be taken with respect to the coefficient-vector of P.
For a number field K and a nonzero linear form

$$L = \alpha_1 X_1 + \ldots + \alpha_n X_n$$

with algebraic coefficients α_i, the extension $K(L)$ is defined by

$$K(L) = K\Big(\frac{\alpha_1}{\alpha_i}, \ldots, \frac{\alpha_n}{\alpha_i}\Big),$$

where i is a subscript having $\alpha_i \neq 0$.
Now suppose that S is a finite subset of $\mathfrak{M}(K)$. Assume that for each $v \in S$ we are given linearly independent linear forms $L_1^{(v)}, \ldots, L_n^{(v)}$ in $\boldsymbol{X} = (X_1, \ldots, X_n)$

with coefficients in $\overline{\mathbb{Q}}$. Let D be a quantity such that for $v \in S$ and for $i = 1, \dots, n$

$$[K(L_i^{(v)}) : K] \le D. \tag{1.11}$$

Moreover let H be a quantity with

$$H(L_i^{(v)}) \le H \quad (v \in S,\ i = 1, \dots, n). \tag{1.12}$$

The following theorem extends theorem 1.2 to the case when we ask for approximation by vectors $\boldsymbol{x} \in \overline{\mathbb{Q}}^n$. It also makes a statement in the sense of problem II:

Theorem 1.3. (J.-H. Evertse and H. P. Schlickewei, 2002, [4]). *Let K, S, $L_1^{(v)}, \dots, L_n^{(v)}$ $(v \in S)$ be as above. Let D and H be given by* (1.11) *and* (1.12) *respectively.*
Suppose that S has cardinality s. Assume that for each $v \in S$ the normalized absolute value $\| \ \|_v$ on K is extended to $\overline{\mathbb{Q}}$ as in (1.6). *Let $0 < \delta < 1$.*
Then there exist proper linear subspaces $T_1, \dots, T_{t_1}$ of $\overline{\mathbb{Q}}^n$, all defined over K, where

$$t_1 = t_1(n, s, D, \delta) \le (3n)^{2ns}\, 2^{3(n+9)^2} \delta^{-ns-n-4} \log(4D) \log\log(4D) \tag{1.13}$$

with the following property:
The set of solutions $\boldsymbol{x} \in \overline{\mathbb{Q}}^n$ of the inequalities

$$\prod_{v \in S} \prod_{i=1}^{n} \max_{\sigma \in \mathrm{Gal}(\overline{\mathbb{Q}}|K)} \frac{\left\| L_i^{(v)}(\sigma(\boldsymbol{x})) \right\|_v}{\|\sigma(\boldsymbol{x})\|_v}$$
$$\le \left(\prod_{v \in S} \left\| \det\left(L_1^{(v)}, \dots, L_n^{(v)}\right) \right\|_v \right) H(\boldsymbol{x})^{-n-\delta} \tag{1.14}$$

and

$$H(\boldsymbol{x}) > \max\{n^{4n/\delta}, H\} \tag{1.15}$$

is contained in the union

$$T_1 \cup \dots \cup T_{t_1}.$$

We remark that the assertion of theorem 1.2 remains true if we replace (1.3) by

$$\frac{|L_1(\boldsymbol{x})|}{|\boldsymbol{x}|} \cdots \frac{|L_n(\boldsymbol{x})|}{|\boldsymbol{x}|} < |\det(L_1, \dots, L_n)|\, H(\boldsymbol{x})^{-n-\delta} \quad \text{in } \boldsymbol{x} \in \mathbb{Q}^n. \tag{1.3$'$}$$

Thus theorem 1.2 in a qualitative sense treats the special case of theorem 1.3 when $K = \mathbb{Q}$, $S = \{\infty\}$ (the archimedean prime) and when the solutions $\boldsymbol{x}$

are taken from $\mathbb{Q}^n$ instead of $\overline{\mathbb{Q}}^n$.
The first quantitative result for the Subspace Theorem 1.2 is due to W.M. Schmidt [11]. For $S = \{\infty\}$ and for solutions $\boldsymbol{x}$ in $\mathbb{Q}^n$ he has shown that the number of subspaces needed does not exceed

$$(2D)^{2^{26n}\delta^{-2}}.$$

In theorem 1.3 we are dealing with certain *particular* nonnormalized *extensions* of the absolute values $\| \ \|_v$, $v \in S$, to $\overline{\mathbb{Q}}$ (cf. (1.6)). Nevertheless, in view of the maximum over $\sigma \in \mathrm{Gal}(\overline{\mathbb{Q}}\,|K)$ in (1.14) implicitly we have information of the same type on *all* the *extensions* of the absolute values $\| \ \|_v$, $v \in S$, to $\overline{\mathbb{Q}}$. We will discuss this aspect in more detail in section 4 when we study the proof of theorem 1.3.
The picture becomes quite different, when we ask for results where on the left hand side of (1.14) we omit the maximum over $\sigma \in \mathrm{Gal}(\overline{\mathbb{Q}}\,|K)$. Clearly, to get a finiteness result we have to impose *some* extra hypothesis. Our additional hypothesis will be to ask for approximations by elements of *bounded degree.*
For an algebraic number β of degree d let

$$P(X) = b_0X^d + \ldots + b_d = b_0(X - \beta^{(1)}) \ldots (X - \beta^{(d)})$$

be its minimal polynomial with coprime integral coefficients $b_0, \ldots, b_d$. The Mahler measure $M(\beta)$ then is given by

$$M(\beta) = |\,b_0| \prod_{i=1}^{d} \max\{1, |\,\beta^{(i)}|\}. \tag{1.16}$$

It is easily seen that the height $H((1,\beta))$ and the Mahler measure are related by

$$M(\beta) \le H((1,\beta))^d \le d^{d/2} M(\beta). \tag{1.17}$$

A generalization of Roth's Theorem to approximation by algebraic numbers of degree $\le d$ is

Theorem 1.4. (Wirsing, 1971, [13]). *Let α be an algebraic number. Suppose $\delta > 0$. Let d be a natural number. Then there are only finitely many algebraic numbers β of degree $\le d$ satisfying*

$$|\alpha - \beta| < M(\beta)^{-2d-\delta}. \tag{1.18}$$

Notice that for $d = 1$ theorem 1.4 coincides with Roth's Theorem 1.1.
W.M. Schmidt [8] (1970) applying his Subspace Theorem has shown that theorem 1.4 remains true with the exponent $-2d - \delta$ replaced by

$$-d - 1 - \delta. \tag{1.19}$$

In the opposite direction, Wirsing [12] (1961) has shown that *for any real number* α *which is not algebraic of degree* $\leq d$ *there exists a constant* $c > 0$ *such that the inequality*

$$|\alpha - \beta| < cM(\beta)^{-d-1}$$

has infinitely many solutions in algebraic numbers β *of degree* $\leq d$. Therefore Schmidt's result (1.19) is best possible.
In section 5 (theorem 5.3) we will discuss a generalization of Wirsing's theorem 1.4 which may be considered as a first step towards a Subspace Theorem for approximation by elements of bounded degree.

2 Roth's proof of theorem 1.1

We begin with an outline of Roth's approach. His proof of theorem 1.1 is indirect. Assuming that inequality (1.1) has infinitely many solutions, he picks a *suitable* sequence among these solutions, $x_1/y_1, \ldots, x_m/y_m$, say.
For these m solutions he constructs a nonzero auxiliary polynomial P with integral coefficients in m variables $X_1, \ldots, X_m$ such that

(i) on the one hand P vanishes with high order at the point $\left(\frac{x_1}{y_1}, \ldots, \frac{x_m}{y_m}\right)$,

(ii) on the other hand P cannot vanish with too high order at $\left(\frac{x_1}{y_1}, \ldots, \frac{x_m}{y_m}\right)$.

He then obtains a contradiction between **(i)** and **(ii)**, provided m is large enough. Consequently m is bounded.
To be more specific on **(i)** and **(ii)** we first define what we mean by "vanishing with high order". Suppose that the polynomial P has degree $\leq r_h$ in X_h $(h = 1, \ldots, m)$. For the order of vanishing of P at a point $\boldsymbol{z} = (z_1, \ldots, z_m)$ Roth introduces a normalization with respect to the m-tuple $(r_1, \ldots, r_m)$.
The Roth-Index I *of* P *at the point* $\boldsymbol{z}$ *with respect to* $\boldsymbol{r} = (r_1, \ldots, r_m)$ *is defined as the maximum value* t *such that for every tuple of nonnegative integers* $\boldsymbol{i} = (i_1, \ldots, i_m)$ *with*

$$\frac{\boldsymbol{i}}{\boldsymbol{r}} = \frac{i_1}{r_1} + \ldots + \frac{i_m}{r_m} < t \tag{2.1}$$

we have

$$P^{\boldsymbol{i}}(\boldsymbol{z}) = \frac{1}{i_1! \ldots i_m!} \frac{\partial^{i_1+\ldots+i_m}}{\partial X_1^{i_1} \ldots \partial X_m^{i_m}} P(\boldsymbol{z}) = 0. \tag{2.2}$$

With this notion we now go back to the two statements **(i)** and **(ii)** we want to derive.

2.1 Vanishing

We begin with (**i**). Our task is to construct a suitable auxiliary polynomial P. By (1.1), the points x_h/y_h under consideration are good approximations of α. Therefore the strategy is as follows. We try to find a nonzero polynomial P with integral coefficients and with partial degrees $\leq r_h$ $(h = 1, \ldots, m)$ whose index at the point $(\alpha, \ldots, \alpha)$ with respect to $(r_1, \ldots, r_m)$ is large. Using (1.1) and Taylor expansion about $(\alpha, \ldots, \alpha)$, we might hope to be able to prove that the index at $(x_1/y_1, \ldots, x_m/y_m)$ will be large as well.
Let us check what we can expect for the index at $(\alpha, \ldots, \alpha)$: We seek a nonzero polynomial

$$P(X_1, \ldots, X_m) = \sum_{j_1=0}^{r_1} \cdots \sum_{j_m=0}^{r_m} c(j_1, \ldots, j_m) X_1^{j_1} \ldots X_m^{j_m}$$

with integral coefficients and of degree r_h in X_h such that for a value I which is as large as possible the partial derivatives $P^{\boldsymbol{i}}(\boldsymbol{\alpha}) = P^{\boldsymbol{i}}(\alpha, \ldots, \alpha)$ vanish for all $\boldsymbol{i}$ with

$$\frac{i_1}{r_1} + \ldots + \frac{i_m}{r_m} < I. \tag{2.3}$$

To construct P we have to find the

$$(r_1 + 1) \ldots (r_m + 1) = N \tag{2.4}$$

integral coefficients $c(j_1, \ldots, j_m)$. Each relation $P^{\boldsymbol{i}}(\boldsymbol{\alpha}) = 0$ is a linear equation in the $c(j_1, \ldots, j_m)$. The coefficients of these equations are products of binomial coefficients and powers of α. Assuming (for simplicity) that α is an algebraic integer of degree d we see that each power of α is a linear combination of $1, \alpha, \ldots, \alpha^{d-1}$ with rational integral coefficients. Therefore, $P^{\boldsymbol{i}}(\boldsymbol{\alpha}) = 0$ is equivalent to d linear equations in the $c(j_1, \ldots, j_m)$ with rational integral coefficients.
The quotients i_h/r_h may be interpreted as random variables with expectation $1/2$. Thus by the law of large numbers $i_1/r_1 + \ldots + i_m/r_m \approx m/2$ with high probability. More precisely we obtain

Lemma 2.1. *Suppose* $0 < \varepsilon < 1$. *Then the number of tuples* $(i_1, \ldots, i_m)$ *with* $0 \leq i_h \leq r_h$ *and with* $i_1/r_1 + \ldots + i_m/r_m < m(1/2 - \varepsilon)$ *is at most*

$$(r_1 + 1) \ldots (r_m + 1) e^{-\varepsilon^2 m/4}. \tag{2.5}$$

This motivates a construction where we ask for the index I of P at $(\alpha, \ldots, \alpha)$ to satisfy

$$I \geq m(1/2 - \varepsilon). \tag{2.6}$$

Here ε is a small parameter depending on δ. In fact an appropriate choice is $\varepsilon = c\delta$ with a small absolute constant c. With I as in (2.6) in view of lemma 2.1 we may conclude that altogether we obtain at most

$$d(r_1+1)\dots(r_m+1)e^{-\varepsilon^2 m/4} = M \qquad (2.7)$$

linear equations with integral coefficients to be satisfied by the coefficients $c(j_1,\dots,j_m)$ of the polynomial P.
Let M and N be as in (2.7) and (2.4) respectively, then we have M relations in N variables. If

$$m \geq 4\varepsilon^{-2}\log 2d \qquad (2.8)$$

we get $M \leq (1/2)N$, so that the number of equations does not exceed half the number of variables. Under this assumption we apply Siegel's Lemma. It follows that our system admits a nontrivial solution in integers $c(j_1,\dots,j_m)$ $(0 \leq j_h \leq r_h)$ whose absolute values are bounded by the maximum absolute value of the coefficients of the equations. It may be shown that these in turn are bounded by a constant B which depends only upon α and $r_1,\dots,r_m$. But clearly, once we have found suitable coefficients $c(j_1,\dots,j_m)$ we have found our desired polynomial P. The bound for the $c(j_1,\dots,j_m)$ then essentially also is a bound for $H(P)$.
We summarize what we have discussed so far in order to prove an assertion of type (**i**):

Theorem 2.2 (Index Theorem). *Suppose $\varepsilon > 0$. Let α be an algebraic number of degree d and let m be an integer satisfying* (2.8). *Let $r_1,\dots,r_m$ be natural numbers.*
Then there exists a nonzero polynomial $P(X_1,\dots,X_m)$ with rational integral coefficients having the following properties:

(*a*) *P has degree $\leq r_h$ in X_h.*

(*b*) *P has index $\geq m(1/2-\varepsilon)$ at the point $\boldsymbol{\alpha}$ with respect to $(r_1,\dots,r_m)$.*

(*c*) $H(P) \leq B_1(\alpha)^{r_1+\dots+r_m}$.

Now let P be the polynomial obtained with the Index Theorem 2.2. Studying the Taylor expansion of P about $(\alpha,\dots,\alpha)$ we show that P has an index at $(x_1/y_1,\dots,x_m/y_m)$ which is comparatively large as well. We state at once the conclusion we obtain:

Assertion (i). *Suppose $\delta > 0$. Put $\varepsilon = \delta/40$. Let m be a natural number satisfying* (2.8).
Let $x_1/y_1,\dots,x_m/y_m$ be solutions of (1.1), *i.e. of*

$$|\alpha - x_h/y_h| < y_h^{-2-\delta} \qquad (2.9)$$

with $0 < y_1 < y_2 < \dots < y_m$.
Suppose moreover that

$$y_h^\delta \geq B_2, \tag{2.10}$$

where $B_2 = B_2(\alpha)$. Let $r_1, \ldots, r_m$ be natural numbers such that

$$y_1^{r_1} \leq y_h^{r_h} \leq y_1^{r_1(1+\varepsilon)} \qquad (h = 1, \ldots, m). \tag{2.11}$$

Let P be the polynomial of the Index Theorem. Then the index of P at the point $(x_1/y_1, \ldots, x_m/y_m)$ with respect to $(r_1, \ldots, r_m)$ is $\geq \varepsilon m$.

To illustrate the deduction of Assertion (**i**) from the Index Theorem, for simplicity let us consider only $P(x_1/y_1, \ldots, x_m/y_m)$. Since P has integral coefficients and partial degrees $\leq r_h$ $(h = 1, \ldots, m)$, $P(x_1/y_1, \ldots, x_m/y_m)$ is a rational number with denominator $\leq y_1^{r_1} \ldots y_m^{r_m}$. Therefore, in order to show that $P(x_1/y_1, \ldots, x_m/y_m) = 0$ it suffices to verify that

$$|P(x_1/y_1, \ldots, x_m/y_m)| < \frac{1}{y_1^{r_1} \ldots y_m^{r_m}}. \tag{2.12}$$

By Taylor

$$P(x_1/y_1, \ldots, x_m/y_m) = \sum_{i_1=0}^{r_1} \ldots \sum_{i_m=0}^{r_m} P^{\boldsymbol{i}}(\alpha)(x_1/y_1 - \alpha)^{i_1} \ldots (x_m/y_m - \alpha)^{i_m}. \tag{2.13}$$

Using (c) of the Index Theorem we get

$$|P^{\boldsymbol{i}}(\alpha)| \leq (4B_1(\alpha)\max\{1, |\alpha|\})^{r_1+\ldots+r_m} = B_3(\alpha)^{r_1+\ldots+r_m}. \tag{2.14}$$

On the other hand, assertion (b) of the Index Theorem implies

$$P^{\boldsymbol{i}}(\alpha) = 0 \ \text{ unless } \ \frac{i_1}{r_1} + \ldots + \frac{i_m}{r_m} \geq m(1/2 - \varepsilon). \tag{2.15}$$

The term

$$(x_1/y_1 - \alpha)^{i_1} \ldots (x_m/y_m - \alpha)^{i_m}$$

in (2.13) by (2.9) has absolute value

$$< \left(y_1^{-i_1} \ldots y_m^{-i_m}\right)^{2+\delta}. \tag{2.16}$$

Now, disregarding the ε (which we will indicate by $\underset{\varepsilon}{=}$), we get with (2.11)

$$y_1^{r_1} \underset{\varepsilon}{=} y_h^{r_h} \qquad (h = 1, \ldots, m).$$

Therefore by (2.15), (2.16) the monomials $(x_1/y_1 - \alpha)^{i_1} \ldots (x_m/y_m - \alpha)^{i_m}$ that really do show up in (2.13) may be estimated by

$$
\begin{aligned}
\left(y_1^{-i_1}\dots y_m^{-i_m}\right)^{2+\delta} &= \left(y_1^{-r_1\frac{i_1}{r_1}}\dots y_m^{-r_m\frac{i_m}{r_m}}\right)^{2+\delta} \\
&\underset{\varepsilon}{=} y_1^{-r_1\left(\frac{i_1}{r_1}+\dots+\frac{i_m}{r_m}\right)(2+\delta)} \\
&\underset{\varepsilon}{\leq} y_1^{-r_1\frac{m}{2}(2+\delta)} \qquad (2.17)\\
&= y_1^{-r_1 m(1+\delta/2)} \\
&\underset{\varepsilon}{=} y_1^{-r_1(1+\delta/2)}\dots y_m^{-r_m(1+\delta/2)}.
\end{aligned}
$$

Computing all the details we finally obtain combining (2.14), (2.16) and (2.17)

$$
\left|\sum P^{\boldsymbol{i}}(\alpha)(x_1/y_1-\alpha)^{i_1}\dots(x_m/y_m-\alpha)^{i_m}\right| \leq \left(\prod_{h=1}^{m}\left(2B_3(\alpha)y_h^{-\delta/4}\right)^{r_h}\right) y_1^{-r_1}\dots y_m^{-r_m}.
$$

So we get the desired relation (2.12), provided that the constant $B_2(\alpha)$ in (2.10) is large enough.
We can summarize as follows: A crucial point in the deduction of (2.12) is estimate (2.17). We rely first on (2.15) in order to guarantee that we have only to worry about tuples $i_1,\dots,i_m$ with

$$
\frac{i_1}{r_1}+\dots+\frac{i_m}{r_m} \geq m(1/2-\varepsilon). \qquad (2.18)
$$

A second important hypothesis we use in the deduction of (2.17) is (2.11), i.e.,

$$
y_1^{r_1} \underset{\varepsilon}{=} y_h^{r_h}. \qquad (2.19)
$$

We will see in section 5 that the analogue of (2.18) requires considerably more work when we study approximations by algebraic numbers of bounded degree.

2.2 Non-Vanishing

We now turn to step (**ii**) in Roth's method. It contains the main difficulty in his approach as in general it is extremely hard to prove that a given polynomial P does not vanish at a specific point $(x_1/y_1,\dots,x_m/y_m)$. This difficulty is overcome in the famous Roth's Lemma. It says the following:

Lemma 2.3 (Roth's Lemma). *Let $\varepsilon < 1/12$ be a positive number. Let m be a natural number. Put $\omega = \omega(m,\varepsilon) = 24\cdot 2^{-m}(\varepsilon/12)^{2^{m-1}}$. Suppose that P is a nonzero polynomial with integral coefficients of degree $\leq r_h$ in X_h, where*

$$
\omega r_h \geq r_{h+1} \quad (h=1,\dots,m-1). \qquad (2.20)
$$

Suppose that $(x_1, y_1), \ldots, (x_m, y_m)$ *are pairs of coprime integers with* $0 < y_1 < \ldots < y_m$ *satisfying*

$$y_h^{r_h} \geq y_1^{r_1} \text{ and } y_h^{\omega} \geq 2^{3m} \quad (h = 1, \ldots, m) \tag{2.21}$$

as well as

$$H(P) \leq y_1^{\omega r_1}. \tag{2.22}$$

Then P *has index* $\leq \varepsilon$ *at the point* $(x_1/y_1, \ldots, x_m/y_m)$ *with respect to* $(r_1, \ldots, r_m)$.

Since in these notes we focus rather on other aspects of the method, we will not discuss here the difficult proof of Roth's Lemma.
Notice that (2.20) means that the sequence $r_1, \ldots, r_m$ is very rapidly decreasing. Accordingly (2.21) says that the sequence of the denominators of $x_1/y_1, \ldots, x_m/y_m$ is very rapidly increasing.

2.3 Conclusion

To reach the desired contradiction, in the final step we compare the vanishing-result from section 2.1 with the non-vanishing-result from section 2.2. To do so we have to find a sequence of solutions $x_1/y_1, \ldots, x_m/y_m$ which allows a choice of parameters such that simultaneously the hypotheses of Assertion (**i**) on the one hand and those of Roth's Lemma 2.3 on the other hand are satisfied.
In (1.1) we may assume that $\delta < 1$. We then pick $\varepsilon = \delta/40$, m according to (2.8) and ω as in Roth's Lemma. We let x_1/y_1 be a solution of (1.1) such that $y_1^{\omega} > B_1^m$ where B_1 is the constant from the Index Theorem 2.2. Next we choose successively solutions $x_2/y_2, \ldots, x_m/y_m$ of (1.1) such that

$$y_{h+1}^{\omega} > y_h^2 \quad (h = 1, \ldots, m-1).$$

Finally we let r_1 be a natural number with

$$y_1^{\varepsilon r_1} \geq y_m,$$

and for $2 \leq h \leq m$ we define r_h by

$$r_h = \left[\frac{r_1 \log y_1}{\log y_h}\right] + 1.$$

It is easily checked that with this choice of parameters (2.10), (2.11), (2.20), (2.21) are true. Given ε, m and $r_1, \ldots, r_m$ as above, we construct the polynomial P of the Index Theorem. It turns out that then also hypothesis (2.22) of Roth's Lemma is fulfilled.
Consequently the index I of P at the point $(x_1/y_1, \ldots, x_m/y_m)$ with respect to $(r_1, \ldots, r_m)$ satisfies

$$I \geq m\varepsilon \quad \text{by Assertion (i)},$$
$$I < \varepsilon \quad \text{by Roth's Lemma},$$

and we have reached the desired contradiction.
A closer analysis of the proof implies that *with* m *as in* (2.8) *and with* ω *as in Roth's Lemma* we get the following statement:
The set of solutions x/y *of* (1.1) *with*

$$y^{\omega} > B(\alpha)^{m}$$

is contained in the union of less than m *intervals, each of the type*

$$(Q, Q^{\omega^{-1}}]. \tag{2.23}$$

We will see in section 6 that this last statement is crucial when we want to obtain quantitative results in the sense of problem II of section 1.

3 Schmidt's proof of theorem 1.2

We next give an outline of Schmidt's proof of theorem 1.2. We can divide the proof into three main parts. In the first part the assertion on the set of solutions of (1.3) is reduced to an assertion on integral points in certain *parallelepipeds.* In the second part Schmidt applies Roth's method assuming that the successive minima of these parallelepipeds satisfy a suitable hypothesis. We will call it the *approximation part* of the proof. In the third part he manipulates the parallelepipeds in such a way that the hypotheses needed in the approximation argument are satisfied. For this purpose he applies transference principles from the geometry of numbers. We call this third part the *geometry part* of the proof.

3.1 Parallelepipeds

We begin by reducing the assertion of theorem 1.2 to an assertion on parallelepipeds. In (1.2) we have seen that in homogeneous form Roth's Theorem 1.1 may be viewed as the two-dimensional case of the Subspace Theorem 1.2. Instead of (1.2), in fact (1.1) even gives the simultaneous inequalities

$$|y| \leq \max\{|x|, |y|\}, \qquad |y\alpha - x| < \max\{|x|, |y|\}^{-1-\delta}.$$

Now suppose $n \geq 2$, and let $\boldsymbol{x}$ be a solution of (1.3) with $L_i(\boldsymbol{x}) \neq 0$ $(i = 1, \ldots, n)$. There exist real numbers $c_1, \ldots, c_n$ with

$$c_1 + \ldots + c_n = 0$$

such that

$$|L_1(\boldsymbol{x})| \leq |\boldsymbol{x}|^{c_1-\delta/n}, \ \dots \ , |L_n(\boldsymbol{x})| \leq |\boldsymbol{x}|^{c_n-\delta/n}. \tag{3.1}$$

A compactness argument shows that indeed it suffices to study (3.1) for a *fixed* tuple $c_1, \dots, c_n$ which does not depend upon $\boldsymbol{x}$.
Let $Q > 1$ be a real parameter (to be thought of as being $|\boldsymbol{x}|$). We consider the set of points $\boldsymbol{x} \in \mathbb{Z}^n$ defined by the inequalities

$$|L_1(\boldsymbol{x})| \leq Q^{c_1-\delta}, \ \dots \ , |L_n(\boldsymbol{x})| \leq Q^{c_n-\delta} \tag{3.2}$$

where the c_i are fixed real numbers satisfying

$$c_1 + \ldots + c_n = 0 \ \text{ and } \ |c_i| \leq 1 \quad (i = 1, \dots, n). \tag{3.3}$$

The assertion of theorem 1.2 may be reduced to

Theorem 3.1 (Parametric Subspace Theorem). *There exist finitely many proper linear subspaces $T_1, \dots, T_t$ of $\mathbb{Q}^n$ with the following property: As Q runs through the set of values*

$$Q > c(n, \delta, L_1, \dots, L_n), \tag{3.4}$$

the set of solutions $\boldsymbol{x} \in \mathbb{Z}^n$ of (3.2) is contained in the union

$$T_1 \cup \ldots \cup T_t.$$

Notice that initially, when introducing Q, we thought of it as being $|\boldsymbol{x}|$. However in the above statement Q has become an independent parameter. So at least at first glance, given Q, inequalities (3.2) may have many more solutions than (3.1) for a given value $|\boldsymbol{x}|$. The miracle is that nevertheless, the proof goes through even with the additional free parameter Q we encounter in (3.2).
We suppose for simplicity that our forms L_i have real algebraic coefficients and that

$$\det(L_1, \dots, L_n) = 1. \tag{3.5}$$

Let $\Pi(Q, \boldsymbol{c})$ be the parallelepiped in $\mathbb{R}^n$ determined by

$$|L_1(\boldsymbol{x})| \leq Q^{c_1}, \ \dots \ , |L_n(\boldsymbol{x})| \leq Q^{c_n}. \tag{3.6}$$

In view of (3.3) and (3.5), $\Pi(Q, \boldsymbol{c})$ has volume

$$V(\Pi) = 2^n. \tag{3.7}$$

For $\lambda > 0$, let $\lambda \, \Pi(Q, \boldsymbol{c})$ be the subset of $\mathbb{R}^n$ obtained with the inequalities

$$|L_1(\boldsymbol{x})| \leq Q^{c_1}\lambda, \ \dots \ , |L_n(\boldsymbol{x})| \leq Q^{c_n}\lambda. \tag{3.8}$$

In (3.2) we study $\mathbb{Z}^n \cap Q^{-\delta}\Pi(Q, \boldsymbol{c})$.
Minkowski's successive minima $\lambda_1, \dots, \lambda_n$ of $\Pi(Q, \boldsymbol{c})$ are defined as follows:

λ_i is the smallest value λ such that there exist i linearly independent points in $\mathbb{Z}^n$ satisfying (3.8). Minkowski's Theorem on successive minima says

$$\frac{2^n}{n!} \leq \lambda_1 \dots \lambda_n V(\Pi) \leq 2^n$$

(cf. Schmidt [10], chapter IV, theorem 1A).
With our normalization (3.7) this gives

$$\frac{1}{n!} \leq \lambda_1 \dots \lambda_n \leq 1. \tag{3.9}$$

Now if (3.2) has a solution $\boldsymbol{x} \in \mathbb{Z}^n$, $\boldsymbol{x} \neq \mathbf{0}$, at all, then by definition

$$\lambda_1 = \lambda_1(Q) \leq Q^{-\delta}. \tag{3.10}$$

3.2 The approximation part

In the approximation part of the proof Schmidt treats parameters Q such that instead of (3.10) we have the much stronger hypothesis

$$\lambda_{n-1} = \lambda_{n-1}(Q) \leq Q^{-\delta}. \tag{3.11}$$

(Notice that for $n = 2$ hypotheses (3.10) and (3.11) coincide.) In view of (3.9), for each Q with (3.11) which is large enough we have $\lambda_n(Q) > 1$. Thus the definition of successive minima implies that for large Q satisfying (3.11) the set of integral points in $Q^{-\delta}\Pi(Q, \mathbf{c})$ generates a linear subspace $T(Q)$ of $\mathbb{Q}^n$ of codimension 1.
The procedure for Roth's Theorem described in section 2 now changes as follows. We pick m suitable parameters $Q_1 < \dots < Q_m$ with (3.11). They will play the rôle of the $y_1, \dots, y_m$ we had in the one-dimensional case. For each of these parameters Q_h we let $T_h = T(Q_h)$ be the $(n-1)$-dimensional subspace corresponding to $\lambda_{n-1}(Q_h)$ $(h = 1, \dots, m)$. These subspaces replace the solutions $x_1/y_1, \dots, x_m/y_m$ we had in the one-dimensional case. However as illustrated in (1.2), we should rather make a comparison with the homogenized version of Roth's Theorem. Thus in dimension n the $(n-1)$-dimensional subspaces $T_1, \dots, T_m$ are the analogues of the one-dimensional subspaces of $\mathbb{Q}^2$ generated by the points $(x_1, y_1), \dots, (x_m, y_m)$ respectively. Roth's auxiliary polynomial $P(X_1, \dots, X_m)$ in homogenized form becomes $R(X_1, Y_1; \dots; X_m, Y_m) = Y_1^{r_1} \dots Y_m^{r_m} P(X_1/Y_1, \dots, X_m/Y_m)$. Then R is homogeneous in X_h, Y_h of degree r_h $(h = 1, \dots, m)$. This explains why in the current context, when we are dealing with linear forms in dimension n, it is natural to ask for an auxiliary polynomial P in the variables $X_{11}, \dots, X_{1n}; \dots; X_{m1}, \dots, X_{mn}$ such that P is homogeneous in the block of variables $X_{h1}, \dots, X_{hn}$ of degree r_h $(h = 1, \dots, m)$.
The requirements corresponding to (**i**) and (**ii**) in section 2 now become:

(i)$_n$ P vanishes with high order on the m-fold product of subspaces $T_1 \times \cdots \times T_m$,

(ii)$_n$ P cannot vanish with too high order on the product $T_1 \times \cdots \times T_m$.

Again we will see that for m large enough **(i)**$_n$ and **(ii)**$_n$ will contradict each other.

We need a generalization of the index. The suitable definition in the current context is as follows:

Suppose that the polynomial P has degree $\le r_h$ in the block of variables $X_{h1}, \ldots, X_{hn}$ $(h = 1, \ldots, m)$. Let $V_1, \ldots, V_m$ be subspaces of $\mathbb{R}^n$ of codimension 1.

The index I of P on the product $V_1 \times \cdots \times V_m$ with respect to $\boldsymbol{r} = (r_1, \ldots, r_m)$ is the maximum value t such that for every mn-tuple of nonnegative integers $\mathfrak{I} = (i_{11}, \ldots, i_{1n}; \ldots; i_{m1}, \ldots, i_{mn})$ with

$$\frac{\mathfrak{I}}{\boldsymbol{r}} = \frac{i_{11} + \ldots + i_{1n}}{r_1} + \ldots + \frac{i_{m1} + \ldots + i_{mn}}{r_m} < t \tag{3.12}$$

we have

$$P^{\mathfrak{I}}(\boldsymbol{z}) = \frac{1}{i_{11}! \ldots i_{mn}!} \frac{\partial^{i_{11}+\ldots+i_{mn}}}{\partial X_{11}^{i_{11}} \ldots \partial X_{mn}^{i_{mn}}} P(\boldsymbol{z}) = 0 \tag{3.13}$$

for any point $\boldsymbol{z} \in V_1 \times \cdots \times V_m$.

3.2.1 Vanishing

We indicate the essential modifications to be made in the deduction of **(i)**$_n$ as compared with **(i)** in section 2.1. Each subspace $T_h = T(Q_h)$ has integral basis vectors $\boldsymbol{g}_{h1}, \ldots, \boldsymbol{g}_{h,n-1}$ lying in $Q_h^{-\delta} \Pi(Q_h, \mathbf{c})$ $(h = 1, \ldots, m)$. Thus each T_h is not very far from any of the subspaces U_i $(i = 1, \ldots, n)$ given by the relation

$$L_i(\boldsymbol{x}) = 0. \tag{3.14}$$

We therefore try to construct a nonzero polynomial P with integral coefficients which is homogeneous of degree r_h in $X_{h1}, \ldots, X_{hn}$ $(h = 1, \ldots, m)$ and whose index with respect to $U_i \times \cdots \times U_i$ is as large as possible $(i = 1, \ldots, n)$.

Since P is homogeneous in the blocks it may be written as

$$P(X_{11}, \ldots, X_{mn}) = \sum_{j_{11}, \ldots, j_{mn}} c(j_{11}, \ldots, j_{mn}) X_{11}^{j_{11}} \ldots X_{mn}^{j_{mn}}$$

where the sum is over mn-tuples $j_{11}, \ldots, j_{mn}$ of nonnegative integers satisfying

$$j_{h1} + \ldots + j_{hn} = r_h \quad (h = 1, \ldots, m).$$

Suppose that $L(\boldsymbol{X}) = \alpha_1 X_1 + \ldots + \alpha_n X_n$ has $\alpha_1 \neq 0$ and that P has index I on the m-fold product of subspaces $U \times \cdots \times U$, where U is given by $L = 0$. Then it may be shown that there exists a tuple $\mathfrak{I}$ of type

$$\mathfrak{I} = (i_{11}, 0, \ldots, 0; i_{21}, 0, \ldots, 0; \ldots ; i_{m1}, 0, \ldots, 0) \tag{3.15}$$

with

$$\frac{i_{11}}{r_1} + \ldots + \frac{i_{m1}}{r_m} = I$$

such that $P^{\mathfrak{I}} \not\equiv 0$ on $U \times \cdots \times U$. Therefore, to guarantee that P has index $\geq I$ on $U \times \cdots \times U$ it will suffice to guarantee that $P^{\mathfrak{I}} \equiv 0$ on $U \times \cdots \times U$ for all tuples $\mathfrak{I}$ of type (3.15) with

$$\frac{i_{11}}{r_1} + \ldots + \frac{i_{m1}}{r_m} < I. \tag{3.16}$$

For $k = 1, \ldots, n$ we interpret the ratios i_{hk}/r_h with $0 \leq i_{hk} \leq r_h$ and with $i_{h1} + \ldots + i_{hn} = r_h$ as random variables. They will have expectation $1/n$. Consequently $\sum_{h=1}^m i_{hk}/r_h \approx m/n$ with high probability. Thus, a realistic value for a large index will be $m(1/n - \varepsilon)$.
The number of $(n-1)$-tuples of nonnegative integers $(i_2, \ldots, i_n)$ with $i_2 + \ldots + i_n = r - i$ equals $\binom{r-i+n-2}{n-2}$. Therefore each relation $P^{\mathfrak{I}} \equiv 0$ on $U \times \cdots \times U$ with $\mathfrak{I}$ as in (3.15) may be expressed by

$$\binom{r_1 - i_{11} + n - 2}{n-2} \cdots \binom{r_m - i_{m1} + n - 2}{n-2} \tag{3.17}$$

linear relations in the coefficients $c(j_{11}, \ldots, j_{mn})$ of P. Consequently, to ensure that P has index $\geq m(1/n - \varepsilon)$ on $U \times \cdots \times U$ by (3.15), (3.16), (3.17) we have to guarantee that

$$M' = \sum \binom{r_1 - i_{11} + n - 2}{n-2} \cdots \binom{r_m - i_{m1} + n - 2}{n-2}$$

linear relations in the coefficients of P are fulfilled. Here the summation is over $i_{11}, \ldots, i_{m1}$ with

$$\frac{i_{11}}{r_1} + \ldots + \frac{i_{m1}}{r_m} < m(1/n - \varepsilon).$$

Now for the quantity M' the following lemma is true:

Lemma 3.2. *Let $n \geq 2$ be an integer. Suppose $r_1, \ldots, r_m$ are natural numbers and let $0 < \varepsilon < 1$. Then the number M' of mn-tuples of nonnegative integers $(i_{11}, \ldots, i_{1n}; \ldots ; i_{m1}, \ldots, i_{mn})$ with*

$$i_{h1} + \ldots + i_{hn} = r_h \qquad (h = 1, \ldots, m)$$

and with

$$\frac{i_{11}}{r_1} + \frac{i_{21}}{r_2} + \ldots + \frac{i_{m1}}{r_m} < m(1/n - \varepsilon)$$

satisfies

$$M' \le \binom{r_1 + n - 1}{n - 1} \cdots \binom{r_m + n - 1}{n - 1} e^{-\varepsilon^2 m/4}. \tag{3.18}$$

Taking into account each of our linear forms $L_1, \ldots, L_n$ we see that with nM' linear relations in the coefficients we can guarantee that for $i = 1, \ldots, n$ the polynomial P has index $\ge m(1/n - \varepsilon)$ on each of the products $U_i \times \cdots \times U_i$. On the other hand P has

$$N' = \binom{r_1 + n - 1}{n - 1} \cdots \binom{r_m + n - 1}{n - 1} \tag{3.19}$$

hypothetical coefficients $c(j_{11}, \ldots, j_{mn})$. Comparing (3.18) and (3.19) it is clear that for m large enough the number of conditions nM' will be not larger than half the number N' of variables. Thus using Siegel's Lemma, similarly as in the proof of Roth's Theorem we get a nonzero approximation polynomial P whose coefficients are bounded as in the Index Theorem 2.2 and which has

$$\operatorname{ind} P \ge m(1/n - \varepsilon) \quad \text{on} \quad U_i \times \cdots \times U_i \qquad (i = 1, \ldots, n). \tag{3.20}$$

We perform this construction with $\varepsilon = c\delta/n^2$ where c is a small absolute constant.

In step $(\mathbf{i})_n$ our goal is to prove that P vanishes with high order on the product of subspaces $T_1 \times \cdots \times T_m = T(Q_1) \times \cdots \times T(Q_m)$ corresponding to $\lambda_{n-1}(Q_h)$ $(h = 1, \ldots, m)$ respectively. In fact we want to show that

$$\operatorname{ind} P \ge m\varepsilon \quad \text{on} \quad T_1 \times \cdots \times T_m. \tag{3.21}$$

Let $\boldsymbol{g}_{h1}, \ldots, \boldsymbol{g}_{h,n-1}$ be a basis of T_h $(h = 1, \ldots, m)$. It may be shown that (3.21) is verified, once we have proved that

$$P^{\mathfrak{J}}(\boldsymbol{y}_1, \ldots, \boldsymbol{y}_m) = 0 \tag{3.22}$$

for all $\mathfrak{J}$ with

$$\mathfrak{J}/\boldsymbol{r} < 2\varepsilon m \tag{3.23}$$

and for all mn-tuples $(\boldsymbol{y}_1, \ldots, \boldsymbol{y}_m)$, where $\boldsymbol{y}_h$ runs through the linear combinations of the basis vectors $\boldsymbol{g}_{h1}, \ldots, \boldsymbol{g}_{h,n-1}$ of T_h given by

$$\boldsymbol{y}_h = s_{h1}\boldsymbol{g}_{h1} + \ldots + s_{h,n-1}\boldsymbol{g}_{h,n-1} \qquad (h = 1, \ldots, m). \tag{3.24}$$

Here the s_{hj} are integers with

$$1 \le s_{hj} \le [\varepsilon^{-1}] + 1 \quad (j = 1, \dots, n-1;\ h = 1, \dots, m). \tag{3.25}$$

To verify (3.22) under the hypotheses (3.23)–(3.25) we write uniquely

$$\begin{aligned} P^{\mathfrak{J}}(\boldsymbol{y}_1, \dots, \boldsymbol{y}_m) &= P^{\mathfrak{J}}(y_{11}, \dots, y_{1n}; \dots; y_{m1}, \dots, y_{mn}) \\ &= \sum_{j_{11}, \dots, j_{mn}} d^{\mathfrak{J}}(j_{11}, \dots, j_{mn}) \prod_{h=1}^{m} \left(L_1(\boldsymbol{y}_h)^{j_{h1}} \dots L_n(\boldsymbol{y}_h)^{j_{hn}} \right). \end{aligned} \tag{3.26}$$

The basis vectors $\boldsymbol{g}_{hj}$ $(j = 1, \dots, n-1)$ of T_h according to (3.11) may be chosen in $\mathbb{Z}^n$ such as to satisfy

$$|L_i(\boldsymbol{g}_{hj})| \le Q_h^{c_i - \delta} \quad (i = 1, \dots, n;\ j = 1, \dots, n-1;\ h = 1, \dots, m). \tag{3.27}$$

Since moreover P has integral coefficients, to prove (3.22) it suffices to show that

$$\left| P^{\mathfrak{J}}(\boldsymbol{y}_1, \dots, \boldsymbol{y}_m) \right| < 1. \tag{3.28}$$

Now in our construction of the polynomial P, similarly as in the Index Theorem 2.2 we can show that the coefficients $d^{\mathfrak{J}}(j_{11}, \dots, j_{mn})$ satisfy

$$\left| d^{\mathfrak{J}}(j_{11}, \dots, j_{mn}) \right| \le C^{r_1 + \dots + r_m} \tag{3.29}$$

where C is a constant which depends only upon the forms $L_1, \dots, L_n$. Moreover using (3.20), for $\mathfrak{J}$ with (3.23) we obtain in analogy with (2.15)

$$d^{\mathfrak{J}}(j_{11}, \dots, j_{mn}) = 0$$

unless

$$\left| \sum_{h=1}^{m} \frac{j_{hk}}{r_h} - \frac{m}{n} \right| \le 3nm\varepsilon \quad \text{for all } k \ (k = 1, \dots, n). \tag{3.30}$$

Consequently, when estimating the right hand side of (3.26) we have only to worry about monomials with

$$\frac{m}{n} - 3nm\varepsilon \le \frac{j_{1k}}{r_1} + \dots + \frac{j_{mk}}{r_m} \le \frac{m}{n} + 3nm\varepsilon \quad (k = 1, \dots, n). \tag{3.31}$$

Assuming that our parameters $Q_1, \dots, Q_m$ satisfy the analogue of (2.19), i.e., assuming that we have

$$Q_1^{r_1} \underset{\varepsilon}{=} Q_h^{r_h} \quad (h = 1, \dots, m), \tag{3.32}$$

we get for the typical monomial in (3.26) using (3.3), (3.24), (3.25), (3.27), (3.31), (3.32)

$$
\begin{aligned}
&\left|L_1(y_1)^{j_{11}} \dots L_n(y_1)^{j_{1n}} \dots L_1(y_m)^{j_{m1}} \dots L_n(y_m)^{j_{mn}}\right| \\
&\underset{\varepsilon}{<} Q_1^{(c_1-\delta)j_{11}} \dots Q_1^{(c_n-\delta)j_{1n}} \dots Q_m^{(c_1-\delta)j_{m1}} \dots Q_m^{(c_n-\delta)j_{mn}} \\
&= Q_1^{(c_1-\delta)r_1\frac{j_{11}}{r_1}} \dots Q_1^{(c_n-\delta)r_1\frac{j_{1n}}{r_1}} \dots Q_m^{(c_1-\delta)r_m\frac{j_{m1}}{r_m}} \dots Q_m^{(c_n-\delta)r_m\frac{j_{mn}}{r_m}} \\
&\underset{\varepsilon}{=} Q_1^{(c_1-\delta)r_1\frac{j_{11}}{r_1}} \dots Q_1^{(c_n-\delta)r_1\frac{j_{1n}}{r_1}} \dots Q_1^{(c_1-\delta)r_1\frac{j_{m1}}{r_m}} \dots Q_1^{(c_n-\delta)r_1\frac{j_{mn}}{r_m}} \\
&= Q_1^{(c_1-\delta)r_1(\frac{j_{11}}{r_1}+\dots+\frac{j_{m1}}{r_m})+\dots+(c_n-\delta)r_1(\frac{j_{1n}}{r_1}+\dots+\frac{j_{mn}}{r_m})} \\
&\underset{\varepsilon}{=} Q_1^{(c_1-\delta)r_1\frac{m}{n}+\dots+(c_n-\delta)r_1\frac{m}{n}} \\
&\underset{\varepsilon}{=} Q_1^{-r_1 m\delta} \\
&\underset{\varepsilon}{=} Q_1^{-r_1\delta} \dots Q_m^{-r_m\delta}.
\end{aligned}
\tag{3.33}
$$

Estimate (3.33) is the n-dimensional homogeneous analogue of (2.17). (Notice that in section 2 we have considered the inhomogeneous problem. In a homogeneous approach, in the last term of formula (2.17) we have to replace the parentheses $(1+\delta/2)$ by $\delta/2$.) Combining (3.33) with (3.29) we get the desired (3.28) provided that $Q_1, \dots, Q_m$ are large enough.
We underline the fact that the proof of (3.22) depends heavily upon (3.30) and (3.32). These two hypotheses correspond to hypotheses (2.18) and (2.19) in Roth's construction.

3.2.2 Non-Vanishing

As for step $(\mathbf{ii})_n$, Schmidt derives a homogeneous version of Roth's Lemma 2.3 for polynomials in mn variables with an inductive argument, again assuming that the sequence $r_1, \dots, r_m$ is very rapidly decreasing and that $Q_1, \dots, Q_m$ is very rapidly increasing similarly to (2.20), (2.21). He replaces hypothesis (2.22) by a similar hypothesis to be satisfied by the heights of the subspaces $T_1, \dots, T_m$. Evertse [3] on the basis of Faltings' Product Theorem has given a much better version of Roth's Lemma. We quote it without discussing the proof:

Lemma 3.3 (Evertse's Lemma). *Let $0 < \varepsilon \leq 1$. Let m, n be integers ≥ 2. Put $\theta = \theta(m, \varepsilon) = \varepsilon/(2m^2)$. Let $r_1, \dots, r_m$ be natural numbers satisfying*

$$\theta r_h \geq r_{h+1} \quad (h = 1, \dots, m-1).$$

Suppose that $P = P(X_{11}, \dots, X_{1n}; \dots; X_{m1}, \dots, X_{mn})$ is a polynomial with coefficients in $\overline{\mathbb{Q}}$, homogeneous in the block of variables $X_{h1}, \dots, X_{hn}$ of degree r_h $(h = 1, \dots, m)$. Let $T_1, \dots, T_m$ be $(n-1)$-dimensional linear subspaces of $\overline{\mathbb{Q}}^n$ with

$$H(T_h)^{(2\theta/3)^m (n-1)^{-1} r_h} \geq e^{r_1+\dots+r_m} H(P) \quad (h = 1, \dots, m).$$

Then P has index $< m\varepsilon$ on the product $T_1 \times \cdots \times T_m$ with respect to $(r_1, \ldots, r_m)$.

In Evertse's Lemma 3.3 the m-th power θ^m should be compared with the parameter ω in Roth's Lemma. θ^m is only simply exponential in ε, whereas ω is doubly exponential. In a qualitative sense, i.e., if we are only interested in finiteness results, both Schmidt's extension of Roth's Lemma and Evertse's Lemma eventually lead to the same conclusion. If however we are interested in *counting* the number of subspaces, Evertse's Lemma yields much better bounds. As in these notes we focus rather on step $(\mathbf{i})_n$, we will not discuss this aspect further.

3.2.3 Conclusion

We combine the assertion about non-vanishing in Evertse's Lemma 3.3 with the result (3.21) on vanishing from section 3.2.1. Similarly as in section 2.3 we choose the parameters such that simultaneously the hypotheses of Evertse's Lemma as well as those needed for (3.21) (cf. in particular (3.32)) are satisfied. The final outcome of the approximation part of the proof then may be formulated as follows:

Theorem 3.4 (on the Penultimate Minimum). *Suppose $0 < \delta < 1$. Let $m > c(n, \delta, L_1, \ldots, L_n)$.*
Let Q run through values

$$Q > Q_0 = Q_0(m, \delta, L_1, \ldots, L_n)$$

and such that the penultimate minimum of $\Pi(Q, \mathbf{c})$ satisfies

$$\lambda_{n-1}(Q) \leq Q^{-\delta}. \tag{3.34}$$

Then either for all values Q under consideration the subspace $T(Q)$ corresponding to $\lambda_{n-1}(Q)$ will be the same, or the values Q are contained in fewer than m intervals of type

$$Q_h < Q \leq Q_h^E \quad (h = 1, \ldots, m) \tag{3.35}$$

where $E = E(m)$.

We remark that in the above theorem the constants $c(n, \delta, L_1, \ldots, L_n)$, $Q_0(m, \delta, L_1, \ldots, L_n)$ as well as $E(m)$ may be determined explicitly. Notice that the assertion of the Theorem on the Penultimate Minimum 3.4 contains two alternatives. In our discussion in sections 3.2.1 and 3.2.2 we have dealt only with the steps needed for the second alternative. We then get the intervals in (3.35). The first alternative comes from subspaces $T(Q)$ whose height is not bounded from below by some fixed positive power of Q. For such subspaces we cannot guarantee the second hypothesis in Evertse's Lemma. Schmidt handles

such subspaces with an argument from the geometry of numbers which does not enter into the general scheme illustrated above.
In section 6 we will see that we can determine an explicit upper bound for the *number* of subspaces $T(Q)$ needed, when Q runs through an interval of type (3.35). Consequently, the Theorem on the Penultimate Minimum allows it to handle problem II from section 1 for parameters Q satisfying (3.34). For such parameters it clearly also implies the assertion of the Parametric Subspace Theorem 3.1.

3.3 The geometry part

In the *approximation part* treated in section 3.2 we have studied values Q such that the penultimate minimum of the parallelepiped $\Pi(Q,\boldsymbol{c})$ in (3.6) satisfies $\lambda_{n-1}(Q) \le Q^{-\delta}$ (cf. (3.11), (3.34)). However, as we have seen, in general for a value Q such that (3.2) has a solution $\boldsymbol{x} \in \mathbb{Z}^n$, $\boldsymbol{x} \neq \boldsymbol{0}$, at all, we only have (3.10), i.e., $\lambda_1(Q) \le Q^{-\delta}$.
If Q is large we may conclude with (3.9) that there exists p with $1 \le p \le n-1$ such that

$$\lambda_p(Q) \le Q^{-\delta/(n-1)}\lambda_{p+1}(Q). \tag{3.36}$$

We suppose that p with (3.36) is chosen maximal.
Let $\boldsymbol{g}_1(Q),\dots,\boldsymbol{g}_n(Q)$ be linearly independent integral points such that

$$\boldsymbol{g}_i \in \lambda_i(Q)\Pi(Q,\boldsymbol{c}). \tag{3.37}$$

Since p in (3.36) is chosen maximal, every solution $\boldsymbol{x}$ of (3.2) is contained in the p-dimensional subspace $T_{(p)}(Q)$ of $\mathbb{Q}^n$ generated by $\boldsymbol{g}_1,\dots,\boldsymbol{g}_p$. As in (3.36) we have a gap between λ_p and λ_{p+1}, although there may be many choices for $\boldsymbol{g}_1,\dots,\boldsymbol{g}_p$, the subspace $T_{(p)}$ will be unique. However, clearly in general $T_{(p)}$ will depend upon the parameter Q.
Our goal is to derive a statement of the type given in the Theorem on the Penultimate Minimum 3.4, but now for the collection of subspaces $T_{(p)}(Q)$ where Q runs through the set of values satisfying (3.36). In fact the union over p with $1 \le p \le n-1$ of the family of subspaces $T_{(p)}$ then finally gives us the subspaces $T_1,\dots,T_t$ in the assertion of the Parametric Subspace Theorem 3.1. We actually want to apply the Theorem on the Penultimate Minimum to the subspaces $T_{(p)}$. Therefore our goal is to transform (3.36) into a hypothesis of type (3.34).
Schmidt achieves this goal with two transference principles. The first one is based on a method given by Davenport. Evertse [2] has proved a nice variant of Davenport's Lemma. It is this variant that we will quote here.

Lemma 3.5 (Davenport's Lemma). *Let $\boldsymbol{g}_1,\dots,\boldsymbol{g}_n$ be linearly independent integral points satisfying*

$$|L_i(\boldsymbol{g}_j)| \le Q^{c_i}\lambda_j \quad (1 \le i,j \le n). \tag{3.38}$$

Then there exist a permutation π of $\{1,\ldots,n\}$ and linearly independent integral points $\boldsymbol{h}_1,\ldots,\boldsymbol{h}_n$ with the following two properties:

(*a*) $\operatorname{span}\{\boldsymbol{h}_1,\ldots,\boldsymbol{h}_j\} = \operatorname{span}\{\boldsymbol{g}_1,\ldots,\boldsymbol{g}_j\}$ *for* $j=1,\ldots,n$,

(*b*) $|L_{\pi(i)}(\boldsymbol{h}_j)| \le 2^{i+j} Q^{c_{\pi(i)}} \min\{\lambda_i,\lambda_j\} \quad (1\le i,j\le n)$.

Notice that the permutation π in the above lemma in general will depend upon Q.

Schmidt's second tool relies on ideas introduced by Mahler and uses multilinear algebra. Given $1\le k\le n$ write $N=\binom{n}{k}$. Let $C(n,k)$ denote the set of k-tuples $\sigma=\{1\le i_1<i_2<\ldots<i_k\le n\}$. For n vectors $\boldsymbol{a}_i=(a_{i1},\ldots,a_{in})$ we let $\boldsymbol{A}_\sigma=\boldsymbol{a}_{i_1}\wedge\ldots\wedge\boldsymbol{a}_{i_k}$ $(\sigma\in C(n,k))$ be the N vectors in N-dimensional space whose components are the $(k\times k)$-determinants formed from the components of $\boldsymbol{a}_{i_1},\ldots,\boldsymbol{a}_{i_k}$ (in the lexicographical ordering).

For our linear forms L_i we define $L_\sigma=L_{i_1}\wedge\ldots\wedge L_{i_k}$ as the linear form in N variables obtained with the coefficient vectors of the forms $L_{i_1},\ldots,L_{i_k}$. Moreover, we write $c_\sigma=c_{i_1}+\ldots+c_{i_k}$ and $\lambda_\sigma=\lambda_{i_1}\ldots\lambda_{i_k}$. Let $\sigma_1,\ldots,\sigma_N$ be the lexicographical ordering of the elements in $C(n,k)$. By (3.5) we have

$$\det(L_{\sigma_1},\ldots,L_{\sigma_N})=\det(L_1,\ldots,L_n)^{\frac{k}{n}N}=1. \tag{3.39}$$

Using (3.3) we obtain

$$c_{\sigma_1}+\ldots+c_{\sigma_N}=0 \text{ and } |c_{\sigma_i}|\le k \quad (i=1,\ldots,N). \tag{3.40}$$

Laplace's identity says that for $\sigma=\{i_1<\ldots<i_k\}$ and $\tau=\{j_1<\ldots<j_k\}$

$$L_\sigma(\boldsymbol{A}_\tau)=\det\begin{pmatrix} L_{i_1}(\boldsymbol{a}_{j_1}) & \ldots & L_{i_1}(\boldsymbol{a}_{j_k}) \\ \vdots & & \vdots \\ L_{i_k}(\boldsymbol{a}_{j_1}) & \ldots & L_{i_k}(\boldsymbol{a}_{j_k}) \end{pmatrix}. \tag{3.41}$$

We combine Davenport's Lemma 3.5 with multilinear algebra. To simplify the notation let us restrict ourselves to values Q where the permutation π in Davenport's Lemma is the identity.

We apply the above considerations with

$$k=n-p, \tag{3.42}$$

where p is given by (3.36).

In view of (3.41) and by assertion (*b*) of Davenport's Lemma we obtain for any pair $\sigma,\tau\in C(n,k)$

$$|L_\sigma(\boldsymbol{H}_\tau)|\le k!\,2^{2n}Q^{c_\sigma}\min\{\lambda_\sigma,\lambda_\tau\} \tag{3.43}$$

where $\boldsymbol{H}_\tau=\boldsymbol{h}_{j_1}\wedge\ldots\wedge\boldsymbol{h}_{j_k}$.

We use (3.43) with the points $\boldsymbol{H}_{\sigma_1},\ldots,\boldsymbol{H}_{\sigma_{N-1}}$. Accordingly, we get for $j=1,\ldots,N-1$

$$\begin{aligned}|L_{\sigma_i}(\boldsymbol{H}_{\sigma_j})| &\le k!\,2^{2n}Q^{c_{\sigma_i}}\lambda_{\sigma_i} \qquad (i=1,\dots,N-1)\\ |L_{\sigma_N}(\boldsymbol{H}_{\sigma_j})| &\le k!\,2^{2n}Q^{c_{\sigma_N}}\lambda_{\sigma_{N-1}}.\end{aligned} \tag{3.44}$$

Write

$$\begin{aligned}\rho_i &= \frac{\lambda_{\sigma_i}}{(\lambda_{\sigma_1}\dots\lambda_{\sigma_{N-1}}\lambda_{\sigma_{N-1}})^{1/N}} \qquad (i=1,\dots,N-1)\\ \rho_N &= \frac{\lambda_{\sigma_{N-1}}}{(\lambda_{\sigma_1}\dots\lambda_{\sigma_{N-1}}\lambda_{\sigma_{N-1}})^{1/N}}.\end{aligned} \tag{3.45}$$

Then clearly

$$\rho_1\cdots\rho_N = 1. \tag{3.46}$$

(3.44) and (3.45) yield

$$|L_{\sigma_i}(\boldsymbol{H}_{\sigma_j})| \le k!\,2^{2n}Q^{c_{\sigma_i}}\rho_i(\lambda_{\sigma_1}\dots\lambda_{\sigma_{N-1}}\lambda_{\sigma_{N-1}})^{1/N} \tag{3.47}$$

for $i=1,\dots,N$ and for $j=1,\dots,N-1$. However

$$\lambda_{\sigma_1}\dots\lambda_{\sigma_{N-1}}\lambda_{\sigma_{N-1}} = \lambda_{\sigma_1}\dots\lambda_{\sigma_N}\cdot\frac{\lambda_{\sigma_{N-1}}}{\lambda_{\sigma_N}} = (\lambda_1\dots\lambda_n)^{\frac{k}{n}N}\cdot\frac{\lambda_{\sigma_{N-1}}}{\lambda_{\sigma_N}}.$$

Combining this with (3.9) we get

$$\lambda_{\sigma_1}\dots\lambda_{\sigma_{N-1}}\lambda_{\sigma_{N-1}} \le \lambda_{\sigma_{N-1}}/\lambda_{\sigma_N}. \tag{3.48}$$

With our value of k in (3.42) we have $\sigma_{N-1} = \{p,\ p+2,\dots,n\}$, $\sigma_N = \{p+1,\dots,n\}$. Thus

$$\lambda_{\sigma_{N-1}}/\lambda_{\sigma_N} = \frac{\lambda_p\lambda_{p+2}\dots\lambda_n}{\lambda_{p+1}\lambda_{p+2}\dots\lambda_n} = \lambda_p/\lambda_{p+1}.$$

So by (3.36) and (3.48) we may infer that

$$(\lambda_{\sigma_1}\dots\lambda_{\sigma_{N-1}}\lambda_{\sigma_{N-1}})^{1/N} \le Q^{-\delta/(n-1)N}. \tag{3.49}$$

Comparing (3.47) and (3.49), we see that for large enough Q the points $\boldsymbol{H}_{\sigma_1},\dots,\boldsymbol{H}_{\sigma_{N-1}}$ are solutions of the simultaneous inequalities

$$|L_{\sigma_i}(\boldsymbol{X})| \le Q^{c_{\sigma_i}}\rho_i Q^{-\delta/nN} \qquad (i=1,\dots,N). \tag{3.50}$$

Notice that by (3.45) the parameters ρ_i in (3.50) in general will depend upon Q. However, with the bound for the c_{σ_i}-s from (3.40) we can show that the ρ_i-s lie between fixed powers of Q. Using a compactness argument, we may replace the ρ_i-s by powers Q^{d_i}, say, where $d_1,\dots,d_N$ are taken from a finite set whose cardinality depends only upon n and δ. Hence there will be no loss of generality, if we restrict ourselves to a fixed tuple $d_1,\dots,d_N$. Writing

$$e_i = c_{\sigma_i} + d_i \quad (i = 1, \ldots, N)$$

by (3.40) and (3.46) we get

$$e_1 + \ldots + e_N = 0, \quad |e_i| \le c(n) \quad (i = 1, \ldots, N). \tag{3.51}$$

We define the parallelepiped $\Pi^{(k)}(Q, \boldsymbol{e})$ in $\mathbb{R}^N$ by the inequalities

$$|L_{\sigma_i}(\boldsymbol{X})| \le Q^{e_i} \quad (i = 1, \ldots, N). \tag{3.52}$$

Then (3.50) says that

$$\boldsymbol{H}_{\sigma_1}, \ldots, \boldsymbol{H}_{\sigma_{N-1}} \in Q^{-\delta/nN} \Pi^{(k)}(Q, \boldsymbol{e}). \tag{3.53}$$

Let $\mu_1 = \mu_1(Q), \ldots, \mu_N = \mu_N(Q)$ be the successive minima of $\Pi^{(k)}(Q, \boldsymbol{e})$. In view of (3.53) we have

$$\mu_{N-1}(Q) \le Q^{-\delta/nN}. \tag{3.54}$$

We want to apply the Theorem on the Penultimate Minimum 3.4 to the parallelepipeds $\Pi^{(k)}(Q, \boldsymbol{e})$ in N-dimensional space. To do so we have to guarantee our normalizations (3.3) and (3.5). (3.5) becomes (3.39). The analogue of (3.3) by (3.51) will be satisfied if we replace the parameter Q by $Q^{c(n)}$. Then in view of (3.54) we may apply the Theorem on the Penultimate Minimum with $\delta/(nc(n)N)$ instead of δ.
Consequently, for the parameters Q under consideration the $(N-1)$-dimensional subspaces $T^{(k)}(Q)$ in $\mathbb{Q}^N$ corresponding to $\mu_{N-1}(Q)$ satisfy the assertion of this theorem.
The subspaces $T^{(k)}(Q)$ are generated by the vectors $\boldsymbol{H}_{\sigma_1}(Q), \ldots, \boldsymbol{H}_{\sigma_{N-1}}(Q)$. However, in the context of (3.36) we wanted information on the subspaces $T_{(p)}(Q)$ of dimension p in n-dimensional space. These are generated by the points $\boldsymbol{g}_1(Q), \ldots, \boldsymbol{g}_p(Q)$ from (3.37). By Davenport's Lemma 3.5 they are also generated by the points $\boldsymbol{h}_1(Q), \ldots, \boldsymbol{h}_p(Q)$ found with that lemma. Now with k as in (3.42) and with our definition of the points $\boldsymbol{H}_\sigma$, it is a fact from multilinear algebra that the span of $\boldsymbol{H}_{\sigma_1}, \ldots, \boldsymbol{H}_{\sigma_{N-1}}$ determines the span of $\boldsymbol{h}_1, \ldots, \boldsymbol{h}_p$ uniquely. Therefore, since we have control over the subspaces $T^{(k)}(Q)$ in $\mathbb{Q}^N$ generated by $\boldsymbol{H}_{\sigma_1}, \ldots, \boldsymbol{H}_{\sigma_{N-1}}$, we have control over the subspaces $T_{(p)}(Q)$ in $\mathbb{Q}^n$ as well. This finishes the proof of the Parametric Subspace Theorem 3.1 and hence the proof of theorem 1.2.

4 The proof of theorem 1.3

In this section we give an outline of the proof of a *qualitative* version of theorem 1.3. In other words we discuss how we can prove that we do need only finitely many subspaces to cover the set of solutions of (1.14), (1.15). But we do not focus on an explicit upper bound for the number of subspaces as in (1.13).

To avoid technicalities, we assume that the set $S \subset \mathfrak{M}(K)$ consists of a single place $v_0 \in \mathfrak{M}(K)$ (archimedean or nonarchimedean, whatsoever).
So we are considering a single system of linearly independent linear forms $L_1, \ldots, L_n$ with algebraic coefficients.
In view of the determinant factor on the right hand side of (1.14) we may suppose that

$$\det(L_1, \ldots, L_n) = 1. \tag{4.1}$$

We then have to study the inequality

$$\prod_{i=1}^{n} \max_{\sigma \in \mathrm{Gal}(\overline{\mathbb{Q}}|K)} \frac{\|L_i(\sigma(\boldsymbol{x}))\|_{v_0}}{\|\sigma(\boldsymbol{x})\|_{v_0}} \leq H(\boldsymbol{x})^{-n-\delta}.$$

To further simplify our exposition we suppose that the forms $L_1, \ldots, L_n$ have coefficients in our ground field K. So we have to study

$$\prod_{i=1}^{n} \max_{\sigma \in \mathrm{Gal}(\overline{\mathbb{Q}}|K)} \frac{\|\sigma(L_i(\boldsymbol{x}))\|_{v_0}}{\|\sigma(\boldsymbol{x})\|_{v_0}} \leq H(\boldsymbol{x})^{-n-\delta}. \tag{4.2}$$

4.1 Parallelepipeds

Our first goal will be to construct in analogy with (3.2), (3.6) a family of parallelepipeds corresponding to the set of solutions $\boldsymbol{x}$ in $\overline{\mathbb{Q}}^n$ of (4.2).
For each solution $\boldsymbol{x}$ with $L_i(\boldsymbol{x}) \neq 0$ $(i = 1, \ldots, n)$ there exist real numbers $e_1, \ldots, e_n$ with

$$e_1 + \ldots + e_n = -n \tag{4.3}$$

such that

$$\max_{\sigma \in \mathrm{Gal}(\overline{\mathbb{Q}}|K)} \frac{\|\sigma(L_i(\boldsymbol{x}))\|_{v_0}}{\|\sigma(\boldsymbol{x})\|_{v_0}} \leq H(\boldsymbol{x})^{e_i - \delta/n}. \tag{4.4}$$

Given $\boldsymbol{x} \in \overline{\mathbb{Q}}^n$ with (4.4), let E be a finite extension of K such that $\boldsymbol{x} \in E^n$. Recall from (1.6) that our absolute value $\| \ \|_{v_0}$ is normalized on K. Write $\mathfrak{M}_{v_0}(E)$ for the set of places $w \in \mathfrak{M}(E)$ with $w \mid v_0$. Since in (4.4) we take the maximum over $\sigma \in \mathrm{Gal}(\overline{\mathbb{Q}}|K)$, by (1.7), (1.8) the normalized absolute values $\| \ \|_w$ on E with $w \in \mathfrak{M}_{v_0}(E)$ satisfy

$$\frac{\|L_i(\boldsymbol{x})\|_w}{\|\boldsymbol{x}\|_w} \leq H(\boldsymbol{x})^{(e_i - \delta/n)d(w|v_0)} \quad (i = 1, \ldots, n;\ w \in \mathfrak{M}_{v_0}(E)). \tag{4.5}$$

To transform inequalities (4.5) into inequalities related with a parallelepiped, we use the following lemma. It is an easy consequence of Dirichlet's Unit Theorem.

Lemma 4.1. *Let E be a number field. Suppose that for $w \in \mathfrak{M}(E)$ we are given positive real numbers A_w satisfying*

$$A_w = 1 \quad \textit{for all but finitely many } w \in \mathfrak{M}(E), \tag{4.6}$$

$$\prod_{w \in \mathfrak{M}(E)} A_w > 1. \tag{4.7}$$

Then there exist a finite extension F of E and an element $\beta \in F^*$ with the following property:

$$\|\beta\|_u \le A_w^{d(u|w)} \quad \textit{for all pairs } (u,w) \textit{ with } u \in \mathfrak{M}(F),\ w \in \mathfrak{M}(E),\ u\,|\,w. \tag{4.8}$$

We apply lemma 4.1 to (4.5).
We put

$$\begin{aligned} A_w &= H(\boldsymbol{x})^{(1+\delta/2n)d(w|v_0)}\|\boldsymbol{x}\|_w^{-1} && \text{for } w \in \mathfrak{M}_{v_0}(E) \\ A_w &= \|\boldsymbol{x}\|_w^{-1} && \text{for } w \in \mathfrak{M}(E)\backslash\mathfrak{M}_{v_0}(E). \end{aligned}$$

Since $\sum_{w \in \mathfrak{M}_{v_0}(E)} d(w|v_0) = 1$ and since $\|\boldsymbol{x}\|_w = 1$ for all but finitely many $w \in \mathfrak{M}(E)$, hypotheses (4.6) and (4.7) are satisfied, provided $H(\boldsymbol{x}) > 1$. Let F be a finite extension of E and let $\beta \in F^*$ be as in the assertion of lemma 4.1. Notice that for $u \in \mathfrak{M}(F)$, $w \in \mathfrak{M}(E)$, $v \in \mathfrak{M}(K)$ such that $u\,|\,w$ and $w\,|\,v$

$$d(u|w)\, d(w|v) = d(u|v).$$

Therefore (4.5) implies for the point $\beta\boldsymbol{x}$

$$\begin{aligned} \|L_i(\beta\boldsymbol{x})\|_u &\le H(\boldsymbol{x})^{(e_i+1-\delta/2n)d(u|v_0)} && (i = 1,\ldots,n;\ u \in \mathfrak{M}_{v_0}(F)) \\ \|\beta x_i\|_u &\le 1 && (i = 1,\ldots,n;\ u \in \mathfrak{M}(F)\backslash\mathfrak{M}_{v_0}(F)). \end{aligned}$$

Writing $c_i = e_i + 1$ we obtain in view of (4.3)

$$c_1 + \ldots + c_n = 0 \tag{4.9}$$

and

$$\begin{aligned} \|L_i(\beta\boldsymbol{x})\|_u &\le H(\boldsymbol{x})^{(c_i-\delta/2n)d(u|v_0)} && (i = 1,\ldots,n;\ u \in \mathfrak{M}_{v_0}(F)) \\ \|\beta x_i\|_u &\le 1 && (i = 1,\ldots,n;\ u \in \mathfrak{M}(F)\backslash\mathfrak{M}_{v_0}(F)), \end{aligned} \tag{4.10}$$

similarly as in (3.1). Let $\boldsymbol{c} = (c_1,\ldots,c_n)$ be a tuple of real numbers having (4.9) and

$$|c_i| \le 1 \quad (i = 1,\ldots,n), \tag{4.11}$$

and let $Q > 1$ be a real parameter (to be thought of as being a fixed power of $H(\boldsymbol{x})$ in the applications). In connection with (4.10) we introduce the parallelepiped $\Pi(Q,\boldsymbol{c}) \subset \overline{\mathbb{Q}}^n$ as follows:

If E is a finite extension of K, we let $\Pi_E(Q, \boldsymbol{c})$ be the set of points $\boldsymbol{x} \in E^n$ satisfying

$$\begin{aligned} \|L_i(\boldsymbol{x})\|_w &\le Q^{c_i d(w|v_0)} && (i = 1, \ldots, n;\ w \in \mathfrak{M}_{v_0}(E)) \\ \|x_i\|_w &\le 1 && (i = 1, \ldots, n;\ w \in \mathfrak{M}(E) \backslash \mathfrak{M}_{v_0}(E)). \end{aligned} \tag{4.12}$$

We write

$$\Pi(Q, \boldsymbol{c}) = \bigcup_E \Pi_E(Q, \boldsymbol{c}) \tag{4.13}$$

where the union is over all finite extensions E of K. Since in particular $\Pi_K(Q, \boldsymbol{c})$ is given by

$$\begin{aligned} \|L_i(\boldsymbol{x})\|_{v_0} &\le Q^{c_i} && (i = 1, \ldots, n) \\ \|x_i\|_v &\le 1 && (i = 1, \ldots, n;\ v \in \mathfrak{M}(K),\ v \ne v_0), \end{aligned}$$

we say that $\Pi(Q, \boldsymbol{c})$ is *defined over* K. This reflects the fact that in (4.12) for any pair w, v with $v \in \mathfrak{M}(K)$ and $w \in \mathfrak{M}(E)$ such that w lies above v, the normalized absolute value $\| \ \|_w$ of E gets its "proper share". We call $\Pi(Q, \boldsymbol{c})$ the *algebraic closure* of $\Pi_K(Q, \boldsymbol{c})$.
Given $\lambda \in \mathbb{R}$, $\lambda > 0$ we define $\lambda * \Pi(Q, \boldsymbol{c})$ as follows:
For a finite extension E of K we let $\lambda * \Pi_E(Q, \boldsymbol{c})$ be the set of points $\boldsymbol{x} \in E^n$ with

$$\begin{aligned} \|L_i(\boldsymbol{x})\|_w &\le (Q^{c_i}\lambda)^{d(w|v_0)} && (i = 1, \ldots, n;\ w \in \mathfrak{M}_{v_0}(E)) \\ \|x_i\|_w &\le 1 && (i = 1, \ldots, n;\ w \in \mathfrak{M}(E) \backslash \mathfrak{M}_{v_0}(E)). \end{aligned} \tag{4.14}$$

The set $\lambda * \Pi(Q, \boldsymbol{c})$ then is given by

$$\lambda * \Pi(Q, \boldsymbol{c}) = \bigcup_E \lambda * \Pi_E(Q, \boldsymbol{c}), \tag{4.15}$$

where the union is over all finite extensions E of K.
In (4.10), we replace $\delta/2n$ by δ. So we are interested in the set of points $\boldsymbol{x} \in \overline{\mathbb{Q}}^n$ lying in

$$Q^{-\delta} * \Pi(Q, \boldsymbol{c}). \tag{4.16}$$

Here in view of (1.15) Q runs through the set of parameters satisfying

$$Q \ge c(n, \delta, L_1, \ldots, L_n).$$

As in section 3, in order to prove theorem 1.3 it will suffice to establish the following statement about the family of sets $Q^{-\delta} * \Pi(Q, \boldsymbol{c})$:

Theorem 4.2 (Absolute Parametric Subspace Theorem). *There exist finitely many proper linear subspaces of* $\overline{\mathbb{Q}}^n$*, all defined over* K*, with the following property:*
As Q *runs through the set of values*

$$Q > c(n, \delta, L_1, \ldots, L_n), \tag{4.17}$$

the sets $Q^{-\delta} * \Pi(Q, \mathbf{c})$ *defined by* (4.14), (4.15) *are contained in the union*

$$T_1 \cup \ldots \cup T_t.$$

The structure of our proof of the Absolute Parametric Subspace Theorem 4.2 is quite similar to that of the Parametric Subspace Theorem 3.1. Following Roy and Thunder [7], we introduce successive minima of $\Pi(Q, \mathbf{c})$: For $\lambda > 0$ we let $T(\lambda)$ be the $\overline{\mathbb{Q}}$-vector space generated by the elements $\boldsymbol{x} \in \lambda * \Pi(Q, \mathbf{c})$. For $i = 1, \ldots, n$ we define

$$\lambda_i = \inf\{\lambda \mid \dim T(\lambda) \geq i\}.$$

Notice that our definition implies that for any $\eta > 0$ there exists a finite extension E of K and there exist linearly independent points $\boldsymbol{g}_1, \ldots, \boldsymbol{g}_n \in E^n$ with

$$\boldsymbol{g}_i \in \lambda_i (1 + \eta) * \Pi_E(Q, \mathbf{c}). \tag{4.18}$$

The following proposition essentially is a special case of theorem 6.3 of Roy and Thunder [7].

Proposition 4.3. *The successive minima* $\lambda_1, \ldots, \lambda_n$ *of* $\Pi(Q, \mathbf{c})$ *satisfy*

$$n^{-n/2} \leq \lambda_1 \ldots \lambda_n \leq 2^{n(n-1)/2}. \tag{4.19}$$

We remark that in (4.19) we have already used the normalizations (4.1) and (4.9). In a more general setting we should replace (4.19) by

$$n^{-n/2} \leq \lambda_1 \ldots \lambda_n V \leq 2^{n(n-1)/2} \tag{4.20}$$

with $V = \| \det(L_1, \ldots, L_n) \|_{v_0}^{-1} Q^{c_1 + \ldots + c_n}$, allowing for the volume of $\Pi(Q, \mathbf{c})$. The theorem of Roy and Thunder is the absolute generalization of Minkowski's Theorem as quoted in (3.9). This absolute generalization is one of the main ingredients in our proof of theorem 1.3.
Bombieri and Vaaler [1] have proved an extension of Minkowski's Theorem for parallelepipeds of type $\Pi_E(Q, \mathbf{c})$, where E is a finite extension of K. They consider for λ the E vector space $T_E(\lambda)$ generated by the points $\boldsymbol{x} \in \lambda * \Pi_E(Q, \mathbf{c})$. Since for a number field E the nonarchimedean absolute values are discrete they actually suppose that the factor λ in the definition of $\lambda * \Pi_E(Q, \mathbf{c})$ shows up at the archimedean places. (In our more general setting, where we

are considering the algebraic closure $\Pi(Q, \boldsymbol{c})$, in view of lemma 4.1, essentially it does not matter at which place we put the factor λ.) At any rate, with

$$\lambda_{iE} = \inf\{\lambda \mid \dim T_E(\lambda) \geq i\}$$

Bombieri and Vaaler prove an inequality of type

$$1 \ll \lambda_{1E} \ldots \lambda_{nE} V \ll |\Delta_E|^{n/(2e)}. \tag{4.21}$$

Here Δ_E is the discriminant of E and $e = [E : \mathbb{Q}]$.
If we try to apply (4.21) in our proof, due to the discriminant on the right hand side of (4.21) we can derive a result about parallelepipeds $Q^{-\delta} * \Pi_E(Q, \boldsymbol{c})$, but there is no hope to get a result for the set of solutions $\boldsymbol{x} \in \overline{\mathbb{Q}}^n$ of (4.2) on this basis. A further advantage of proposition 4.3 comes into evidence when we ask for an explicit upper bound for the number of subspaces (as in (1.13)). It is only with this proposition that we are able to derive a result with a uniform lower bound for the height of the solutions as in (1.15). An application of (4.21) to the family of parallelepipeds $Q^{-\delta} * \Pi_E(Q, \boldsymbol{c})$ would yield a lower bound in (1.15) which involves the degree e of the field E. This might appear as only a minor difference. However, when we use our theorems to count the number of solutions of diophantine equations this difference matters.
Let us go back to (4.16). For values Q such that $Q^{-\delta} * \Pi(Q, \boldsymbol{c}) \neq \varnothing$ we have

$$\lambda_1(Q) \leq Q^{-\delta}. \tag{4.22}$$

Hence if Q is large, proposition 4.3 implies $\lambda_n(Q) > 1$. Then again we obtain (3.36), i.e.,

$$\lambda_p(Q) \leq Q^{-\delta/(n-1)} \lambda_{p+1}(Q) \quad \text{for some } p \text{ with } 1 \leq p \leq n-1, \tag{4.23}$$

and we have to get control over the subspaces corresponding to λ_p, where p is maximal with (4.23).

4.2 The approximation part

As in section 3.2, here we deal with values Q, where instead of (4.23) we have

$$\lambda_{n-1} = \lambda_{n-1}(Q) \leq Q^{-\delta}. \tag{4.24}$$

Again, we pick m suitable parameters $Q_1 < \ldots < Q_m$ with (4.24). For each h $(1 \leq h \leq m)$, by (4.24) and (4.18) we can find a finite extension E_h of K and linearly independent points $\boldsymbol{g}_{h1}(Q_h), \ldots, \boldsymbol{g}_{h,n-1}(Q_h)$ satisfying

$$\boldsymbol{g}_{hj} \in E_h^n \cap Q_h^{-\delta/2} * \Pi(Q_h, \boldsymbol{c}) \quad (h = 1, \ldots, m;\ j = 1, \ldots, n-1). \tag{4.25}$$

Write $T_h = T(Q_h)$ for the $(n-1)$-dimensional $\overline{\mathbb{Q}}$-vector space generated by the points $\boldsymbol{g}_{h1}, \ldots, \boldsymbol{g}_{h,n-1}$ $(h = 1, \ldots, m)$.

We construct an auxiliary polynomial $P(X_{11},\dots,X_{1n};\dots;X_{m1},\dots,X_{mn})$ as in section 3 whose index is

$$\geq m(1/n-\varepsilon)$$

on the m-fold product $U_i\times\cdots\times U_i$ $(i=1,\dots,n)$, where U_i is defined as in (3.14). Here we choose $\varepsilon=c\delta/n^2$ with a small absolute constant c.
Our goal is to prove that

(i)$_{n,\,\mathrm{abs}}$ P vanishes with high order on the m-fold product of subspaces $T_1\times\cdots\times T_m$,

(ii)$_{n,\,\mathrm{abs}}$ P cannot vanish with too high order on the product $T_1\times\cdots\times T_m$.

Again for m large enough **(i)**$_{n,\,\mathrm{abs}}$ and **(ii)**$_{n,\,\mathrm{abs}}$ will contradict each other provided that the parameters $Q_1,\dots,Q_m$ are suitably chosen.
Here we will discuss only certain aspects of **(i)**$_{n,\,\mathrm{abs}}$.

4.2.1 Vanishing

Similarly as in (3.21), we want to show that

$$\operatorname{ind}P\geq m\varepsilon \quad \text{on} \quad T_1\times\cdots\times T_m. \tag{4.26}$$

To verify (4.26), – mutatis mutandis – it will suffice to prove (3.22) under the assumptions (3.23)–(3.25).
In section 3 (as in section 2) we have constructed an auxiliary polynomial with rational integral coefficients. The only reason that we asked for integral coefficients was relation (3.28) (or (2.12) respectively). In fact we applied the "Fundamental Theorem of Diophantine Approximation": an integer whose absolute value is less than one equals zero.
In the current context the linear combinations $\boldsymbol{y}_h$ obtained with the points $\boldsymbol{g}_{h1},\dots,\boldsymbol{g}_{h,n-1}$ from (4.25) in analogy with (3.24) have components in the field E_h $(h=1,\dots,m)$. Therefore, to construct a polynomial P with integral coefficients would be of no big help. We use the version of Siegel's Lemma for number fields proved by Bombieri and Vaaler [1] to get a nonzero auxiliary polynomial P with coefficients in K which however otherwise essentially has the same properties as the auxiliary polynomial in section 3.
To prove (3.22), i.e., to prove

$$P^{\mathfrak{J}}(\boldsymbol{y}_1,\dots,\boldsymbol{y}_m)=0, \tag{4.27}$$

we use the product formula.
Let E be the compositum of the fields $E_1,\dots,E_m$ in (4.25). Then $\boldsymbol{y}_1,\dots,\boldsymbol{y}_m$ will lie in E. Consequently, to verify (4.27) is suffices to prove that

$$\prod_{w\in\mathfrak{M}(E)}\left\|P^{\mathfrak{J}}(\boldsymbol{y}_1,\dots,\boldsymbol{y}_m)\right\|_w<1. \tag{4.28}$$

In our proof of (4.28) it will be crucial that for $h = 1, \dots, m$ the parallelepipeds $\Pi(Q_h, \mathbf{c})$ from (4.12) and (4.13) are defined over K. In fact combining (3.24), (3.25) and (4.25) we see that the points $\mathbf{y}_1, \dots, \mathbf{y}_m$ satisfy the inequalities

$$\begin{aligned} &\|L_i(\mathbf{y}_h)\|_w \leq Q_h^{(c_i-\delta/2)d(w|v_0)}(2n\varepsilon^{-1})^{s(w)} && (w \in \mathfrak{M}_{v_0}(E)) \\ &\|y_{hi}\|_w \quad\ \leq (2n\varepsilon^{-1})^{s(w)} && (w \in \mathfrak{M}(E)\backslash\mathfrak{M}_{v_0}(E)) \end{aligned} \tag{4.29}$$

for $i = 1, \dots, n$ and $h = 1, \dots, m$. Here we have put

$$s(w) = \begin{cases} d(w|\infty) & \text{for } w \mid \infty,\ \infty \text{ the archimedean place on } \mathbb{Q}, \\ 0 & \text{for } w \text{ nonarchimedean.} \end{cases} \tag{4.30}$$

In analogy with estimate (3.33) in section 3.2.1 for the monomials in (3.26), using (4.29) we obtain the estimate

$$\begin{aligned} &\left\|L_1(\mathbf{y}_1)^{j_{11}} \dots L_n(\mathbf{y}_1)^{j_{1n}} \dots L_1(\mathbf{y}_m)^{j_{m1}} \dots L_n(\mathbf{y}_m)^{j_{mn}}\right\|_w \\ &\qquad \underset{\varepsilon,n}{<} \left(Q_1^{-r_1\delta/2} \dots Q_m^{-r_m\delta/2}\right)^{d(w|v_0)} \quad \text{for } w \in \mathfrak{M}_{v_0}(E), \end{aligned} \tag{4.31}$$

provided that $Q_1, \dots, Q_m$ satisfy (3.32). Notice that in (4.31) essentially we have the same estimate for each w which lies over v_0. This is due to the fact that our parallelepipeds are defined over K.
Moreover similarly as in (3.29), in the current context we get

$$H(P^{\mathfrak{J}}) < C(L_1, \dots, L_n)^{r_1+\dots+r_m}.$$

Therefore, using $\sum_{w\in\mathfrak{M}_{v_0}(E)} d(w|v_0) = \sum_{w|v_0} d(w|v_0) = 1$ we finally end up with

$$\begin{aligned} &\prod_{w\in\mathfrak{M}(E)} \|P^{\mathfrak{J}}(\mathbf{y}_1, \dots, \mathbf{y}_m)\|_w \\ &\qquad \leq C(L_1, \dots, L_n)^{r_1+\dots+r_m} \left(Q_1^{-r_1\delta/2} \dots Q_m^{-r_m\delta/2}\right)^{\sum_{w|v_0} d(w|v_0)} \\ &\qquad = C(L_1, \dots, L_n)^{r_1+\dots+r_m} \left(Q_1^{-r_1} \dots Q_m^{-r_m}\right)^{\delta/2} < 1 \end{aligned}$$

provided $Q_1, \dots, Q_m$ are large enough.
The remainder of the approximation part of the proof does not change, and therefore we will not discuss it further. The final result is a Theorem on the Penultimate Minimum of exactly the same type as theorem 3.4.

4.3 The geometry part

Checking the material from section 3.3, at first glance we might think that the transference principles given there in Davenport's Lemma 3.5 as well as with multilinear algebra easily go through also in our current context. However some care is needed when it comes to Davenport's Lemma: as pointed out in section 4.2, our vanishing argument in its central place depends on the fact that the parallelepipeds $\Pi(Q_h, \mathbf{c})$ $(h = 1, \ldots, m)$ are *defined over* K. Therefore it will be essential that we have a version of Davenport's Lemma at our disposal, where the analogue of hypothesis (b) yields parallelepipeds *defined over* K.

Such a situation being reached, the proof illustrated in section 3.3 goes through also in our more general context.

Therefore in the following we restrict ourselves to a discussion of the generalization of Davenport's Lemma.

We need the following

Lemma 4.4. *For $\lambda > 0$ let $T(\lambda)$ be the $\overline{\mathbb{Q}}$-vector space generated by the points $\boldsymbol{x} \in \lambda * \Pi(Q, \mathbf{c})$. Then $T(\lambda)$ is defined over K.*

Proof. Write $t = \dim T(\lambda)$. Then we can find a basis $\boldsymbol{b}_1, \ldots, \boldsymbol{b}_t$ of $T(\lambda)$ with

$$\boldsymbol{b}_j \in \lambda * \Pi(Q, \mathbf{c}) \quad (j = 1, \ldots, t). \tag{4.32}$$

Let E be a finite extension of K such that $\boldsymbol{b}_1, \ldots, \boldsymbol{b}_t \in E^n$. We may suppose without loss of generality that $E \,|\, K$ is a normal extension.

By (4.14) and (4.32) we have, for $i = 1, \ldots, n$ and $j = 1, \ldots, t$,

$$\begin{aligned} &||L_i(\boldsymbol{b}_j)||_w \le (Q^{c_i}\lambda)^{d(w|v_0)} && (w \in \mathfrak{M}_{v_0}(E)) \\ &||b_{ji}||_w \quad\; \le 1 && (w \in \mathfrak{M}(E) \backslash \mathfrak{M}_{v_0}(E)). \end{aligned} \tag{4.33}$$

For a place $w \in \mathfrak{M}(E)$ and for $\sigma \in \mathrm{Gal}(\overline{\mathbb{Q}} \,|\, E)$ we let w_σ be the place in $\mathfrak{M}(E)$ such that for $x \in E$ the non-normalized absolute values $|\;|_w$ and $|\;|_{w_\sigma}$ satisfy the relation

$$|x|_{w_\sigma} = |\sigma x|_w.$$

If w runs through the set of places in $\mathfrak{M}(E)$ lying above some $v \in \mathfrak{M}(K)$ then so does w_σ. Moreover since $E \,|\, K$ is normal, in (1.8) we have $d(w|v) = d(w_\sigma|v)$ for any $\sigma \in \mathrm{Gal}(\overline{\mathbb{Q}} \,|\, K)$. Therefore the normalized absolute values $||\;||_w$ and $||\;||_{w_\sigma}$ are related by

$$||x||_{w_\sigma} = ||\sigma x||_w \quad \text{for each } x \in E \text{ and for each } \sigma \in \mathrm{Gal}(\overline{\mathbb{Q}} \,|\, K).$$

Hence (4.33) implies for each $\sigma \in \mathrm{Gal}(\overline{\mathbb{Q}} \,|\, K)$, for $i = 1, \ldots, n$ and for $j = 1, \ldots, t$

$$\begin{aligned} ||L_i(\sigma(\boldsymbol{b}_j))||_w &= ||\sigma(L_i(\boldsymbol{b}_j))||_w = ||L_i(\boldsymbol{b}_j)||_{w_\sigma} \\ &\le (Q^{c_i}\lambda)^{d(w_\sigma|v_0)} = (Q^{c_i}\lambda)^{d(w|v_0)} \qquad (w \in \mathfrak{M}_{v_0}(E)) \end{aligned}$$

and

$$\|\sigma(b_{ji})\|_w = \|b_{ji}\|_{w_\sigma} \le 1 \qquad (w \in \mathfrak{M}(E)\backslash\mathfrak{M}_{v_0}(E)).$$

We may conclude that for each $\sigma \in \mathrm{Gal}(\overline{\mathbb{Q}}|K)$ we have $\sigma(\boldsymbol{b}_1), \ldots, \sigma(\boldsymbol{b}_t) \in T(\lambda)$. Thus $\sigma(T(\lambda)) = T(\lambda)$ and the lemma follows. □

Now let $\lambda_1 = \lambda_1(Q), \ldots, \lambda_n = \lambda_n(Q)$ be the successive minima of the parallelepiped $\Pi(Q, \boldsymbol{c})$ given in (4.12), (4.13). Define the integers

$$1 \le r_1 < r_2 < \ldots < r_s = n$$

by

$$\lambda_1 = \ldots = \lambda_{r_1} < \lambda_{r_1+1} = \ldots = \lambda_{r_2} < \ldots < \lambda_{r_{s-1}+1} = \ldots = \lambda_n. \tag{4.34}$$

Let $\varepsilon < 1$ be a positive number, small enough such that

$$\lambda_{r_t}(1+\varepsilon)^2 < \lambda_{r_{t+1}} \qquad (t = 1, \ldots, s-1). \tag{4.35}$$

Moreover let $\boldsymbol{g}_1, \ldots, \boldsymbol{g}_n$ be linearly independent points in $\overline{\mathbb{Q}}^n$ with

$$\boldsymbol{g}_j \in (1+\varepsilon)\lambda_j * \Pi(Q, \boldsymbol{c}) \qquad (j = 1, \ldots, n). \tag{4.36}$$

For $t = 1, \ldots, s$ let V_{r_t} be the $\overline{\mathbb{Q}}$-vector space generated by $\boldsymbol{g}_1, \ldots, \boldsymbol{g}_{r_t}$. Combining lemma 4.4, (4.35), (4.36) and the definition of successive minima we may infer that for $t = 1, \ldots, s$ the space V_{r_t} is defined over K.

Lemma 4.5 (Absolute Davenport's Lemma). *Let $v_1 \in \mathfrak{M}(K)$, $v_1 \ne v_0$ be a nonarchimedean place. Then there exists a permutation π of $\{1, \ldots, n\}$ and there are vectors*

$$\boldsymbol{h}_1 = (h_{11}, \ldots, h_{1n}), \ \ldots, \ \boldsymbol{h}_n = (h_{n1}, \ldots, h_{nn}) \in \overline{\mathbb{Q}}^n$$

with the following properties:

(*a*) $\mathrm{span}\{\boldsymbol{h}_1, \ldots, \boldsymbol{h}_j\} = \mathrm{span}\{\boldsymbol{g}_1, \ldots, \boldsymbol{g}_j\}$ *for* $j = 1, \ldots, n$.

(*b*) *If E is a finite extension of K such that $\boldsymbol{h}_1, \ldots, \boldsymbol{h}_n \in E^n$ then, for $1 \le i, j \le n$,*

$$\begin{aligned} &\|L_i(\boldsymbol{h}_j)\|_w \le Q^{c_i d(w|v_0)} && (w \in \mathfrak{M}_{v_0}(E)) \\ &\|h_{j\pi(i)}\|_w \le (2^{3n^2} \min\{\lambda_i, \lambda_j\})^{d(w|v_1)} && (w \in \mathfrak{M}_{v_1}(E)) \\ &\|h_{ji}\|_w \le 1 && (w \in \mathfrak{M}(E),\ w \nmid v_0,\ w \nmid v_1). \end{aligned}$$

Proof. We first determine the permutation π.
For $t = 1, \ldots, s$ we consider the space V_{r_t}. By definition $\boldsymbol{g}_1, \ldots, \boldsymbol{g}_{r_t}$ are a basis of V_{r_t}. However, since V_{r_t} is defined over K it also has a basis $\boldsymbol{y}_1, \ldots, \boldsymbol{y}_{r_t}$ with $\boldsymbol{y}_j \in K^n$ $(j = 1, \ldots, r_t)$. We pick linearly independent points $\boldsymbol{y}_1, \ldots, \boldsymbol{y}_n$ in K^n such that

$$\text{for } t = 1, \ldots, s \text{ the points } \boldsymbol{y}_1, \ldots, \boldsymbol{y}_{r_t} \text{ are a basis of } V_{r_t}. \tag{4.37}$$

Let V_i be the $\overline{\mathbb{Q}}$-vector space generated by $\boldsymbol{y}_1, \ldots, \boldsymbol{y}_i$ $(i = 1, \ldots, n)$.
The points $\boldsymbol{y} \in V_{n-1}$ satisfy a nontrivial relation

$$a_1 y_1 + \ldots + a_n y_n = 0 \quad (\boldsymbol{y} \in V_{n-1}) \tag{4.38}$$

with coefficients $a_1, \ldots, a_n \in K$. After reordering the variables, if necessary, we may suppose that in (4.38)

$$\|a_n\|_{v_1} = \max\{\|a_1\|_{v_1}, \ldots, \|a_n\|_{v_1}\}.$$

Dividing by a_n, we may rewrite (4.38) as

$$a'_1 y_1 + \ldots + a'_{n-1} y_{n-1} = y_n \quad (\boldsymbol{y} \in V_{n-1}) \tag{4.39}$$

with $a'_1, \ldots, a'_{n-1} \in K$ satisfying

$$\|a'_i\|_{v_1} \leq 1 \quad (i = 1, \ldots, n-1). \tag{4.40}$$

The points $\boldsymbol{y} \in V_{n-2}$ apart from (4.39) will satisfy a second relation which is independent of (4.39). Indeed, we may find such a second relation of type

$$b_1 y_1 + \ldots + b_{n-1} y_{n-1} = 0 \quad (\boldsymbol{y} \in V_{n-2}) \tag{4.41}$$

with $b_1, \ldots, b_{n-1} \in K$.
Again we may reorder the variables such that

$$\|b_{n-1}\|_{v_1} = \max\{\|b_1\|_{v_1}, \ldots, \|b_{n-1}\|_{v_1}\}.$$

And so (4.41) leads to

$$b'_1 y_1 + \ldots + b'_{n-2} y_{n-2} = y_{n-1} \quad (\boldsymbol{y} \in V_{n-2}) \tag{4.42}$$

with $b'_1, \ldots, b'_{n-2} \in K$ satisfying

$$\|b'_i\|_{v_1} \leq 1 \quad (i = 1, \ldots, n-2). \tag{4.43}$$

For points $\boldsymbol{y} \in V_{n-2}$, we may substitute (4.42) into (4.39) to get

$$a''_1 y_1 + \ldots + a''_{n-2} y_{n-2} = y_n \tag{4.44}$$

with $a''_1, \ldots, a''_{n-2} \in K$ and

$$\|a''_i\|_{v_1} \leq 1 \quad (i = 1, \ldots, n-2). \tag{4.45}$$

(At this point it is convenient that we have chosen a nonarchimedean place v_1.)
We may continue in this way. Finally we obtain:

After a suitable permutation of the variables, for each i ($i = 1, \ldots, n$) the points $\boldsymbol{y} \in V_i$ satisfy $n - i$ independent relations of type

$$y_k = \sum_{j=1}^{i} a_{kj}^{(i)} y_j \quad \text{for } k = i+1, \ldots, n \tag{4.46}$$

with coefficients $a_{kj}^{(i)} \in K$ having

$$\left\| a_{kj}^{(i)} \right\|_{v_1} \leq 1 \quad (i+1 \leq k \leq n;\ 1 \leq j \leq i). \tag{4.47}$$

This finishes the construction of the permutation π. For simplicity of notation, in the sequel we assume that π is the identity.
Before we start with the construction of the points $\boldsymbol{h}_j$, we apply lemma 4.1 to shift the factors λ_j from the places lying above v_0 to the places lying above v_1. Initially our points $\boldsymbol{g}_j$ satisfy (4.36) (cf. also (4.14)). With lemma 4.1 we can find a finite extension F of K and elements $\beta_1, \ldots, \beta_n \in F^*$ such that the points

$$\boldsymbol{f}_j = \beta_j \boldsymbol{g}_j \quad (j = 1, \ldots, n) \tag{4.48}$$

have, for $1 \leq i, j \leq n$,

$$\begin{aligned} \|L_i(\boldsymbol{f}_j)\|_w &\leq Q^{c_i d(w|v_0)} && (w \in \mathfrak{M}_{v_0}(F)) \\ \|f_{ji}\|_w &\leq ((1+\varepsilon)^2 \lambda_j)^{d(w|v_1)} && (w \in \mathfrak{M}_{v_1}(F)) \\ \|f_{ji}\|_w &\leq 1 && (w \in \mathfrak{M}(F),\ w \nmid v_0,\ w \nmid v_1). \end{aligned} \tag{4.49}$$

Here we have assumed that F is chosen so large as to contain also the components of $\boldsymbol{g}_1, \ldots, \boldsymbol{g}_n$. Clearly, $\operatorname{span}\{\boldsymbol{f}_1, \ldots, \boldsymbol{f}_j\} = \operatorname{span}\{\boldsymbol{g}_1, \ldots, \boldsymbol{g}_j\}$ for $j = 1, \ldots, n$.
Our construction of the points $\boldsymbol{h}_j$ will be by induction on t with $1 \leq t \leq s$. Indeed, the procedure to find $\boldsymbol{h}_q$ for $r_{t-1} + 1 \leq q \leq r_t$ will follow the same pattern.
We start with $t = 1$ and take $\boldsymbol{h}_1 = \boldsymbol{f}_1, \ldots, \boldsymbol{h}_{r_1} = \boldsymbol{f}_{r_1}$. Then by (4.48) assertion (a) is satisfied for $j = 1, \ldots, r_1$. Moreover, in view of (4.34) and (4.49) assertion (b) is true as well for $\boldsymbol{h}_1, \ldots, \boldsymbol{h}_{r_1}$.
We now suppose that $1 < t \leq s$ and that $\boldsymbol{h}_1, \ldots, \boldsymbol{h}_{r_{t-1}}$ have already been constructed such that assertions (a) and (b) are true for $j = 1, \ldots, r_{t-1}$ and for $i = 1, \ldots, n$. For the vectors $\boldsymbol{h}_{r_{t-1}+1}, \ldots, \boldsymbol{h}_{r_t}$ we try

$$\boldsymbol{h}_q = \gamma_{q0} \boldsymbol{f}_q + \sum_{j=1}^{r_{t-1}} \gamma_{qj} \boldsymbol{h}_j \quad (q = r_{t-1} + 1, \ldots, r_t). \tag{4.50}$$

Our goal is to determine suitable coefficients $\gamma_{q0}, \gamma_{q1}, \ldots, \gamma_{qr_{t-1}} \in \overline{\mathbb{Q}}$ with $\gamma_{q0} \neq 0$. Suppose for the moment that we have found such coefficients. Then

by the induction hypothesis, by (4.48) and since $\gamma_{q0} \neq 0$ we see that assertion (a) will be true for $j = 1, \ldots, r_t$.
There remains assertion (b). We fix q with $r_{t-1} + 1 \leq q \leq r_t$. Instead of (4.50) we write

$$\boldsymbol{h}_q = \gamma_0 \boldsymbol{f}_q + \sum_{j=1}^{r_{t-1}} \gamma_j \boldsymbol{h}_j. \tag{4.51}$$

With $\boldsymbol{h}_q = (h_{q1}, \ldots, h_{qn})$ relation (4.51) reads as

$$h_{qi} = \gamma_0 f_{qi} + \sum_{j=1}^{r_{t-1}} \gamma_j h_{ji} \quad (i = 1, \ldots, n). \tag{4.52}$$

To get an idea how to choose $\gamma_0, \gamma_1, \ldots, \gamma_{r_{t-1}}$ we first manipulate (4.52) for i in the range $1 \leq i \leq r_{t-1}$.
By the induction hypothesis $\boldsymbol{h}_1, \ldots, \boldsymbol{h}_{r_{t-1}}$ are a basis of $V_{r_{t-1}}$. Therefore relations (4.46) for $V_{r_{t-1}}$ imply that the truncated vectors

$$\left(h_{11}, \ldots, h_{1r_{t-1}}\right), \ldots, \left(h_{r_{t-1},1}, \ldots, h_{r_{t-1},r_{t-1}}\right)$$

are linearly independent. Hence we can find $\vartheta_1, \ldots, \vartheta_{r_{t-1}} \in \overline{\mathbb{Q}}$ such that

$$f_{qi} = \vartheta_1 h_{1i} + \ldots + \vartheta_{r_{t-1}} h_{r_{t-1},i} \quad (i = 1, \ldots, r_{t-1}). \tag{4.53}$$

The field F in (4.48), (4.49) may be chosen so large that also the points $\boldsymbol{h}_1, \ldots, \boldsymbol{h}_{r_{t-1}}$ lie in F^n. Then the coefficients $\vartheta_1, \ldots, \vartheta_{r_{t-1}}$ in (4.53) will lie in F as well.
Substituting (4.53) into (4.52) we get

$$h_{qi} = \sum_{j=1}^{r_{t-1}} (\gamma_0 \vartheta_j + \gamma_j) h_{ji} \quad (i = 1, \ldots, r_{t-1}). \tag{4.54}$$

Next we treat (4.52) for i in the range $r_{t-1} + 1 \leq i \leq n$. We now use relations (4.46) for $V_{r_{t-1}}$ in a more specific way. Accordingly for $j = 1, \ldots, r_{t-1}$, the components h_{ji} with $r_{t-1} + 1 \leq i \leq n$ of $\boldsymbol{h}_j$ satisfy equations of type

$$h_{ji} = \sum_{k=1}^{r_{t-1}} a_{ik} h_{jk} \tag{4.55}$$

with coefficients $a_{i1}, \ldots, a_{ir_{t-1}} \in K$ having

$$\|a_{ik}\|_{v_1} \leq 1 \quad (k = 1, \ldots, r_{t-1}). \tag{4.56}$$

Substituting (4.55) into (4.52) we obtain

$$h_{qi} = \gamma_0 f_{qi} + \sum_{k=1}^{r_{t-1}} a_{ik} \sum_{j=1}^{r_{t-1}} \gamma_j h_{jk} \quad (r_{t-1} + 1 \leq i \leq n). \tag{4.57}$$

In (4.57) the inner sum by (4.52) satisfies

$$\sum_{j=1}^{r_{t-1}} \gamma_j h_{jk} = h_{qk} - \gamma_0 f_{qk}.$$

Substituting this into (4.57) we finally arrive at

$$h_{qi} = \gamma_0 f_{qi} + \sum_{k=1}^{r_{t-1}} a_{ik}(h_{qk} - \gamma_0 f_{qk}) \quad (r_{t-1}+1 \le i \le n). \tag{4.58}$$

Comparing (4.54), (4.56), (4.58) and using the induction hypothesis we may now get a clearer idea how to choose the coefficients $\gamma_0, \gamma_1, \ldots, \gamma_{r_{t-1}}$ in (4.51). In fact an application of proposition 4.3 shows that we can find a finite extension E of F and elements $\gamma_0, \gamma_1, \ldots, \gamma_{r_{t-1}} \in E$ with $\gamma_0 \ne 0$ and

$$\begin{aligned} \|\gamma_j\|_u &\le (r_{t-1}+1)^{-s(u)} && (u \in \mathfrak{M}(E)\backslash\mathfrak{M}_{v_1}(E);\ j = 0,\ldots,r_{t-1}) \\ \|\gamma_0\vartheta_j + \gamma_j\|_u &\le 1 && (u \in \mathfrak{M}_{v_1}(E);\ j = 1,\ldots,r_{t-1}) \\ \|\gamma_0\|_u &\le 2^{2(r_{t-1}+1)^2 d(u|v_1)} && (u \in \mathfrak{M}_{v_1}(E)). \end{aligned} \tag{4.59}$$

Here $s(u)$ is defined by (4.30).
We prove that with this choice of $\gamma_0, \gamma_1, \ldots, \gamma_{r_{t-1}}$ the point $\boldsymbol{h}_q$ in (4.51) will satisfy assertion (b).
We begin with $u \in \mathfrak{M}_{v_1}(E)$.
For i with $1 \le i \le r_{t-1}$ we combine (4.54) and (4.59) to obtain

$$\|h_{qi}\|_u \le \max_{1\le j\le r_{t-1}} \{\|\gamma_0\vartheta_j + \gamma_j\|_u \|h_{ji}\|_u\} \le \max_{1\le j\le r_{t-1}} \|h_{ji}\|_u.$$

By the induction hypothesis for $1 \le j \le r_{t-1}$ and for $i = 1,\ldots,n$

$$\|h_{ji}\|_u \le \left(2^{3n^2}\min\{\lambda_i,\lambda_j\}\right)^{d(u|v_1)} \le \left(2^{3n^2}\lambda_i\right)^{d(u|v_1)}.$$

Since $r_{t-1}+1 \le q \le r_t$ we may conclude that

$$\|h_{qi}\|_u \le \left(2^{3n^2}\min\{\lambda_i,\lambda_q\}\right)^{d(u|v_1)} \quad \text{for } i = 1,\ldots,r_{t-1} \tag{4.60}$$

as desired in (b).
We next treat for $u \in \mathfrak{M}_{v_1}(E)$ the range $r_{t-1}+1 \le i \le n$.
Here we use (4.58). Accordingly we get for such i

$$\|h_{qi}\|_u \le \max\{\|\gamma_0\|_u \|f_{qi}\|_u;\ \|a_{ik}\|_u \|h_{qk} - \gamma_0 f_{qk}\|_u\ (k = 1,\ldots,r_{t-1})\}.$$

Now (4.34), (4.49), (4.59) and $r_{t-1}+1 \le q \le r_t$ give

$$\|\gamma_0\|_u \|f_{qi}\|_u \le \left(2^{2n^2}(1+\varepsilon)^2\lambda_{r_{t-1}+1}\right)^{d(u|v_1)}.$$

On the other hand, by (4.56) and by (4.60) (which already is proved) we may infer that for $1 \leq k \leq r_{t-1}$

$$\|a_{ik}\|_u \, \|h_{qk} - \gamma_0 f_{qk}\|_u \leq \max\{\|h_{qk}\|_u, \|\gamma_0\|_u \, \|f_{qk}\|_u\}$$
$$\leq \left(2^{3n^2} \max\{\lambda_{r_{t-1}}, \lambda_{r_{t-1}+1}\}\right)^{d(u|v_1)}.$$

So altogether we obtain for $r_{t-1} + 1 \leq i \leq n$

$$\|h_{qi}\|_u \leq \left(2^{3n^2} \lambda_{r_{t-1}+1}\right)^{d(u|v_1)} \leq \left(2^{3n^2} \min\{\lambda_i, \lambda_q\}\right)^{d(u|v_1)}. \tag{4.61}$$

Inequalities (4.60), (4.61) say that assertion (b) is satisfied for $\boldsymbol{h}_q$ and for $u \in \mathfrak{M}_{v_1}(E)$.
The remaining part of assertion (b) for $\boldsymbol{h}_q$ follows from an easy combination of (4.49), (4.51) and (4.59).
Since q with $r_{t-1} + 1 \leq q \leq r_t$ is arbitrary, our construction together with the induction hypothesis now implies the truth of assertion (b) for $\boldsymbol{h}_1, \ldots, \boldsymbol{h}_{r_t}$.
□

If we combine Davenport's Lemma 4.5 with multilinear algebra, we obtain assuming (4.23) a parallelepiped $\Pi^{(n-p)}$ in N-dimensional space with

$$N = \binom{n}{n-p}$$

exactly as in section 3.3. Assertion (b) of Davenport's Lemma implies that $\Pi^{(n-p)}$ again is defined over K. The penultimate minimum μ_{N-1} of this parallelepiped satisfies

$$\mu_{N-1} \leq Q^{-\delta/nN} \tag{4.62}$$

in complete analogy with (3.54). However, (4.62) is of the same type as hypothesis (4.24) in section 4.2. Since moreover all other hypotheses we have made in section 4.2 will be satisfied, the material developed there applies and the Absolute Parametric Subspace Theorem 4.2 follows just in the same way as explained in section 3.3.

5 Generalization of theorem 1.4

We begin by formulating a conjecture which would generalize the Subspace Theorem in the direction indicated by Wirsing's theorem 1.4.
Suppose that in (1.18) β is the quotient of algebraic numbers ζ and ξ: $\beta = \zeta/\xi$. Then clearly we can find positive constants c_1 and c_2 depending only on α such that

$$c_1|\zeta| \leq |\xi| \leq c_2|\zeta|.$$

Moreover, by (1.17)

$$M(\beta) \gg\ll H((\zeta,\xi))^d.$$

Thus inequality (1.18) essentially is the same as

$$\frac{|\xi|}{\max\{|\zeta|,|\xi|\}} \frac{|\xi\alpha-\zeta|}{\max\{|\zeta|,|\xi|\}} < H((\zeta,\xi))^{-2d^2-d\delta}. \tag{5.1}$$

Here we ask for solution vectors $(\zeta,\xi) \in \overline{\mathbb{Q}}^2$ with $\xi \neq 0$ and such that

$$[\mathbb{Q}(\xi/\xi,\ \zeta/\xi):\mathbb{Q}] = [\mathbb{Q}(1,\beta):\mathbb{Q}] = [\mathbb{Q}(\beta):\mathbb{Q}] \leq d. \tag{5.2}$$

So we can say that the vector (ζ,ξ) has degree $\leq d$ over $\mathbb{Q}$. In general, we define for a vector $\boldsymbol{x} \in \overline{\mathbb{Q}}^n\backslash\{\mathbf{0}\}$ and for a number field K the extension $K(\boldsymbol{x})$ by

$$K(\boldsymbol{x}) = K\left(\frac{x_1}{x_i},\ldots,\frac{x_n}{x_i}\right), \tag{5.3}$$

where i is a subscript with $x_i \neq 0$. In analogy with (5.2) we say that $\boldsymbol{x}$ has degree $\leq d$ over K if $[K(\boldsymbol{x}):K] \leq d$. Obviously, the degree of $\boldsymbol{x}$ depends only on the one-dimensional subspace of $\overline{\mathbb{Q}}^n$ generated by $\boldsymbol{x}$.
More generally, a *linear subspace* T of the $\overline{\mathbb{Q}}$-vector space $\overline{\mathbb{Q}}^n$ is said to have *degree* $\leq d$ over K, if there exists an extension E of K of degree $\leq d$ such that T is defined over E. In other words a subspace T has degree $\leq d$ over K if it may be defined by a system of linear equations

$$c_{i1}x_1 + \ldots + c_{in}x_n = 0 \quad (i = 1,\ldots,m)$$

whose coefficients c_{ij} $(1 \leq i \leq m,\ 1 \leq j \leq n)$ are contained in an extension of K of degree $\leq d$.
In view of (5.1)–(5.3) we formulate the following conjecture on a generalization of theorem 1.4 to n dimensions:

Conjecture 5.1. Let $n \geq 2$ and d be natural numbers. Then there exists a positive function $c(n,d)$ depending only on n and d with the following property:
Let $\delta > 0$. Let K be a number field. Let S be a finite subset of $\mathfrak{M}(K)$. Assume that for each $v \in S$ we are given linearly independent linear forms $L_1^{(v)}(\boldsymbol{X}),\ldots,L_n^{(v)}(\boldsymbol{X})$ in $\boldsymbol{X} = (X_1,\ldots,X_n)$ with algebraic coefficients. Suppose that for each $v \in S$ the normalized absolute value $\|\ \|_v$ on K is extended to $\overline{\mathbb{Q}}$ as in (1.6). Let $\mathcal{L}$ be the set of points $\boldsymbol{x} \in \overline{\mathbb{Q}}^n\backslash\{\mathbf{0}\}$ satisfying

$$\prod_{v\in S} \frac{\|L_1^{(v)}(\boldsymbol{x})\|_v}{\|\boldsymbol{x}\|_v} \cdots \frac{\|L_n^{(v)}(\boldsymbol{x})\|_v}{\|\boldsymbol{x}\|_v} < H(\boldsymbol{x})^{-c(n,d)-\delta} \tag{5.4}$$

and

$$x \text{ has degree} \leq d \text{ over } K. \tag{5.5}$$

Then there exist finitely many proper linear subspaces $T_1, \ldots, T_{t_2}$ of $\overline{\mathbb{Q}}^n$, all of degree $\leq d$ over K, such that $\mathcal{L}$ is contained in the union

$$T_1 \cup \ldots \cup T_{t_2}.$$

In this section we will give an argument which justifies the hope that a possible function $c(n,d)$ might be $2(n-1)d^2$.
We try to attack conjecture 5.1 along the path chosen in sections 3 and 4. To simplify the exposition we will treat here only the special case when S consists of a single place $v_0 \in \mathfrak{M}(K)$. $\| \;\|_{v_0}$ will be the associated absolute value, normalized on K but extended to $\overline{\mathbb{Q}}$. We then have a single system of linearly independent linear forms $L_1, \ldots, L_n$. Again to avoid technicalities, we assume that $L_1, \ldots, L_n$ have coefficients in K and we suppose that

$$\det(L_1, \ldots, L_n) = 1. \tag{5.6}$$

So (5.4) reads as

$$\frac{\|L_1(\boldsymbol{x})\|_{v_0}}{\|\boldsymbol{x}\|_{v_0}} \cdots \frac{\|L_n(\boldsymbol{x})\|_{v_0}}{\|\boldsymbol{x}\|_{v_0}} < H(\boldsymbol{x})^{-c(n,d)-\delta}. \tag{5.7}$$

5.1 Parallelepipeds

Notice that (5.7) is invariant under replacing $\boldsymbol{x}$ by $\alpha\boldsymbol{x}$ for any $\alpha \in \overline{\mathbb{Q}}^*$. Therefore without loss of generality we may restrict ourselves to solutions $\boldsymbol{x}$ of (5.7) which have some component equal to 1. Recall the definition of $K(\boldsymbol{x})$ in (5.3). With this normalization we get

$$[K(\boldsymbol{x}) : K] = [K(x_1, \ldots, x_n) : K] \leq d. \tag{5.8}$$

We fix such a solution $\boldsymbol{x}$ of (5.7), (5.8) and we write $K(\boldsymbol{x}) = F$. Let $w_0 \in \mathfrak{M}(F)$ be the place of F lying above v_0 such that the normalized absolute value $\| \;\|_{w_0}$ on F satisfies

$$\|x\|_{w_0} = \|x\|_{v_0}^{d(w_0|v_0)} \quad \text{for } x \in F, \tag{5.9}$$

where $d(w_0|v_0) = [F_{w_0} : K_{v_0}]/[F : K]$. Here (5.8) implies

$$d(w_0|v_0) \geq 1/d. \tag{5.10}$$

With w_0, (5.7) for our particular $\boldsymbol{x}$ becomes

$$\frac{\|L_1(\boldsymbol{x})\|_{w_0}}{\|\boldsymbol{x}\|_{w_0}} \cdots \frac{\|L_n(\boldsymbol{x})\|_{w_0}}{\|\boldsymbol{x}\|_{w_0}} < H(\boldsymbol{x})^{-d(w_0|v_0)(c(n,d)+\delta)}. \tag{5.11}$$

The factors on the left hand side of (5.11) satisfy

$$\frac{\|L_i(\boldsymbol{x})\|_{w_0}}{\|\boldsymbol{x}\|_{w_0}} \leq \frac{\|L_i\|_{w_0}\,\|\boldsymbol{x}\|_{w_0}}{\|\boldsymbol{x}\|_{w_0}} = \|L_i\|_{w_0} \ll 1 \quad (i = 1, \ldots, n).$$

Assuming that $L_i(\boldsymbol{x}) \neq 0$ for $i = 1, \ldots, n$ we can find real numbers $e_1, \ldots, e_n$ with

$$e_1 + \ldots + e_n = -\big(c(n,d) + \delta\big), \quad e_i \leq 0 \quad (i = 1, \ldots, n) \tag{5.12}$$

such that

$$\frac{\|L_i(\boldsymbol{x})\|_{w_0}}{\|\boldsymbol{x}\|_{w_0}} \ll H(\boldsymbol{x})^{d(w_0|v_0)e_i} \quad (i = 1, \ldots, n), \tag{5.13}$$

where the constants in $\ll$ depend only on $L_1, \ldots, L_n$. Using a compactness argument (and replacing δ by $\delta/2$), we may restrict ourselves to $e_1, \ldots, e_n$ from a finite set which depends only upon d, δ and $L_1, \ldots, L_n$. Thus we may even suppose that $e_1, \ldots, e_n$ in (5.12), (5.13) are fixed.
Apart from (5.13) we have the trivial inequalities

$$\|x_i\|_w \leq \|\boldsymbol{x}\|_w \quad (i = 1, \ldots, n;\ w \in \mathfrak{M}(F),\ w \neq w_0). \tag{5.14}$$

We now apply lemma 4.1 to (5.13), (5.14). We pick a place $v_1 \in \mathfrak{M}(K)$, $v_1 \neq v_0$, and we define for $w \in \mathfrak{M}(F)$

$$A_w = \begin{cases} \|\boldsymbol{x}\|_w^{-1} & \text{for } w \in \mathfrak{M}(F),\ w \nmid v_1 \\ \|\boldsymbol{x}\|_w^{-1} H(\boldsymbol{x})^{d(w|v_1)(1+\delta/(2c(n,d)))} & \text{for } w \in \mathfrak{M}(F),\ w \mid v_1. \end{cases} \tag{5.15}$$

By lemma 4.1 there exist a finite extension E of F and an element $\beta \in E^*$ such that

$$\|\beta\|_u \leq A_w^{d(u|w)} \quad \text{for all pairs } (u, w),\ u \in \mathfrak{M}(E),\ w \in \mathfrak{M}(F),\ u|w. \tag{5.16}$$

The point $\beta\boldsymbol{x}$ then satisfies the inequalities

$$\begin{aligned} \|L_i(\beta\boldsymbol{x})\|_u &\ll H(\beta\boldsymbol{x})^{d(u|v_0)e_i} && (u \in \mathfrak{M}(E),\ u \mid w_0) \\ \|\beta x_i\|_u &\leq H(\beta\boldsymbol{x})^{d(u|v_1)(1+\delta/(2c(n,d)))} && (u \in \mathfrak{M}(E),\ u \mid v_1) \\ \|\beta x_i\|_u &\leq 1 && (u \in \mathfrak{M}(E),\ u \nmid w_0,\ u \nmid v_1) \end{aligned} \tag{5.17}$$

for $i = 1, \ldots, n$. The third inequality in (5.17) implies in particular

$$\|L_i(\beta\boldsymbol{x})\|_u \ll 1 \quad (i = 1, \ldots, n;\ u \in \mathfrak{M}(E),\ u \mid v_0,\ u \nmid w_0). \tag{5.18}$$

Define real numbers $c_1, \ldots, c_n$ by

$$c_i = e_i/(1 + \delta/c(n,d)) \quad (i = 1, \ldots, n).$$

It is easily seen that a function $c(n,d)$ which might be suitable in conjecture 5.1 satisfies

$$c(n,d) \geq nd. \tag{5.19}$$

In fact for $d > 1$ the lower bound (5.19) in view of (1.17), (1.19) is even far from the truth. At any rate, by (5.12), (5.19) we get

$$c_1 + \ldots + c_n = -c(n,d) \leq -nd, \quad c_i \leq 0 \quad (i = 1, \ldots, n). \tag{5.20}$$

Given $(c_1, \ldots, c_n)$ with (5.20), we define the tuple $\boldsymbol{c} = (c_{wi})$ for $w \in \mathfrak{M}(F)$, $w \mid v_0$ and for $i = 1, \ldots, n$ as follows:

$$c_{wi} = \begin{cases} c_i & (i = 1, \ldots, n;\ w = w_0) \\ 0 & (i = 1, \ldots, n;\ w \mid v_0,\ w \neq w_0). \end{cases} \tag{5.21}$$

For $Q > 1$, for $\boldsymbol{c} = (c_{wi})$ as in (5.20), (5.21) and for our field F we define the parallelepiped $\Pi_F(F, Q, \boldsymbol{c})$ as the set of points $\boldsymbol{x} \in F^n$ satisfying the inequalities

$$\begin{aligned} \|L_i(\boldsymbol{x})\|_w &\leq Q^{d(w|v_0)c_{wi}} && (w \in \mathfrak{M}(F),\ w \mid v_0) \\ \|x_i\|_w &\leq Q^{d(w|v_1)} && (w \in \mathfrak{M}(F),\ w \mid v_1) \\ \|x_i\|_w &\leq 1 && (w \in \mathfrak{M}(F),\ w \nmid v_0,\ w \nmid v_1) \end{aligned} \tag{5.22}$$

for $i = 1, \ldots, n$. For a finite extension E of F we let $\Pi_E(F, Q, \boldsymbol{c})$ be the set of points $\boldsymbol{x} \in E^n$ having

$$\begin{aligned} \|L_i(\boldsymbol{x})\|_u &\leq Q^{d(u|v_0)c_{wi}} && (u \in \mathfrak{M}_w(E),\ w \in \mathfrak{M}_{v_0}(F)) \\ \|x_i\|_u &\leq Q^{d(u|v_1)} && (u \in \mathfrak{M}(E),\ u \mid v_1) \\ \|x_i\|_u &\leq 1 && (u \in \mathfrak{M}(E),\ u \nmid v_0,\ u \nmid v_1) \end{aligned} \tag{5.23}$$

for $i = 1, \ldots, n$. Finally we write

$$\Pi(F, Q, \boldsymbol{c}) = \bigcup_E \Pi_E(F, Q, \boldsymbol{c}), \tag{5.24}$$

where the union is over all finite extensions E of F.
Given $\lambda > 0$, we define $\lambda * \Pi_F(F, Q, \boldsymbol{c})$ as the set of points $\boldsymbol{x} \in F^n$ with

$$\begin{aligned} \|L_i(\boldsymbol{x})\|_w &\leq Q^{d(w|v_0)c_{wi}} && (w \in \mathfrak{M}(F),\ w \mid v_0) \\ \|x_i\|_w &\leq (Q\lambda)^{d(w|v_1)} && (w \in \mathfrak{M}(F),\ w \mid v_1) \\ \|x_i\|_w &\leq 1 && (w \in \mathfrak{M}(F),\ w \nmid v_0,\ w \nmid v_1) \end{aligned} \tag{5.25}$$

for $i = 1, \ldots, n$. We define the algebraic closure $\lambda * \Pi(F, Q, \boldsymbol{c})$ starting from (5.25) in analogy with (5.23), (5.24). (For $\lambda * \Pi_E(F, Q, \boldsymbol{c})$ the right hand

side in the analogue of the second inequality (5.25) becomes $(Q\lambda)^{d(u|v_1)}$ for $u \in \mathfrak{M}(E)$, $u \mid v_1$.)
Write

$$Q = H(\boldsymbol{x})^{1+\delta/c(n,d)} \quad \text{and} \quad \eta = \delta/3c(n,d).$$

Disregarding the $\ll$ in (5.17), (5.18), we have to treat the points

$$\boldsymbol{x} \in Q^{-\eta} * \Pi(F, Q, \boldsymbol{c}). \tag{5.26}$$

Notice that in (5.23) the inequalities for u with $u \nmid v_0$ are defined over the ground field K. The situation is different for the first inequality corresponding to the places $u \in \mathfrak{M}(E)$ with $u \mid v_0$. By (5.21), here the exponents on the right hand side are $c_i d(u|v_0)$ for $u \mid w_0$ but they are equal to zero for $u \mid v_0$, $u \nmid w_0$. In our construction, we started with a solution $\boldsymbol{x}$ of (5.7). With this solution $\boldsymbol{x}$ we obtained the field F, the tuple $c_1, \ldots, c_n$ in (5.20) and finally the parallelepiped $\Pi(F, Q, \boldsymbol{c})$ with (5.21), (5.22), (5.23), (5.24). As may be seen from (5.21), (5.23) our parallelepiped $\Pi(F, Q, \boldsymbol{c})$ is defined over F. However, F depends upon the solution $\boldsymbol{x}$ of (5.7) we started with and in general different solutions will lead to different fields. Therefore, in order to study conjecture 5.1 we have to investigate the two-parameter family of parallelepipeds $\Pi(F, Q, \boldsymbol{c})$ where Q is a free parameter and where F runs through the extensions of K of degree $\leq d$.
In contrast with the current situation, in section 4 we have dealt with one-parameter families of parallelepipeds $\Pi(Q, \boldsymbol{c})$. Here only Q was a free parameter. But all parallelepipeds $\Pi(Q, \boldsymbol{c})$ were defined over the same field K.
For the family $\Pi(F, Q, \boldsymbol{c})$ we formulate the following parametric version of conjecture 5.1:

Conjecture 5.2 (Parametric Conjecture). Let $n \geq 2$ and d be natural numbers. Then there exists a positive function $c(n,d)$ as in (5.19) *with the following properties:*
Suppose $0 < \eta < 1$. Let $c_1, \ldots, c_n$ be real numbers satisfying (5.20). *For an extension F of K of degree $\leq d$ define the tuple $\boldsymbol{c} = (c_{wi})$ ($w \in \mathfrak{M}(F)$, $w \mid v_0$; $i = 1, \ldots, n$) by* (5.21). *Let $\Pi(F, Q, \boldsymbol{c})$ be the parallelepiped given by* (5.22)–(5.24).
Then there exist finitely many proper linear subspaces $T_1, \ldots, T_t$ of $\overline{\mathbb{Q}}^n$, all of degree $\leq d$ over K, with the following property:
As Q runs through the set of values

$$Q \geq c(n, \eta, d, L_1, \ldots, L_n) \tag{5.27}$$

and as F runs through the extensions of K with

$$[F : K] \leq d \tag{5.28}$$

the union of the sets

$$Q^{-\eta} * \Pi(F, Q, \mathbf{c}) \tag{5.29}$$

is contained in

$$T_1 \cup \ldots \cup T_t.$$

Similarly as in the preceding sections we attack the parametric conjecture 5.2 with geometry of numbers. In proposition 4.3 we have quoted the theorem of Roy and Thunder [7] in a very special case. For our current application we need the more general version as in (4.20). We have $\sum_{w|v_1} d(w|v_1) = 1$. Moreover by (5.10), (5.20) we see that

$$\sum_{i=1}^{n} d(w_0|v_0)c_i = -d(w_0|v_0)\, c(n,d) \leq -n.$$

Therefore the parallelepiped $\Pi(F, Q, \mathbf{c})$ in (5.22)–(5.24) has volume

$$\begin{aligned} V = Q^{n-d(w_0|v_0)c(n,d)} \, \|\det(L_1, \ldots, L_n)\|_w^{-1} \\ = Q^{n-d(w_0|v_0)c(n,d)} \leq 1. \end{aligned} \tag{5.30}$$

Thus by (4.20)

$$\lambda_1 \ldots \lambda_n \gg\ll Q^{d(w_0|v_0)c(n,d)-n}. \tag{5.31}$$

Notice that here $\lambda_i = \lambda_i(F, Q)$. Now for a pair (F, Q) such that $Q^{-\eta} * \Pi(F, Q, \mathbf{c}) \neq \varnothing$ we clearly have $\lambda_1(F, Q) \leq Q^{-\eta}$. So if Q is large, (5.31) implies

$$\lambda_n(F, Q) > Q^{(d(w_0|v_0)c(n,d)-n)/(n-1)+\eta/n} \geq Q^{\eta/n}. \tag{5.32}$$

The picture seems to be quite similar to the one we had in sections 3 and 4. Indeed, to prove the parametric conjecture 5.2, by (5.32) in analogy with (3.36) it would suffice to deal with the set of pairs (F, Q) such that

$$\lambda_p(F, Q) < Q^{-\eta/(n-1)} \lambda_{p+1}(F, Q), \tag{5.33}$$

or by (5.32) even

$$\lambda_p(F, Q) < Q^{-(d(w_0|v_0)c(n,d)-n)/(n-1)^2 - \eta/(n-1)} \lambda_{p+1}(F, Q),$$

for some p with $1 \leq p \leq n-1$.

5.2 The approximation part

As in the preceding sections we replace (5.33) by the much stronger hypothesis

$$\lambda_{n-1} = \lambda_{n-1}(F, Q) \leq Q^{-\eta}. \tag{5.34}$$

Given a pair (F, Q) with (5.34) similarly as in (4.25) we can find a finite extension E of F and linearly independent points $\boldsymbol{g}_1, \ldots, \boldsymbol{g}_{n-1}$ satisfying

$$\boldsymbol{g}_j \in E^n \cap Q^{-\eta/2} * \Pi(F, Q, \mathbf{c}) \quad (j = 1, \ldots, n-1). \tag{5.35}$$

Let $T = T(F, Q)$ be the $(n-1)$-dimensional subspace of $\overline{\mathbb{Q}}^n$ generated by $\boldsymbol{g}_1, \ldots, \boldsymbol{g}_{n-1}$.
We now stop making conjectures. Put

$$c(n, d) = 2(n-1)d^2. \tag{5.36}$$

We prove:

Theorem 5.3 (on the Penultimate Minimum). (H. Locher and H. P. Schlickewei, [5]). *Let $c(n, d)$ be given by* (5.36). *Let $c_1, \ldots, c_n$ be real numbers satisfying* (5.20). *For an extension F of K with $[F : K] \leq d$ define the tuple $\mathbf{c} = (c_{wi})$ by* (5.21). *Let $\Pi(F, Q, \mathbf{c})$ be the parallelepiped given by* (5.22)–(5.24). *Write $\lambda_i(F, Q)$ $(i = 1, \ldots, n)$ for its successive minima. Suppose $0 < \eta < 1$. Let $m > c_1(n, \eta, d)$.*
Consider the set $\mathfrak{R}$ of pairs (F, Q) such that

$$Q \geq Q_0 = Q_0(m, L_1, \ldots, L_n) \tag{5.37}$$

$$[F : K] \leq d \tag{5.38}$$

$$\lambda_{n-1}(F, Q) \leq Q^{-\eta}. \tag{5.39}$$

Then either for all pairs $(F, Q) \in \mathfrak{R}$ the subspace $T(F, Q)$ corresponding to $\lambda_{n-1}(F, Q)$ will be the same, or the values Q under consideration are contained in fewer than m intervals of type

$$Q_h < Q \leq Q_h^E \quad (h = 1, \ldots, m-1). \tag{5.40}$$

Here $E = E(m, n, \eta, d)$.

We remark that for $n = 2$ hypothesis (5.39) becomes

$$\lambda_1(F, Q) \leq Q^{-\eta}.$$

This hypothesis is certainly satisfied for the sets $Q^{-\eta} * \Pi(F, Q, \mathbf{c})$ in the parametric conjecture 5.2. So in particular theorem 5.3 implies the conjecture in the case $n = 2$ with $c(n, d) = 2(n-1)d^2$ which then becomes $2d^2$. As explained in (5.1), this means that for $n = 2$ theorem 5.3 coincides with Wirsing's theorem 1.4.
We also remark that all constants in the above theorem may be determined explicitly.
It follows from lemma 4.4 that for each Q the subspace $T(F, Q)$ is defined over F and so it has degree $\leq d$ over K. Now it is easily shown that the

height of $T(F,Q)$ is bounded by Q^c with a constant c that depends only upon $c_1,\dots,c_n$, n, d, and $L_1,\dots,L_n$. But then we may infer from (5.40) that the heights of the subspaces $T(F,Q)$ under consideration are below some fixed bound. Since the degree of $T(F,Q)$ over K is $\le d$ we may conclude that the number of these subspaces is finite.
To prove theorem 5.3 we pick m suitable pairs

$$(F_1,Q_1),\ \dots\ ,(F_m,Q_m) \tag{5.41}$$

with (5.37)–(5.39) and such that $Q_1 < \dots < Q_m$. For $h = 1,\dots,m$ we write $T_h = T(F_h,Q_h)$.
As in sections 3 and 4, again our strategy will be to construct an auxiliary polynomial $P(X_{11},\dots,X_{1n};\ \dots\ ;X_{m1},\dots,X_{mn})$ such that

$(\mathbf{i})_{n,\,\mathrm{bound}}$ P vanishes with high order on the m-fold product of subspaces $T_1 \times \cdots \times T_m$,

$(\mathbf{ii})_{n,\,\mathrm{bound}}$ P cannot vanish with too high order on the product $T_1\times\cdots\times T_m$.

Similarly as in sections 3 and 4, our goal is to construct P in such a way that for m large enough and for a suitable choice of $Q_1,\dots,Q_m$ we obtain a contradiction between $(\mathbf{i})_{n,\,\mathrm{bound}}$ and $(\mathbf{ii})_{n,\,\mathrm{bound}}$.
The main problem in our proof occurs with $(\mathbf{i})_{n,\,\mathrm{bound}}$. The difficulty comes from the fact that in our sequence (5.41) the fields $F_1,\dots,F_m$ in general will not coincide. So we have to deal simultaneously with parallelepipeds which are defined over distinct fields.
In the sequel we will discuss only the ideas needed for $(\mathbf{i})_{n,\,\mathrm{bound}}$. The other parts of our proof of theorem 5.3 are similar to their analogues in the proof of the Theorem on the Penultimate Minimum in sections 3 and 4.

5.2.1 Vanishing

Our purpose is to construct a nonzero auxiliary polynomial P such that

$$P \text{ has index } m\varepsilon \text{ on } T_1 \times \cdots \times T_m, \tag{5.42}$$

where $\varepsilon = c\eta/n^2$ with a small absolute constant c. Let us discuss where we end up with if we proceed in the same way as in sections 3 and 4. So we assume that we have a polynomial P in $X_{11},\dots,X_{1n};\ \dots\ ;X_{m1},\dots,X_{mn}$ which is homogeneous in the block of variables $X_{h1},\dots,X_{hn}$ of degree r_h $(h = 1,\dots,m)$. We assume moreover that P has index $\ge m(1/n-\varepsilon)$ on the m-fold product of subspaces $U_i \times \cdots \times U_i$ where U_i is given by $L_i = 0$ $(i = 1,\dots,n)$.
We suppose (similarly as in section 3.2.1) that P has been constructed with parameters $r_1,\dots,r_m$ such that

$$Q_1^{r_1} \underset{\varepsilon}{=} Q_h^{r_h} \quad (h = 1,\dots,m). \tag{5.43}$$

Since $(F_1, Q_1), \ldots, (F_m, Q_m)$ fulfill (5.37)–(5.39) we know that for each h $(1 \le h \le m)$ there exist a finite extension E_h of F_h and linearly independent points $\boldsymbol{g}_{h1}, \ldots, \boldsymbol{g}_{h,n-1}$ having

$$\boldsymbol{g}_{hj} \in E_h^n \cap Q_h^{-\eta/2} * \Pi(F_h, Q_h, \boldsymbol{c}) \quad (h = 1, \ldots, m;\ j = 1, \ldots, n-1). \quad (5.44)$$

Define the linear combinations $\boldsymbol{y}_h$ of $\boldsymbol{g}_{h1}, \ldots, \boldsymbol{g}_{h,n-1}$ $(h = 1, \ldots, m)$ in analogy with (3.24), (3.25). In order to prove (5.42), we have to verify the analogue of (3.22). Let us restrict ourselves to an investigation of

$$P(\boldsymbol{y}_1, \ldots, \boldsymbol{y}_m) = 0. \quad (5.45)$$

Write E for the compositum of the fields $E_1, \ldots, E_m$ from (5.44). To establish (5.45) it suffices to prove

$$\prod_{u \in \mathfrak{M}(E)} \|P(\boldsymbol{y}_1, \ldots, \boldsymbol{y}_m)\|_u < 1,$$

or equivalently

$$\prod_{v \in \mathfrak{M}(K)} \|N_{E|K}(P(\boldsymbol{y}_1, \ldots, \boldsymbol{y}_m))\|_v < 1. \quad (5.46)$$

In view of (5.25) it is plausible that in estimating the left hand side of (5.46) we will get the main contribution from the places v_0 and v_1. We begin with v_1 and study

$$\|N_{E|K}(P(\boldsymbol{y}_1, \ldots, \boldsymbol{y}_m))\|_{v_1}. \quad (5.47)$$

Let D be the degree of the compositum E over K. So we have D embeddings $\tau_1, \ldots, \tau_D$ of E over K into $\overline{\mathbb{Q}}$. We write

$$P(\boldsymbol{y}_1, \ldots, \boldsymbol{y}_m) = \sum_{j_{11}, \ldots, j_{mn}} c(j_{11}, \ldots, j_{mn})\, y_{11}^{j_{11}} \cdots y_{1n}^{j_{1n}} \cdots y_{m1}^{j_{m1}} \cdots y_{mn}^{j_{mn}}.$$

In order to derive an upper bound for (5.47) we have to estimate

$$\prod_{k=1}^{D} \max_{j_{11}, \ldots, j_{mn}} \left\| \tau_k\left(y_{11}^{j_{11}} \cdots y_{1n}^{j_{1n}} \cdots y_{m1}^{j_{m1}} \cdots y_{mn}^{j_{mn}}\right) \right\|_{v_1}. \quad (5.48)$$

Here we suppose that the absolute value $\| \ \|_{v_1}$ has been extended to $\overline{\mathbb{Q}}$. Using (5.25), (5.44) together with (3.24) we see that

$$\begin{aligned} &\left\| \tau_k\left(y_{11}^{j_{11}} \cdots y_{1n}^{j_{1n}} \cdots y_{m1}^{j_{m1}} \cdots y_{mn}^{j_{mn}}\right) \right\|_{v_1} \\ &\qquad \underset{\varepsilon, n}{\le} Q_1^{(j_{11} + \ldots + j_{1n})(1 - \eta/2)} \cdots Q_m^{(j_{m1} + \ldots + j_{mn})(1 - \eta/2)} \\ &\qquad = Q_1^{r_1(1 - \eta/2)} \cdots Q_m^{r_m(1 - \eta/2)}. \end{aligned}$$

Consequently we obtain

$$\prod_{k=1}^{D} \max_{j_{11},\ldots,j_{mn}} \|\tau_k(y_{11}^{j_{11}}\ldots y_{1n}^{j_{1n}}\ldots y_{m1}^{j_{m1}}\ldots y_{mn}^{j_{mn}})\|_{v_1}$$
$$\leq Q_1^{Dr_1(1-\eta/2)}\ldots Q_m^{Dr_m(1-\eta/2)}. \tag{5.49}$$

Next we consider v_0 and estimate

$$\|N_{E|K}(P(\boldsymbol{y}_1,\ldots,\boldsymbol{y}_m))\|_{v_0}. \tag{5.50}$$

We write (similarly as in (3.26))

$$\begin{aligned} P(\boldsymbol{y}_1,\ldots,\boldsymbol{y}_m) &= P(y_{11},\ldots,y_{1n};\ldots;y_{m1},\ldots,y_{mn}) \\ &= \sum_{j_{11},\ldots,j_{mn}} d(j_{11},\ldots,j_{mn}) \prod_{h=1}^{m} \big(L_1(\boldsymbol{y}_h)^{j_{h1}}\ldots L_n(\boldsymbol{y}_h)^{j_{hn}}\big). \end{aligned} \tag{5.51}$$

To estimate (5.50), by (5.51) we have to estimate the term

$$\prod_{k=1}^{D} \max_{j_{11},\ldots,j_{mn}} \left\|\tau_k\left(\prod_{h=1}^{m}\big(L_1(\boldsymbol{y}_h)^{j_{h1}}\ldots L_n(\boldsymbol{y}_h)^{j_{hn}}\big)\right)\right\|_{v_0}. \tag{5.52}$$

In contrast with the situation encountered in sections 3 and 4, in our current context given k with $1 \leq k \leq D$ we do not have for each of the factors $\tau_k\big(L_1(\boldsymbol{y}_h)^{j_{h1}}\ldots L_n(\boldsymbol{y}_h)^{j_{hn}}\big)$ $(h = 1,\ldots,m)$ in (5.52) an estimate of the same type.

For h with $1 \leq h \leq m$ we consider the field of definition F_h of the parallelepiped $\Pi(F_h, Q_h, \mathbf{c})$. To simplify the notation we assume that $[F_h : K] = d$ $(1 \leq h \leq m)$. Let τ_{v_0} be the embedding of $\overline{\mathbb{Q}}$ into $\overline{K}_{v_0}$ from (1.6). Then for each h $(1 \leq h \leq m)$ there will be $D' = D/d$ embeddings $\tau_{h1},\ldots,\tau_{hD'}$, say, among $\tau_1,\ldots,\tau_D$ whose restrictions to F_h coincide with τ_{v_0} restricted to F_h. Consequently using (5.21), (5.25) and (3.24) we obtain for $h = 1,\ldots,m$ and for $\ell = 1,\ldots,D'$

$$\|\tau_{h\ell}(L_i(\boldsymbol{y}_h))\|_{v_0} \underset{\varepsilon,n}{\leq} Q_h^{c_i}, \tag{5.53}$$

whereas for $\tau_k \in \{\tau_1,\ldots,\tau_D\}$, $\tau_k \notin \{\tau_{h1},\ldots,\tau_{hD'}\}$ we can only guarantee the estimate

$$\|\tau_k(L_i(\boldsymbol{y}_h))\|_{v_0} \underset{\varepsilon,n}{\leq} Q_h^0. \tag{5.54}$$

For $h = 1,\ldots,m$ and for $k = 1,\ldots,D$ we define ε_{hk} by

$$\varepsilon_{hk} = \begin{cases} 1 & \text{if } \tau_k \in \{\tau_{h1},\ldots,\tau_{hD'}\} \\ 0 & \text{if } \tau_k \notin \{\tau_{h1},\ldots,\tau_{hD'}\}. \end{cases} \tag{5.55}$$

Then (5.53), (5.54) may be written as

$$\|\tau_k(L_i(\mathbf{y}_h))\|_{v_0} \underset{\varepsilon,n}{\leq} Q_h^{\varepsilon_{hk} c_i} \tag{5.56}$$

for $i = 1, \dots, n$, for $h = 1, \dots, m$ and for $k = 1, \dots, D$.
In the situation studied in section 4, the fields F_h would satisfy $F_h = K$ for each h ($h = 1, \dots, m$). Therefore, with our current notation, in section 4 we would get $\varepsilon_{hk} = 1$ for any pair (h, k) with $1 \leq h \leq m$ and $1 \leq k \leq D$.
Now with (5.56), (5.43) we obtain for the typical monomial in (5.52)

$$\begin{aligned}
&\left\|\tau_k\left(L_1(\mathbf{y}_1)^{j_{11}} \dots L_n(\mathbf{y}_1)^{j_{1n}} \dots L_1(\mathbf{y}_m)^{j_{m1}} \dots L_n(\mathbf{y}_m)^{j_{mn}}\right)\right\|_{v_0} \\
&\underset{\varepsilon,n}{\leq} Q_1^{j_{11}\varepsilon_{1k}c_1 + \dots + j_{1n}\varepsilon_{1k}c_n} \dots Q_m^{j_{m1}\varepsilon_{mk}c_1 + \dots + j_{mn}\varepsilon_{mk}c_n} \\
&= Q_1^{r_1(\frac{j_{11}}{r_1}\varepsilon_{1k}c_1 + \dots + \frac{j_{1n}}{r_1}\varepsilon_{1k}c_n)} \dots Q_m^{r_m(\frac{j_{m1}}{r_m}\varepsilon_{mk}c_1 + \dots + \frac{j_{mn}}{r_m}\varepsilon_{mk}c_n)} \\
&\underset{\varepsilon,n}{=} Q_1^{r_1(\frac{j_{11}}{r_1}\varepsilon_{1k} + \dots + \frac{j_{m1}}{r_m}\varepsilon_{mk})c_1 + \dots + r_1(\frac{j_{1n}}{r_1}\varepsilon_{1k} + \dots + \frac{j_{mn}}{r_m}\varepsilon_{mk})c_n}.
\end{aligned} \tag{5.57}$$

If in (5.57) we had $\varepsilon_{hk} = 1$ for each pair (h, k) (as is the case in sections 3 and 4), we could use relation (3.31), i.e.,

$$\frac{j_{1\ell}}{r_1} + \dots + \frac{j_{m\ell}}{r_m} \underset{\varepsilon,n}{=} \frac{m}{n} \qquad (\ell = 1, \dots, n).$$

(3.31) in turn is a consequence of condition (3.20) for the index.
However in (5.52) or (5.57) the situation is more complicated. We only know that in the weighted sums $\frac{j_{1\ell}}{r_1}\varepsilon_{1k} + \dots + \frac{j_{m\ell}}{r_m}\varepsilon_{mk}$ for each h ($h = 1, \dots, m$) exactly D/d among the D weights ε_{hk} ($k = 1, \dots, D$) are equal to 1 whereas the others are equal to 0. But in general, as h varies and k is fixed, the weights ε_{hk} will not have the same value.
To overcome the difficulty in (5.57) we will prove a refinement of lemmata 2.1 and 3.2. Our procedure extends the argument given by Wirsing [13].
Lemma 3.2 makes a statement on the distribution of the quotients i_{hk}/r_h corresponding to the exponent tuples $\mathfrak{I} = (i_{11}, \dots, i_{1n}; \dots; i_{m1}, \dots, i_{mn})$ of the auxiliary polynomial $P(X_{11}, \dots, X_{1n}; \dots; X_{m1}, \dots, X_{mn})$. Here

$$i_{11} + \dots + i_{1n} = r_1, \; \dots \;, i_{m1} + \dots + i_{mn} = r_m,$$

as we assumed that P is homogeneous of degree r_h in $X_{h1}, \dots, X_{hn}$. In the sequel we will view the quotients i_{hj}/r_h as continuous variables x_{hj}. Then clearly for $h = 1, \dots, m$ we have

$$x_{h1} + \dots + x_{hn} = 1, \quad x_{hi} \geq 0 \quad (i = 1, \dots, n). \tag{5.58}$$

For $0 \leq \gamma \leq 1$ we write

$$\Delta_{n-1}(\gamma) = \{(x_1,\dots,x_{n-1}) \mid x_1,\dots,x_{n-1} \geq 0,\ x_1+\dots+x_{n-1} \leq \gamma\}. \tag{5.59}$$

We observe that $\Delta_{n-1}(1)$ has volume equal to $1/(n-1)!$. More generally, $\Delta_{n-1}(\gamma)$ has volume $\gamma^{n-1}/(n-1)!$. Therefore the expectation for a point $(x_1,\dots,x_{n-1}) \in \Delta_{n-1}(1)$ to be contained in the subset $\Delta_{n-1}(\gamma)$ equals γ^{n-1}. For $m \in \mathbb{N}$ we put

$$\mathcal{A}_m = \{\boldsymbol{x} \in \mathbb{R}^{m(n-1)} \mid x_{h1},\dots,x_{h,n-1} \geq 0,\ x_{h1}+\dots+x_{h,n-1} \leq 1\ (1 \leq h \leq m)\}.$$

For $0 \leq \gamma \leq 1$ and for $\boldsymbol{x} \in \mathcal{A}_m$ we define the function

$$\begin{aligned}\xi_\gamma(\boldsymbol{x}) &= \big|\{h \mid h \in \{1,\dots,m\},\ x_{h1}+\dots+x_{h,n-1} \leq \gamma\}\big| \\ &= \sum_{h=1}^{m} \chi_{\Delta_{n-1}(\gamma)}((x_{h1},\dots,x_{h,n-1})),\end{aligned}$$

where for a set $M \subset \mathbb{R}^{n-1}$ we write χ_M for its characteristic function. We remark that the function $\xi_\gamma(\boldsymbol{x})$ has expectation $m\gamma^{n-1}$. Given $\varepsilon > 0$, we let

$$M_\varepsilon(m,\gamma) = \{\boldsymbol{x} \in \mathcal{A}_m \mid \xi_\gamma(\boldsymbol{x}) \geq m(\gamma^{n-1}-\varepsilon)\} \tag{5.60}$$

and

$$M_\varepsilon(m) = \{\boldsymbol{x} \in \mathcal{A}_m \mid \xi_\gamma(\boldsymbol{x}) \geq m(\gamma^{n-1}-\varepsilon) \text{ for all } \gamma \in [0,1]\}. \tag{5.61}$$

Thus

$$M_\varepsilon(m) = \bigcap_{\gamma\in[0,1]} M_\varepsilon(m,\gamma).$$

Now suppose $\boldsymbol{x} = (x_{11},\dots,x_{1n};\ \dots\ ; x_{m1},\dots,x_{mn})$ is a point in $\mathbb{R}^{mn}$ satisfying (5.58). For $\ell = 1,\dots,n$ we put

$$\boldsymbol{x}_{(\ell)} = (x_{11},\dots,x_{1,\ell-1},x_{1,\ell+1},\dots,x_{1n};\ \dots\ ; x_{m1},\dots,x_{m,\ell-1},x_{m,\ell+1},\dots,x_{mn}).$$

We write $\boldsymbol{x} \in\in M_\varepsilon(m)$ if $\boldsymbol{x}_{(\ell)} \in M_\varepsilon(m)$ for each ℓ ($\ell = 1,\dots,n$). If $\boldsymbol{x} \in\in M_\varepsilon(m)$ is not satisfied we write $\boldsymbol{x} \notin\notin M_\varepsilon(m)$.

Lemma 5.4. *Suppose $\varepsilon > 0$ and $m > c(d,\varepsilon)$. Let $I_1,\dots,I_D$ be subsets of $\{1,\dots,m\}$ such that*

$$\sum_{k=1}^{D} |I_k| = \frac{Dm}{d}. \tag{5.62}$$

Then we have for each ℓ with $1 \leq \ell \leq n$

$$\sum_{k=1}^{D} \inf_{\boldsymbol{x}\in\in M_\varepsilon(m)} \sum_{h\in I_k} x_{h\ell} \geq \frac{Dm}{2(n-1)d^2}(1-\varepsilon). \tag{5.63}$$

Proof. We treat the case $\ell = n$. Given $\boldsymbol{x} \in\in M_\varepsilon(m)$ there exists a permutation π of $\{1, \ldots, m\}$ such that

$$x_{\pi(1),1} + \ldots + x_{\pi(1),n-1} \leq \ldots \leq x_{\pi(m),1} + \ldots + x_{\pi(m),n-1}.$$

In view of (5.58) we may infer that

$$1 - x_{\pi(1),n} \leq \ldots \leq 1 - x_{\pi(m),n}.$$

For $t \in \{1, \ldots, m\}$ we put $\gamma_t = \big((x_{\pi(t),1} + \ldots + x_{\pi(t),n-1})^{n-1} - \varepsilon\big)^{1/(n-1)}$. The definitions of $\xi_\gamma(\boldsymbol{x})$ and $M_\varepsilon(m)$ imply

$$t \geq \xi_{\gamma_t}(\boldsymbol{x}) \geq m\big((x_{\pi(t),1} + \ldots + x_{\pi(t),n-1})^{n-1} - 2\varepsilon\big)$$

and therefore

$$t \geq m\big((1 - x_{\pi(t),n})^{n-1} - 2\varepsilon\big).$$

We may conclude that for $t = 1, \ldots, m$

$$x_{\pi(t),n} \geq 1 - \left(\frac{t}{m} + 2\varepsilon\right)^{1/(n-1)}. \tag{5.64}$$

Combining (5.62) and (5.64) we get

$$\begin{aligned}\sum_{k=1}^{D} \inf_{\boldsymbol{x} \in\in M_\varepsilon(m)} \sum_{h \in I_k} x_{hn} &\geq \sum_{k=1}^{D} \sum_{t=m-|I_k|+1}^{m} \left(1 - \left(\frac{t}{m} + 2\varepsilon\right)^{1/(n-1)}\right) \\ &= \frac{Dm}{d} - \sum_{k=1}^{D} \sum_{t=m-|I_k|+1}^{m} \left(\frac{t}{m} + 2\varepsilon\right)^{1/(n-1)}. \end{aligned} \tag{5.65}$$

However for $m \geq c(\varepsilon)$

$$\sum_{t=m-|I_k|+1}^{m} \left(\frac{t}{m} + 2\varepsilon\right)^{1/(n-1)} \leq \frac{n-1}{n}(m+3) - \frac{n-1}{n} m \left(1 - \frac{|I_k|}{m}\right)^{n/(n-1)}.$$

On the other hand Hölder's inequality and (5.62) imply

$$\sum_{k=1}^{D} \left(1 - \frac{|I_k|}{m}\right)^{n/(n-1)} \geq D \left(\frac{d-1}{d}\right)^{n/(n-1)}.$$

Thus (5.65) yields

$$\begin{aligned}&\sum_{k=1}^{D} \inf_{\boldsymbol{x} \in\in M_\varepsilon(m)} \sum_{h \in I_k} x_{hn} \\ &\geq \frac{Dm}{d} - \frac{n-1}{n} D(m+3) + \left(1 - \frac{1}{d}\right)^{n/(n-1)} \frac{n-1}{n} Dm. \end{aligned} \tag{5.66}$$

But

$$\left(1-\frac{1}{d}\right)^{n/(n-1)} = 1 - \frac{n}{n-1}\frac{1}{d} + \frac{n}{2(n-1)^2}\frac{1}{d^2} + \dots$$
$$\geq 1 - \frac{n}{n-1}\frac{1}{d} + \frac{n}{2(n-1)^2}\frac{1}{d^2}. \tag{5.67}$$

Combining (5.66) and (5.67) we obtain the assertion provided $m \geq c(d, \varepsilon)$.
□

Remark 5.5. The first part of inequalities (5.65) is true independently of hypothesis (5.62). Therefore, replacing (5.62) by

$$I_k = \{1, \dots, m\} \text{ for } k = 1, \dots, D$$

(this is exactly the case of sections 3 and 4) we get

$$\sum_{k=1}^{D} \inf_{\boldsymbol{x} \in\in M_\varepsilon(m)} \sum_{h \in I_k} x_{hn} \geq Dm - \sum_{k=1}^{D} \sum_{t=1}^{m} \left(\frac{t}{m} + 2\varepsilon\right)^{1/(n-1)}$$

and thus

$$\sum_{k=1}^{D} \inf_{\boldsymbol{x} \in\in M_\varepsilon(m)} \sum_{h \in I_k} x_{hn} \geq D\frac{m}{n}(1 - 5\varepsilon).$$

Let us go back to (5.52). Suppose for the moment that we have constructed the auxiliary polynomial P such that in (5.51) the coefficients $d(j_{11}, \dots, j_{mn})$ vanish unless we have

$$\left(\frac{j_{11}}{r_1}, \dots, \frac{j_{1n}}{r_1}; \dots; \frac{j_{m1}}{r_m}, \dots, \frac{j_{mn}}{r_m}\right) \in\in M_\varepsilon(m). \tag{5.68}$$

Our goal had been to derive an upper bound for the expression in (5.52). By (5.57) we obtain

$$\prod_{k=1}^{D} \max_{j_{11}, \dots, j_{mn}} \left\| \tau_k \left(\prod_{h=1}^{m} \left(L_1(\boldsymbol{y}_h)^{j_{h1}} \dots L_n(\boldsymbol{y}_h)^{j_{hn}}\right)\right)\right\|_{v_0}$$
$$\underset{\varepsilon,n}{\leq} \prod_{k=1}^{D} \max_{j_{11}, \dots, j_{mn}} Q_1^{r_1\left(\sum_{h=1}^{m} \frac{j_{h1}}{r_h}\varepsilon_{hk}\right)c_1 + \dots + r_1\left(\sum_{h=1}^{m} \frac{j_{hn}}{r_h}\varepsilon_{hk}\right)c_n}. \tag{5.69}$$

By (5.20), none of the exponents c_i $(i = 1, \dots, n)$ in (5.69) is positive. Therefore in order to get the desired upper bound we have to derive a lower bound for

$$\sum_{k=1}^{D} \min_{j_{11},\dots,j_{mn}} \left(\frac{j_{1\ell}}{r_1} \varepsilon_{1k} + \dots + \frac{j_{m\ell}}{r_m} \varepsilon_{mk} \right) \quad (\ell = 1, \dots, n),$$

where the minimum is taken over $(j_{11}, \dots, j_{mn})$ satisfying (5.68). Moreover we need an estimate which does not depend upon the particular distribution of the weights ε_{hk} from (5.55).

We apply lemma 5.4. For $k = 1, \dots, D$ we let $I_k = \{h \mid 1 \le h \le m,\ \varepsilon_{hk} = 1\}$. By (5.55) and since $D' = D/d$ the sets I_k then fulfill (5.62). Therefore lemma 5.4 in conjunction with (5.68) implies for $\ell = 1, \dots, n$

$$\sum_{k=1}^{D} \min_{j_{11},\dots,j_{mn}} \left(\frac{j_{1\ell}}{r_1} \varepsilon_{1k} + \dots + \frac{j_{m\ell}}{r_m} \varepsilon_{mk} \right) \underset{\varepsilon}{\geq} \frac{Dm}{2(n-1)d^2}.$$

Combination with (5.69), (5.20), (5.36) and (5.43) yields

$$\prod_{k=1}^{D} \max_{j_{11},\dots,j_{mn}} \left\| \tau_k \left(\prod_{h=1}^{m} \left(L_1(y_h)^{j_{h1}} \dots L_n(y_h)^{j_{hn}} \right) \right) \right\|_{v_0}$$

$$\underset{\varepsilon,n}{\leq} Q_1^{r_1 \frac{Dm}{2(n-1)d^2}(c_1+\dots+c_n)} \underset{\varepsilon}{=} Q_1^{-r_1 Dm} \underset{\varepsilon}{\leq} Q_1^{-r_1 D} \dots Q_m^{-r_m D}. \tag{5.70}$$

Comparing (5.49) and (5.70) we get

$$\prod_{i=0}^{1} \prod_{k=1}^{D} \max_{j_{11},\dots,j_{mn}} \left\| \tau_k \left(\prod_{h=1}^{m} \left(L_1(y_h)^{j_{h1}} \dots L_n(y_h)^{j_{hn}} \right) \right) \right\|_{v_i}$$

$$\underset{\varepsilon,n}{\leq} Q_1^{-r_1 D\eta/2} \dots Q_m^{-r_m D\eta/2}. \tag{5.71}$$

Notice that (5.71) stands in perfect analogy with (2.17) and (3.33). Thus it becomes plausible that with our assumption (5.68) we can derive an estimate of type

$$\prod_{v \in \mathfrak{M}(K)} \left\| N_{E|K}(P(y_1, \dots, y_m)) \right\|_v$$

$$\underset{\varepsilon,n}{\leq} C(L_1, \dots, L_n)^{D(r_1+\dots+r_m)} \left(Q_1^{Dr_1} \dots Q_m^{Dr_m} \right)^{-\eta/2}.$$

But then the desired relation (5.46) follows provided that the parameters $Q_1, \dots, Q_m$ are large enough.

The above considerations depend on the assumption that we can construct a nonzero approximation polynomial P satisfying (5.68). We will now justify this assumption.

For this purpose we verify that with a suitable choice of the parameters the number of potential coefficients in our approximation polynomial P will be

at least twice as large as the number of conditions imposed by hypothesis (5.68). Since the conditions again may be expressed as linear relations in the coefficients of the polynomial, an application of Siegel's Lemma then will deliver a nonzero polynomial P. The height of P will be bounded in terms of our linear forms and of the partial degrees $r_1, \ldots, r_m$. Moreover we will have (5.68). By (5.71) this will suffice to guarantee (5.46) and finally (5.42).
We return to the set $M_\varepsilon(m)$ in (5.61). For a set $M \subset \mathcal{A}_m$ we write $\complement M$ for its complement in $\mathcal{A}_m$, in other words $\complement M = \mathcal{A}_m \backslash M$. In analogy with (2.5) and (3.18) we can prove the following version of the law of large numbers:

Lemma 5.6. *Suppose* $0 < \varepsilon \leq 1$. *Then for each* $\gamma \in [0,1]$ *we have*

$$\int\limits_{\complement M_\varepsilon(m,\gamma)} \mathrm{d}\boldsymbol{x} \leq e^{-\frac{1}{4}\varepsilon^2 m} \int\limits_{(\Delta_{n-1}(1))^m} \mathrm{d}\boldsymbol{x} = e^{-\frac{1}{4}\varepsilon^2 m} \int\limits_{\mathcal{A}_m} \mathrm{d}\boldsymbol{x}.$$

Lemma 5.6 may be interpreted as a continuous version of lemma 4C, chapter V in W. M. Schmidt [10].
Combining lemma 5.6 and the definition of $M_\varepsilon(m)$ in (5.61) we get

Lemma 5.7. *Let* $m, n \in \mathbb{N}$ *with* $n \geq 2$. *Suppose* $0 < \varepsilon \leq 1$. *Then we have*

$$\int\limits_{\complement M_\varepsilon(m)} \mathrm{d}\boldsymbol{x} < \frac{2^n}{\varepsilon} e^{-2^{-(2n+1)}\varepsilon^2 m} \int\limits_{\mathcal{A}_m} \mathrm{d}\boldsymbol{x}.$$

To derive lemma 5.7 from lemma 5.6 we proceed as follows. We cover the interval $\{\gamma \mid 0 \leq \gamma \leq 1\}$ by $\ell \leq 2^n/\varepsilon$ subintervals of type $J_k = [\gamma_k - \varepsilon^*, \gamma_k]$ $(k = 1, \ldots, \ell)$ where $\varepsilon^* = \varepsilon/(2^{n-1}+1)$. It follows that

$$\complement M_\varepsilon(m) \subseteq \bigcup_{k=1}^{\ell} \complement M_{\varepsilon^*}(m, \gamma_k).$$

We then apply lemma 5.6 to the sets $M_{\varepsilon^*}(m, \gamma_k)$. Adding up we get the assertion.
The following corollary gives the discrete version of lemma 5.7.

Corollary 5.8. *Let* $0 < \varepsilon \leq 1$. *Suppose* $m, n \in \mathbb{N}$, $n \geq 2$. *Let* $r_1, \ldots, r_m$ *be natural numbers satisfying* $\min\{r_1, \ldots, r_m\} \geq c(n, m, \varepsilon)$. *Let* $\mathcal{M}$ *be the set of tuples of nonnegative integers* $\mathfrak{I} = (i_{11}, \ldots, i_{1n}; \ldots; i_{m1}, \ldots, i_{mn})$ *with*

$$i_{h1} + \ldots + i_{hn} = r_h \quad (h = 1, \ldots, m) \tag{5.72}$$

and with

$$\left(\frac{i_{11}}{r_1}, \ldots, \frac{i_{1n}}{r_1}; \ldots; \frac{i_{m1}}{r_m}, \ldots, \frac{i_{mn}}{r_m}\right) \notin M_\varepsilon(m). \tag{5.73}$$

Then $\mathcal{M}$ *has cardinality*

$$|\mathcal{M}| \leq \frac{2^{2n}}{\varepsilon} e^{-2^{-(2n+1)}\varepsilon^2 m} \binom{r_1+n-1}{n-1} \cdots \binom{r_m+n-1}{n-1}. \tag{5.74}$$

Our remark 5.5 combined with the assertions of lemma 5.4 and corollary 5.8 imply that altogether we have a generalization of lemmata 2.1 and 3.2.
In view of corollary 5.8 it is clear that for large m the number of conditions imposed by our assumption (5.68) will be much smaller than the total number

$$\binom{r_1+n-1}{n-1}\cdots\binom{r_m+n-1}{n-1}$$

of hypothetical coefficients of our polynomial to be constructed. Consequently an application of Siegel's Lemma will suffice to guarantee condition (5.68) provided the parameters $m, r_1, \ldots, r_m$ are suitably chosen. The remainder of the proof of theorem 5.3 goes along the same lines as explained in section 3.2.

In contrast with the preceding sections 3 and 4 we do not include here a subsection "geometry part". The reason is as follows: With $c(n,d) = 2(n-1)d^2$ as in (5.36) the right hand side in (5.31) will be a positive power of Q. If we use (5.31) in (3.48) of section 3.3 we get an analogue of (3.49), however with a positive power of Q as upper bound. Therefore the "geometry part" as applied in sections 3 and 4 fails in our current context. In fact we do not see how to reduce (5.33) to a hypothesis of type (5.34).

6 Gap principles

In this section we want to describe methods to count the number of solutions or more generally the number of subspaces showing up in our theorems.

6.1 Vanishing determinants

We begin by studying the case of Roth's Theorem 1.1. Suppose we have two rational approximations $x_1/y_1 \neq x_2/y_2$ in their lowest terms and with $0 < y_1 \leq y_2$. So both are solutions of

$$\left|\alpha - \frac{x}{y}\right| < y^{-2-\delta}. \tag{6.1}$$

We obtain

$$\frac{1}{y_1y_2} \leq \left|\frac{x_1}{y_1} - \frac{x_2}{y_2}\right| \leq \left|\frac{x_1}{y_1} - \alpha\right| + \left|\alpha - \frac{x_2}{y_2}\right| < y_1^{-2-\delta} + y_2^{-2-\delta} \leq 2y_1^{-2-\delta}. \tag{6.2}$$

Therefore, assuming $y_1^{\delta/2} \geq 2$, we may infer that

$$y_1^{1+\delta/2} < y_2. \tag{6.3}$$

Now suppose that we have a sequence $x_1/y_1, x_2/y_2, \dots$ of distinct solutions of (6.1).
Assume moreover that for given values Q and E with

$$Q > 2^{2/\delta}, \quad E > 1 \tag{6.4}$$

the denominators y_i all satisfy

$$Q < y_i \le Q^E. \tag{6.5}$$

The interval $(Q, Q^E]$ in (6.5) may be covered by

$$\le 1 + \frac{\log E}{\log(1+\delta/2)} < 1 + 4\delta^{-1}\log E$$

intervals of type $(Q, Q^{1+\delta/2}]$. We conclude with (6.3) that our sequence does not contain more than $1 + 4\delta^{-1}\log E$ elements x_i/y_i satisfying (6.5).
To generalize the argument to n dimensions it is convenient to consider instead of (6.2) the homogeneous analogue

$$|x_1 y_2 - x_2 y_1| < y_1^{-1-\delta} y_2 + y_2^{-1-\delta} y_1$$

which then gives for $y_1^{\delta/2} \ge 2$ and $y_1 < y_2 \le y_1^{1+\delta/2}$

$$\left| \det \begin{pmatrix} x_1 & y_1 \\ x_2 & y_2 \end{pmatrix} \right| < 1,$$

so that

$$\det \begin{pmatrix} x_1 & y_1 \\ x_2 & y_2 \end{pmatrix} = 0. \tag{6.6}$$

In other words, if $x_1/y_1, x_2/y_2$ are solutions of (6.1) satisfying

$$Q < y_i \le Q^{1+\delta/2} \quad (i = 1, 2)$$

for some $Q \ge 2^{2/\delta}$ then the vectors (x_1, y_1) and (x_2, y_2) are linearly dependent.
This simple argument may be generalized to n dimensions as follows:

Lemma 6.1. *Let $\Pi(Q, \mathbf{c})$ be the parallelepiped in $\overline{\mathbb{Q}}^n$ defined in* (4.13). *Suppose $0 < \delta < 1$. Let Q_0 be a quantity satisfying*

$$Q_0 \ge n^{2/\delta}. \tag{6.7}$$

Suppose that $E > 1$.
Then, as Q runs through values satisfying

$$Q_0 < Q \le Q_0^E, \tag{6.8}$$

the union of the sets $Q^{-\delta} * \Pi(Q, \mathbf{c})$ may be covered by a collection of not more than

$$1 + 4\delta^{-1}\log E \tag{6.9}$$

proper linear subspaces of $\overline{\mathbb{Q}}^n$.

Remark 6.2. Notice that our hypothesis (6.8) is exactly of the same shape as assertion (3.35) in the Theorem on the Penultimate Minimum 3.4. Therefore, once in that theorem we have control over the parameters m and E we can determine an explicit upper bound for the number of subspaces needed.

Proof of lemma 6.1. Similarly to the argument used above it will suffice to prove that to cover points $\boldsymbol{x}$ corresponding to parameters Q in an interval of type

$$Q_0 < Q \leq Q_0^{1+\delta/2} \tag{6.10}$$

a single subspace will suffice.
Let $\boldsymbol{x}_1, \ldots, \boldsymbol{x}_n$ be points in $\overline{\mathbb{Q}}^n$ with

$$\boldsymbol{x}_i \in Q_i^{-\delta} * \Pi(Q_i, \boldsymbol{c}) \quad (i = 1, \ldots, n), \tag{6.11}$$

where the parameters Q_i satisfy (6.10). We want to prove that

$$\det(\boldsymbol{x}_1, \ldots, \boldsymbol{x}_n) = 0.$$

Let G be a finite extension of K such that

$$\boldsymbol{x}_i \in G^n \quad (i = 1, \ldots, n). \tag{6.12}$$

It suffices to show that

$$\prod_{w \in \mathfrak{M}(G)} \|\det(\boldsymbol{x}_1, \ldots, \boldsymbol{x}_n)\|_w < 1. \tag{6.13}$$

By (6.11), (6.12) and (4.14) we have for $1 \leq i, j \leq n$

$$\begin{aligned} \|L_j(\boldsymbol{x}_i)\|_w &\leq Q_i^{(c_j - \delta) d(w|v_0)} && (w \in \mathfrak{M}(G),\ w \mid v_0) \\ \|x_{ij}\|_w &\leq 1 && (w \in \mathfrak{M}(G),\ w \nmid v_0). \end{aligned} \tag{6.14}$$

We estimate $\|\det(\boldsymbol{x}_1, \ldots, \boldsymbol{x}_n)\|_w$. Recall that we had assumed

$$\det(L_1, \ldots, L_n) = 1.$$

Therefore, for any $w \in \mathfrak{M}(G)$

$$\|\det(\boldsymbol{x}_1, \ldots, \boldsymbol{x}_n)\|_w = \left\| \det \begin{pmatrix} L_1(\boldsymbol{x}_1) & \ldots & L_n(\boldsymbol{x}_1) \\ \vdots & & \vdots \\ L_1(\boldsymbol{x}_n) & \ldots & L_n(\boldsymbol{x}_n) \end{pmatrix} \right\|_w . \tag{6.15}$$

Now

$$\|\det(L_i(\boldsymbol{x}_j))\|_w \leq (n!)^{s(w)} \max_{i_1, \ldots, i_n} \|L_{i_1}(\boldsymbol{x}_1) \ldots L_{i_n}(\boldsymbol{x}_n)\|_w$$

where the maximum is taken over all permutations $i_1, \ldots, i_n$ of $1, \ldots, n$. Here $s(w)$ is defined in (4.30).
For $w \mid v_0$ we obtain in view of (6.14)

$$\| \det(L_i(\boldsymbol{x}_j)) \|_w \leq (n!)^{s(w)} \max_{i_1, \ldots, i_n} Q_1^{(c_{i_1} - \delta) d(w|v_0)} \cdots Q_n^{(c_{i_n} - \delta) d(w|v_0)}. \tag{6.16}$$

For $i = 1, \ldots, n$ we define

$$b_i = \begin{cases} c_i & \text{if } c_i < 0 \\ (1 + \delta/2) c_i & \text{if } c_i \geq 0. \end{cases} \tag{6.17}$$

Combination of (6.16), (6.17) and (6.10) yields for w with $w \mid v_0$

$$\begin{aligned} \| \det(L_i(\boldsymbol{x}_j)) \|_w &\leq (n!)^{s(w)} Q_0^{(b_1 + \ldots + b_n - n\delta) d(w|v_0)} \\ &\leq (n!)^{s(w)} Q_0^{(c_1 + \ldots + c_n + (|c_1| + \ldots + |c_n|)\delta/2 - n\delta) d(w|v_0)}. \end{aligned}$$

By (4.9) and (4.11), $\sum_{i=1}^n c_i = 0$ and $\sum_{i=1}^n |c_i| \leq n$. Therefore we get for w with $w \mid v_0$

$$\| \det(L_i(\boldsymbol{x}_j)) \|_w \leq (n!)^{s(w)} Q^{-d(w|v_0) n\delta/2}. \tag{6.18}$$

For w with $w \nmid v_0$ inequalities (6.14) imply

$$\| \det(\boldsymbol{x}_1, \ldots, \boldsymbol{x}_n) \|_w \leq (n!)^{s(w)}. \tag{6.19}$$

Combination of (6.18) and (6.19) yields in conjunction with (6.7) and (6.15)

$$\prod_{w \in \mathfrak{M}(G)} \| \det(\boldsymbol{x}_1, \ldots, \boldsymbol{x}_n) \|_w \leq n! \, Q_0^{-n\delta/2} < \left(n Q_0^{-\delta/2}\right)^n \leq 1,$$

and so

$$\det(\boldsymbol{x}_1, \ldots, \boldsymbol{x}_n) = 0.$$

Thus indeed $\boldsymbol{x}_1, \ldots, \boldsymbol{x}_n$ are linearly dependent. The lemma follows. □

We now study the parallelepipeds of section 5. Again, to simplify the exposition we restrict ourselves to the case when we start with a single absolute value $\| \; \|_{v_0}$ on K.
We pick a place $v_1 \in \mathfrak{M}(K)$, $v_1 \neq v_0$ and we let $(c_1, \ldots, c_n)$ be an n-tuple of real numbers. For an extension F of K of degree $\leq d$ the place $w_0 \in \mathfrak{M}(F)$ is determined by (5.9). Given $Q > 1$ the set $Q^{-\delta} * \Pi(F, Q, \boldsymbol{c})$ will be the algebraic closure of the parallelepiped $Q^{-\delta} * \Pi_F(F, Q, \boldsymbol{c})$ defined by (5.21), (5.25), i.e., defined by

$$\begin{aligned} &\|L_i(\boldsymbol{x})\|_{w_0} \le Q^{d(w_0|v_0)c_i} && \\ &\|L_i(\boldsymbol{x})\|_{w} \le 1 && (w \in \mathfrak{M}(F),\ w \mid v_0,\ w \ne w_0) \\ &\|x_i\|_{w} \le Q^{d(w|v_1)(1-\delta)} && (w \in \mathfrak{M}(F),\ w \mid v_1) \\ &\|x_i\|_{w} \le 1 && (w \in \mathfrak{M}(F),\ w \nmid v_0,\ w \nmid v_1) \end{aligned} \tag{6.20}$$

for $i = 1, \dots, n$.
Again we assume that (5.6) is satisfied. We prove:

Lemma 6.3. *Let* $0 < \delta < 1$. *Suppose that in* (6.20) *we have*

$$c_1 + \dots + c_n = -nd^n \quad \textit{and} \quad c_i \le 0 \quad (i = 1, \dots, n). \tag{6.21}$$

Then as F *runs through the set of extensions of* K *of degree* $\le d$ *and as* Q *runs through the set of values*

$$Q_0 < Q \le Q_0^{1+\delta/2}, \tag{6.22}$$

where

$$Q_0 \ge n^{2/\delta}, \tag{6.23}$$

the union of the sets $Q^{-\delta} * \Pi(F, Q, \mathbf{c})$ *is contained in a single proper linear subspace of* $\overline{\mathbb{Q}}^n$.

Proof. We prove that any n points $\boldsymbol{x}_1, \dots, \boldsymbol{x}_n$ in the above union are linearly dependent. Given $\boldsymbol{x}_1, \dots, \boldsymbol{x}_n$, let $F_1, \dots, F_n$ and $Q_1, \dots, Q_n$ be the corresponding fields and parameters respectively, i.e., $\boldsymbol{x}_i \in Q_i^{-\delta} * \Pi(F_i, Q_i, \mathbf{c})$ $(i = 1, \dots, n)$. Let E be the compositum of $F_1, \dots, F_n$.
Write u_0 for the place on E such that the normalized absolute value $\|\ \|_{u_0}$ on E satisfies

$$\|\ \|_{u_0} = \|\ \|_{v_0}^{d(u_0|v_0)}.$$

Since the F_i have degree $\le d$ over K, the degree of E over K is $\le d^n$. Consequently we obtain

$$d(u_0|v_0) \ge 1/d^n. \tag{6.24}$$

Combining the second part of (6.21) with (6.24) we infer from (6.20) that any point $\boldsymbol{x} \in Q_i^{-\delta} * \Pi(F_i, Q_i, \mathbf{c})$ $(i = 1, \dots, n)$ lies in the algebraic closure of the parallelepiped defined over E and given by

$$\begin{aligned} &\|L_i(\boldsymbol{x})\|_{u_0} \le Q_i^{d^{-n}c_i} && \\ &\|L_i(\boldsymbol{x})\|_{u} \le 1 && (u \in \mathfrak{M}(E),\ u \mid v_0,\ u \ne u_0) \\ &\|x_i\|_{u} \le Q_i^{d(u|v_1)(1-\delta)} && (u \in \mathfrak{M}(E),\ u \mid v_1) \\ &\|x_i\|_{u} \le 1 && (u \in \mathfrak{M}(E),\ u \nmid v_0,\ u \nmid v_1) \end{aligned} \tag{6.25}$$

for $i = 1, \ldots, n$. Let G be a finite extension of E such that $\boldsymbol{x}_1, \ldots, \boldsymbol{x}_n \in G^n$. Using again the second part of (6.21) we get in conjunction with (6.22) and (6.25)

$$\begin{array}{lll} \|L_i(\boldsymbol{x}_j)\|_w \leq Q_0^{d(w|u_0)d^{-n}c_i} & & (w \in \mathfrak{M}(G),\ w \mid u_0) \\ \|L_i(\boldsymbol{x}_j)\|_w \leq 1 & & (w \in \mathfrak{M}(G),\ w \mid v_0,\ w \nmid u_0) \\ \|x_{ji}\|_w \leq Q_0^{d(w|v_1)(1-\delta)(1+\delta/2)} & & (w \in \mathfrak{M}(G),\ w \mid v_1) \\ \|x_{ji}\|_w \leq 1 & & (w \in \mathfrak{M}(G),\ w \nmid v_0,\ w \nmid v_1) \end{array}$$

for $1 \leq i, j \leq n$. This implies (similarly as in lemma 6.1)

$$\begin{array}{ll} \|\det(\boldsymbol{x}_1, \ldots, \boldsymbol{x}_n)\|_w \leq (n!)^{s(w)} Q_0^{d(w|u_0)d^{-n}(c_1+\ldots+c_n)} & (w \in \mathfrak{M}(G),\ w \mid u_0) \\ \|\det(\boldsymbol{x}_1, \ldots, \boldsymbol{x}_n)\|_w \leq (n!)^{s(w)} Q_0^{d(w|v_1)(1-\delta/2)n} & (w \in \mathfrak{M}(G),\ w \mid v_1) \\ \|\det(\boldsymbol{x}_1, \ldots, \boldsymbol{x}_n)\|_w \leq (n!)^{s(w)} & (w \in \mathfrak{M}(G),\ w \nmid u_0,\ w \nmid v_1). \end{array}$$

Altogether this gives

$$\prod_{w \in \mathfrak{M}(G)} \|\det(\boldsymbol{x}_1, \ldots, \boldsymbol{x}_n)\|_w \leq n!\, Q_0^{d^{-n}(c_1+\ldots+c_n)+n(1-\delta/2)}.$$

So by (6.21), (6.23)

$$\begin{aligned} \prod_{w \in \mathfrak{M}(G)} \|\det(\boldsymbol{x}_1, \ldots, \boldsymbol{x}_n)\|_w &< n^n Q_0^{-d^{-n} n d^n + n - n\delta/2} \\ &= n^n Q_0^{-n\delta/2} = \left(n Q_0^{-\delta/2}\right)^n \leq 1. \end{aligned}$$

We may conclude that $\boldsymbol{x}_1, \ldots, \boldsymbol{x}_n$ are linearly dependent. □

An analysis of the proof of lemma 6.3 shows that the method does not allow it to weaken the hypothesis

$$c_1 + \ldots + c_n = -nd^n$$

made in (6.21).

6.2 Application of Minkowski's Theorem

Let us go back to lemma 6.1. It can be proved with a completely different method based on Minkowski's geometry of numbers. Since this approach is very useful to obtain quantitative results which are suitable for section 5, we will briefly sketch the argument it gives for lemma 6.1.
Let us consider inequalities

$$|L_1(\boldsymbol{x})| \leq Q^{c_1-\delta}, \ \ldots, \ |L_n(\boldsymbol{x})| \leq Q^{c_n-\delta}. \tag{6.26}$$

Here we assume that $L_1, \ldots, L_n$ are real linear forms with $\det(L_1, \ldots, L_n) = 1$. Moreover we let $c_1, \ldots, c_n$ be real numbers having

$$c_1 + \ldots + c_n = 0 \quad \text{and} \quad |c_i| \leq 1 \quad (i = 1, \ldots, n). \tag{6.27}$$

We are interested in the set of solutions $\boldsymbol{x} \in \mathbb{Z}^n$ of (6.26) where Q runs through an interval

$$Q_0 < Q \leq Q_0^{1+\delta/2} \tag{6.28}$$

with

$$Q_0 \geq n^{2/\delta}. \tag{6.29}$$

For subscripts i with $c_i - \delta \leq 0$ we get with (6.28)

$$|L_i(\boldsymbol{x})| \leq Q_0^{c_i - \delta}.$$

On the other hand for subscripts i with $c_i - \delta > 0$ hypotheses (6.27) and (6.28) yield

$$|L_i(\boldsymbol{x})| \leq Q_0^{(c_i-\delta)(1+\delta/2)} \leq Q_0^{c_i-\delta/2}.$$

Combining these two inequalities, we see that for any pair $(\boldsymbol{x}, Q)$ under consideration the point $\boldsymbol{x}$ satisfies the inequalities

$$|L_1(\boldsymbol{x})| \leq Q_0^{c_1-\delta/2}, \ \ldots, \ |L_n(\boldsymbol{x})| \leq Q_0^{c_n-\delta/2}. \tag{6.30}$$

Let $\Pi(Q_0, \boldsymbol{c})$ be the parallelepiped in $\mathbb{R}^n$ given by the simultaneous inequalities

$$|L_1(\boldsymbol{x})| \leq Q_0^{c_1}, \ \ldots, \ |L_n(\boldsymbol{x})| \leq Q_0^{c_n}.$$

With our hypotheses made on the determinant and with (6.27) we may infer that the successive minima $\lambda_1, \ldots, \lambda_n$ of $\Pi(Q_0, \boldsymbol{c})$ have

$$\frac{1}{n!} \leq \lambda_1 \ldots \lambda_n \leq 1 \tag{6.31}$$

(cf. (3.9)). Therefore $\lambda_n > 1/n$. However by (6.29)

$$Q_0^{-\delta/2} \leq 1/n.$$

Thus by (6.30) any point $\boldsymbol{x}$ under consideration satisfies

$$|L_1(\boldsymbol{x})| < Q_0^{c_1}\lambda_n, \ \ldots, \ |L_n(\boldsymbol{x})| < Q_0^{c_n}\lambda_n. \tag{6.32}$$

Since in (6.32) we have strict inequalities, the definition of successive minima implies that our set of points $\boldsymbol{x}$ spans a subspace of dimension $< n$. Consequently the set of points $\boldsymbol{x} \in \mathbb{Z}^n \cap Q^{-\delta} * \Pi(Q, \boldsymbol{c})$ where Q runs through the values with (6.28), (6.29) is contained in a single proper linear subspace of $\mathbb{Q}^n$.

We now apply this approach to the parallelepipeds studied in lemma 6.3. Let $Q^{-\delta} * \Pi(F, Q, \boldsymbol{c})$ be the parallelepiped from (6.20). We prove:

Lemma 6.4. *Let $0 < \delta < 1$. Suppose that instead of (6.21) we have*

$$c_1 + \ldots + c_n = -nd^2 \quad \text{and} \quad c_i \le 0 \quad \text{for} \quad i = 1, \ldots, n. \tag{6.33}$$

Consider the set M of pairs (F, Q) satisfying the following three properties:

(i) *F is an extension of K with*

$$[F : K] \le d. \tag{6.34}$$

(ii) *Q is a real parameter with*

$$Q_0 < Q \le Q_0^{1+\delta/2}, \tag{6.35}$$

where

$$Q_0 \ge n^{1/\delta}. \tag{6.36}$$

(iii) *The parallelepiped $Q^{-\delta} * \Pi(F, Q, \mathbf{c})$ contains $n-1$ linearly independent points.*

Then the union $\bigcup_{(F,Q)\in M} Q^{-\delta} * \Pi(F, Q, \mathbf{c})$ is contained in a single proper linear subspace of $\overline{\mathbb{Q}}^n$.

The significant feature in lemma 6.4 is hypothesis (6.33), i.e.,

$$c_1 + \ldots + c_n = -nd^2.$$

This should be compared with the much stronger assumption

$$c_1 + \ldots + c_n = -nd^n$$

in lemma 6.3. But notice that on the other hand in lemma 6.3 we are not restricted by any hypothesis like (iii) of lemma 6.4.
The gap principle of lemma 6.4 allows us to give an explicit upper bound for the number of subspaces needed in theorem 5.3.

Proof of lemma 6.4. For a pair $(F, Q) \in M$ we let $\lambda_1, \ldots, \lambda_n$ be the successive minima of $\Pi(F, Q, \mathbf{c})$. Since $d(w_0|v_0) \ge 1/d$ the volume V of $\Pi(F, Q, \mathbf{c})$ by (6.33) equals

$$V = Q^{n-d(w_0|v_0)nd^2} \le Q^{n-nd}.$$

In view of (4.20) we may infer that

$$\lambda_1 \ldots \lambda_n \ge n^{-n/2} Q^{nd-n} \ge n^{-n/2}.$$

In particular we get

$$\lambda_n \ge n^{-1/2}.$$

On the other hand hypothesis (iii) implies that for any pair $(F, Q) \in M$ we have

$$\lambda_{n-1} \leq Q^{-\delta}.$$

Using (6.36) we may infer that $Q^{-\delta} * \Pi(F, Q, \mathbf{c})$ generates a subspace of $\overline{\mathbb{Q}}^n$ of codimension 1. We denote this subspace by $U(F, Q)$. We prove that for any two pairs (F_1, Q_1) and (F_2, Q_2) in M we have $U(F_1, Q_1) = U(F_2, Q_2)$.
Let E be the compositum $F_1 F_2$. Let u_0 be the place on E satisfying the analogue of (5.9). By (6.34) we obtain

$$[F_1 F_2 : K] \leq d^2, \tag{6.37}$$

and therefore $d(u_0|v_0) \geq 1/d^2$. In view of hypothesis (iii), we can find for $k = 1, 2$ linearly independent vectors $\boldsymbol{g}_1^{(k)}, \ldots, \boldsymbol{g}_{n-1}^{(k)}$ and a finite extension G of E with $\boldsymbol{g}_1^{(k)}, \ldots, \boldsymbol{g}_{n-1}^{(k)} \in G^n$ such that for $i = 1, \ldots, n$, for $j = 1, \ldots, n-1$ and for $k = 1, 2$ we have

$$\begin{aligned}
&\left\|L_i(\boldsymbol{g}_j^{(k)})\right\|_w \leq Q_k^{d(w|u_0)d^{-2}c_i} && (w \in \mathfrak{M}(G),\ w \mid u_0)\\
&\left\|L_i(\boldsymbol{g}_j^{(k)})\right\|_w \leq 1 && (w \in \mathfrak{M}(G),\ w \mid v_0,\ w \nmid u_0)\\
&\left\|g_{ji}^{(k)}\right\|_w \leq Q_k^{d(w|v_1)(1-\delta)} && (w \in \mathfrak{M}(G),\ w \mid v_1)\\
&\left\|g_{ji}^{(k)}\right\|_w \leq 1 && (w \in \mathfrak{M}(G),\ w \nmid v_1,\ w \nmid v_0).
\end{aligned} \tag{6.38}$$

Notice that for the second inequality in (6.38) our hypothesis $c_i \leq 0$ is crucial. We now assume that $Q_1 \leq Q_2$. Then by (6.35) and (6.38) we have for $i = 1, \ldots, n$, for $j = 1, \ldots, n-1$ and for $k = 1, 2$

$$\begin{aligned}
&\left\|L_i(\boldsymbol{g}_j^{(k)})\right\|_w \leq Q_1^{d(w|u_0)d^{-2}c_i} && (w \in \mathfrak{M}(G),\ w \mid u_0)\\
&\left\|L_i(\boldsymbol{g}_j^{(k)})\right\|_w \leq 1 && (w \in \mathfrak{M}(G),\ w \mid v_0,\ w \nmid u_0)\\
&\left\|g_{ji}^{(k)}\right\|_w \leq Q_1^{d(w|v_1)(1-\delta/2)} && (w \in \mathfrak{M}(G),\ w \mid v_1)\\
&\left\|g_{ji}^{(k)}\right\|_w \leq 1 && (w \in \mathfrak{M}(G),\ w \nmid v_1,\ w \nmid v_0).
\end{aligned} \tag{6.39}$$

Let $\Pi(E, Q_1, \mathbf{c})$ be the algebraic closure of the parallelepiped defined over E by

$$\begin{aligned}
&\|L_i(\boldsymbol{x})\|_{u_0} \leq Q_1^{d^{-2}c_i} && \\
&\|L_i(\boldsymbol{x})\|_u \leq 1 && (u \in \mathfrak{M}(E),\ u \mid v_0,\ u \neq u_0)\\
&\|x_i\|_u \leq Q_1^{d(u|v_1)} && (u \in \mathfrak{M}(E),\ u \mid v_1)\\
&\|x_i\|_u \leq 1 && (u \in \mathfrak{M}(E),\ u \nmid v_1,\ u \nmid v_0)
\end{aligned} \tag{6.40}$$

for $i = 1, \ldots, n$. Write $\mu_1, \ldots, \mu_n$ for the successive minima of $\Pi(E, Q_1, \mathbf{c})$. By (6.39) we have

$$\mu_{n-1} \le Q_1^{-\delta/2}.$$

On the other hand, similarly as for λ_n above we can show that

$$\mu_n \ge n^{-1/2}.$$

But by (6.36)

$$Q_1^{-\delta/2} < n^{-1/2}.$$

Therefore $Q_1^{-\delta/2} * \Pi(E, Q_1, \mathbf{c})$ generates a subspace U of $\overline{\mathbb{Q}}^n$ of dimension $n-1$. However (6.39) implies

$$\boldsymbol{g}_1^{(1)}, \dots, \boldsymbol{g}_{n-1}^{(1)}, \boldsymbol{g}_1^{(2)}, \dots, \boldsymbol{g}_{n-1}^{(2)} \in U.$$

We may conclude that

$$U(F_1, Q_1) = U(F_2, Q_2) = U$$

as desired. □

References

1. E. Bombieri and J. Vaaler, *On Siegel's Lemma*, Invent. Math. 73 (1983), 11–32.
2. J.-H. Evertse, *An improvement of the quantitative Subspace Theorem*, Compos. Math. 101 (1996), 225–311.
3. J.-H. Evertse, *An explicit version of Faltings' Product Theorem and an improvement of Roth's Lemma*, Acta Arith. 73 (1995), 215–248.
4. J.-H. Evertse and H. P. Schlickewei, *A quantitative version of the Absolute Subspace Theorem*, J. reine angew. Math. 548 (2002), 21–127.
5. H. Locher and H. P. Schlickewei, *The Theorem on the Penultimate Minimum with algebraic numbers of bounded degree*, unpublished manuscript.
6. K.F. Roth, *Rational approximations to algebraic numbers*, Mathematika 2 (1955), 1–20.
7. D. Roy and J. L. Thunder, *An absolute Siegel's Lemma*, J. reine angew. Math. 476 (1996), 1–26.
8. W. M. Schmidt, *Simultaneous approximation to algebraic numbers by rationals*, Acta Math. 125 (1970), 189–201.
9. W. M. Schmidt, *Norm form equations*, Ann. Math. 96 (1972), 526–551.
10. W. M. Schmidt, *Diophantine Approximation*, Lect. Notes Math. 785, Springer-Verlag, Berlin-Heidelberg-New York, 1980.
11. W. M. Schmidt, *The Subspace Theorem in Diophantine approximations*, Compos. Math. 69 (1989), 121–173.
12. E. Wirsing, *Approximation mit algebraischen Zahlen beschränkten Grades*, J. reine angew. Math. 206 (1961), 67–77.
13. E. Wirsing, *On approximation of algebraic numbers by algebraic numbers of bounded degree*, Proc. Symp. Pure Math. 20 (1969 Number Theory Institute), D. J. Lewis ed., Amer. Math. Soc., Providence, R. I., 1971, 213–247.

Linear Recurrence Sequences

Wolfgang M. Schmidt

Department of Mathematics, University of Colorado

1 Introduction

The best known linear recurrence sequence is the Fibonacci Sequence 1, 1, 2, 3, 5, 8, ..., where each term (after the first two terms) is a sum of the two preceding terms. We may extend it to become a "doubly infinite" sequence

$$\ldots, -8, 5, -3, 2, -1, 1, 0, 1, 1, 2, 3, 5, 8, \ldots .$$

More generally, a *linear recurrence sequence of order* $\leqq t$ is a sequence $\ldots, u_{-3}, u_{-2}, u_{-1}, u_0, u_1, u_2, u_3, \ldots$ of complex numbers where each term is a linear combination of the t preceding terms with fixed coefficients $c_0, \ldots, c_{t-1}$:

$$u_{k+t} = c_{t-1}u_{k+t-1} + \cdots + c_0 u_k \qquad (k \in \mathbb{Z}). \tag{1.1}$$

Let V be the space of doubly infinite sequences, i.e., of functions $u : \mathbb{Z} \to \mathbb{C}$. It is a complex vector space under component-wise addition. We denote the value of $u \in V$ at $k \in \mathbb{Z}$ by u_k. Sequences $u \in V$ will sometimes be defined by saying "let $u_k = f(k)$", or similar. Let R be the ring $\mathbb{C}[Z, 1/Z]$ where Z is a variable. We let R act on V as follows. When

$$\mathcal{P} = \sum_m c_m Z^m \in R$$

(with finitely many nonzero coefficients c_m) and $u \in V$, we let $\mathcal{P}u$ be the doubly infinite sequence with

$$(\mathcal{P}u)_k = \sum_m c_m u_{k+m} \qquad (k \in \mathbb{Z}).$$

The relation (1.1) may be written as $\mathcal{P}u = 0$ (i.e., the zero element of V) with $\mathcal{P}(Z) = Z^t - c_{t-1}Z^{t-1} - \cdots - c_0$. It is now easily seen that u is a linear recurrence sequence precisely if there is a nonzero $\mathcal{P} \in R$ with $\mathcal{P}u = 0$.

More generally, let V_n be the vector space of functions $\mathbb{Z}^n \to \mathbb{C}$. Its elements will sometimes be called *multisequences*. When $u \in V_n$, its value at $\mathbf{k} \in \mathbb{Z}^n$ will be denoted $u_{\mathbf{k}}$. Further let R_n be the ring $\mathbb{C}[Z_1, \ldots, Z_n, Z_1^{-1}, \ldots, Z_n^{-1}]$ in variables $Z_1, \ldots, Z_n$. Elements of R_n will be referred to as *generalized polynomials*, and particular elements

$$\mathbf{Z}^{\mathbf{m}} = Z_1^{m_1} \ldots Z_n^{m_n}$$

as *monomials*. Note that R_n is a vector space over $\mathbb{C}$, and an ideal $\mathcal{I} \subset R_n$ is a subspace, so that also $R_n/\mathcal{I}$ is a vector space. Therefore for ideals $\mathcal{I}_1, \mathcal{I}_2$,

$$\dim(R_n/\mathcal{I}_1) + \dim(R_n/\mathcal{I}_2) \\ = \dim(R_n/(\mathcal{I}_1 \cap \mathcal{I}_2)) + \dim(R_n/(\mathcal{I}_1 + \mathcal{I}_2)). \quad (1.2)$$

Again R_n acts on V_n: when

$$\mathcal{P}(\mathbf{Z}) = \sum_{\mathbf{m}} c_{\mathbf{m}} \mathbf{Z}^{\mathbf{m}} \in R_n$$

and $u \in V_n$, we let $\mathcal{P}u$ be the multisequence with

$$(\mathcal{P}u)_{\mathbf{k}} = \sum_{\mathbf{m}} c_{\mathbf{m}} u_{\mathbf{k}+\mathbf{m}}.$$

In particular, $(\mathbf{Z}^{\mathbf{m}} u)_{\mathbf{k}} = u_{\mathbf{k}+\mathbf{m}}$. Clearly $\mathcal{P}u$ is bilinear in the spaces R_n and V_n. When $\mathcal{P}_1, \mathcal{P}_2$ are in R_n,

$$(\mathcal{P}_1\mathcal{P}_2)u = \mathcal{P}_1(\mathcal{P}_2 u) = \mathcal{P}_2(\mathcal{P}_1 u);$$

by linearity it suffices to check this when $\mathcal{P}_1, \mathcal{P}_2$ are monomials. Thus V_n is an R_n-module.

Given $u \in V_n$, the elements $\mathcal{P} \in R_n$ with $\mathcal{P}u = 0$ make up an ideal $\mathcal{I}(u)$ of R_n. An element $u \in V_n$ will be called a *multi-recurrence sequence* if $R_n/\mathcal{I}(u)$ is of finite dimension. This dimension will be called the *order* of u. When $\mathcal{P}$ annihilates both u and v (i.e., $\mathcal{P}u = \mathcal{P}v = 0$), then $\mathcal{P}$ annihilates $u + v$. Therefore

$$\mathcal{I}(u) \cap \mathcal{I}(v) \subset \mathcal{I}(u + v). \quad (1.3)$$

Hence by (1.2), $\dim(R_n/\mathcal{I}(u+v)) \leqq \dim(R_n/\mathcal{I}(u) \cap \mathcal{I}(v)) \leqq \dim(R_n/\mathcal{I}(u)) + \dim(R_n/\mathcal{I}(v))$. The multi-recurrence sequences form a subspace of V_n, and

$$\text{order}\,(u + v) \leqq \text{order}\,(u) + \text{order}\,(v).$$

Given an ideal $\mathcal{I} \subset R_n$, the multisequences u which are annihilated by $\mathcal{I}$ (i.e., have $\mathcal{P}u = 0$ for every $\mathcal{P} \in \mathcal{I}$) make up a subspace $V(\mathcal{I})$ of V_n. When u is annihilated by either $\mathcal{I}_1$ or by $\mathcal{I}_2$, then it is annihilated by $\mathcal{I}_1 \cap \mathcal{I}_2$, so that

$$V(\mathcal{I}_1 \cap \mathcal{I}_2) \supset V(\mathcal{I}_1) + V(\mathcal{I}_2). \quad (1.4)$$

Lemma 1.1. $\dim V(\mathcal{I}) = \dim(R_n/\mathcal{I})$.

Proof. For ease of notation, we will suppose that $\dim(R_n/\mathcal{I})$ is finite. When $\mathcal{I} = R_n$, clearly $V(\mathcal{I}) = \{0\}$. We may therefore suppose that $d = \dim(R_n/\mathcal{I}) > 0$. There is a basis of $R_n/\mathcal{I}$ consisting of images of monomials, say images of $\mathbf{Z}^{\mathbf{m}_1}, \ldots, \mathbf{Z}^{\mathbf{m}_d}$. Every monomial $\mathbf{Z}^{\boldsymbol{\ell}}$ may be uniquely written as

$$\mathbf{Z}^{\boldsymbol{\ell}} = r_{\boldsymbol{\ell}1}\mathbf{Z}^{\mathbf{m}_1} + \cdots + r_{\boldsymbol{\ell}d}\mathbf{Z}^{\mathbf{m}_d} + \mathcal{P}_{\boldsymbol{\ell}} \tag{1.5}$$

with coefficients $r_{\boldsymbol{\ell}i} \in \mathbb{C}$ and $\mathcal{P}_{\boldsymbol{\ell}} \in \mathcal{I}$. As is easily seen, the generalized polynomials $\mathcal{P}_{\boldsymbol{\ell}}$ ($\boldsymbol{\ell} \in \mathbb{Z}^n$) are a set of generators of $\mathcal{I}$ (as a vector space over $\mathbb{C}$).

Given constants $c_1, \ldots, c_d$, we wish to construct $u \in V(\mathcal{I})$ with

$$u_{\mathbf{m}_i} = c_i \qquad (i = 1, \ldots, d).$$

The condition $u \in V(\mathcal{I})$ means that $(\mathcal{P}u)_{\mathbf{q}} = 0$ for each $\mathcal{P} \in \mathcal{I}$ and $\mathbf{q} \in \mathbb{Z}^n$. In particular, we need

$$(\mathcal{P}u)_{\mathbf{0}} = 0 \qquad (\mathcal{P} \in \mathcal{I}). \tag{1.6}$$

But with $\mathcal{P} \in \mathcal{I}$, also $\mathbf{Z}^{\mathbf{q}}\mathcal{P} \in \mathcal{I}$, so that (1.6) yields $(\mathcal{P}u)_{\mathbf{q}} = (\mathbf{Z}^{\mathbf{q}}\mathcal{P}u)_{\mathbf{0}} = 0$. Hence (1.6) is a necessary and sufficient condition for u to be in $V(\mathcal{I})$. Since the $\mathcal{P}_{\boldsymbol{\ell}}$ are a set of generators for $\mathcal{I}$, it will be enough that $(\mathcal{P}_{\boldsymbol{\ell}}u)_{\mathbf{0}} = 0$ for each $\boldsymbol{\ell}$, which by (1.5) means that

$$u_{\boldsymbol{\ell}} = r_{\boldsymbol{\ell}1}u_{\mathbf{m}_1} + \cdots + r_{\boldsymbol{\ell}d}u_{\mathbf{m}_d} = r_{\boldsymbol{\ell}1}c_1 + \cdots + r_{\boldsymbol{\ell}d}c_d.$$

This choice for each $\boldsymbol{\ell}$ is certainly possible, and uniquely determines u. Since $c_1, \ldots, c_d$ were arbitrary, indeed $\dim V(\mathcal{I}) = d = \dim(R_n/\mathcal{I})$. □

Now suppose $\mathcal{I}_1 + \mathcal{I}_2 = R_n$. A multisequence annihilated by both $\mathcal{I}_1$ and $\mathcal{I}_2$ is annihilated by R_n, hence is 0. Thus $V(\mathcal{I}_1) \cap V(\mathcal{I}_2) = V(\mathcal{I}_1 + \mathcal{I}_2) = \{0\}$ in this case, and (1.4) yields $V(\mathcal{I}_1 \cap \mathcal{I}_2) \supset V(\mathcal{I}_1) \oplus V(\mathcal{I}_2)$. A comparison of dimension using (1.2), and our Lemma, gives

$$V(\mathcal{I}_1 \cap \mathcal{I}_2) = V(\mathcal{I}_1) \oplus V(\mathcal{I}_2). \tag{1.7}$$

2 Functions of Polynomial-Exponential Type

We will describe $V(\mathcal{I})$ when $\mathcal{I}$ is an ideal in R_n with finite $\dim(R_n/\mathcal{I})$. We begin with the case $n = 1$.

When some $\mathcal{Q}(Z) \in \mathcal{I}$, then $Z^m\mathcal{Q}(Z) \in \mathcal{I} \cap \mathbb{C}[Z]$ for suitable m, hence $\mathcal{Q}(Z) \in Z^{-m}(\mathcal{I} \cap \mathbb{C}[Z])$. Here $\mathcal{I} \cap \mathbb{C}[Z]$ is an ideal in $\mathbb{C}[Z]$, therefore a principal ideal, generated by a polynomial $\mathcal{P}(Z) = Z^d + \cdots + c_0$. Note that unless $\mathcal{P} = 0$, we have $c_0 \neq 0$, for otherwise $Z^{-1}\mathcal{P}(Z) \in \mathcal{I} \cap \mathbb{C}[Z]$. It is now clear that $\mathcal{I}$ is generated by $\mathcal{P}(Z)$, and $u \in V(\mathcal{I})$ iff $\mathcal{P}u = 0$. When $\mathcal{P}$ generates $\mathcal{I}$, we will set $V(\mathcal{P}) = V(\mathcal{I})$.

If u is a linear recurrence sequence and $\mathcal{I} = \mathcal{I}(u)$, the polynomial $\mathcal{P}(Z)$ described above is called the *companion polynomial* of u. Clearly, the order of u is the degree of the companion polynomial.

When $u_k = P(k)\alpha^k$ with $\alpha \neq 0$ and $P \in \mathbb{C}[X]$, then

$$((Z-\alpha)u)_k = u_{k+1} - \alpha u_k = P(k+1)\alpha^{k+1} - \alpha P(k)\alpha^k = ((P(k+1)-P(k))\alpha \cdot \alpha^k.$$

Here the degree of the polynomial in front of α^k is reduced. We may infer that when $\deg P < a$, then $(Z-\alpha)^a u = 0$. A comparison of dimensions utilizing Lemma 1.1 shows that

$V((Z-\alpha)^a)$ consists of sequences u with $u_k = P(k)\alpha^k$ where $\deg P < a$.

When $\mathcal{P}(Z) = (Z-\alpha_1)^{a_1} \cdots (Z-\alpha_q)^{a_q}$ with distinct $\alpha_1, \dots, \alpha_q$, then (1.7) and induction on q shows that $V(\mathcal{P})$ is the direct sum of the $V((Z-\alpha_i)^{a_i})$ $(i = 1, \dots, q)$. Therefore

$V((Z-\alpha_1)^{a_1} \cdots (Z-\alpha_q)^{a_q})$ consists of sequences

$u_k = P_1(k)\alpha_1^k + \cdots + P_q(k)$ where $\deg P_i < a_i$ $(i = 1, \dots, q)$,

and every such sequence can be written uniquely in this way. $V(\mathcal{P})$ has dimension $d = a_1 + \cdots + a_q$. It is clear that given constants $c_1, \dots, c_d$, there is exactly one recurrence sequence $u \in V(\mathcal{P})$ with $u_k = c_k$ for $k = 1, \dots, d$. Thus $V(\mathcal{P})$ has a basis $u^{(1)}, \dots, u^{(d)}$ with

$$u_k^{(i)} = \delta_{ik} \qquad (1 \leqq i, k \leqq d), \tag{2.1}$$

where δ is the Kronecker symbol.

Now let $n \in \mathbb{N}$ be arbitrary. A function of *polynomial–exponential type* is a function $\mathbb{Z}^n \to \mathbb{C}$ of the form

$$f(\mathbf{k}) = f(k_1, \dots, k_n) = \sum_{i=1}^{q} P_i(\mathbf{k})\boldsymbol{\alpha}_i^{\mathbf{k}}$$

where $P_i \in \mathbb{C}[X_1, \dots, X_n]$ and

$$\boldsymbol{\alpha}_i^{\mathbf{k}} = \alpha_{i1}^{k_1} \cdots \alpha_{in}^{k_n}$$

with nonzero α_{ij} $(1 \leqq i \leqq q,\ 1 \leqq j \leqq n)$. An element $u \in V_n$ will be said to be of polynomial–exponential type if $u_{\mathbf{k}} = f(\mathbf{k})$ with f a function of this type.

Theorem 2.1. *An element $u \in V_n$ of polynomial–exponential type is a multi-recurrence sequence.*

When $\mathcal{I} \subset R_n$ is an ideal with finite $\dim(R_n/\mathcal{I})$, *then each element of $V(\mathcal{I})$ is of polynomial–exponential type. Therefore every multi-recurrence sequence is of this type.*

Proof. Suppose

$$u_{\mathbf{k}} = \mathbf{k}^{\mathbf{a}-1}\boldsymbol{\alpha}^{\mathbf{k}} = k_1^{a_1-1}\cdots k_n^{a_n-1}\alpha_1^{k_1}\cdots\alpha_n^{k_n}$$

where $a_1,\ldots,a_n$ are in $\mathbb{N}$. Then $(Z_i-\alpha_i)^{a_i}u = 0$ $(i=1,\ldots,n)$, so that $\mathcal{I}(u)$ contains each $(Z_i-\alpha_i)^{a_i}$, and $\dim(R_n/\mathcal{I}(u)) \leqq a_1\cdots a_n$, hence is finite, so that u is a multi-recurrence sequence. The first assertion of the Theorem follows from the fact, noted above, that the multi-recurrence sequences form a subspace of V_n, so that a sum of such sequences is again multi-recurrent.

The second assertion will be done by induction on n. The case $n=1$ has already been established. Since $\dim(R_n/\mathcal{I})<\infty$, there are only finitely many powers of Z_1 independent mod $\mathcal{I}$, so some polynomial $\mathcal{P}_1=\mathcal{P}_1(Z_1)\neq 0$ is in $\mathcal{I}$. Let $V_1(\mathcal{P}_1)$ consist of sequences $v\in V_1$ with $\mathcal{P}_1 v=0$. Then $V_1(\mathcal{P}_1)$ is spanned by certain sequences $u^{(i)}$ $(1\leqq i\leqq d)$ with (2.1). Every $u\in V(\mathcal{I})$ has $\mathcal{P}_1 u=0$. Therefore when $k_2,\ldots,k_n$ are given, the sequence $u_{k_1,k_2,\ldots,k_n}$ with subscript k_1 lies in $V_1(\mathcal{P}_1)$. We may infer that

$$u_{k_1,k_2,\ldots,k_n} = \sum_{i=1}^{d} c_{k_2,\ldots,k_n}(i)u_{k_1}^{(i)} \tag{2.2}$$

with certain coefficients $c_{k_2,\ldots,k_n}(i)$.

There are only finitely many monomials $Z_2^{k_2}\cdots Z_n^{k_n}$ independent mod $\mathcal{I}$. Hence if $\mathcal{I}'=\mathcal{I}\cap\mathbb{C}[Z_2,\ldots,Z_n,Z_2^{-1},\ldots,Z_n^{-1}]=\mathcal{I}\cap R'$, say, then $\dim(R'/\mathcal{I}')$ is finite. By induction, $V(\mathcal{I}')\subset R'$ is spanned by finitely many elements $v^{(1)},\ldots,v^{(\ell)}$ of polynomial–exponential type in $n-1$ variables. For given k_1, our $u\in V(\mathcal{I})$ with subscripts $k_2,\ldots,k_n$ is in $V(\mathcal{I}')$. Thus

$$u_{k_1,\ldots,k_n} = \sum_{m=1}^{\ell} c_{k_1}(m)v_{k_2,\ldots,k_n}^{(m)} \tag{2.3}$$

with certain coefficients $c_{k_1}(m)$.

For $1\leqq i\leqq d$, we may infer from (2.1), (2.2), (2.3) that

$$c_{k_2,\ldots,k_n}(i) = u_{i,k_2,\ldots,k_n} = \sum_{m=1}^{\ell} c_i(m)v_{k_2,\ldots,k_n}^{(m)}.$$

Since by induction the $u^{(i)}$ and $v^{(m)}$ are of polynomial–exponential type, we see from (2.2) that so is u. □

Example. Let $n=2$, and for ease of notation, write Z,W in place of Z_1,Z_2, and k,ℓ in place of k_1,k_2. Let $\mathcal{I}\subset R_2$ be generated by $\mathcal{P}_1=Z^2-1$, $\mathcal{P}_2=(Z-1)(W^2+W+1)$, $\mathcal{P}_3=(Z+1)(W^3-4W^2+W+6)$. An element $u\in V(\mathcal{I})$ has $\mathcal{P}_1u=0$, hence has $u_{k,\ell}=c_\ell+c'_\ell(-1)^k$ with coefficients c_ℓ, c'_ℓ. The relation $\mathcal{P}_2u=0$ yields $2(-1)^{k+1}(c'_{\ell+2}+c'_{\ell+1}+c'_\ell)=0$. The roots of W^2+W+1 are the primitive cube roots $\zeta_3,\bar{\zeta}_3$ of 1, so that c'_ℓ is a linear combination of

$\zeta_3^\ell, \bar{\zeta}_3^\ell$. Finally $\mathcal{P}_3 u = 0$ yields $2(c_{\ell+3} - 4c_{\ell+2} + c_{\ell+1} + 6c_\ell) = 0$. Since the roots of $W^3 - 4W^2 + W + 6$ are $-1, 2, 3$, we see that c_ℓ is a linear combination of $(-1)^\ell, 2^\ell, 3^\ell$. Therefore $V(\mathcal{I})$ is spanned by the multi-sequences

$$(-1)^k \zeta_3^\ell,\ (-1)^k \bar{\zeta}_3^\ell,\ (-1)^\ell,\ 2^\ell,\ 3^\ell,$$

and $\dim V(\mathcal{I}) = 5$.

Observe that

$$(W^3-4W^2+W+6)\mathcal{P}_2-(W^2+W+1)\mathcal{P}_3 = -2(W^2+W+1)(W^3-4W^2+W+6)$$

and

$$(5W^2-31W+55)\mathcal{P}_2-(5W-6)\mathcal{P}_3 = 91Z-10W^4+52W^3-58W^2-48W-19$$

are in $\mathcal{I}$, so that $1, W, W^2, W^3, W^4 \bmod \mathcal{I}$ are a basis of $R_2/\mathcal{I}$.

Lemma 2.2. *Suppose $\boldsymbol{\alpha}_1, \ldots, \boldsymbol{\alpha}_q$ are distinct elements of $(\mathbb{C}^\times)^n$. Then the sequences u with*

$$u_{\mathbf{k}} = \mathbf{k}^{\mathbf{a}} \boldsymbol{\alpha}_i^{\mathbf{k}} = k_1^{a_1} \cdots k_n^{a_n} \alpha_{i1}^{k_1} \cdots \alpha_{in}^{k_n} \tag{2.4}$$

for $1 \leqq i \leqq q$ and any $a_1, \ldots, a_n$ in $\mathbb{Z}_{\geqq 0}$ are linearly independent.

Proof. The case $n = 1$ follows from the fact that for any $b_1, \ldots, b_q$ in $\mathbb{N}$, the sequences $k^{a_i} \alpha_i^k$ with $1 \leqq i \leqq q$, $0 \leqq a_i < b_i$ are a basis of $V(\mathcal{P})$ with $\mathcal{P} = (Z - \alpha_1)^{b_1} \cdots (Z - \alpha_q)^{b_q}$. We will proceed by induction on n. Suppose there are t distinct numbers among $\alpha_{1n}, \ldots, \alpha_{qn}$. If these are $\beta_1, \ldots, \beta_t$, we may change our notation and ordering, so that $\boldsymbol{\alpha}_1, \ldots, \boldsymbol{\alpha}_q$ become $\boldsymbol{\alpha}_1^{(i)}, \ldots, \boldsymbol{\alpha}_{q_i}^{(i)}$ with $i = 1, \ldots, t$, where $q_1 + \cdots + q_t = q$ and

$$\boldsymbol{\alpha}_j^{(i)} = (\boldsymbol{\beta}_{ij}, \beta_i)$$

with $\boldsymbol{\beta}_{ij} \in (\mathbb{C}^\times)^{n-1}$, and $\boldsymbol{\beta}_{i1}, \ldots, \boldsymbol{\beta}_{i,q_i}$ distinct for each given i. Set $\tilde{\mathbf{k}} = (k_1, \ldots, k_{n-1})$, $\tilde{\mathbf{a}} = (a_1, \ldots, a_{n-1})$ and $\tilde{\mathbf{k}}^{\tilde{\mathbf{a}}} = k_1^{a_1} \cdots k_{n-1}^{a_{n-1}}$, as well as

$$\boldsymbol{\beta}_{ij}^{\tilde{\mathbf{k}}} = \beta_{ij1}^{k_1} \cdots \beta_{ij,n-1}^{k_{n-1}}.$$

The multisequences (2.4) are of the form

$$\tilde{\mathbf{k}}^{\tilde{\mathbf{a}}} \boldsymbol{\beta}_{ij}^{\tilde{\mathbf{k}}} k_n^{a_n} \beta_i^{k_n}$$

with $1 \leqq i \leqq t$, $1 \leqq j \leqq q_i$ and $\tilde{\mathbf{a}} \in (\mathbb{Z}_{\geqq 0})^{n-1}$, $a_n \in \mathbb{Z}_{\geqq 0}$. Suppose we had a linear dependence relation, so that a finite sum

$$\sum_{i=1}^{t} \sum_{a_n} \left(\sum_{j=1}^{q_i} \sum_{\tilde{\mathbf{a}}} c(i, a_n, j, \tilde{\mathbf{a}}) \tilde{\mathbf{k}}^{\tilde{\mathbf{a}}} \boldsymbol{\beta}_{ij}^{\tilde{\mathbf{k}}} \right) k_n^{a_n} \beta_i^{k_n} = 0.$$

By the case $n = 1$, we have for each i and each a_n,

$$\sum_{j=1}^{q_i} \sum_{\tilde{\mathbf{a}}} c(i, a_n, j, \tilde{\mathbf{a}}) \tilde{\mathbf{k}}^{\tilde{\mathbf{a}}} \beta_{ij}^{\tilde{\mathbf{k}}} = 0.$$

Since $\boldsymbol{\beta}_{i1}, \ldots, \boldsymbol{\beta}_{i,q_i}$ are distinct, the case $n-1$ shows that each coefficient $c(i, a_n, j, \tilde{\mathbf{a}})$ is zero. The Lemma follows. □

Lemma 2.3. *Suppose $u \in V(\mathcal{I})$ where $u = u^{(1)} + \cdots + u^{(q)}$, and*

$$u_{\mathbf{k}}^{(i)} = P_i(\mathbf{k}) \boldsymbol{\alpha}_i^{\mathbf{k}}$$

with distinct $\boldsymbol{\alpha}_1, \ldots, \boldsymbol{\alpha}_q$. Then each $u^{(i)} \in V(\mathcal{I})$.

Proof. $(\mathbf{Z}^{\mathbf{m}} u^{(i)})_{\mathbf{k}} = \boldsymbol{\alpha}_i^{\mathbf{m}} P_i(\mathbf{k}+\mathbf{m}) \boldsymbol{\alpha}_i^{\mathbf{k}}$, and therefore given $\mathcal{P} \in R_n$, we have $(\mathcal{P} u^{(i)})_{\mathbf{k}} = Q_i(\mathbf{k}) \boldsymbol{\alpha}_i^{\mathbf{k}}$ with some polynomial Q_i. Thus

$$(\mathcal{P} u)_{\mathbf{k}} = \sum_{i=1}^{q} Q_i(\mathbf{k}) \boldsymbol{\alpha}_i^{\mathbf{k}}.$$

By Lemma 2.2, the vanishing of this implies that each $Q_i = 0$, so that $\mathcal{P} u^{(i)} = 0$. Therefore $\mathcal{P} u = 0$ yields $\mathcal{P} u^{(i)} = 0$ $(i = 1, \ldots, q)$, which clearly entails the Lemma. □

Let $P \in \mathbb{C}[X_1, \ldots, X_n]$ be given. The following vector spaces are easily seen to be the same.

(a) The space generated by all partial derivatives

$$\partial^{i_1 + \cdots + i_n} P / \partial X_1^{i_1} \cdots \partial X_n^{i_n},$$

including P itself.

(b) The space generated by the polynomials

$$\Delta_1^{i_1} \cdots \Delta_n^{i_n} P \qquad (i_1, \ldots, i_n \in \mathbb{Z}_{\geqq 0})$$

where Δ_j is the difference operator with

$$\Delta_j Q = Q(X_1, \ldots, X_j + 1, \ldots, X_n) - Q(X_1, \ldots, X_j, \ldots, X_n).$$

(c) The space generated by the polynomials

$$P(X_1 + a_1, \ldots, X_n + a_n)$$

as $(a_1, \ldots, a_n)$ run through $\mathbb{Z}^n$.

We will denote this space by $\mathcal{S}(P)$.

Lemma 2.4. *Suppose $u_{\mathbf{k}} = P(\mathbf{k}) \boldsymbol{\alpha}^{\mathbf{k}}$ with $\boldsymbol{\alpha} \in (\mathbb{C}^{\times})^n$. Then $V(\mathcal{I}(u))$ consists of the sequences $Q(\mathbf{k}) \boldsymbol{\alpha}^{\mathbf{k}}$ with $Q \in \mathcal{S}(P)$, and therefore u is of order* $\dim \mathcal{S}(P)$.

Proof. Set $\mathcal{I} = \mathcal{I}(u)$. Suppose a exceeds the total degree of P. Then $(Z_i - \alpha_i)^a u = 0$, so that each $(Z_i - \alpha_i)^a$ $(i = 1, \ldots, n)$ lies in $\mathcal{I}$.

When $Q_1(\mathbf{k})\boldsymbol{\beta}_1^{\mathbf{k}} + \cdots + Q_q(\mathbf{k})\boldsymbol{\beta}_q^{\mathbf{k}}$ with distinct $\boldsymbol{\beta}_1, \ldots, \boldsymbol{\beta}_q$ is in $V(\mathcal{I})$, then each $Q_i(\mathbf{k})\boldsymbol{\beta}_i^{\mathbf{k}}$ is in $V(\mathcal{I})$. Now suppose some $Q(\mathbf{k})\boldsymbol{\beta}^{\mathbf{k}}$ is in $V(\mathcal{I})$. Then $(Z_i - \alpha_i)^a(Q(\mathbf{k})\boldsymbol{\beta}^{\mathbf{k}}) = 0$. By considering $Q(\mathbf{k})\boldsymbol{\beta}^{\mathbf{k}}$ as a sequence in k_i (with fixed $k_1, \ldots, k_{i-1}, k_{i+1}, \ldots, k_n$), we see from our analysis of the case $n = 1$ at the beginning of this Section that $\beta_i = \alpha_i$. Therefore $\boldsymbol{\beta} = \boldsymbol{\alpha}$, and the elements of $V(\mathcal{I})$ are of the form $Q(\mathbf{k})\boldsymbol{\alpha}^{\mathbf{k}}$.

It is clear that not only is $u \in V(\mathcal{I})$, but also $\mathcal{P}u \in V(\mathcal{I})$ for any $\mathcal{P} \in R_n$. Observe that

$$\begin{aligned}((\alpha_i^{-1} Z_i - 1)u)_{\mathbf{k}} &= (P(k_1, \ldots, k_i + 1, \ldots) - P(k_1, \ldots, k_i, \ldots))\boldsymbol{\alpha}^{\mathbf{k}} \\ &= (\Delta_i P)(\mathbf{k})\boldsymbol{\alpha}^{\mathbf{k}}.\end{aligned}$$

Therefore each $(\Delta_i P)(\mathbf{k})\boldsymbol{\alpha}^{\mathbf{k}}$ is in $V(\mathcal{I})$, and in fact $Q(\mathbf{k})\boldsymbol{\alpha}^{\mathbf{k}} \in V(\mathcal{I})$ for every $Q \in \mathcal{S}(P)$. In particular, $\dim V(\mathcal{I}) \geqq \dim \mathcal{S}(P)$.

Since $(Z_i - \alpha_i)^a \in \mathcal{I}$, the monomials $\mathbf{Z}^{\mathbf{m}}$ with $\mathbf{m}$ in the cube $C(a)$ given by $0 \leqq m_i < a$ $(i = 1, \ldots, n)$ generate the factor space $R_n/\mathcal{I}$. But some of these monomials may be linearly dependent mod $\mathcal{I}$. Which linear combinations of these monomials lie in $\mathcal{I}$? It will be convenient to write such a linear combination as

$$\mathcal{Q}(\mathbf{Z}) = \sum_{\mathbf{m}} \boldsymbol{\alpha}^{-\mathbf{m}} c_{\mathbf{m}} \mathbf{Z}^{\mathbf{m}}$$

with coefficients $c_{\mathbf{m}}$ $(\mathbf{m} \in C(a))$. Then

$$\mathcal{Q}(P(\mathbf{k})\boldsymbol{\alpha}^{\mathbf{k}}) = \sum_{\mathbf{m}} c_{\mathbf{m}} P(\mathbf{k} + \mathbf{m})\boldsymbol{\alpha}^{\mathbf{k}},$$

and $\mathcal{Q} \in \mathcal{I}$ precisely if $\sum_{\mathbf{m}} c_{\mathbf{m}} P(\mathbf{k} + \mathbf{m}) = 0$ for each $\mathbf{k} \in \mathbb{Z}^n$, so that in fact

$$\sum_{\mathbf{m}} c_{\mathbf{m}} P(\mathbf{X} + \mathbf{m}) = 0. \tag{2.5}$$

This last sum here is in $\mathcal{S}(P)$. The vector $\mathbf{c} \in \mathbb{C}^{a^n}$ with coordinates $c_{\mathbf{m}}$ $(\mathbf{m} \in C(a))$ has to satisfy the conditions (2.5), of which at most $\dim \mathcal{S}(P)$ are independent. We obtain at least $a^n - \dim \mathcal{S}(P)$ independent solution vectors $\mathbf{c}$ of (2.5). Therefore at most $\dim \mathcal{S}(P)$ of the monomials $\mathbf{Z}^{\mathbf{m}}$ $(\mathbf{m} \in C(a))$ are independent mod $\mathcal{I}$, and $\dim(R_n/\mathcal{I}) \leqq \dim \mathcal{S}(P)$. Since $\dim(R_n/\mathcal{I}) = \dim V(\mathcal{I}) \geqq \dim \mathcal{S}(P)$ by what was said above, $V(\mathcal{I}(u))$ is as claimed, and u is indeed a multi-recurrent sequence of order $\dim \mathcal{S}(P)$. □

Example. Let $n = 2$, and write Z, W for Z_1, Z_2, further k, ℓ for k_1, k_2, and α, β for α_1, α_2. Set $u_{k,\ell} = P(k, \ell)\alpha^k\beta^\ell$ with $P(k, \ell) = k^2 + \ell^2$. Then $\mathcal{S}(P)$ is spanned by $1, k, \ell, k^2+\ell^2$, hence $V(\mathcal{I}(u))$ by $\alpha^k\beta^\ell, k\alpha^k\beta^\ell, \ell\alpha^k\beta^\ell, (k^2+\ell^2)\alpha^k\beta^\ell$, so that u is of order 4. The polynomials $(Z - \alpha)^3$, $(W - \beta)^3$, $(Z - \alpha)(W - \beta)$, $\beta^2(Z - \alpha)^2 - \alpha^2(W - \beta)^2$ lie in $\mathcal{I}$, and it may be checked that $1, Z, Z^2, W$ form a basis of $R_2/\mathcal{I}$.

3 Generating Functions

Write $\mathbb{N}_0$ (up to now $\mathbb{Z}_{\geqq 0}$) for the set of nonnegative integers. Further let V_n' be the vector space of functions $v : \mathbb{N}_0^n \to \mathbb{C}$. The value of such v at $\mathbf{k} \in \mathbb{N}_0^n$ will be denoted by $v_{\mathbf{k}}$. We will say that v is the *restriction* of $u \in V_n$ if $v_{\mathbf{k}} = u_{\mathbf{k}}$ for every $\mathbf{k} \in \mathbb{N}_0^n$. The *generating function* of v is

$$\mathcal{F}(Z_1, \ldots, Z_n) = \sum_{\mathbf{k} \in \mathbb{N}_0^n} v_{\mathbf{k}} \mathbf{Z}^{\mathbf{k}}.$$

When $\mathcal{P}(Z) = c_m Z^m + \cdots + c_0$ with $c_m c_0 \neq 0$, set $\widehat{\mathcal{P}}(Z) = Z^m \mathcal{P}(1/Z) = c_0 Z^m + \cdots + c_m$.

Theorem 3.1. *Let $\mathcal{P}_1(Z_1), \ldots, \mathcal{P}_n(Z_n)$ be polynomials with nonzero constant terms. Then $v \in V_n'$ is the restriction of a multi-recurrence sequence u with*

$$\mathcal{P}_i u = 0 \qquad (i = 1, \ldots, n) \tag{3.1}$$

precisely if $\mathcal{F}$ is a rational function of the form

$$\mathcal{F}(Z_1, \ldots, Z_n) = \mathcal{G}(Z_1, \ldots, Z_n) / (\widehat{\mathcal{P}}_1(Z_1) \cdots \widehat{\mathcal{P}}_n(Z_n))$$

where $\mathcal{G}$ is a polynomial with

$$\deg_{Z_i} \mathcal{G} < \deg \mathcal{P}_i \qquad (i = 1, \ldots, n).$$

Note that every multi-recurrence sequence u has (3.1) for some $\mathcal{P}_1(Z_1), \ldots, \mathcal{P}_n(Z_n)$ with nonzero constant terms.

Proof. We begin with the case $n = 1$. Say $\mathcal{P}(Z) = \mathcal{P}_1(Z) = c_m Z^m + \cdots + c_0$. Then

$$\widehat{\mathcal{P}}(Z)\mathcal{F}(Z) = \mathcal{G}(Z) + Z^m \sum_{k=0}^{\infty} (c_0 v_k + c_1 v_{k+1} + \cdots + c_m v_{k+m}) Z^k, \tag{3.2}$$

where $\mathcal{G}$ is a polynomial of degree $< m$.

When v is the restriction of u, then the coefficient of Z^k is

$$c_0 u_k + \cdots + c_m u_{k+m} = (\mathcal{P}u)_k,$$

and this is zero if u is recurrent with $\mathcal{P}u = 0$. Then indeed $\mathcal{F} = \mathcal{G}/\widehat{\mathcal{P}}$. Conversely, when this is the case, so that $\widehat{\mathcal{P}}\mathcal{F} = \mathcal{G}$, then the coefficient of each Z^k in (3.2) vanishes. We set $u_k = v_k$ for $k \geqq 0$, and define u_k recursively for $k < 0$ by $c_0 u_k + \cdots + c_m u_{k+m} = 0$. Then $\mathcal{P}u = 0$, and v is the restriction of u.

We note that linear recurrences can be defined over any field $\mathbb{F}$ in place of $\mathbb{C}$. The Theorem is valid for any field, and our proof of the case $n = 1$ goes through. The general case can now be done by induction on n. For ease of notation, we will just do the typical case $n = 2$, and we will write Z, W in place of Z_1, Z_2. Say $\mathcal{P}_1(Z) = c_m Z^m + \cdots + c_0$, $\mathcal{P}_2(W) = d_\ell W^\ell + \cdots + d_0$, so that $\widehat{\mathcal{P}}_1(Z) = c_m + \cdots + c_0 Z^m$, $\widehat{\mathcal{P}}_2(W) = d_\ell + \cdots + d_0 W^\ell$.

Suppose v is the restriction of u with $\mathcal{P}_1 u = \mathcal{P}_2 u = 0$. Then

$$\sum_i c_i v_{k+i,h} = 0 \qquad (k,h) \in \mathbb{N}_0^2, \tag{3.3}$$

$$\sum_j d_j v_{k,h+j} = 0 \qquad (h,k) \in \mathbb{N}_0^2. \tag{3.4}$$

The coefficient of $Z^k W^n$ in

$$\widehat{\mathcal{P}}_1(Z)\widehat{\mathcal{P}}_2(W)\mathcal{F}(Z,W) \tag{3.5}$$

is

$$\sum_{i=\max(0,m-k)}^{m} \quad \sum_{j=\max(0,\ell-h)}^{\ell} \qquad c_i d_j v_{k-m+i,h-\ell+j}.$$

By (3.3), (3.4), this coefficient vanishes when $k \geqq m$ or $h \geqq \ell$. Therefore (3.5) is a polynomial $\mathcal{G}$ of degree $< m$ in Z and $< \ell$ in W, and $\mathcal{F} = \mathcal{G}/(\widehat{\mathcal{P}}_1\widehat{\mathcal{P}}_2)$.

Conversely, suppose this is the case, with the above bounds for the degrees of $\mathcal{G}$. Interpreting $\mathcal{F}(Z,W)$ as being a series $\mathcal{F}_1(Z)$ with coefficients

$$\tilde{v}_k = \sum_{h \geqq 0} v_{kh} W^h$$

in $\mathbb{F}((W))$, we have $\mathcal{F}_1(Z) = \mathcal{G}_1(Z)/\widehat{\mathcal{P}}_1(Z)$ where $\mathcal{G}_1$ is a polynomial with coefficients in $\mathbb{F}((W))$. By the case $n = 1$ we have $\sum_i c_i \tilde{v}_{k+i} = 0$ for $k \geqq 0$, and since the vanishing of a power series is the same as the vanishing of all of its coefficients, we have (3.3). Similarly, (3.4) holds. We can extend v inductively from k to $k-1$ and from h to $h-1$ to a function $v : \mathbb{Z}^2 \to \mathbb{F}$ such that the equations (3.3), (3.4) hold for any k, h. Therefore v is the restriction of a multisequence u with $\mathcal{P}_1 u = \mathcal{P}_2 u = 0$. □

4 Factorization of Polynomial-Exponential Functions

The contents of this Section are essentially due to Ritt [26]. See also the exposition [8]. When considering polynomial–exponential functions $\mathbb{Z}^n \to \mathbb{C}$,

$$f(\mathbf{k}) = \sum_{i=1}^{q} P_i(\mathbf{k})\boldsymbol{\alpha}_i^{\mathbf{k}}, \tag{4.1}$$

note that they form a ring with zero divisors. For instance $(1^k + (-1)^k)(1^k - (-1)^k) = 0$. We therefore will now deal with polynomial–exponential functions $\mathbb{C}^n \to \mathbb{C}$, given by

$$f(\boldsymbol{\xi}) = \sum_{i=1}^{q} P_i(\boldsymbol{\xi}) e(\boldsymbol{\varphi}_i \boldsymbol{\xi}) \tag{4.2}$$

where $e(\xi) = e^{2\pi i \xi}$, and $\boldsymbol{\varphi}_i \boldsymbol{\xi}$ denotes an inner product $\varphi_{i1}\xi_1 + \cdots + \varphi_{in}\xi_n$. These functions form a ring $\mathcal{R}$ without divisors of zero. The restriction of such f to $\mathbb{Z}^n$ gives the function (4.1) with $\boldsymbol{\alpha}_i = (e(\varphi_{i1}), \ldots, e(\varphi_{in}))$.

We may write nonzero f as in (4.2) with distinct vectors $\boldsymbol{\varphi}_1, \ldots, \boldsymbol{\varphi}_q$ and with nonzero polynomials $P_1, \ldots, P_q$. Then f may be written uniquely in this way; to see this it will suffice to show that

$$\sum_{i=1}^{q} P_i(\boldsymbol{\xi}) e(\boldsymbol{\varphi}_i \boldsymbol{\xi}) = 0 \tag{4.3}$$

with distinct $\boldsymbol{\varphi}_1, \ldots, \boldsymbol{\varphi}_q$ implies $P_1 = \cdots = P_q = 0$. For $n = 1$, and assuming $P_1 \neq 0$, we divide by $P_1(\xi) e(\varphi_1 \xi)$ to obtain

$$1 + \sum_{i=2}^{q} \widehat{P}_i(\xi) e(\widehat{\varphi}_i \xi) = 0$$

where the $\widehat{P}_i$ are rational functions, and $\widehat{\varphi}_i = \varphi_i - \varphi_1 \neq 0$ $(i = 2, \ldots, q)$. After differentiation we get

$$\sum_{i=2}^{q} (\widehat{P}_i'(\xi) + 2\pi i \widehat{\varphi}_i \widehat{P}_i(\xi)) e(\widehat{\varphi}_i \xi) = 0, \tag{4.4}$$

so that an inductive assumption yields $\widehat{P}_i' + 2\pi i \widehat{\varphi}_i \widehat{P}_i = 0$, and the only rational solution of such a differential equation is $\widehat{P}_i = 0$. Therefore also $P_i = 0$ for $i = 2, \ldots, q$, and thus by (4.3) in fact for $i = 1, \ldots, q$. To reduce the general case to the case $n = 1$, note that by our hypothesis, $\boldsymbol{\gamma}\boldsymbol{\varphi}_i \neq \boldsymbol{\gamma}\boldsymbol{\varphi}_j$ $(i \neq j)$ for almost every $\boldsymbol{\gamma} \in \mathbb{C}^n$, and substituting $\boldsymbol{\xi} = \zeta\boldsymbol{\gamma}$ into (4.3), we see that each $P_i(\zeta\boldsymbol{\gamma})$ vanishes as a function of ζ for almost every $\boldsymbol{\gamma}$, so that again $P_1 = \cdots = P_q = 0$.

Because of the uniqueness in (4.2), $\mathcal{R}$ is a vector space over $\mathbb{C}$ with basis vectors

$$\xi_1^{a_1} \cdots \xi_n^{a_n} e(\varphi_1 \xi_1) \cdots e(\varphi_n \xi_n) \tag{4.5}$$

where $a_i \in \mathbb{N}_0$ and $\varphi_i \in \mathbb{C}$ $(i = 1, \ldots, n)$.

Let Φ be a semigroup under addition. Define a generalized monomial in $Y_1, \ldots, Y_\ell$ as a formal expression

$$\mathbf{Y}^{\boldsymbol{\varphi}} = Y_1^{\varphi_1} \cdots Y_\ell^{\varphi_\ell} \tag{4.6}$$

with $\boldsymbol{\varphi} = (\varphi_1, \ldots, \varphi_\ell) \in \Phi^\ell$. When D is an integral domain, the finite linear combinations of generalized monomials with coefficients in D form a ring under obvious definitions of addition and multiplication, which we will denote by $D[\mathbf{Y}^{\Phi}]$. Nonzero elements of this ring may be written uniquely as

$$f = \sum_{i=1}^{q} c_i \mathbf{Y}^{\varphi_i} \tag{4.7}$$

with distinct $\varphi_1, \ldots, \varphi_q$, and nonzero $c_1, \ldots, c_q$ in D. We will be particularly interested in the case when $D = \mathbb{C}[\mathbf{X}]$, $\Phi = \mathbb{C}$, and we will write $\mathbb{C}[\mathbf{X}][\mathbf{Y}^{\mathbb{C}}] = \mathbb{C}[\mathbf{X}, \mathbf{Y}^{\mathbb{C}}]$. The elements of $\mathbb{C}[\mathbf{X}, \mathbf{Y}^{\mathbb{C}}]$ are uniquely written as linear combinations of expressions

$$X_1^{a_1} \cdots X_n^{a_n} Y_1^{\varphi_1} \cdots Y_\ell^{\varphi_\ell} \tag{4.8}$$

with $a_i \in \mathbb{N}_0$ $(1 \leqq i \leqq n)$, $\varphi_j \in \mathbb{C}$ $(1 \leqq j \leqq m)$, and coefficients in $\mathbb{C}$. When $n = \ell$, our domain $\mathcal{R}$ is isomorphic to $\mathbb{C}[\mathbf{X}, \mathbf{Y}^{\mathbb{C}}]$ under the linear map which sends (4.5) into (4.8). Therefore to study factorization in $\mathcal{R}$, we will study factorization in $\mathbb{C}[\mathbf{X}, \mathbf{Y}^{\mathbb{C}}]$.

We begin with an investigation of $D[\mathbf{Y}^{\mathbb{C}}]$. Given f as in (4.7), let $\mathcal{F} \subset \mathbb{C}^n$ be the $\mathbb{Z}$-module generated by the "frequencies" $\varphi_1, \ldots, \varphi_q$. Given any $\mathbb{Z}$-module $\mathcal{M} \subset \mathbb{C}^n$, let $\mathcal{M}_{\mathbb{Q}}$ be the $\mathbb{Q}$-module (i.e., vector space) it generates.

Lemma 4.1. *Let f, g, h be nonzero elements of $D[\mathbf{Y}^{\mathbb{C}}]$, and $\mathcal{F}, \mathcal{G}, \mathcal{H}$ the $\mathbb{Z}$-modules generated by their respective frequencies. If*

$$f = gh,$$

then up to normalization, $\mathcal{G}, \mathcal{H}$ are contained in $\mathcal{F}_{\mathbb{Q}}$. More precisely, there is some $\mathbf{Y}^{\rho}$ such that the $\mathbb{Z}$-modules $\overline{\mathcal{G}}$, $\overline{\mathcal{H}}$ generated by the frequencies of $\mathbf{Y}^{-\rho} g, \mathbf{Y}^{\rho} h$ are contained in $\mathcal{F}_{\mathbb{Q}}$.

Proof. We have $\mathcal{F} \subset \mathcal{G} + \mathcal{H}$, therefore $\mathcal{F}_{\mathbb{Q}} \subset (\mathcal{G} + \mathcal{H})_{\mathbb{Q}}$. These are $\mathbb{Q}$-vector spaces, and we can choose a basis $\varphi_1, \ldots, \varphi_u$ of $\mathcal{F}_{\mathbb{Q}}$, and extend it to a basis $\varphi_1, \ldots, \varphi_u, \ldots, \varphi_v$ of $(\mathcal{G} + \mathcal{H})_{\mathbb{Q}}$. We introduce an ordering on $(\mathcal{G} + \mathcal{H})_{\mathbb{Q}}$ by setting $c_1\varphi_1 + \cdots + c_v\varphi_v < c_1'\varphi_1 + \cdots + c_v'\varphi_v$ if $c_s < c_s'$, $c_{s+1} = c_{s+1}', \ldots,$ $c_v = c_v'$ for some s. This ordering respects addition: if $\boldsymbol{\psi} \leqq \boldsymbol{\psi}'$ and $\boldsymbol{\chi} \leqq \boldsymbol{\chi}'$, then $\boldsymbol{\psi} + \boldsymbol{\chi} \leqq \boldsymbol{\psi}' + \boldsymbol{\chi}'$, with equality iff $\boldsymbol{\psi} = \boldsymbol{\psi}'$, $\boldsymbol{\chi} = \boldsymbol{\chi}'$. Let $\boldsymbol{\psi}^+$ and $\boldsymbol{\chi}^+$ be the maximal frequency of g and h with respect to this ordering. Then since $gh = f$, the frequency $\boldsymbol{\psi}^+ + \boldsymbol{\chi}^+$ is in $\mathcal{F}$. Similarly, if $\boldsymbol{\psi}^-$ and $\boldsymbol{\chi}^-$ are the minimal frequencies of g and h, then $\boldsymbol{\psi}^- + \boldsymbol{\chi}^-$ is in $\mathcal{F}$. Therefore $\boldsymbol{\psi}^+ - \boldsymbol{\psi}^- + \boldsymbol{\chi}^+ - \boldsymbol{\chi}^-$ is in $\mathcal{F}_{\mathbb{Q}}$. Writing

$$\boldsymbol{\psi}^+ - \boldsymbol{\psi}^- = \sum_{i=1}^{v} c_i \varphi_i, \quad \boldsymbol{\chi}^+ - \boldsymbol{\chi}^- = \sum_{i=1}^{v} d_i \varphi_i,$$

we have $c_i + d_i = 0$ for $u < i \leqq v$. But since $\boldsymbol{\psi}^+ - \boldsymbol{\psi}^- \geqq \mathbf{0}, \boldsymbol{\chi}^+ - \boldsymbol{\chi}^- \geqq \mathbf{0}$, we have in particular $\sum_{i=u+1}^{v} c_i \varphi_i \geqq \mathbf{0}$, $\sum_{i=u+1}^{v} d_i \varphi_i \geqq \mathbf{0}$, and it follows that $c_i = d_i =$

0 for $u < i \leqq v$, so that $\boldsymbol{\psi}^+ - \boldsymbol{\psi}^-, \boldsymbol{\chi}^+ - \boldsymbol{\chi}^-$ are in $\mathcal{F}_{\mathbb{Q}}$. Now if $\boldsymbol{\psi}$ is any frequency of $\mathcal{G}$, then $\boldsymbol{\psi}^+ - \boldsymbol{\psi}, \boldsymbol{\psi} - \boldsymbol{\psi}^-$ are $\geqq \mathbf{0}$, and $(\boldsymbol{\psi}^+ - \boldsymbol{\psi}) + (\boldsymbol{\psi} - \boldsymbol{\psi}^-) \in \mathcal{F}_{\mathbb{Q}}$, which by the argument just given entails in particular that $\boldsymbol{\psi} - \boldsymbol{\psi}^- \in \mathcal{F}_{\mathbb{Q}}$. Therefore $\mathcal{G} \subset \mathcal{F}_{\mathbb{Q}} + \boldsymbol{\psi}^-$, and similarly $\mathcal{H} \subset \mathcal{F}_{\mathbb{Q}} + \boldsymbol{\chi}^- = \mathcal{F}_{\mathbb{Q}} + (\boldsymbol{\psi}^- + \boldsymbol{\chi}^-) - \boldsymbol{\psi}^- = \mathcal{F}_{\mathbb{Q}} - \boldsymbol{\psi}^-$. It is now clear that the Lemma holds with $\boldsymbol{\rho} = \boldsymbol{\psi}^-$. □

Corollary 4.2. *The units of $D[\mathbf{Y}^{\mathbb{C}}]$ are the expressions*

$$u\mathbf{Y}^{\boldsymbol{\psi}}$$

where u is a unit of D.

Proof. Clearly, these are units. On the other hand, $f = 1$ has only the frequency $\mathbf{0}$, so that $\mathcal{F}_{\mathbb{Q}} = \{\mathbf{0}\}$. When $1 = gh$, then by the Lemma, each of g, h has a single frequency. If $g = c\mathbf{Y}^{\boldsymbol{\psi}}$, $h = d\mathbf{Y}^{\boldsymbol{\chi}}$, then clearly $\boldsymbol{\chi} = -\boldsymbol{\psi}$ and $cd = 1$, so that c, d are units of D. □

Observe that when $D = \mathbb{C}[\mathbf{X}]$, the latter units are nonzero constants.

Given $f \neq 0$ in $D[\mathbf{Y}^{\mathbb{C}}]$, we ask about its factorization.

Define $\mathcal{F}_{\mathbb{Q}}$ as above. Since the $\mathbf{Y}^{\boldsymbol{\rho}}$ are units, it will in view of Lemma 4.1 suffice to consider only factors in the subring $D[\mathbf{Y}^{\mathbb{C}}]_f$ of $D[\mathbf{Y}^{\mathbb{C}}]$ consisting of elements with frequencies in $\mathcal{F}_{\mathbb{Q}}$. Now if $\boldsymbol{\theta}_1, \dots, \boldsymbol{\theta}_k$ constitute a basis of $\mathcal{F}_{\mathbb{Q}}$, the frequencies will be $b_1\boldsymbol{\theta}_1 + \dots + b_k\boldsymbol{\theta}_k$ with rational $b_1, \dots, b_k$. Thus $D[\mathbf{Y}^{\mathbb{C}}]_f$ consists of linear combinations of expressions

$$\mathbf{Y}^{b_1\boldsymbol{\theta}_1} \dots \mathbf{Y}^{b_k\boldsymbol{\theta}_k} \tag{4.9}$$

with coefficients in D. Writing $\mathbf{Z} = (Z_1, \dots, Z_k)$, it is clear that $D[\mathbf{Y}^{\mathbb{C}}]_f$ is isomorphic to $D[\mathbf{Z}^{\mathbb{Q}}]$ via the isomorphism which sends (4.9) to $Z_1^{b_1} \cdots Z_k^{b_k}$. It remains for us to deal with factorization in $D[\mathbf{Z}^{\mathbb{Q}}]$.

We will now specialize to $D = \mathbb{C}$. We will suppose that $f \in \mathbb{C}[\mathbf{Z}^{\mathbb{Z}}]$, and we are interested in its factorization in $\mathbb{C}[\mathbf{Z}^{\mathbb{Q}}]$. If there is an $s \in \mathbb{N}$ such that all the factors of f in $\mathbb{C}[\mathbf{Z}^{\mathbb{Q}}]$ lie in $\mathbb{C}[\mathbf{Z}^{(1/s)\mathbb{Z}}]$, then since the last ring is isomorphic to the unique factorization domain $\mathbb{C}[\mathbf{Z}^{\mathbb{Z}}]$, we have unique factorization of f into irreducible factors.

We may first factor in $\mathbb{C}[\mathbf{Z}^{\mathbb{Z}}]$, hence may restrict ourselves to f which is irreducible in $\mathbb{C}[\mathbf{Z}^{\mathbb{Z}}]$. Write

$$f = \sum_{i=1}^{q} c_i \mathbf{Z}^{\mathbf{b}_i} \tag{4.10}$$

with distinct $\mathbf{b}_1, \dots, \mathbf{b}_q$ and nonzero $c_1, \dots, c_q$ in $\mathbb{C}$.

When $q = 1$, then f is a unit in $\mathbb{C}[\mathbf{Z}^{\mathbb{Q}}]$.

When $q = 2$, up to a unit factor,

$$f = \mathbf{Z}^{\mathbf{b}} - c \tag{4.11}$$

with $\mathbf{b} \neq \mathbf{0}$, $c \in \mathbb{C}^{\times}$. In $\mathbb{C}[\mathbf{Z}^{(1/m)\mathbb{Z}}]$ this factors as

$$\prod_{i=1}^{m}(\mathbf{Z}^{(1/m)\mathbf{b}} - \zeta_m^i c^{1/m}) \tag{4.12}$$

where ζ_m is a primitive m-th root of 1. Therefore f has no irreducible factor in $\mathbb{C}[\mathbf{Z}^{\mathbb{Q}}]$.

When $q \geqq 3$, then since $f \in \mathbb{C}[\mathbf{Z}^{\mathbb{Z}}]$ is irreducible, and $\mathbb{C}$ algebraically closed, the results of the next Section will show that all the factors of f lie in $\mathbb{C}[\mathbf{Z}^{(1/s)\mathbb{Z}}]$ for some s.

We dealt with factorization of $f \in \mathbb{C}[\mathbf{Z}^{\mathbb{Z}}]$. Any $f \in \mathbb{C}[\mathbf{Z}^{\mathbb{Q}}]$ is in $\mathbb{C}[\mathbf{Z}^{(1/m)\mathbb{Z}}]$ for some $m \in \mathbb{N}_0$, and the last ring is isomorphic to $\mathbb{C}[\mathbf{Z}^{\mathbb{Z}}]$. Therefore the factorization of f into factors in $\mathbb{C}[\mathbf{Z}^{\mathbb{Q}}]$ is as above. To summarize: *except for factors of the type* (4.11), *elements of* $\mathbb{C}[\mathbf{Z}^{\mathbb{Q}}]$ *have unique factorization into irreducible elements in this ring.*

The ring $\mathbb{C}[\mathbf{X}^{\mathbb{Q}}, \mathbf{Z}^{\mathbb{Q}}]$ is like the ring $\mathbb{C}[\mathbf{Z}^{\mathbb{Q}}]$, except that there is a different number of variables. In particular, when $f \in \mathbb{C}[\mathbf{X}, \mathbf{Z}^{\mathbb{Q}}]$, then in $\mathbb{C}[\mathbf{X}^{\mathbb{Q}}, \mathbf{Z}^{\mathbb{Q}}]$ it factors into expressions with 2 summands, and other factors which lie in some $\mathbb{C}[\mathbf{X}^{(1/s)\mathbb{Z}}, \mathbf{Z}^{(1/s)\mathbb{Z}}]$. Similarly, in $\mathbb{C}[\mathbf{X}, \mathbf{Z}^{\mathbb{Q}}]$ it factors into expressions with 2 summands (i.e., $\mathbf{X}^{\mathbf{a}}\mathbf{Z}^{\mathbf{b}} - c$ with $\mathbf{a} \in \mathbb{N}_0^n$, $\mathbf{b} \in \mathbb{Q}^k$), while all the other factors lie in some $\mathbb{C}[\mathbf{X}, \mathbf{Z}^{(1/s)\mathbb{Z}}]$. But what about factors with just 2 summands?

$$f = aM(\mathbf{X}) + bN(\mathbf{X})$$

with monomials M, N decomposes uniquely into irreducible factors. So consider

$$f = aM(\mathbf{X})\mathbf{Z}^{\mathbf{b}} + bN(\mathbf{X})$$

with $\mathbf{b} \neq \mathbf{0}$. After dividing by common factors, we may suppose M, N to be coprime. If both are constants, we have essentially $\mathbf{Z}^{\mathbf{b}} - c$, which factors, e.g., as (4.12). In this case, f has no irreducible factors in $\mathbb{C}[\mathbf{Z}^{\mathbb{Q}}]$, hence not in $\mathbb{C}[\mathbf{X}, \mathbf{Z}^{\mathbb{Q}}]$. If not both M, N are constants, then in $\mathbb{F}(\mathbf{X})$, where $\mathbb{F}$ is the algebraic closure of $\mathbb{C}(\mathbf{Z})$, our f factors into irreducible factors of the form $a'M'(\mathbf{X})\mathbf{Z}^{\mathbf{b}'} + b'N'(\mathbf{X})$ where M', N' are monomials (not both constants). There is an $s \in \mathbb{N}$ such that for each factor, $\mathbf{b}' \in (1/s)\mathbb{Z}^k$. We may conclude that the factors of f in $\mathbb{C}[\mathbf{X}, \mathbf{Z}^{\mathbb{Q}}]$ are in $\mathbb{C}[\mathbf{X}, \mathbf{Z}^{(1/s)\mathbb{Z}}]$. This covers (except for unit factors) all elements with 2 summands.

In summary, we see that only factors of the type $\mathbf{Z}^{\mathbf{b}} - c$ do not have unique factorization into irreducibles in $\mathbb{C}[\mathbf{X}, \mathbf{Z}^{\mathbb{Q}}]$. Multiplying such factors together, we get an element

$$\prod_{j=1}^{t}(\mathbf{Z}^{\mathbf{b}_j} - c_j). \tag{4.13}$$

Therefore *in* $\mathbb{C}[\mathbf{X}, \mathbf{Z}^{\mathbb{Q}}]$, *any* $f \neq 0$ *may be factored, uniquely up to units and associates, as a product of an element* (4.13), *a product of irreducible polynomials in* $\mathbb{C}[\mathbf{X}]$, *and further irreducible elements.* Going back to $\mathbb{C}[\mathbf{X}, \mathbf{Y}^{\mathbb{C}}]$, we have

Theorem 4.3. *In* $\mathbb{C}[\mathbf{X}, \mathbf{Y}^{\mathbb{C}}]$*, any* $f \neq 0$ *can be factored, essentially uniquely, into an element*

$$\prod_{j=1}^{t} (\mathbf{Y}^{\varphi_j} - c_j), \tag{4.14}$$

and a product of irreducible elements.

Some of these irreducible elements may be irreducible polynomials in $\mathbb{C}[\mathbf{X}]$, others may involve $\mathbf{Y}$.

Finally in $\mathcal{R}$, the following holds.

Theorem 4.4. *Every nonzero polynomial–exponential function* (4.2) *may be factored, uniquely up to units and associates, into a function*

$$\prod_{j=1}^{t} (e(\varphi_j \xi) - c_j), \tag{4.15}$$

and a product of irreducible elements, some of which may be ordinary polynomials, and some may be genuinely of polynomial–exponential type.

Remark. It is important to observe that a factorization such as (4.14) or (4.15) is NOT unique.

Example. Let $n = 1$,

$$f(\xi) = (e(3\xi) - 1)(e(i\xi) - 2)\xi(e(2\xi) + \xi e(\xi) + 1).$$

Here $(e(3\xi)-1)(e(i\xi)-2)$ is of the form (4.15), and ξ as well as $e(2\xi)+\xi e(\xi)+1$ are irreducible.

5 Gourin's Theorem

Let K be an algebraically closed field of characteristic 0. A polynomial in $K[\mathbf{Z}] = K[Z_1, \ldots, Z_k]$ will be called *primary* in Z_i if the exponents of Z_i occurring in it have no common factor > 1. It will be termed *primary* if it is primary in each Z_i. Note that a polynomial primary in Z_i actually depends on Z_i.

Theorem 5.1. *Suppose* $f \in K[\mathbf{Z}]$ *is irreducible, primary, and has at least three terms. Let* δ *be the maximal degree of* f *in* $Z_1, \ldots, Z_k$*. Suppose* $\mathbf{t} = (t_1, \ldots, t_k) \in \mathbb{N}^k$*. Then there are* k*-tuples* $\mathbf{s} = (s_1, \ldots, s_k)$, $\mathbf{u} = (u_1, \ldots, u_k) \in \mathbb{N}^k$ *such that*

$$t_i = s_i u_i \qquad (i = 1, \ldots, k), \tag{5.1}$$

$$0 < s_i \leqq \delta^2 \qquad (i = 1, \dots, k), \tag{5.2}$$

and the canonical factorization of $f(Z_1^{s_1}, \dots, Z_k^{s_k})$ *into irreducible factors is*

$$f(Z_1^{s_1}, \dots, Z_k^{s_k}) = c \cdot \prod_{j=1}^{N} f_j(Z_1, \dots, Z_k) \tag{5.3}$$

with non-proportional[1] *polynomials* $f_1, \dots, f_N$*, and we also have the canonical factorization*

$$f(Z_1^{t_1}, \dots, Z_k^{t_k}) = c \cdot \prod_{j=1}^{N} f_j(Z_1^{u_1}, \dots, Z_k^{u_k}). \tag{5.4}$$

The Theorem is due to Gourin [15]. Our presentation in this Section is indebted to Schinzel [30], who gave a more general version where f need not be primary, and where K need not be of characteristic 0 or algebraically closed.

Corollary 5.2. *Suppose* $g \in K[\mathbf{Z}^{\mathbb{Z}}]$ *is irreducible and has at least three terms. Let* $h \in K[\mathbf{Z}^{\mathbb{Q}}]$ *be a divisor of* g*. Then* $h \in K[\mathbf{Z}^{(1/m)\mathbb{Z}}]$ *where* $m \leqq m_0(g)$*.*

Proof. After multiplication by a suitable unit of $K[\mathbf{Z}^{\mathbb{Z}}]$ we may suppose that $g \in K[\mathbf{Z}]$, and is not divisible by any Z_i. Now if $h \in K[\mathbf{Z}^{\mathbb{Q}}]$ is a divisor, we may suppose that all of its exponents are nonnegative. Choose m minimal with $h \in K[Z_1^{1/m}, \dots, Z_k^{1/m}]$. We may suppose that each variable Z_i actually occurs in g, and that $g = f(Z_1^{v_1}, \dots, Z_k^{v_k})$ where f is primary. Then f, like g, is irreducible and has at least 3 terms. We will apply the Theorem with $t_i = v_i m$. We get a canonical factorization

$$f(Z_1^{v_1 m}, \dots, Z_k^{v_k m}) = c \cdot \prod_{j=1}^{N} f_j(Z_1^{u_1}, \dots, Z_k^{u_k})$$

in $K[\mathbf{Z}]$ where $v_i m = s_i u_i$ with $s_i \in \mathbb{N}$, $s_i \leqq \delta^2$. Then

$$g(Z_1, \dots, Z_k) = f(Z_1^{v_1}, \dots, Z_k^{v_k}) = c \cdot \prod_{j=1}^{N} f_j(Z_1^{u_1/m}, \dots, Z_k^{u_k/m})$$

is a canonical factorization in $K[Z_1^{1/m}, \dots, Z_k^{1/m}]$.

Thus every divisor of g in $K[Z_1^{1/m}, \dots, Z_k^{1/m}]$ lies in $K[Z_1^{1/d}, \dots, Z_k^{1/d}]$ where $d = m/(\gcd(m, u_1, \dots, u_k))$. Since h is such a divisor, and since m was minimal for h, we get $d = m$, hence $\gcd(m, u_1, \dots, u_k) = 1$. But $(m/m_i) \mid u_i$ where $m_i = \gcd(m, s_i)$, so that $(m/n) \mid u_i$ $(i = 1, \dots, k)$ with

[1] Two polynomials are *proportional* if they differ by a constant factor.

$n = \text{lcm}(m_1, \ldots, m_k) \leqq s_1 \cdots s_k \leqq \delta^{2k}$. Therefore $(m/n) \mid \gcd(m, u_1, \ldots, u_k)$, which yields $m/n = 1$, and $m = n \leqq \delta^{2k}$. □

Our Corollary establishes a claim made in the last Section.

The proof of Theorem 5.1 will depend on several Lemmas. We will write ζ_t for a primitive t-th root of 1.

Lemma 5.3. *Let $f(Z_1, \ldots, Z_k)$ be the polynomial of Theorem 5.1. In the factorization of $f(Z_1^{t_1}, \ldots, Z_k^{t_k})$ into irreducible factors, say*

$$f(Z_1^{t_1}, \ldots, Z_k^{t_k}) = c \cdot \prod_{j=1}^{N} \tilde{f}_j(Z_1, \ldots, Z_k)^{\tilde{e}_j}, \tag{5.5}$$

each $\tilde{e}_j = 1$, and the $\tilde{f}_j$ $(j = 1, \ldots, N)$ form a complete set of nonproportional polynomials obtained from $\tilde{f}_1$ by substitutions of the form

$$(Z_1, \ldots, Z_k) \mapsto (\zeta_{t_1}^{i_1} Z_1, \ldots, \zeta_{t_k}^{i_k} Z_k) \tag{5.6}$$

with integers $i_1, \ldots, i_k$.

Proof. All polynomials obtained from $\tilde{f}_1$ by the mentioned substitutions are irreducible and, when nonproportional, are coprime. Hence $f(Z_1^{t_1}, \ldots, Z_k^{t_k}) = gh$, where g is the product of all nonproportional polynomials obtained from $\tilde{f}_1$. Every substitution

$$(Z_1, \ldots, Z_k) \mapsto (\zeta_{t_1}^{i_1} Z_1, Z_2, \ldots, Z_k) \tag{5.7}$$

maps g into a proportional polynomial. If we had $Z_1 \mid g$, then $Z_1 \mid f$, so that f is reducible or proportional to Z_1, with both alternatives ruled out by the hypothesis.

Hence g contains a term not depending on Z_1. Therefore a substitution (5.7) not only maps g into a proportional polynomial, but in fact leaves g invariant. We may infer that g is a polynomial in $Z_1^{t_1}, Z_2, \ldots, Z_k$. By symmetry, $g = g_0(Z_1^{t_1}, \ldots, Z_k^{t_k})$ with a polynomial g_0. Then

$$h = f(Z_1^{t_1}, \ldots, Z_k^{t_k}) / g_0(Z_1^{t_1}, \ldots, Z_k^{t_k})$$

lies in $K[Z_1^{t_1}, \ldots, Z_k^{t_k}]$; say it equals $h_0(Z_1^{t_1}, \ldots, Z_k^{t_k})$. We have

$$f(Z_1^{t_1}, \ldots, Z_k^{t_k}) = g_0(Z_1^{t_1}, \ldots, Z_k^{t_k}) h_0(Z_1^{t_1}, \ldots, Z_k^{t_k}),$$

hence $f = g_0 h_0$. This contradicts the irreducibility of f unless h_0 is constant. Further $g_0(Z_1^{t_1}, \ldots, Z_k^{t_k}) = g(Z_1, \ldots, Z_k)$ is of the form $\prod_{j=1}^{N} \tilde{f}_j(Z_1, \ldots, Z_k)$ as specified in the Lemma. □

Since f is primary, it depends on every variable Z_j, and hence so does $\tilde{f}_1$. Write $\tilde{f}_1(Z_1, \ldots, Z_k) = f_1(Z_1^{u_1}, \ldots, Z_k^{u_k})$ where f_1 is primary. By Lemma 5.3, $\tilde{f}_j = f_j(Z_1^{u_1}, \ldots, Z_k^{u_k})$ where f_j is primary. Since f is primary, $u_i \mid t_i$, say $t_i = s_i u_i$ $(i = 1, \ldots, k)$. Now (5.5) with $\tilde{e}_j = 1$ yields (5.4), hence (5.3). The polynomials $f_1, \ldots, f_N$ are irreducible and not proportional to each other. Only (5.2) remains to be proved.

Lemma 5.4. *f_1 has more than two terms.*

Proof. Clearly f_1 has more than one term. It it had two terms, then since it is primary, depends on each Z_j and is irreducible, we have after suitable numbering of variables

$$f_1 = aZ_1 \cdots Z_\ell - bZ_{\ell+1} \cdots Z_k$$

for some ℓ, $1 \leqq \ell < k$. Then

$$\tilde{f}_1 = aZ_1^{u_1} \cdots Z_\ell^{u_\ell} - bZ_{\ell+1}^{u_{\ell+1}} \cdots Z_k^{u_k}.$$

The substitutions (5.6) map $\tilde{f}_1$ into polynomials

$$\zeta_{t_1}^{u_1 i_1} \cdots \zeta_{t_\ell}^{u_\ell i_\ell}(aZ_1^{u_1} \cdots Z_\ell^{u_\ell} - \zeta_{t_1}^{-u_1 i_1} \cdots \zeta_{t_\ell}^{-u_\ell i_\ell} \zeta_{t_{\ell+1}}^{u_{\ell+1} i_{\ell+1}} \cdots \zeta_{t_k}^{u_k i_k} bZ_{\ell+1}^{u_{\ell+1}} \cdots Z_k^{u_k}),$$

proportional to

$$aZ_1^{u_1} \cdots Z_\ell^{u_\ell} - \zeta bZ_{\ell+1}^{u_{\ell+1}} \cdots Z_k^{u_k},$$

where ζ runs through the r-th roots of unity, with r the order of the group generated by $\zeta_{t_1}^{u_1}, \ldots, \zeta_{t_k}^{u_k}$. By Lemma 5.3, $f(Z_1^{t_1}, \ldots, Z_k^{t_k})$ is proportional to the product of the latter polynomials, which equals

$$(aZ_1^{u_1} \cdots Z_\ell^{u_\ell})^r - (bZ_{\ell+1}^{u_{\ell+1}} \cdots Z_k^{u_k})^r.$$

But this is impossible, since f has more than two terms. □

Lemma 5.5. *Suppose*

$$f_1 = \sum_{j=1}^{q} a_j \mathbf{Z}^{\mathbf{b}_j}, \tag{5.8}$$

where $q \geqq 3$, the a_j are nonzero, and $\mathbf{b}_1, \ldots, \mathbf{b}_q$ are distinct. Set $\mathbf{b}_j' = (1, b_{j1}, \ldots, b_{jk})$ $(j = 1, \ldots, q)$. Then there are (at least) three independent ones among the vectors $\mathbf{b}_1', \ldots, \mathbf{b}_q'$.

Proof. If not, all the $\mathbf{b}_j'$ are linear combinations of some integer vectors $\mathbf{u}', \mathbf{v}'$. Say $\mathbf{b}_j' = c_j\mathbf{u}' + d_j\mathbf{v}'$ $(j = 1, \ldots, q)$ with integers c_j, d_j. Certainly $\mathbf{u}', \mathbf{v}'$ may be chosen such that their first coordinate is 1, say $\mathbf{u}' = (1, \mathbf{u})$, $\mathbf{v}' = (1, \mathbf{v})$, and then $c_j + d_j = 1$ $(j = 1, \ldots, q)$. Now

$$f_1 = \sum_{j=1}^{q} a_j (\mathbf{Z}^{\mathbf{u}})^{c_j} (\mathbf{Z}^{\mathbf{v}})^{d_j} = \mathbf{Z}^{\mathbf{v}} \sum_{j=1}^{q} a_j (\mathbf{Z}^{\mathbf{u}-\mathbf{v}})^{c_j}.$$

Since $q \geqq 3$ and K is algebraically closed,

$$\sum_{j=1}^{q} a_j X^{c_j} = X^p g_1(X) g_2(X)$$

with $p \in \mathbb{Z}$, and polynomials g_1, g_2 with positive degree and nonzero constant term. But then

$$f_1 = \mathbf{Z}^{\mathbf{v}+p(\mathbf{u}-\mathbf{v})} g_1(\mathbf{Z}^{\mathbf{u}-\mathbf{v}}) g_2(\mathbf{Z}^{\mathbf{u}-\mathbf{v}}),$$

contradicting the fact that f_1 is irreducible. □

Lemma 5.6. *Suppose f_1 is given by* (5.8). *Set*

$$n_i = \deg_{Z_i} f_1, \qquad m_i = \deg_{Z_i} f \qquad (i = 1, \dots, k). \tag{5.9}$$

There is some i, $1 < i \leqq k$, and there are $u < v < w$, such that the determinant

$$D = \begin{vmatrix} 1 & b_{u1} & b_{ui} \\ 1 & b_{v1} & b_{vi} \\ 1 & b_{w1} & b_{wi} \end{vmatrix} \tag{5.10}$$

has $0 < |D| \leqq n_1 n_i$.

Proof. By the preceding Lemma, the matrix

$$\begin{pmatrix} 1 & b_{11} & \cdots & b_{1k} \\ & & \cdots & \\ 1 & b_{q1} & \cdots & b_{qk} \end{pmatrix} \tag{5.11}$$

has rank at least 3. Since f_1 is primary and not divisible by Z_1, the exponents $b_{11}, \dots, b_{q1}$ of Z_1 in (5.8) are not all equal. Therefore the first two columns in (5.11) are independent, and there is a third column independent of them. Hence some determinant D is $\neq 0$. If we let $b_{u1}, \dots, b_{wi}$ in (5.10) vary in [0,1], the maximum modulus of D is reached when in fact these numbers are 0 or 1, and is easily checked to be 1. Now in our situation, where $0 \leqq b_{j1} \leqq n_1$, $0 \leqq b_{ji} \leqq n_i$, we have $|D| \leqq n_1 n_i$ by homogeneity. □

Proof of Theorem 5.1. Only (5.2) still needs to be established. In Lemma 5.6, suppose $i = 2$, without loss of generality. We claim that when $d \not\equiv 0 \pmod{s_1}$, then

$$f_1(\zeta_{s_1}^d Z_1, Z_2, \dots, Z_k) \tag{5.12}$$

is not proportional to $f_1(Z_1, \dots, Z_k)$. For since f_1 is not divisible by Z_1, it has a term independent of Z_1. So if the above polynomials were proportional, they would in fact be equal. But then

$$a_j \zeta_{s_1}^{d b_{j1}} = a_j \qquad (j = 1, \dots, q)$$

and $d b_{j1} \equiv 0 \pmod{s_1}$. But since (because f_1 is primary) $\gcd(b_{11}, \dots, b_{q1}) = 1$, this yields $d \equiv 0 \pmod{s_1}$, establishing the claim.

In fact, any two polynomials (5.12) with $d \not\equiv d' \pmod{s_1}$ are not proportional. As a consequence, $N \geqq s_1$ in (5.3), so that the right hand side has degree $\geqq s_1 n_1$ in Z_1. The left hand side has degree $s_1 m_1$, so that

$$n_1 \leqq m_1.$$

We next claim that when d is arbitrary, and $e \not\equiv 0 \pmod{s_2/(s_2, D)}$,[2] then

$$f_1(\zeta_{s_1}^d Z_1, \zeta_{s_2}^e Z_2, Z_3, \dots, Z_k) \tag{5.13}$$

is not proportional to $f_1(Z_1, \dots, Z_k)$. For if this were so, then

$$a_j \zeta_{s_1}^{db_{j1}} \zeta_{s_2}^{eb_{j2}} = (\text{const}) \cdot a_j \qquad (j = 1, \dots, q).$$

The constant must be a root of 1. We may suppose the argument of ζ_{s_2} is $2\pi/s_2$. Writing $2\pi\rho$, $2\pi\sigma$ for the argument of the constant, and of $\zeta_{s_1}^d$, respectively, we obtain

$$\rho - b_{j1}\sigma - b_{j2}(e/s_2) \equiv 0 \pmod 1.$$

This is true in particular for $j = u, v$ or w. Solving this set of 3 linear congruences for ρ, σ, e/s_2, we see that they are integer multiples of $1/D$. In particular, $(e/s_2)D \in \mathbb{Z}$, therefore $eD \equiv 0 \pmod{s_2}$, so that $e \equiv 0 \pmod{s_2/(s_2, D)}$, establishing the claim.

In fact any two polynomials (5.13) with respective exponents d, e and d', e' will be nonproportional unless $e \equiv e' \pmod{s_2/(s_2, D)}$. On the other hand when $e = e'$ is given, they will by our first claim (with $\zeta_{s_2}^e Z_2$ in place of Z_2) be nonproportional unless $d \equiv d' \pmod{s_1}$. Hence as (d, e) runs through the pairs with $0 \leqq d < s_1$, $0 \leqq e < s_2/(s_2 D)$, the polynomials (5.13) will be nonproportional to each other. Since f_1 divides $f(Z_1^{s_1}, \dots, Z_k^{s_k})$, so does each polynomial in (5.13). Therefore, $N \geqq s_1 s_2/(s_2, D) \geqq s_1 s_2/D \geqq s_1 s_2/(n_1 n_2)$ in (5.3). The degree in Z_2 of the right hand side is $n_2 N \geqq s_1 s_2/n_1$. The degree of the left hand side is $s_2 m_2$, so that $s_1 \leqq n_1 m_2 \leqq m_1 m_2 \leqq \delta^2$. By symmetry, (5.2) holds. □

6 Hadamard Products, Quotients and Roots

Most of the results of this Section will be stated without proof. Suppose $\mathcal{F} \in \mathbb{C}[[Z]]$, so that

$$\mathcal{F}(Z) = \sum_{k=0}^{\infty} a_k Z^k \tag{6.1}$$

is a formal power series with complex coefficients. It is an almost immediate consequence of Theorem 3.1 that

$\mathcal{F}(Z)$ is a rational function, i.e., it belongs to $\mathbb{C}(Z)$, precisely if there is a linear recurrence sequence u and a k_0 such that $a_k = u_k$ for $k > k_0$.

[2] We set $(s_2, D) = \gcd(s_2, D)$.

The *Hadamard Product* of $\mathcal{F}(Z)$ and $\mathcal{G}(Z) = \sum_{k=0}^{\infty} b_k Z^k$ is defined to be

$$(\mathcal{F} \circ \mathcal{G})(Z) = \sum_{k=0}^{\infty} a_k b_k Z^k.$$

Lemma 6.1. *Suppose $\mathcal{F}$, $\mathcal{G}$ are rational functions. Then so is $\mathcal{F} \circ \mathcal{G}$.*

Proof. By Section 2, $a_k = u_k = f(k)$ for $k > k_0$, where f is a function of polynomial–exponential type. Similary $b_k = g(k)$ for $k > k_1$ with g of this type. But then also $f(k)g(k)$ is of polynomial–exponential type, and $a_k b_k = f(k)g(k)$ for $k > \max(k_0, k_1)$. The Lemma follows. □

Let $\mathcal{J}$ be the set of subscripts k with $b_k \neq 0$. A *Hadamard Quotient* of $\mathcal{F}$ over $\mathcal{G}$ is any series

$$\mathcal{H}(Z) = \sum_{k=0}^{\infty} h_k Z^k \tag{6.2}$$

such that $h_k = a_k / b_k$ for $k \in \mathcal{J}$. There is a unique Hadamard Quotient when $\mathcal{J} = \mathbb{N}_0$. Two rational, nonzero series $\mathcal{F}$, $\mathcal{G}$ need not have a rational Hadamard Quotient. For when u is a linear recurrence with recurrence relation $u_{k+t} = c_{t-1} u_{k+t-1} + \cdots + c_0 u_k$ $(k \in \mathbb{Z})$, then its members u_k with $k \geqq 0$ will lie in the ring $A \subset \mathbb{C}$ generated over $\mathbb{Z}$ by $u_0, \ldots, u_{t-1}, c_0, \ldots, c_{t-1}$. This ring is finitely generated over $\mathbb{Z}$. But in general, when a_k, b_k come from linear recurrence sequences, there is no reason why the quotients a_k / b_k $(k \in \mathcal{J})$ should lie in such a finitely generated ring. (Take, e.g., $a_k = 1$, $b_k = 2^k - 1$, and observe that for every odd prime p there are infinitely many k with $p \mid b_k$.) However, there is

Theorem 6.2 (Hadamard Quotient Theorem). *Let $\mathcal{F}$, $\mathcal{G}$ be nonzero series which are rational functions. Suppose there is a ring $A \subset \mathbb{C}$, finitely generated over $\mathbb{Z}$, such that $a_k / b_k \in A$ for $k \in \mathcal{J}$. Then there is a Hadamard Quotient of $\mathcal{F}$ over $\mathcal{G}$ which is a rational function.*

The first complete proof may be found in the papers [28] by Rumely. They use an outline by van der Poorten [23]. When $b_k \neq 0$ for large k, i.e., when $\mathbb{N}_0 \backslash \mathcal{J}$ is finite, the coefficients h_k with $k \in \mathbb{N}_0 \backslash \mathcal{J}$ of the Hadamard Quotients do not affect rationality, and therefore in this case every Hadamard Quotient is rational. This happens in particular when $\mathcal{G}$ comes from a "non-degenerate" linear recurrence, i.e., when $b_k = v_k$ for large k, where v is non-degenerate. (See Corollary 7.2 below.)

In terms of linear recurrences, we may infer that if *u, v are linear recurrences with $v_k \neq 0$ for $k > k_0$, and if the quotients u_k / v_k $(k > k_0)$ lie in a finitely generated ring over $\mathbb{Z}$, then there is a linear recurrence w such that $u_k / v_k = w_k$ for $k > k_0$.*

Let $\mathcal{H}^{\circ d} = \mathcal{H} \circ \mathcal{H} \circ \cdots \circ \mathcal{H}$ with d factors. A *Hadamard d-th root* of $\mathcal{F}$ is a series $\mathcal{H}$ with $\mathcal{H}^{\circ d} = \mathcal{F}$. With $\mathcal{F}$, $\mathcal{H}$ given by (6.1), (6.2), this means that

$$a_k = h_k^d \qquad (k \in \mathbb{N}_0).$$

Theorem 6.3 (Hadamard Root Theorem). *Suppose a field $K \subset \mathbb{C}$ is finitely generated over $\mathbb{Q}$, and each coefficient a_k of a series $\mathcal{F}$ is the d-th power of an element of K. If $\mathcal{F}$ is a rational function, it has a d-th Hadamard root $\mathcal{H} \in K[[Z]]$ which is also a rational function, i.e., belongs to $K(Z)$.*

An ingenious proof was recently given by Zannier [48], thereby settling a long standing conjecture usually attributed to Pisot. Rumely and van der Poorten [29] had reduced the general case to the case when K is a number field.

Corollary 6.4. *Suppose u is a linear recurrence such that u_k is a perfect d-th power in K for each $k \in \mathbb{N}_0$. Then there is a linear recurrence v such that $u_k = v_k^d$ and $v_k \in K$ for each $k \in \mathbb{Z}$.*

Proof. By the Theorem, $\mathcal{F} = \sum_{k=0}^{\infty} u_k Z^k$ has a rational Hadamard d-th root $\mathcal{H} = \sum_{k=0}^{\infty} h_k Z^k$ with $h_k^d = u_k$, $h_k \in K$ for $k \in \mathbb{N}_0$. Then there is a k_0 and a linear recurrence v having $v_k = h_k$ for $k > k_0$. Thus $v_k^d = u_k$, $v_k \in K$ for $k > k_0$. Since v_k^d, u_k are given by polynomial–exponential functions, $v_k^d = u_k$ for $k \in \mathbb{Z}$. Further it is easily seen that if $v_k \in K$ for $k > k_0$, then the companion polynomial has coefficients in K, and therefore $v_k \in K$ for $k \in \mathbb{Z}$. □

As indicated at the beginning, Theorems 6.2, 6.3 will not be proved here. For a general discussion of related topics, see also [24].

7 The Zero-Multiplicity, and Polynomial-Exponential Equations

The *zero-multiplicity* (also called *number of zeros*) of a multi-linear recurrence sequence u is the number of $\mathbf{k} \in \mathbb{Z}^n$ with $u_{\mathbf{k}} = 0$.

A linear recurrence sequence $u \in V_1$ is called *non-degenerate* if the quotients α_i/α_j $(i \neq j)$ of the roots $\alpha_1, \ldots, \alpha_q$ of its companion polynomial are not roots of 1. Recall that we may write

$$u_k = P_1(k)\alpha_1^k + \cdots + P_q(k)\alpha_q^k$$

with polynomials $P_1, \ldots, P_q$, none of which will be zero if $\alpha_1, \ldots, \alpha_q$ are roots of the companion polynomial. Let us call u *weakly non-degenerate* if there is an i_0 such that α_{i_0}/α_j is not a root of 1 for $1 \leqq j \leqq q$, $j \neq i_0$. The Skolem–Mahler–Lech Theorem ([45], [20], [18]) says that a non-degenerate linear recurrence has finite zero-multiplicity. In fact, this is true more generally for weakly non-degenerate sequences. That some condition on the sequence is necessary is seen by the example

$$u_k = 2^k + (-2)^k + 3^k - (3\zeta_3)^k, \tag{7.1}$$

where ζ_3 is a primitive cube root of 1. Here $u_k = 0$ when $k \equiv 3 \pmod 6$, and in fact

$$2^k + (-2)^k = 3^k - (3\zeta_3)^k = 0 \tag{7.2}$$

for such k. In this example, the quotients $2/(-2)$ and $3/(3\zeta_3)$ are roots of 1.

In general, when u is a multi-linear recurrence sequence, $u_{\mathbf{k}} = 0$ leads to an equation

$$\sum_{i=1}^{q} P_i(\mathbf{k})\boldsymbol{\alpha}_i^{\mathbf{k}} = 0, \tag{7.3}$$

i.e., an equation of polynomial–exponential type. The example (7.1) with (7.2) suggests that we may often have to deal with a "splitting up" of the equation, more precisely, with a partition π of $\{1,\ldots,q\}$. The sets $\lambda \subset \{1,\ldots,q\}$ belonging to π will be considered to be elements of π: we write $\lambda \in \pi$. Given π, the set of equations

$$\sum_{i\in\lambda} P_i(\mathbf{k})\boldsymbol{\alpha}_i^{\mathbf{k}} = 0 \qquad (\lambda \in \pi) \tag{7.3π}$$

implies (7.3). When π' is a refinement of (7.3π), then (7.3π') implies (7.3π). Let $\mathcal{S}(\pi)$ consist of solutions of (7.3π) which are not solutions of (7.3π') for any proper refinement π' of π. Every solution of (7.3) lies in $\mathcal{S}(\pi)$ for some π, but π is not necessarily determined by the solution.

Write $i \overset{\pi}{\sim} j$ if i, j lie in the same subset λ belonging to π. Finally let $G(\pi)$ be the subgroup of $\mathbb{Z}^n$ consisting of $\mathbf{k}$ having $\boldsymbol{\alpha}_i^{\mathbf{k}} = \boldsymbol{\alpha}_j^{\mathbf{k}}$ for every pair i, j with $i \overset{\pi}{\sim} j$.

Theorem 7.1 (M. Laurent, [17]). *Suppose $G(\pi) = \{\mathbf{0}\}$. Then $\mathcal{S}(\pi)$ is finite.*

In fact, Laurent proves a more general result, to the effect that when writing $\mathbf{k} = \mathbf{k}_G + \mathbf{k}_G^{\perp}$, where $\mathbf{k}_G$ lies in the subspace of $\mathbb{Q}^n$ generated by G, and $\mathbf{k}_G^{\perp}$ in the orthogonal complement, then the solutions in $\mathcal{S}(\pi)$ have $|\mathbf{k}_G^{\perp}| \ll \log^+ |\mathbf{k}|$, where $\log^+ x = \log x$ when $x \geqq e$, but $= 1$ when $0 < x < e$.

Corollary 7.2. *A weakly non-degenerate linear recurrence sequence $u \in V_1$ has finite zero-multiplicity. In particular, the Skolem–Mahler–Lech Theorem holds.*

Deduction of Corollary. We may suppose that the hypothesis holds with $i_0 = 1$, so that α_1/α_j for $1 < j \leqq q$ is not a root of 1. It will suffice to show that $\mathcal{S}(\pi)$ is finite for every partition π. Given π, let λ be the set belonging to π which contains 1. When the cardinality $|\lambda| = 1$, then (7.3π) yields

$P_1(k)\alpha_1^k = 0$. Since α_1 is a root of the companion polynomial of u, the polynomial $P_1 \neq 0$, and there are only finitely many k with $P_1(k) = 0$. When $|\lambda| > 1$, there is a $j \neq 1$ in λ, and since α_1/α_j is not a root of 1, the equation $\alpha_1^k = \alpha_j^k$ yields $k = 0$. Therefore $G(\pi) = \{0\}$, and $\mathcal{S}(\pi)$ is finite by the Theorem. □

For points $\mathbf{x} = (x_1, \ldots, x_n)$, $\mathbf{y} = (y_1, \ldots, y_n)$ in $\mathbb{C}^n$ with nonzero coordinates, set

$$\mathbf{x} * \mathbf{y} = (x_1 y_1, \ldots, x_n y_n).$$

Then $(\mathbb{C}^\times)^n$ (often denoted $\mathbb{G}_m^n$) becomes a group under the operation $*$. The symbol Γ will denote a subgroup of this group, and we set $\Gamma_D = \Gamma \cap D$, where D is the "diagonal", consisting of elements $(x, \ldots, x) \in (\mathbb{C}^\times)^n$. We further set $\mathbb{P}(\Gamma) = \Gamma/\Gamma_D$. A solution of an equation

$$x_1 + \cdots + x_n = 0 \tag{7.4}$$

will be called *nondegenerate*[3] if no subsum vanishes, i.e., if $\sum_{i\in\mathcal{I}} x_i \neq 0$ for every proper, nonempty subset $\mathcal{I}$ of $\{1, \ldots, n\}$.

Theorem 7.3. *Suppose Γ is finitely generated. Then up to factors of proportionality,* (7.4) *has only finitely many nondegenerate solutions $\mathbf{x} \in \Gamma$. Put differently,* (7.4) *has only finitely many nondegenerate solutions in $\mathbb{P}(\Gamma)$.*

This Theorem was first proved by Evertse [10] and by van der Poorten and Schlickewei [25].

Proof of Theorem 7.3. Also $\mathbb{P}(\Gamma)$ is finitely generated. Say $\mathbb{P}(\Gamma) = T \otimes \bar{\Gamma}_1$ where T is the torsion subgroup and $\bar{\Gamma}_1$ a free abelian group of some rank r. Here T is finite, say containing $\bar{\boldsymbol{\zeta}}_1, \ldots, \bar{\boldsymbol{\zeta}}_m$, and say $\bar{\Gamma}_1$ is generated by $\bar{\boldsymbol{\gamma}}_1, \ldots, \bar{\boldsymbol{\gamma}}_r$, where generally $\bar{\mathbf{x}}$ denotes the image of $\mathbf{x} \in \Gamma$ in $\mathbb{P}(\Gamma) = \Gamma/\Gamma_D$. Every $\mathbf{x} \in \Gamma$ may be written as $\mathbf{x} = \mathbf{d} * \mathbf{y}$ where $\mathbf{d} \in D$ and

$$\mathbf{y} = \boldsymbol{\zeta}_i \boldsymbol{\gamma}_1^{k_1} \cdots \boldsymbol{\gamma}_r^{k_r} \tag{7.5}$$

with some i, $1 \leqq i \leqq m$, and $\mathbf{k} = (k_1, \ldots, k_r) \in \mathbb{Z}^r$. It will suffice to show that the equation $y_1 + \cdots + y_n = 0$ with $\mathbf{y}$ of the form (7.5) has only finitely many nondegenerate solutions. We may restrict ourselves to fixed i. When $\boldsymbol{\zeta}_i = (\zeta_1, \ldots, \zeta_n)$ and $\boldsymbol{\gamma}_j = (\gamma_{j1}, \ldots, \gamma_{jn})$ $(1 \leqq j \leqq r)$, we are led to the equation

$$\zeta_1 \gamma_{11}^{k_1} \cdots \gamma_{r1}^{k_r} + \cdots + \zeta_n \gamma_{1n}^{k_1} \cdots \gamma_{rn}^{k_r} = 0.$$

Nondegenerate solutions belong to $\mathcal{S}(\pi)$ where π is the "partition" consisting of the single set $\lambda = \{1, \ldots, n\}$. Hence $G(\pi)$ consists of $\mathbf{k} = (k_1, \ldots, k_r)$ with

$$\gamma_{11}^{k_1} \cdots \gamma_{r1}^{k_r} = \cdots = \gamma_{1n}^{k_1} \cdots \gamma_{rn}^{k_r}.$$

But this is the same as $\boldsymbol{\gamma}_1^{k_1} \cdots \boldsymbol{\gamma}_r^{k_r} \in D$, which by our construction of $\bar{\boldsymbol{\gamma}}_1, \ldots, \bar{\boldsymbol{\gamma}}_r$ yields $\mathbf{k} = \mathbf{0}$. We may conclude that $G(\pi) = \{\mathbf{0}\}$, so that $\mathcal{S}(\pi)$ is finite by Theorem 7.1. □

[3] To distinguish the unfortunate identical terminology for linear recurrences and for solutions of linear equations, I write non-degenerate versus nondegenerate.

8 Proof of Laurent's Theorem in the Number Field Case

We will suppose that the components of each $\boldsymbol{\alpha}_i$, and the coefficients of the polynomials P_i $(i = 1, \ldots, q)$ lie in a number field K.

Let $M(K)$ be the set of places of K, and for $v \in M(K)$ let $|\cdot|_v$ be the associated absolute value, normalized so that it extends the standard or a p-adic absolute value of $\mathbb{Q}$. Let $\|\cdot\|_v$ be the renormalized absolute value, given by $\|x\|_v = |x|_v^{d_v/d}$, where d is the degree of K, and d_v the local degree belonging to v. When $\mathbf{x} = (x_1, \ldots, x_n) \in K^n$, set $\|\mathbf{x}\|_v = \max(\|x_1\|_v, \ldots, \|x_n\|_v)$. The absolute multiplicative Height of $\mathbf{x}$ is given by

$$H(\mathbf{x}) = \prod_{v \in M(K)} \|\mathbf{x}\|_v.$$

By the Product Formula, $H(\lambda\mathbf{x}) = H(\mathbf{x})$ for $\lambda \in K^\times$. Clearly

$$H(\mathbf{x} * \mathbf{y}) \leqq H(\mathbf{x})H(\mathbf{y}). \tag{8.1}$$

Further when $\mathbf{x} \in (K^\times)^n$ and $\mathbf{x}^{-1}$ is the inverse of $\mathbf{x}$ with respect to $*$, we claim that

$$H(\mathbf{x}^{-1}) \leqq H(\mathbf{x})^{n-1}. \tag{8.2}$$

Since $H(\lambda\mathbf{x}) = H(\mathbf{x})$, we may suppose that $x_1 = 1$, and then indeed,

$$\begin{aligned} H(\mathbf{x}^{-1}) &= \prod_v \max(\|x_1\|_v^{-1}, \ldots, \|x_n\|_v^{-1}) \leqq \prod_v \prod_{i=2}^n \max(\|x_1\|_v^{-1}, \|x_i\|_v^{-1}) \\ &= \prod_v \prod_{i=1}^n \max(\|x_i\|_v, \|x_1\|_v) = \prod_{i=1}^n H((x_1, x_i)) \leqq H(\mathbf{x})^{n-1}. \end{aligned}$$

When λ is a subset of $\{1, \ldots, n\}$, say $\lambda = \{i_1, \ldots, i_m\}$, set $H_\lambda(\mathbf{x}) = H((x_{i_1}, \ldots, x_{i_m}))$.

Lemma 8.1. *Let C be a collection of subsets λ of $\{1, \ldots, n\}$. Suppose for any i, j in $1 \leqq i, j \leqq n$, there are sets $\lambda_1, \ldots, \lambda_k$ in C with $i \in \lambda_1$, $j \in \lambda_k$ and $\lambda_i \cap \lambda_{i+1} \neq \varnothing$ for $1 \leqq i < k$. Then there is a collection $C_1 \subset C$ of at most $n - 1$ subsets which has the same property as C. For any $\mathbf{x} \in (K^\times)^n$,*

$$H(\mathbf{x}) \leqq \prod_{\lambda \in C_1} H_\lambda(\mathbf{x}).$$

Proof. We first claim that when $\lambda \cap \mu \neq \varnothing$, then

$$H_{\lambda \cup \mu}(\mathbf{x}) \leqq H_\lambda(\mathbf{x}) H_\mu(\mathbf{x}).$$

Without loss of generality we may suppose that $\lambda \cap \mu = \{1, \ldots, r\}$, $\lambda = \{1, \ldots, r, r+1, \ldots, r+s\}$, $\mu = \{1, \ldots, r, r+s+1, \ldots, r+s+t\}$. We further may suppose that $x_1 = 1$. The claim now follows from

$$\|(1,\ldots,x_{r+s+t})\|_v \leqq \|(1,\ldots,x_r,x_{r+1},\ldots,x_{r+s})\|_v \cdot \|(1,\ldots,x_r,x_{r+s+1},\ldots,x_{r+s+t})\|_v.$$

Pick $\nu_1 \in \mathcal{C}$ of cardinality $|\nu_1| > 1$. If $|\nu_1| < n$, there is a $\nu_2 \in \mathcal{C}$ with $\nu_1 \cap \nu_2 \neq \varnothing$ and $|\nu_1 \cup \nu_2| > |\nu_1|$. If $\nu_1,\ldots,\nu_p$ have been chosen, set $\mu_p = \nu_1 \cup \cdots \cup \nu_p$. If $|\mu_p| < n$, pick $\nu_{p+1} \in \mathcal{C}$ with $\mu_p \cap \nu_{p+1} \neq \varnothing$ and $|\mu_p \cup \nu_{p+1}| > |\mu_p|$. After choosing at most $n-1$ sets ν_i, we get a set $\mu_t = \nu_1 \cup \cdots \cup \nu_t$ equal to $\{1,\ldots,n\}$. The collection $\mathcal{C}_1$ consisting of $\nu_1,\ldots,\nu_t$ has the property in the hypothesis for $\mathcal{C}$. By the claim at the beginning,

$$H_{\mu_{p+1}}(\mathbf{x}) \leqq H_{\mu_p}(\mathbf{x}) H_{\nu_{p+1}}(\mathbf{x}) \qquad (1 \leqq p < t).$$

Therefore

$$H(\mathbf{x}) \leqq \prod_{p=1}^{t} H_{\nu_p}(\mathbf{x}) = \prod_{\nu \in \mathcal{C}_1} H_\nu(\mathbf{x}).$$

□

We will consider an equation of polynomial–exponential type,

$$\sum_{i=1}^{q} P_i(\mathbf{k}) \boldsymbol{\alpha}_i^{\mathbf{k}} = 0 \tag{8.3}$$

defined over K, i.e., the coefficients of the polynomials P_i, as well as the components of the $\boldsymbol{\alpha}_i$, lie in K. When $\mathbf{k}$ is a solution, the vector

$$\mathbf{x} = \left(P_1(\mathbf{k})\boldsymbol{\alpha}_1^{\mathbf{k}},\ldots,P_q(\mathbf{k})\boldsymbol{\alpha}_q^{\mathbf{k}}\right) \tag{8.4}$$

lies in the space E given by $x_1 + \cdots + x_q = 0$.

Lemma 8.2. *Let* (8.3) *be defined over* K. *Let* $C > 1$, *and consider the solutions* $\mathbf{k} \in \mathbb{Z}^n$ *with*

$$H(\boldsymbol{\alpha}_1^{\mathbf{k}},\ldots,\boldsymbol{\alpha}_q^{\mathbf{k}}) > C^{|\mathbf{k}|}, \tag{8.5}$$

where $|\mathbf{k}| = (k_1^2 + \cdots + k_n^2)^{1/2}$. *For such solutions,* $\mathbf{x}$ *lies in finitely many proper subspaces of* E.

Proof. Say $\boldsymbol{\alpha}_i = (\alpha_{i1},\ldots,\alpha_{in})$. There is a finite set $S \subset M(K)$ such that $\|\alpha_{ij}\|_v = 1$ for $1 \leqq i \leqq q$, $1 \leqq j \leqq n$ and $v \notin S$. We may enlarge S if necessary, to guarantee that it contains the set $M_\infty(K)$ of infinite places of K. Each α_{ij} is an S-unit (i.e., $\|\alpha_{ij}\|_v = 1$ for $v \notin S$), and the Product Formula yields

$$\prod_{v \in S} \|\boldsymbol{\alpha}_i^{\mathbf{k}}\|_v = 1 \qquad (i = 1,\ldots,q).$$

Writing $\mathbf{x}$ given by (8.4) as $(x_1,\ldots,x_q)$, we have

$$\prod_{v \in S} \|x_i\|_v \leqq A\,|\mathbf{k}|^B \qquad (i = 1, \ldots, q)$$

for some constants A, B determined by the polynomials $P_1, \ldots, P_q$. Setting $\mathbf{y} = (x_1, \ldots, x_{q-1})$, we have

$$\prod_{v \in S} \prod_{i=1}^{q} (\|x_i\|_v / \|\mathbf{y}\|_v) \leqq A^q |\mathbf{k}|^{qB} H(\mathbf{y})^{-q}. \tag{8.6}$$

For each $v \in S$, choose $i = i(v)$ in $1 \leqq i(v) \leqq q-1$ such that

$$\|\mathbf{y}\|_v = \|x_{i(v)}\|_v.$$

There are $(q-1)^{|S|}$ possibilities for all these choices. We will restrict ourselves to solutions with given $i(v)$ for $v \in S$. For $v \in S$ let L_{vj} $(1 \leqq j \leqq q-1)$ be the set of $q-1$ linear forms

$$\{X_1, \ldots, \dot{X}_{i(v)}, \ldots, X_{q-1}, X_1 + \cdots + X_{q-1}\}$$

in variables $X_1, \ldots, X_{q-1}$, where the dot signifies that $X_{i(v)}$ is omitted. Then, since $\mathbf{x}$ lies in E, (8.6) may be rewritten as

$$\prod_{v \in S} \prod_{j=1}^{q-1} (\|L_{vj}(\mathbf{y})\|_v / \|\mathbf{y}\|_v) \leqq A^q |\mathbf{k}|^{qB} H(\mathbf{y})^{-q}. \tag{8.7}$$

Since $H(P_1(\mathbf{k}), \ldots, P_q(\mathbf{k})) \ll |\mathbf{k}|^{B_1}$, the inequalities (8.1), (8.2), (8.5) yield

$$\begin{aligned} H(x_1, \ldots, x_q) &\geqq H(\boldsymbol{\alpha}_1^{\mathbf{k}}, \ldots, \boldsymbol{\alpha}_q^{\mathbf{k}}) / H(P_1(\mathbf{k})^{-1}, \ldots, P_q(\mathbf{k})^{-1}) \\ &\gg C^{|\mathbf{k}|} |\mathbf{k}|^{-B_2}. \end{aligned} \tag{8.8}$$

We have $y_1 + \cdots + y_{q-1} + x_q = 0$, therefore $H(x_1, \ldots, x_q) \ll H(\mathbf{y})$, so that $H(\mathbf{y}) \gg C^{|\mathbf{k}|} |\mathbf{k}|^{-B_2}$, and

$$A^q |\mathbf{k}|^{qB} < H(\mathbf{y})^{\varepsilon}$$

when $\varepsilon > 0$, and $|\mathbf{k}|$ is large. Thus (8.7) yields

$$\prod_{v \in S} \prod_{j=1}^{q-1} (\|L_{vj}(\mathbf{y})\|_v / \|\mathbf{y}\|_v) < H(\mathbf{y})^{-q+\varepsilon}.$$

By the Subspace Theorem (see the lectures of H. P. Schlickewei), and since we are dealing with points $\mathbf{y} \in K^{q-1}$ (and $q-1 < q-\varepsilon$ for small ε), these points will be contained in the union of a finite number of proper subspaces of K^{q-1}, therefore $\mathbf{x} = (y_1, \ldots, y_{q-1}, x_q)$ in a finite number of proper subspaces of the space E: $x_1 + \cdots + x_q = 0$. □

Lemma 8.3. *Let* (8.3), (8.4) *be as above, and consider nondegenerate solutions of* (8.3), *i.e., solutions where no proper, nonempty subsum of* $x_1+\cdots+x_q$ *vanishes. Then given* $C>1$, *there are only finitely many such solutions with* (8.5).

Proof. Suppose $1 \leqq t \leqq q-1$. We wish to show that in any t-dimensional subspace T of the space $x_1+\cdots+x_q=0$, there are only finitely many solutions in question. Suppose that either $t=1$, or $t>1$ and our assertion is true for $(t-1)$-dimensional subspaces. Without loss of generality we may suppose that T may be parametrized by

$$x_\ell = \sum_{j=1}^{t} c_{\ell j} x_j \qquad (t < \ell \leqq q). \tag{8.9}$$

Let λ_ℓ $(t<\ell\leqq q)$ be the set consisting of ℓ and the subscripts j with $c_{\ell j}\neq 0$. We may suppose that each λ_ℓ contains elements besides ℓ, for otherwise $x_\ell=0$ in T, giving degenerate solutions. We distinguish two cases.

(a) The collection $\mathcal{C}$ of sets λ_ℓ has the property of Lemma 8.1 (with q in place of n). Then by (8.8), and that Lemma, there is an ℓ, $t<\ell\leqq q$, having $H_{\lambda_\ell}(\mathbf{x}) \gg C_1^{|\mathbf{k}|}$ with $C_1>1$, therefore (by an argument as for (8.8)) with $H_{\lambda_\ell}(\mathbf{z}) \gg C_2^{|\mathbf{k}|}$ where $\mathbf{z}=(\boldsymbol{\alpha}_1^{\mathbf{k}},\ldots,\boldsymbol{\alpha}_q^{\mathbf{k}})$ and $C_2>1$. Let $\mathbf{x}_{\lambda_\ell}\in K^{|\lambda_\ell|}$ be the point with components having subscripts in λ_ℓ. By the preceding Lemma, $\mathbf{x}_{\lambda_\ell}$ for our solutions will lie in finitely many proper subspaces of the space given by (8.9) (with fixed ℓ). Thus the x_j with $1\leqq j\leqq t$ will satisfy a linear relation, and therefore $\mathbf{x}$ will lie in a proper subspace of T. When $t=1$, then each $x_i=0$, giving a degenerate solution, and when $t>1$, we are finished by induction.

(b) There is a partition of $\{1,\ldots,q\}$ into two proper nonempty sets μ, ν, such that each λ_ℓ $(t<\ell\leqq q)$ is a subset of μ or of ν. Then T is defined by certain linear equations $f(\mathbf{x})=0$, where the support of f (i.e., the set of subscripts i such that X_i occurs with nonzero coefficient) lies either in μ or in ν. By nondegeneracy, we may assume that we don't have the linear equations $\sum_{i\in\mu} x_i=0$ or $\sum_{i\in\nu} x_i=0$ valid in T, but then $\sum_{i=1}^q x_i=0$ is not valid in T, contradicting the fact that T was a subspace of the space E with $x_1+\cdots+x_q=0$. □

Lemma 8.4. *Let* $\psi(\mathbf{k})$ *be a function* $\mathbb{Z}^n\to\mathbb{R}$ *with*

(i) $\psi(\mathbf{0})=0$, $\psi(\mathbf{k})\geqq 1$ *when* $\mathbf{k}\neq\mathbf{0}$,

(ii) $\psi(\ell\mathbf{k})=|\ell|\,\psi(\mathbf{k})$ *for* $\ell\in\mathbb{Z}$,

(iii) $\psi(\mathbf{k}+\mathbf{k}')\leqq\psi(\mathbf{k})+\psi(\mathbf{k}')$.

Then there is a constant $c_1>0$ *with* $\psi(\mathbf{k})\geqq c_1|\mathbf{k}|$.

Proof. (iii) implies that

(iv) $\psi(\mathbf{k})\leqq c_2|\mathbf{k}|$

with a constant c_2. As is easily seen, ψ can be extended to a function on $\mathbb{Q}^n$ by setting $\psi(\ell^{-1}\mathbf{k}) = \ell^{-1}\psi(\mathbf{k})$ for $\ell \in \mathbb{N}$, and this extended function will have (ii), (iii), (iv) for $\ell \in \mathbb{Q}$, $\mathbf{k}, \mathbf{k}' \in \mathbb{Q}^n$. Since $\psi(\mathbf{k}) - \psi(\mathbf{k}') \leqq \psi(\mathbf{k}-\mathbf{k}') \leqq c_2|\mathbf{k}-\mathbf{k}'|$, our function is Lipschitz, hence can be extended to a continuous function on $\mathbb{R}^n$ which again satisfies (ii), (iii), (iv). By Dirichlet's Theorem on simultaneous approximation, for any $\mathbf{x} \in \mathbb{R}^n \backslash \{\mathbf{0}\}$ and $\varepsilon > 0$, there are $\mathbf{k} \in \mathbb{Z}^n \backslash \{\mathbf{0}\}$ and $\ell \in \mathbb{N}$ with $|\ell\mathbf{x} - \mathbf{k}| < \varepsilon$. Taking $\varepsilon < 1/c_2$ we obtain $\psi(\ell\mathbf{x}) \geqq \psi(\mathbf{k}) - \psi(\ell\mathbf{x} - \mathbf{k}) > 1 - c_2\varepsilon > 0$, so that $\psi(\mathbf{x}) > 0$. By continuity, $\psi(\mathbf{x}) \geqq c_1 > 0$ for $\mathbf{x}$ on the sphere $|\mathbf{x}| = 1$, so that $\psi(\mathbf{x}) \geqq c_1|\mathbf{x}|$ for any $\mathbf{x}$, in particular for $\mathbf{x} \in \mathbb{Z}^n$. □

A quantitative version of this Lemma is given by Schlickewei [34].

We now can complete the proof of Theorem 7.1 in the number field case. Suppose $\mathbf{x} = (x_1, \ldots, x_n) \in \left(\overline{\mathbb{Q}}^\times\right)^n$. It is well known that $H(\mathbf{x}) = 1$ if all the ratios x_i/x_j are roots of 1, and $H(\mathbf{x}) > 1$ otherwise. In fact when $\mathbf{x} \in (K^\times)^n$ where K is an algebraic number field and not all ratios x_i/x_j are roots of 1, then $H(\mathbf{x}) \geqq c_3$ where $c_3 = c_3(K) > 1$. As a companion to the Height $H(\mathbf{x})$ we define the absolute logarithmic height to be

$$h(\mathbf{x}) = \log H(\mathbf{x}).$$

Then $h(\mathbf{x} * \mathbf{y}) \leqq h(\mathbf{x}) + h(\mathbf{y})$, and $h(\mathbf{x}) \geqq c_4 = c_4(K) > 0$ if $\mathbf{x} \in (K^\times)^n$ has some ratio x_i/x_j which is not a root of unity.

Given $\boldsymbol{\alpha}_1, \ldots, \boldsymbol{\alpha}_q$ as in the Theorem, and $\lambda \in \pi$, say $\lambda = \{i_1, \ldots, i_m\}$, set

$$\psi_\lambda(\mathbf{k}) = \max\left(h(\boldsymbol{\alpha}_{i_1}^{\mathbf{k}}, \ldots, \boldsymbol{\alpha}_{i_m}^{\mathbf{k}}),\ h(\boldsymbol{\alpha}_{i_1}^{-\mathbf{k}}, \ldots, \boldsymbol{\alpha}_{i_m}^{-\mathbf{k}})\right).$$

Clearly $\psi_\lambda(\ell\mathbf{k}) = |\ell|\,\psi_\lambda(\mathbf{k})$, and $\psi_\lambda(\mathbf{k} + \mathbf{k}') \leqq \psi_\lambda(\mathbf{k}) + \psi_\lambda(\mathbf{k}')$. Further set

$$\psi(\mathbf{k}) = \max_{\lambda \in \pi} \psi_\lambda(\mathbf{k}).$$

Then ψ has properties (ii), (iii) of Lemma 8.4. Now if $\mathbf{k} \in \mathbb{Z}^n \backslash \{\mathbf{0}\}$, and if $\boldsymbol{\alpha}_i^{\mathbf{k}}/\boldsymbol{\alpha}_j^{\mathbf{k}}$ were a root of unity for each pair i, j with $i \overset{\pi}{\sim} j$, then $\boldsymbol{\alpha}_i^{\ell\mathbf{k}} = \boldsymbol{\alpha}_j^{\ell\mathbf{k}}$ for each such pair and some $\ell \in \mathbb{N}$. But this is ruled out by the hypothesis that $G(\pi) = \{\mathbf{0}\}$. Therefore for each $\mathbf{k}$ in question, there is a λ with $\psi_\lambda(\mathbf{k}) \geqq c_4$, and therefore $\psi(\mathbf{k}) \geqq c_4$. Using the last inequality in place of the condition (i) of Lemma 8.4, we may conclude that $\psi(\mathbf{k}) \geqq c_5|\mathbf{k}|$ with $c_5 > 0$, and there is always some $\lambda \in \pi$ with $\psi_\lambda(\mathbf{k}) \geqq c_5|\mathbf{k}|$. Thus if again $\lambda = \{i_1, \ldots, i_m\}$,

$$\max\left(H(\boldsymbol{\alpha}_{i_1}^{\mathbf{k}}, \ldots, \boldsymbol{\alpha}_{i_m}^{\mathbf{k}}),\ H(\boldsymbol{\alpha}_{i_1}^{-\mathbf{k}}, \ldots, \boldsymbol{\alpha}_{i_m}^{-\mathbf{k}})\right) \geqq C_0^{|\mathbf{k}|}$$

where $C_0 = \exp c_5 > 1$. In fact by (8.2),

$$H(\boldsymbol{\alpha}_{i_1}^{\mathbf{k}}, \ldots, \boldsymbol{\alpha}_{i_m}^{\mathbf{k}}) \geqq C^{|\mathbf{k}|} \tag{8.10}$$

with $C = C_0^{1/m} > 1$. To prove the finiteness of $\mathcal{S}(\pi)$, we may concentrate on a particular $\lambda \in \pi$, i.e., we may concentrate on nondegenerate solutions of

$$P_{i_1}(\mathbf{k})\boldsymbol{\alpha}_{i_1}^{\mathbf{k}} + \cdots + P_{i_m}(\mathbf{k})\boldsymbol{\alpha}_{i_m}^{\mathbf{k}} = 0$$

with $\mathbf{k}$ subject to (8.10). By Lemma 8.3, there are only finitely many such solutions. □

9 A Specialization Argument

In fact a more precise version of Laurent's Theorem is known in the number field case. Suppose the number field K is of degree d, and the polynomials in (7.3) have total degrees $\delta_1, \dots, \delta_q$. Then when $G(\pi) = \{\mathbf{0}\}$, the cardinality

$$|\mathcal{S}(\pi)| \leqq c(d, n, q, \delta_1, \dots, \delta_q). \tag{9.1}$$

See Theorem 12.1 below, which however will not be proved in these Notes. Using (9.1) we will derive Laurent's Theorem 7.1 in full generality in this Section. However, in general, a bound for $|\mathcal{S}(\pi)|$ will not follow. But (9.1) can be replaced by a bound independent of d in some cases, and as will be pointed out at the end of this Section, in these cases a general bound for $|\mathcal{S}(\pi)|$ ensues.

Lemma 9.1. *Let $Y \subset \mathbb{C}^m$ be an irreducible algebraic variety of degree δ, and F a Zariski-closed subset of $\mathbb{C}^m$ such that $Y \backslash F$ is not empty. Suppose Y is defined over a field $K \subset \mathbb{C}$. Then there exists an element $\boldsymbol{\beta} = (\beta_1, \dots, \beta_m)$ in $Y \backslash F$ such that $\beta_1, \dots, \beta_m$ are algebraic over K, and in fact*

$$[K(\beta_1, \dots, \beta_m) : K] \leqq \delta. \tag{9.2}$$

Proof. Say $\dim Y = D$. Since $Y \backslash F$ is not empty, $Y \cap F$ is a proper submanifold of Y, each component of which has dimension $< D$. Every sufficiently general linear manifold M of dimension $m - D$ has

$$|M \cap Y| = \delta \tag{9.3}$$

and

$$M \cap Y \cap F = \varnothing.$$

Let M be such a manifold, defined over K, and pick $\boldsymbol{\beta} \in M \cap Y$, so that $\boldsymbol{\beta} \in Y \backslash F$. Since M and Y are defined over K, for any embedding σ of $K(\boldsymbol{\beta})$ into $\mathbb{C}$ leaving elements of K fixed, we have $\sigma(\boldsymbol{\beta}) \in M \cap Y$. Therefore (9.3) yields (9.2). □

Consider a system of equations (7.3π), i.e.,

$$\sum_{i \in \lambda} P_i(\mathbf{k}) \boldsymbol{\alpha}_i^{\mathbf{k}} = 0 \qquad (\lambda \in \pi). \tag{9.4}$$

The point $\langle \boldsymbol{\alpha}_1, \dots, \boldsymbol{\alpha}_q, P_1, \dots, P_q \rangle$ whose coordinates are the α_{ij} $(1 \leqq i \leqq q,$ $1 \leqq j \leqq n)$ and the coefficients of $P_1, \dots, P_q$ lies in a space $W = \mathbb{C}^m$ where m depends on q, n and the total degrees $\delta_1, \dots, \delta_q$ of $P_1, \dots, P_q$. The system of equations (9.4) with $\lambda \in \pi$ and $\mathbf{k} \in \mathcal{S}(\pi)$ defines an algebraic variety $Y \subset W$ defined over $\mathbb{Q}$. Now let $\mathcal{Z}$ be a finite subset of $\mathcal{S}(\pi)$. Define F as the variety in W where some α_{ij} vanishes or where a subsum of a sum in (9.4) vanishes for some $\lambda \in \pi$, $\mathbf{k} \in \mathcal{Z}$. Recall the definition of the group $G(\pi)$; we will now denote it by $G(\pi; \boldsymbol{\alpha}_1, \dots, \boldsymbol{\alpha}_q)$.

Lemma 9.2. *Suppose* $\mathbf{z}_0 = \langle \boldsymbol{\alpha}_1, \ldots, \boldsymbol{\alpha}_q, P_1, \ldots, P_q \rangle$ *is in* $Y \backslash F$ *and has*

$$G(\pi; \boldsymbol{\alpha}_1, \ldots, \boldsymbol{\alpha}_q) = \{\mathbf{0}\}.$$

Then there are n *points*

$$\mathbf{z}_t = \langle \boldsymbol{\alpha}_{t1}, \ldots, \boldsymbol{\alpha}_{tq}, P_{t1}, \ldots, P_{tq} \rangle \qquad (t = 1, \ldots, n) \tag{9.5}$$

in $Y \backslash F$ *with* $[K(\mathbf{z}_t) : K] \leqq \delta$ *and*

$$\bigcap_{t=1}^{n} G(\pi; \boldsymbol{\alpha}_{t1}, \ldots, \boldsymbol{\alpha}_{tq}) = \{\mathbf{0}\}.$$

Proof. Suppose $\mathbf{z}_0 \in Y_0$ where Y_0 is an irreducible component of Y defined over some number field K. Set $d = \deg K$. Let Y_0 be of degree δ. We will construct points (9.5) in $Y_0 \backslash F$ with algebraic coordinates such that for $t = 1, \ldots, n$,

$$G_t := \bigcap_{j=1}^{t} G(\pi, \boldsymbol{\alpha}_{j1}, \ldots, \boldsymbol{\alpha}_{jq})$$

is a sublattice of $\mathbb{Z}^n$ of rank $\leqq n - t$.

Suppose $t = 1$, or $t > 1$ and $\mathbf{z}_1, \ldots, \mathbf{z}_{t-1}$ have already been constructed. When $t = 1$, let $\mathbf{u}_1, \ldots, \mathbf{u}_n$ be a basis of $\mathbb{Z}^n$, and when $t > 1$ let $\mathbf{u}_1, \ldots, \mathbf{u}_{n-t+1}$ be a basis of a lattice of rank $n - t + 1$ containing G_{t-1}. We will construct $\mathbf{z}_t = \langle \boldsymbol{\alpha}_{t1}, \ldots, \boldsymbol{\alpha}_{tq}, P_{t1}, \ldots, P_{tq} \rangle$ such that the system of equations in $k \in \mathbb{Z}$,

$$\boldsymbol{\alpha}_{ti}^{k\mathbf{u}_1} = \boldsymbol{\alpha}_{tj}^{k\mathbf{u}_1} \qquad (\text{any } i, j \text{ with } i \overset{\pi}{\sim} j) \tag{9.6}$$

has the only solution $k = 0$. Then G_t intersects the line spanned by $\mathbf{u}_1$ only in $\mathbf{0}$, and therefore rank $G_t \leqq (n - t + 1) - 1 = n - t$.

Let $V(k, i, j) \subset W$ be defined by (9.6), more precisely by $\boldsymbol{\alpha}_i^{k\mathbf{u}_1} = \boldsymbol{\alpha}_j^{k\mathbf{u}_1}$, and F' the union of F and all the $V(k, i, j)$ with k, i, j having $k > 0$, $\phi(k) \leqq d\delta$ and $i \overset{\pi}{\sim} j$. By hypothesis, since $G(\pi; \boldsymbol{\alpha}_1, \ldots, \boldsymbol{\alpha}_q) = \{\mathbf{0}\}$, we have $\mathbf{z}_0 \in Y_0 \backslash F'$. By Lemma 9.1 there is a $\mathbf{z}_t \in Y_0 \backslash F'$ with $[K(\mathbf{z}_t) : K] \leqq \delta$. We claim that $\mathbf{z}_t$ does not satisfy (9.6) for any $k \neq 0$. Since $\mathbf{z}_t \notin F'$, the multiplicative group Γ generated by the quotients $\boldsymbol{\alpha}_{ti}^{\mathbf{u}_1} / \boldsymbol{\alpha}_{tj}^{\mathbf{u}_1}$ with $i \overset{\pi}{\sim} j$ does not have order k with $\phi(k) \leqq d\delta$. So if (9.6) holds for some k, then the group Γ is cyclic of some order k with $\phi(k) > d\delta$, and $\deg \mathbb{Q}(\Gamma) > d\delta$, contradicting $[\mathbb{Q}(\mathbf{z}_t) : \mathbb{Q}] \leqq [K(\mathbf{z}_t) : \mathbb{Q}] = [K(\mathbf{z}_t) : K]\, d \leqq \delta d$. □

Proof of Laurent's Theorem in the General Case. Each $\mathbf{k} \in \mathcal{Z}$ satisfies

$$\sum_{i \in \lambda} P_{ti}(\mathbf{k}) \boldsymbol{\alpha}_{ti}^{\mathbf{k}} = 0 \qquad (t = 1, \ldots, n, \ \text{ and } \lambda \in \pi). \tag{9.7}$$

This is a system like (9.4), except that q is now replaced by qn. Further π is replaced by the partition π_1 of the set of pairs (t, i) defined by $(t, i) \overset{\pi_1}{\sim} (t', j)$ iff

$t = t'$, $i \overset{\pi}{\sim} j$. Because $\mathbf{z}_t \notin F'$, no subsum of any of the sums in (9.7) vanishes, and therefore each $\mathbf{k} \in \mathcal{Z}$ lies in $\mathcal{S}(\pi_1)$. Here $G(\pi_1) = \{\mathbf{0}\}$ by Lemma 9.2. Since each $[K(\mathbf{z}_t) : K] \leqq \delta$, the system of equations (9.7) is defined over a number field of degree $\leqq d\delta^n$. By (9.1),

$$|\mathcal{Z}| \leqq |\mathcal{S}(\pi_1)| \leqq c(d\delta^n, n, qn, \delta_1, \dots, \delta_q, \dots, \delta_1, \dots \delta_q). \tag{9.8}$$

The quantities occurring in this bound are independent of the finite set $\mathcal{Z} \subset \mathcal{S}(\pi)$. Therefore in fact $|\mathcal{S}(\pi)|$ is bounded by the right hand side of (9.8), so that $\mathcal{S}(\pi)$ is finite. □

Since in general we don't know δ, our method does not give a bound on $|\mathcal{S}(\pi)|$. Suppose that for some values of $n, \delta_1, \dots, \delta_q$, the estimate (9.1) can be replaced by a bound

$$|\mathcal{S}(\pi)| \leqq c'(n, q, \delta_1, \dots, \delta_q)$$

independent of d. Then (9.8) can be replaced by

$$|\mathcal{Z}| \leqq c'(n, qn, \delta_1, \dots, \delta_q, \dots, \delta_1, \dots, \delta_q),$$

and we obtain an explicit bound for $|\mathcal{S}(\pi)|$. This is in fact the case when $n = 1$, or when $\delta_1 = \dots = \delta_q = 0$, so that our polynomials are constants. See Theorems 12.5 and 12.3.

10 A Method of Zannier Using Derivations

Zannier developed a method for polynomial–exponential equations in one variable which often leads to much better results than the specialization approach of the last Section.

Consider

$$f(k) = \sum_{i=1}^{q} P_i(k)\alpha_i^k,$$

and the equation

$$f(k) = 0. \tag{10.1}$$

Call elements α, β of $\mathbb{C}^\times$ *equivalent* if α/β is algebraic. Say $\alpha_1, \dots, \alpha_q$ belong to r equivalence classes, represented by $\beta_1, \dots, \beta_r$. Collecting summands of f with equivalent α's, and changing the notation, we may write

$$f(k) = f_1(k)\beta_1^k + \dots + f_r(k)\beta_r^k \tag{10.2}$$

where

$$f_i(k) = \sum_{n=1}^{q_i} P_{in}(k)\alpha_{in}^k$$

with $q_1 + \cdots + q_r = q$, and with algebraic numbers α_{in}. Given i, let t_i be the dimension of the vector space spanned by the coefficients of $P_{i1}, \ldots, P_{iq_i}$ over the field $\overline{\mathbb{Q}}$ of algebraic numbers. Then $t_i \leqq \sum_{n=1}^{q_i} \langle P_{in} \rangle$, where $\langle P \rangle$ denotes the number of nonzero coefficients of a polynomial P. Therefore

$$A := \sum_{i=1}^{r} t_i \leqq \sum_{i=1}^{q} \langle P_i \rangle.$$

Theorem 10.1 ("First Splitting Lemma"). *Every solution $k \in \mathbb{Z}$ of (10.1), with at most $\frac{1}{2}A(A-1)$ exceptions, has*

$$f_1(k) = \cdots = f_r(k) = 0. \tag{10.3}$$

This Theorem is due to Zannier [47]; see also [4], where a function field analog is proved.

Let $\rho_{i1}, \ldots, \rho_{it_i}$ be a basis of the space spanned by the coefficients of $P_{i1}, \ldots, P_{iq_i}$ over $\overline{\mathbb{Q}}$. Thus

$$P_{in} = \sum_{j=1}^{t_i} \rho_{ij} S_{inj} \qquad (1 \leqq i \leqq r,\ 1 \leqq n \leqq q_i) \tag{10.4}$$

where S_{inj} is a polynomial with algebraic coefficients. Now $f_i(k) = 0$ means that

$$\sum_{j=1}^{t_i} \rho_{ij} \left(\sum_{n=1}^{q_i} S_{inj} \alpha_{in}^k \right) = \sum_{j=1}^{t_i} \rho_{ij} h_{ij}(k) = 0,$$

where

$$h_{ij}(k) = \sum_{n=1}^{q_i} S_{inj}(k) \alpha_{in}^k$$

has all its data, i.e., the coefficients of the S_{inj}, and the α_{in}, in $\overline{\mathbb{Q}}$. Since $\rho_{i1}, \ldots, \rho_{it_i}$ are linearly independent over $\overline{\mathbb{Q}}$, we get

$$h_{ij}(k) = 0 \qquad (1 \leqq i \leqq r,\ 1 \leqq j \leqq t_i).$$

Thus by the Theorem, with at most $\frac{1}{2}A(A-1)$ exceptions, we are reduced to polynomial–exponential equations with algebraic data.

A special case of the Theorem is when each quotient α_i/α_j $(i \neq j)$ is transcendental. Then $r = q$, $q_1 = \cdots = q_r = 1$, and (10.3) becomes

$$P_1(k) = \cdots = P_q(k) = 0.$$

The proof of the Theorem requires some preparations. Let K be a number field containing the α_{ij}, and the coefficients of the polynomials S_{inj} in (10.4). Let L be obtained from K by adjoining the β_i and the ρ_{ij} (i.e., the coefficients of the P_{in}). A *derivation* on L is a map $\delta : L \to L$ with $\delta(a+b) = \delta(a) + \delta(b)$

and $\delta(ab) = a\delta(b) + b\delta(a)$. The elements $a \in L$ with $\delta(a) = 0$ make up a field F, the *constant field* of δ on L. We may extend δ to $L(X)$ by setting $\delta(X) = 0$. When $H \in L(X)$, write $H^\delta = \delta(H)$. It is an easy exercise to show that the constant field of δ on $L(X)$ is $F(X)$. If $a \in F$ such that $H(a)$ is defined, then $\delta(H(a)) = H^\delta(a)$.

Lemma 10.2. *There is a derivation δ of L with constant field $F \supset K$ such that*

$$\delta(\beta_i/\beta_j) \neq 0 \qquad (i \neq j \text{ in } 1 \leqq i, j \leqq r). \tag{10.5}$$

Proof. For fixed $i < j$, since the fields $K(\beta_i/\beta_j)$ and $K(X)$ are isomorphic, there is a unique derivation δ_{ij} on $K(\beta_i/\beta_j)$ which is trivial on K (i.e., has $\delta(a) = 0$ for $a \in K$) and has $\delta_{ij}(\beta_i/\beta_j) = 1$. As is well known, δ_{ij} can be extended to a derivation on L. Let $\{a_{uv}\}$ with $1 \leqq u < v \leqq r$ be an $(r(r-1)/2)$-tuple of elements of K such that $\sum_{u<v} a_{uv}\delta_{uv}(\beta_i/\beta_j) \neq 0$ for $i < j$. Then $\delta = \sum_{u<v} a_{uv}\delta_{uv}$ is a derivation on L which is trivial on K, and has $\delta(\beta_i/\beta_j) \neq 0$ for any $i < j$, hence in fact for any $i \neq j$. □

From now on, δ as in the Lemma will be fixed, and extended to $L(X)$ as indicated above. For given i, $1 \leqq i \leqq r$, let $\gamma_{i1}, \ldots, \gamma_{im_i}$ be a basis of the vector space spanned by the coefficients of $P_{i1}, \ldots, P_{iq_i}$ over the constant field F of δ. Then

$$P_{in} = \sum_{j=1}^{m_i} \gamma_{ij} Q_{inj} \qquad (1 \leqq i \leqq r,\ 1 \leqq n \leqq q_i),$$

where each Q_{inj} is a polynomial with coefficients in F. Since the coefficients of the S_{inj} in (10.4) lie in $K \subset F$, we have $m_i \leqq t_i$, so that

$$B := \sum_{i=1}^{r} m_i \leqq A.$$

Let $\mathcal{S}$ be the set of cardinality B consisting of pairs (i, j) with $1 \leqq i \leqq r$, $1 \leqq j \leqq m_i$, and set

$$g_{ij}(k) = \sum_{n=1}^{q_i} Q_{inj}\alpha_{in}^k \qquad ((i,j) \in \mathcal{S}). \tag{10.6}$$

Proposition 10.3. *With at most $\frac{1}{2}B(B-1)$ exceptions, every solution of* (10.1) *has*

$$g_{ij}(k) = 0 \qquad ((i,j) \in \mathcal{S}). \tag{10.7}$$

This Proposition implies the Theorem, because

$$f_i(k) = \sum_{n=1}^{q_i} P_{in}(k)\alpha_{in}^k = \sum_{j=1}^{m_i} \gamma_{ij}\Big(\sum_{n=1}^{q_i} Q_{inj}(k)\alpha_{in}^k\Big) = \sum_{j=1}^{m_i} \gamma_{ij} g_{ij}(k). \quad (10.8)$$

We now proceed to prove the Proposition, following [47] almost verbatim. Set

$$\lambda_i = \delta(\beta_i)/\beta_i \qquad (1 \leqq i \leqq r). \quad (10.9)$$

We define polynomials $C_{ij\ell}(X) \in L[X]$ inductively on ℓ by

$$C_{ij0}(X) = \gamma_{ij}, \quad C_{ij,\ell+1}(X) = C_{ij\ell}^{\delta}(X) + \lambda_i X C_{ij\ell}(X),$$

where $(i,j) \in \mathcal{S}$. Let us agree on some ordering of $\mathcal{S}$, and let $R(X)$ be the determinant of the $(B \times B)$-matrix whose ℓ-th row consists of the polynomials $C_{ij,\ell-1}(X)$ in the agreed order. Since $C_{ij\ell}$ has degree $\leqq \ell$, we see that

$$\deg R \leqq 1 + 2 + \cdots + (B-1) = \frac{1}{2}B(B-1).$$

From (10.9) we get $\delta(\beta_i^k) = k\delta(\beta_i)\beta_i^{k-1} = k\lambda_i\beta_i^k$. Therefore

$$\delta(C_{ij\ell}(k)\beta_i^k) = C_{ij\ell}^{\delta}(k)\beta_i^k + C_{ij\ell}(k)k\lambda_i\beta_i^k = C_{ij,\ell+1}(k)\beta_i^k,$$

whence

$$\delta^{\ell}(\gamma_{ij}\beta_i^k) = C_{ij\ell}(k)\beta_i^k \qquad ((i,j) \in \mathcal{S},\ \ell \geqq 0). \quad (10.10)$$

In view of (10.2), (10.8), the equation $f(k) = 0$ may be written as

$$\sum_{i=1}^{r}\Big(\sum_{j=1}^{m_i} \gamma_{ij} g_{ij}(k)\Big)\beta_i^k = 0.$$

For given $k \in \mathbb{Z}$, "differentiating" ℓ times we obtain

$$\sum_{i=1}^{r}\Big(\sum_{j=1}^{m_i} C_{ij\ell}(k) g_{ij}(k)\Big)\beta_i^k = 0,$$

on using (10.10). These equations, for $\ell = 0, 1, \ldots, B-1$, make up a system of B linear equations in the B quantities $g_{ij}(k)$ with $(i,j) \in \mathcal{S}$. The determinant of these B equations equals $R(k)$, times a product of powers of $\beta_1, \ldots, \beta_r$. We will see below that $R(X) \neq 0$. Therefore our system of equations has a nontrivial solution $\{g_{ij}(k)\}$ $((i,j) \in \mathcal{S})$ for at most $\deg R \leqq \frac{1}{2}B(B-1)$ values of k. Thus for all but at most $\frac{1}{2}B(B-1)$ solutions of (10.1) we have indeed (10.7).

It remains for us to prove that $R(X) \neq 0$. Otherwise, $R(k) = 0$ for each k. For each such k, the Wronskian (with respect to δ) of the elements

$$\gamma_{ij}\beta_i^k \qquad ((i,j) \in \mathcal{S}), \tag{10.11}$$

in view of (10.10), equals $R(k)$, times some powers of $\beta_1, \ldots, \beta_r$, and therefore this Wronskian vanishes. We may conclude (just as with the standard derivative) that the B quantities (10.11) are linearly dependent over the constant field F.

Lemma 10.4. *Let $T_{ij} \in L(X)$, for $(i,j) \in \mathcal{S}$, have the property that the quantities $T_{ij}(k)\beta_i^k$ are linearly dependent over F for all large integers k. Then there is an h in $1 \leqq h \leqq r$ such that the functions T_{hj} $(1 \leqq j \leqq m_h)$ are linearly dependent over $F(X)$.*

Assuming the Lemma, we may argue as follows. The Lemma certainly applies to $T_{ij} = \gamma_{ij} \in L$. So for some h, the numbers γ_{hj} $(1 \leqq j \leqq m_h)$ are linearly dependent over $F(X)$. If $U_j(X) \in F(X)$ $(1 \leqq j \leqq m_h)$ are the coefficients of a linear dependency relation, substitution of $k \in \mathbb{Z}$ for which the $U_j(k)$ $(1 \leqq j \leqq m_h)$ are defined and not all zero, leads to a linear dependency relation of the γ_{hj} $(1 \leqq j \leqq m_h)$ over F, contradicting their construction.

It remains for us to prove the Lemma. We will use induction on $B = \sum_{i=1}^r m_i$. The assertion is trivial when $B = 1$. We may suppose that $T_{11} \neq 0$. Dividing all terms $T_{ij}(k)\beta_i^k$ by $T_{11}(k)\beta_1^k$, we may suppose that $T_{11} = 1$, $\beta_1 = 1$. Since the quotients β_i/β_j are not altered by this change, our new β_i's again have $\delta(\beta_i/\beta_j) \neq 0$ for $i \neq j$ by (10.5). Therefore $\delta(\beta_1) = \delta(1) = 0$, $\delta(\beta_i) = \delta(\beta_i/\beta_1) \neq 0$ for $1 < i \leqq r$. Again setting $\lambda_i = \delta(\beta_i)/\beta_i$, we have $\lambda_1 = 0$ but $\lambda_i \neq 0$ for $1 < i \leqq r$.

By hypothesis, we have for each large k a nontrivial relation

$$\sum_{(i,j)\in\mathcal{S}} u_{ij}(k)T_{ij}(k)\beta_i^k = 0$$

with coefficients $u_{ij}(k) \in F$. Here $u_{ij}(k) \neq 0$ for some $(i,j) \in \mathcal{S}\backslash(1,1) = \mathcal{S}'$, say. Applying δ, we see that the $B-1$ quantities $\delta(T_{ij}(k)\beta_i^k)$ with $(i,j) \in \mathcal{S}'$ are linearly dependent over F for large k. But these quantities are $(T_{ij}^\delta(k) + k\lambda_i T_{ij}(k))\beta_i^k$, hence are of the same type as the previous quantities $T_{ij}(k)\beta_i^k$. By induction, there is an h, $1 \leqq h \leqq r$, and there are $V_j \in F(X)$ for $j = 1', \ldots, m_h$ (where $1' = 2$ if $h = 1$, but $1' = 1$ when $h > 1$), and not all zero, such that

$$\sum_{j=1'}^{m_h} V_j(X)(T_{hj}^\delta(X) + X\lambda_h T_{hj}(X)) = 0.$$

Substituting $X = k$ where k is an integer such that all our rational functions are defined at k, and multiplying by β_h^k, we obtain

$$\sum_{j=1'}^{m_h} V_j(k)\delta(T_{hj}(k)\beta_h^k) = 0.$$

Setting $H(X) = \sum_{j=1'}^{m_h} V_j(X)T_{hj}(X)$, and recalling that $V_j \in F(X)$, we get $\delta(H(k)\beta_h^k) = 0$, therefore

$$(H^\delta(k) + k\lambda_h H(k))\beta_h^k = 0.$$

Thus $H^\delta(k) + k\lambda_h H(k) = 0$ for large k, so that in fact

$$H^\delta(X) + \lambda_h X H(X) = 0. \tag{10.12}$$

Suppose initially that $h > 1$, so that $\lambda_h \neq 0$. Writing $H(X) = \dfrac{P(X)}{Q(X)}$, and multiplying (10.12) by $Q(X)^2$, we get

$$P^\delta(X)Q(X) - P(X)Q^\delta(X) + \lambda_h X P(X)Q(X) = 0.$$

Comparison of degrees gives $P(X) = 0$, so that $H(X) = 0$. By the definition of $H(X)$, this gives the desired linear dependence of the $T_{hj}(X)$ $(1 \leqq j \leqq m_h)$ over $F(X)$.

When $h = 1$, so that $\lambda_h = 0$, then (10.12) yields $H^\delta(X) = 0$, so that $H(X) \in F(X)$ by the "exercise" mentioned above. But now setting $V_1 = -H$, we have $\sum_{j=1}^{m_1} V_j T_{1j} = 0$, and the $T_{1j}(X)$ $(1 \leqq j \leqq m_1)$ are dependent over $F(X)$.

This completes the proof of the Lemma, hence of Theorem 10.1. □

11 Applications to Linear Recurrences

The recurrence sequences considered in this Section will be non-degenerate and not of the form $\alpha_0^k P(k)$ where α_0 is a root of 1. It will be convenient to write such a sequence u as

$$u_k = \sum_{i=0}^{q} P_i(k)\alpha_i^k, \tag{11.1}$$

with the following convention. If some root of the companion polynomial is a root of unity, let this root be α_0, and $\alpha_1, \ldots, \alpha_q$ the other roots. If no root of the companion polynomial is a root of unity, let these roots be $\alpha_1, \ldots, \alpha_q$, and set $\alpha_0 = 1$, $P_0 = 0$. Let v be another linear recurrence, written as

$$v_h = \sum_{i=0}^{q'} Q_i(h)\beta_i^h, \tag{11.2}$$

with the same convention. The two sequences u, v are said to be *related* if $q = q'$ and after suitable reordering of $\beta_1, \ldots, \beta_q$ we have

$$\alpha_i^a = \beta_i^b \qquad (i = 1, \ldots, q) \tag{11.3}$$

with certain nonzero integers a, b. They are *doubly related* if there is a second reordering of $\beta_1, \ldots, \beta_q$ with this property, i.e., if there is a non-trivial permutation σ of $\{1, \ldots, q\}$ such that we have both (11.3) and

$$\alpha_i^{a'} = \beta_{\sigma(i)}^{b'} \qquad (i = 1, \ldots, q) \tag{11.4}$$

with nonzero integers a', b'. They are called *simply related* if they are related but not doubly related.

The following Lemma comes from work of Evertse [10] and of Laurent [17].

Lemma 11.1. *When u, v are doubly related, then q is even, $a'/b' = -a/b$, and after a suitable ordering of $\alpha_1, \ldots, \alpha_q$ and $\beta_1, \ldots, \beta_q$ we have (11.3) as well as*

$$\alpha_i^{a'} = \beta_{i+1}^{b'}, \qquad \alpha_{i+1}^{a'} = \beta_i^{b'} \qquad (1 \leqq i < q, \ i \text{ odd}). \tag{11.5}$$

(Thus (11.4) holds with σ the permutation interchanging i, $i+1$ for i odd.) The products $\alpha_i\alpha_{i+1}$ and $\beta_i\beta_{i+1}$ for i odd are roots of unity. Conversely, if u, v are related, if q is even and $\alpha_i\alpha_{i+1}$ is a root of unity for i odd, then u, v are doubly related.

There cannot be a third permutation of $\beta_1, \ldots, \beta_q$ with a property like (11.4).

Proof. Let u, v be doubly related with (11.3), (11.4). The permutation σ of $\{1, \ldots, q\}$ is a product of disjoint cycles

$$(i_0, i_1, \ldots, i_{t-1}),$$

where we allow cycles of length $t = 1$. The relation (11.4) yields $\alpha_{i_j}^{a'} = \beta_{i_{j+1}}^{b'}$ for $0 \leqq j < t$ (where we set $i_t = i_0$), and therefore

$$\alpha_{i_j}^{a'b} = \beta_{i_{j+1}}^{b'b} = \alpha_{i_{j+1}}^{b'a} \qquad (0 \leqq j < t).$$

This implies $\alpha_{i_j}^{(a'b)^t} = \alpha_{i_j}^{(b'a)^t}$, and since α_{i_j} is not a root of unity, $(a'b)^t = (b'a)^t$, therefore $a'b = \pm b'a$. If the + sign holds, $\alpha_{i_j}^{a'b} = \alpha_{i_{j+1}}^{a'b}$, so that $\alpha_{i_j}/\alpha_{i_{j+1}}$ is a root of unity, and $i_{j+1} = i_j$ by nondegeneracy. In this case every cycle of σ is of length 1, so that σ is trivial. Therefore the $-$ sign holds, whence $a'/b' = -a/b$, and $\alpha_{i_j}^{a'b} = \alpha_{i_{j+1}}^{-a'b} = \alpha_{i_{j+2}}^{a'b}$, so that $i_{j+2} = i_j$, and every cycle of σ is of length 2. We may conclude that q is even, and σ is a product of $q/2$ cycles of length 2. After reordering, the cycles are $(1,2), (3,4), \ldots, (q-1, q)$. The relation $\alpha_{i_j}^{a'b} = \alpha_{i_{j+1}}^{-a'b}$ now says that $(\alpha_i\alpha_{i+1})^m = 1$ with $m = a'b$, for i odd, so that $\alpha_i\alpha_{i+1}$ is a root of 1. The same is true for $\beta_i\beta_{i+1}$. There cannot be a third permutation, for if it belonged to exponents a'', b'', then $a''/b'' = -a/b$ and $a''/b'' = -a'/b'$, which is impossible.

On the other hand, if u, v are related satisfying (11.3), if q is even and $\alpha_i\alpha_{i+1}$ for i odd is a root of unity, say $(\alpha_i\alpha_{i+1})^m = 1$, then $\alpha_i^{ma}\beta_{i+1}^{mb} = \alpha_i^{ma}\alpha_{i+1}^{ma} = 1$, also $\alpha_{i+1}^{ma}\beta_i^{mb} = \alpha_{i+1}^{ma}\alpha_i^{ma} = 1$, so that (11.5) holds with $a' = ma$, $b' = -mb$, and u, v are doubly related. $\square$

Of course, any recurrence is related to itself. The recurrence is called *symmetric* if it is doubly related with itself. If u, v are related, they are doubly related iff u is symmetric, and also iff v is symmetric. The Fibonacci sequence is symmetric since the roots of its companion polynomial $Z^2 - Z - 1$ have product -1.

From now on, when u, v are related we will suppose that the α's and β's are ordered such that (11.3) holds, and when they are doubly related, such that also (11.5) holds.

Theorem 11.2. *Let u, v be non-degenerate linear recurrence sequences given by* (11.1) *and* (11.2), *and consider the equation*

$$u_k = v_h \tag{11.6}$$

in the integer variables k, h. This equation has only finitely many solutions unless u, v are related. When u, v are simply related, all but finitely many solutions satisfy the system of equations

$$P_i(k)\alpha_i^k = Q_i(h)\beta_i^h \qquad (i = 0, 1, \ldots, q). \tag{11.7}$$

When u, v are doubly related, all but finitely many solutions satisfy either (11.7) *or the system*

$$\begin{cases} P_i(k)\alpha_i^k &= Q_{i+1}(h)\beta_{i+1}^h \qquad (1 \leqq i < q, \;\; i \text{ odd}) \\ P_{i+1}(k)\alpha_{i+1}^k &= Q_i(h)\beta_i^h \qquad (1 \leqq i < q, \;\; i \text{ odd}) \\ P_0(k)\alpha_0^k &= Q_0(h)\beta_0^h. \end{cases} \tag{11.8}$$

This Theorem is essentially a reformulation of Théorème 3 of Laurent [17].

Proof. It will be convenient to introduce the following equivalence relation on $\mathbb{C}^\times$. Set $\alpha \approx \beta$ if α/β is a root of unity. Then α is a root of unity iff $\alpha \approx 1$. Since u is non-degenerate, $\alpha_i \not\approx \alpha_j$ for $i \neq j$. The $q + q' + 2$ summands on the left-hand side of

$$\sum_{i=0}^{q} P_i(k)\alpha_i^k + \sum_{i=0}^{q'}(-Q_i(h))\beta_i^h = 0 \tag{11.9}$$

will be parametrized by symbols

$$0, 1, \ldots, q, \hat{0}, \hat{1}, \ldots, \hat{q}'.$$

Given a partition π with $G(\pi) \neq \{\mathbf{0}\}$ of this set of $q + q' + 2$ symbols, we need to consider solutions $(k, h) \in \mathcal{S}(\pi)$. Let $(r, s) \neq (0, 0)$ be in $G(\pi)$.

CASE A. π contains a singleton $\{i\}$ or $\{\hat{\imath}\}$ with $i \geqq 1$. Say $\{\hat{\imath}\} \in \pi$; then (π) yields $Q_i(h)\beta_i^h = 0$, which gives only finitely many possibilities for h. Given h, (11.9) becomes

$$\sum_{i=1}^{q} P_i(k)\alpha_i^k + (\alpha_0^k P_0(k) + c) = 0$$

with constant c. Since $\alpha_0 \approx 1$ and any two among $1, \alpha_1, \dots \alpha_q$ are $\not\approx$ to each other, it is an easy consequence of the Skolem–Mahler–Lech Theorem that this last equation has only finitely many solutions k.

CASE B. π contains no singleton $\{i\}$ or $\{\hat{i}\}$ with $i \geqq 1$. If we had $i \overset{\pi}{\sim} j$ with some $i \neq j$, then $r = 0$ (again using that $1, \alpha_1, \dots, \alpha_q$ are $\not\approx$ to each other). But then if $\hat{1} \overset{\pi}{\to} \sim\hat{e}$ with some $e \neq 1$, we also get $s = 0$, and if $\hat{1} \overset{\pi}{\to} \sim e$, then $\beta_{\hat{1}}^s = \alpha_e^r = 1$, again yielding $s = 0$, contradicting $(r, s) \neq (0, 0)$. Thus $i \overset{\pi}{\to} \not\sim j$ for $i \neq j$, and similarly $\hat{i} \overset{\pi}{\to} \not\sim \hat{j}$. Further if we had $i \overset{\pi}{\sim} \hat{0}$ for some $i \geqq 1$, then $r = 0$, and now $\hat{1} \overset{\pi}{\sim} j$ for some j leads to $\beta_{\hat{1}}^s = \alpha_j^r = 1$, so that again we get the impossible $r = s = 0$. Therefore $(r, s) \neq (0, 0)$ is only possible if each i ($i \geqq 1$) is paired with a unique $\hat{j}$ ($j \geqq 1$) such that $i \overset{\pi}{\sim} \hat{j}$. Thus $q = q'$, and after suitable ordering, $i \sim \hat{i}$ for $1 \leqq i \leqq q$, and $\alpha_i^r = \beta_{\hat{i}}^s$. Therefore u, v are related. (11.6) can have infinitely many solutions only if u, v are related. Further (π) yields (11.7).

Note that π is *essentially* given by $i \overset{\pi}{\sim} \hat{i}$ ($i = 1, \dots, q$), but it may contain $\{0, \hat{0}\}$, or the singletons $\{0\}$, $\{\hat{0}\}$. If there is another partition π' with infinite $S(\pi')$ which is essentially different from π, we get another pairing of elements of $\{1, \dots, q\}$ and $\{\hat{1}, \dots, \hat{q}\}$. Then u, v are doubly related, and after suitable ordering, π' contains the sets $\{1, \hat{2}\}, \{2, \hat{1}\}, \dots, \{q-1, \hat{q}\}, \{q, \widehat{q-1}\}$, and ($\pi'$) yields (11.8). The Theorem follows easily. □

The *pair-multiplicity* of a linear recurrence u is the number of pairs k, h of integers with $k \neq h$ having

$$u_k = u_h.$$

The Fibonacci sequence has $u_k = u_{-k}$ for k odd, hence has infinite pair-multiplicity. When u is non-degenerate and not symmetric, the pair-multiplicity is a finite number, plus the number of pairs $k \neq h$ with

$$P_i(k)\alpha_i^k = P_i(h)\alpha_i^h \qquad (i = 0, \dots, q).$$

The following result of Evertse [10] is therefore an immediate consequence of Lemma 11.4 below.

Theorem 11.3. *A non-degenerate, and not symmetric, linear recurrence has finite zero-multiplicity.*

Lemma 11.4. *Suppose $\alpha \neq 0$, $\alpha \not\approx 1$, and $P(X)$ is a nonzero polynomial. Then there are only finitely many pairs k, h with $k \neq h$ and*

$$P(k)\alpha^k = P(h)\alpha^h. \tag{11.10}$$

Proof. We may specialize, following the procedure of Section 9. The coefficients of P, and α, make up a vector in a space W of dimension $2 + \deg P$. When $\mathcal{Z}$ is the set of solutions $(h, k) \in \mathbb{Z}^2$, then (11.10) for every $(h, k) \in \mathcal{Z}$ defines a variety $Y \subset W$. Let $F \subset W$ be the variety where $P = 0$, or $\alpha = 0$. The given P, α provide a point $\mathbf{z}_0 \in Y \backslash F$; say $\mathbf{z}_0 \in Y_0 \backslash F$ where Y_0 is an irreducible component of Y, of some degree δ, defined over a number field of some degree d. Let F' be the union of F and the pairs P, α with $\alpha^k = 1$ for some k with $\phi(k) \leqq d\delta$. Then there is an algebraic specialization $\widehat{\alpha}$, $\widehat{P}$ of α, P in $Y_0 \backslash F$, and $\widehat{\alpha}$ again is not a root of 1. We therefore may suppose that our data are algebraic, in some field K.

Now $|\alpha|_v \neq 1$ for some place v, say $|\alpha|_v > 1$. When v is Archimedean and P of degree δ (unrelated to the δ above), then $|P(k)|_v \geqq c_1|k|^\delta > 0$ when $|k| \geqq k_0$. Given k, (11.10) by the Skolem–Mahler–Lech Theorem has only finitely many solutions h. We therefore may suppose that $|k| \geqq k_0$, and also that $|h| \geqq k_0$. Since $|P(h)|_v \leqq c_2|h|^\delta$, we obtain

$$\begin{aligned} |\alpha|_v^{k-h} &= |P(h)/P(k)|_v \leqq (c_2/c_1)|h/k|^\delta \\ &= (c_2/c_1)\left|1 + \frac{h-k}{k}\right|^\delta \leqq (c_2/c_1)(1 + |h-k|)^\delta. \end{aligned}$$

Thus when $k > h$, we see that $|h - k|$ is bounded, and by symmetry this also holds when $h > k$. Therefore it will suffice to study solutions with $k = h + m$ where m is fixed. But then (11.10) yields $P(h) = P(h+m)\alpha^m$. But for $m \in \mathbb{Z}$, $m \neq 0$, the polynomial $P(X) - P(X+m)\alpha^m$ is nonzero, hence has only finitely many zeros.

Suppose $|\alpha|_v = 1$ for v Archimedean. Then (11.10) yields $|P(k)|_v = |P(h)|_v$, from which one easily deduces $\big||k|^\delta - |h|^\delta\big| \leqq c_3(|k|^{\delta-1} + |h|^{\delta-1})$. The left side here is $\big||k| - |h|\big|(|k|^{\delta-1} + \cdots + |h|^{\delta-1})$, so that in fact $\big||k| - |h|\big| \leqq c_3$. Therefore either $k = h + m$ with bounded m, and the proof is completed as before. Or $k = -h + m$ with bounded m. Substitution into (11.10) gives

$$P(h)\alpha^{2h} = P(-h+m)\alpha^m. \tag{11.11}$$

But if $|\alpha|_v = 1$ for every Archimedean place v, there is some non-Archimedean w with $|\alpha|_w > 1$. It is clear that $|P(h)|_w, |P(-h+m)|_w \leqq c_4$ when $h, m \in \mathbb{Z}$, and an argument like Liouville's shows that $|P(h)|_w, |P(-h+m)|_w \gg |h|^{-B}$ for some B, and therefore $|h|^{-B} \ll |P(-h+m)/P(h)|_w \ll |h|^B$, and (11.11) implies that h is bounded. □

When u is a linear recurrence, then an equation

$$Au_k^r = Bu_h^s$$

in $(k, h) \in \mathbb{Z}^2$ with fixed $A \neq 0$, $B \neq 0$, $r \in \mathbb{N}$, $s \in \mathbb{N}$, $r \neq s$ is of the type (11.6), because a power of a linear recurrence (of a function of polynomial–exponential type!) is again a linear recurrence. In [38] it was shown that unless

u is "exceptional" with respect to A, B, r, s, the equation has only finitely many solutions. It is exceptional if $u_k = P(k)\alpha^k$, and if either α is a root of unity, or $P(k)$ is a constant c, and $Ac^r = Bc^s\alpha^n$ for some n divisible by $\gcd(r, s)$.

When u again denotes a non-degenerate sequence, we now turn to an equation

$$Au_k + Bu_h + Cu_\ell = 0 \tag{11.12}$$

with nonzero coefficients A, B, C, in unknowns k, h, ℓ. There may be solutions with $u_k = 0$, $Bu_h + Cu_\ell = 0$. By the Skolem–Mahler–Lech Theorem, there are only finitely many choices for k, and $Bu_h + Cu_\ell = 0$ is of the type dealt with in Theorem 11.2. Therefore we will restrict ourselves to *proper* solutions, i.e., solutions with $u_k u_h u_\ell \neq 0$.

Theorem 11.5. *Suppose u is non-degenerate, and the polynomial P_0 of* (11.1) *is zero. If u is not symmetric, then all but finitely many proper solutions of* (11.12) *satisfy the system*

$$AP_i(k)\alpha_i^k + BP_i(h)\alpha_i^h + CP_i(\ell)\alpha_i^\ell = 0 \qquad (1 \leqq i \leqq q). \tag{11.13}$$

If u is symmetric, all but finitely many solutions satisfy (11.13) *or*

$$\begin{aligned} AP_i(k)\alpha_i^k + BP_i(h)\alpha_i^h + CP_{i+1}(\ell)\alpha_{i+1}^\ell &= 0 \quad (1 \leqq i \leqq q, \\ AP_{i+1}(k)\alpha_{i+1}^k + BP_{i+1}(h)\alpha_{i+1}^h + CP_i(\ell)\alpha_i^\ell &= 0 \qquad\quad i \text{ odd}), \end{aligned} \tag{11.13ℓ}$$

or system (11.13k) *or* (11.13h) *obtained from* (11.13ℓ) *by a permutation of the rôles of the variables.*

See Schlickewei and Schmidt [37]. In that paper an example is given which shows that the condition $P_0 = 0$ cannot be avoided. A proof will not be given here.

In [37] it was further shown that in all cases, the equation (11.6) has finitely many solutions, plus possibly a "linear family" of solutions

$$(k(t), h(t)) = (a_0 t + k_0,\ b_0 t + h_0) \qquad (t \in \mathbb{Z}),$$

and in "exceptional" cases a finite number of "exponential families" of the type

$$(k(t), h(t)) = (cR^t + dt + e,\ c'R^t + d't + e') \qquad (t \in \mathbb{N})$$

with $R > 1$ and coefficients $c \neq 0$, $c' \neq 0$, d, d', e, e'.

The equation (11.12) in variables k, h, ℓ has, at least when all the data are algebraic, finitely many solutions, plus a finite number of linear families

$$(k(t), h(t), \ell(t)) = (a_0 t + k_0,\ b_0 t + h_0,\ c_0 t + \ell_0) \qquad (t \in \mathbb{Z}),$$

plus a finite number of exponential families

$$(k(t), h(t), \ell(t)) = (cR^t + dt + e,\ c'R^t + d't + e',\ c''R^t + d''t + e'') \quad (t \in \mathbb{N}).$$

One is led to the following

Conjecture 11.6. Let $u^{(1)}, \dots, u^{(n)}$ be linear recurrences, and consider the equation

$$u_{k_1}^{(1)} + \cdots + u_{k_n}^{(n)} = 0$$

in integer variables $k_1, \dots, k_n$. The solutions consist of a finite set, together with finitely many families $(k_1(\mathbf{t}), \dots, k_n(\mathbf{t}))$ where $\mathbf{t}$ ranges through $\mathbb{N}^m$ for some m, and each $k_j(\mathbf{t})$ is of polynomial–exponential type, i.e.,

$$k_j(\mathbf{t}) = \sum_{i=1}^{q_j} P_{ji}(\mathbf{t}) \boldsymbol{\alpha}_{ji}^{\mathbf{t}}.$$

Moreover, when no roots of the companion polynomials of $u^{(1)}, \dots, u^{(n)}$ are radicals, i.e., have $\alpha^k = a$ with some $k \in \mathbb{Z}\backslash\{0\}$ and $a \in \mathbb{Z}$, all the functions $k_j(\mathbf{t})$ will be linear in $\mathbf{t}$.

No such simple conjecture can be made for general equations of polynomial–exponential type. The equation

$$\sum_{i=2}^{n} k_i^2 - 2\sum_{i=1}^{n-1} k_{i+1} 2^{k_i} + \sum_{i=1}^{n-1} 4^{k_i} = 0$$

is

$$\sum_{i=1}^{n-1} (k_{i+1} - 2^{k_i})^2 = 0$$

in disguised form, and has the solutions

$$k_1 = t, \ k_2 = 2^t, \ k_3 = 2^{2^t}, \dots, \ k_n = 2^{2^{\cdot^{\cdot^{\cdot^{2^t}}}}} \qquad (t \in \mathbb{Z}, \ t \geqq 0).$$

12 Bounds for the Number of Solutions of Polynomial-Exponential Equations

There are only very special cases where we can make our results *effective*, i.e., where we are able to give a bound for the size $|\mathbf{k}|$ of the solutions to our equations. We will not be concerned with questions of effectiveness here, which sometimes are resolved by Baker's methods or other methods. The reader might consult the book by Shorey and Tijdeman [44] on such questions.

Rather, in these lectures we will be concerned with estimating the *number* of solutions. Consider again a system of equations (π) where π is a partition of $\{1, \dots, q\}$. Put

$$A = \sum_{i=1}^{q} \binom{n + \delta_i}{n}$$

where δ_i is the total degree of the polynomial P_i. Note that A is the number of coefficients of any set of q polynomials of total degrees $\delta_1, \dots, \delta_q$. Further set

$$B = \max(n, A),$$

so that $B = \max(n, q)$ if all the polynomials $P_1, \dots, P_q$ are constants, and $B = A$ otherwise.

Improving on their earlier work [36], Schlickewei and Schmidt [39] obtained the following

Theorem 12.1. *Suppose* $G(\pi) = \{\mathbf{0}\}$, *and* (7.3) *is defined over a number field of degree* d. *Then*

$$|\mathcal{S}(\pi)| < N(d, B) = 2^{35B^3} d^{6B^2}.$$

This result, which will not be proved in the present lectures, yields (9.1). As was pointed out at the end of Section 9, it would be desirable to get a bound independent of d, in fact independent of the equations being defined over a number field. We are able to do this in two situations: when the polynomials are constants, so that we are dealing with purely exponential equations, and when $n = 1$.

As for exponential equations, we will for simplicity restrict ourselves to a sharpening of Theorem 7.3. We will study nondegenerate solutions of

$$a_1 x_1 + \cdots + a_q x_q = 0, \tag{12.1}$$

i.e., solutions where no proper, nonempty subsum of the sum vanishes. Write

$$Z(q, r)$$

for the maximal number of nondegenerate solutions $\mathbf{x} \in \mathbb{P}(\Gamma)$ of (12.1), over all groups $\Gamma \subset (\mathbb{C}^\times)^q$ of rank r, and over all coefficients $a_1, \dots, a_q$. Observe that a group Γ of rank r need not be finitely generated. Rather, the maximal number of multiplicatively independent elements of Γ is r.

We begin with $Z(q, 0)$. This is the number (up to factors of proportionality) of non-degenerate solutions $(x_1, \dots, x_q)$ of (12.1) where each x_i is a root of unity. Improving on earlier work of Schlickewei [33], Evertse [12] established

Theorem 12.2. $Z(q, 0) < q^{3q^2}$.

When $q = p$, a prime, the p-th roots of unity $\xi_1, \dots, \xi_p$ in any order are solutions of

$$\xi_1 + \cdots + \xi_p = 0.$$

Therefore $Z(p, 0) \geqq p!$.

In general, let us remark that instead of a group $\Gamma \subset (\mathbb{C}^\times)^q$ we could have taken a group $\Gamma' \subset \mathbb{C}^\times$ and considered solutions of (12.1) with each component x_i in Γ', i.e., solutions in $(\Gamma')^q$. Write $Z'(q, r)$ for the maximal number of nondegenerate solutions $\mathbf{x} \in \mathbb{P}((\Gamma')^q)$ over all groups Γ' of rank r, and all possible coefficients $a_1, \dots, a_q$. It is clear that

$$Z'(q,r) \leqq Z(q,qr).$$

On the other hand, given $\Gamma \subset (\mathbb{C}^\times)^q$, let Γ_i be the group of numbers x_i occurring in some $(x_1, \dots, x_i, \dots, x_q) \in \Gamma$, and $\Gamma' \subset \mathbb{C}^\times$ the group generated by $\Gamma_1, \dots, \Gamma_q$. When Γ has rank r, each Γ_i has rank $\leqq r$, hence Γ' has rank $\leqq qr$. Since $\Gamma \subset (\Gamma')^q$,

$$Z(q,r) \leqq Z'(q,qr).$$

A very recent result of Evertse, Schlickewei and Schmidt [14], reformulated for our homogeneous equations, is that

$$Z(q,r) \leqq e^{(6(q-1))^{3(q-1)}(r+1)}.$$

Here we will prove the slightly weaker

Theorem 12.3. $Z(q,r) \leqq e^{(4q)^{4q}(r+1)}$.

In fact a bound $e^{e^{cq}(r+1)}$ with an absolute constant c can be obtained if results from David and Philippon's recent paper [5] are used instead of results from Schmidt's [41].

Quite likely, these estimates are far from best possible. But $Z(q,r)$ for fixed r is non-decreasing in q and tends to infinity with q, and for fixed $q \geqq 3$ is non-decreasing in r and tends to infinity with r. We claim that

$$Z(q,0) \geqq (q-1)!\,.$$

For let $\zeta_1, \dots, \zeta_{q-1}$ be distinct roots of unity such that $\zeta_{i_1} + \dots + \zeta_{i_\ell} \neq 0$ for $\ell > 0$ and $i_1 < \dots < i_\ell$. Set $b = -\zeta_1 - \dots - \zeta_{q-1}$. Then the equation

$$x_1 + \dots + x_{q-1} + bx_q = 0$$

has the solution $(\zeta_1, \dots, \zeta_{q-1}, 1)$, in fact the $(q-1)!$ solutions obtained from it by a permutation of $\zeta_1, \dots, \zeta_{q-1}$. These solutions are not proportional to each other and are nondegenerate.

Let $Z_{\mathbb{Q}}(q,r)$ be defined as $Z(q,r)$, but with regard to equations with rational coefficients. We claim that

$$Z_{\mathbb{Q}}(q,0) \geqq (p-1)!$$

for every prime $p \leqq q$. For the equation

$$x_1 + \dots + x_{p-1} + (1/m)x_p + \dots + (1/m)x_q = 0$$

with $m = q-p+1$ has the solution $(\zeta, \zeta^2, \dots, \zeta^{p-1}, 1, \dots, 1)$ where ζ is a primitive p-th root of 1, and the solutions obtained by permuting $\zeta, \zeta^2, \dots, \zeta^{p-1}$.

Again these solutions are non-proportional to each other, and nondegenerate. Since by analytic number theory we may pick p close to q, we have $Z_{\mathbb{Q}}(q,0) \gg (q!)^{1-\varepsilon}$ for $\varepsilon > 0$, and in fact ε may be replaced by a suitable negative power of q. We finally remark that when $p_1, \ldots, p_t$ are primes with $\sum_{i=1}^{t}(p_i - 1) \leqq q-1$, then $Z_{\mathbb{Q}}(q,0) \geqq \prod_{i=1}^{t}(p_i-1)!$.

As for the dependence on r, Erdős, Stewart and Tijdeman [7] had given an example which shows that

$$Z(3,r) \geqq \exp\left(c(r/\log r)^{1/2}\right)$$

with an absolute constant $c > 0$. This can be generalized to $Z(q,r) \geqq \exp\left(c(q)(r/\log r)^{(q-2)/(q-1)}\right)$, but it has been conjectured that the correct lower bound should be

$$\exp\left(c(q)(r/\log r)^{(q-1)/q}\right),$$

or even $\exp\left(c(q)r^{(q-1)/q}\right)$.

For $q = 3$, the assertion of Theorem 12.3 had been proved earlier. Schlickewei [32] showed that $Z(3,r) \leqq c_0(r)$, and Beukers and Schlickewei [3] using the "hypergeometric method" established

$$Z(3,r) \leqq 2^{8(r+2)},$$

which is much better than the case $q = 3$ of Theorem 12.3.

In older work, only the case $\Gamma = (U_S)^q$ had been considered, where U_S is a group of S-units of a number field K. Thus S is a finite subset of $M(K)$ containing all the Archimedean places, and U_S consists of $x \in K$ having $\|x\|_v = 1$ for every place $v \notin S$. When $q = 3$ and $K = \mathbb{Q}$, Mahler [19] had shown that (12.1) has only finitely many solutions with each $x_i \in U_S$ $(i = 1,2,3)$, and Lang [16] extended Mahler's result to arbitrary number fields. (In fact Lang showed a result of this kind for any field K of characteristic 0 and a group $\Gamma \subset (K^{\times})^3$ of finite rank.)

For general q and $\Gamma = (U_S)^q$, Evertse [10] and van der Poorten and Schlickewei [25] had shown that (12.1) has only finitely many nondegenerate solutions $\mathbf{x} \in \mathbb{P}(\Gamma)$. The first quantitative result here is due to Evertse [9], who showed that for $q = 3$ the number of such solutions is

$$\leqq 3 \cdot 7^{4s}$$

where $s = |S|$. For q in general, Schlickewei [31] showed that the number of such solutions is

$$\leqq c(q,s).$$

The best explicit value of $c(q,s)$ again is due to Evertse [11], who established this with

$$c(q,s) = 2^{35q^4 s}.$$

Observe that when $\Gamma \subset (K^\times)^q$ is any group of finite rank r, then $\Gamma \subset (U_S)^q$ for some S, but $s = |S|$ may be much larger than r. At any rate, s has to be at least equal to the number of Archimedean places.

* * *

A linear recurrence sequence $u \in V_1$ is *simple* if its companion polynomial has only simple roots. It is of the form

$$u_k = a_1\alpha_1^k + \cdots + a_t\alpha_t^k \tag{12.2}$$

when of order t. Now if the sequence is weakly non-degenerate, say α_1/α_j not a root of 1 for $j \neq 1$, we observe that when $u_k = 0$, some subsum including $a_1\alpha_1^k$ will vanish, say

$$a_1\alpha_1^k + a_{i_2}\alpha_{i_2}^k + \cdots + a_{i_q}\alpha_{i_q}^k = 0, \tag{12.3}$$

such that no proper nonempty sumsum of (12.3) vanishes. Thus

$$(\alpha_1^k, \alpha_{i_2}^k, \ldots, \alpha_{i_q}^k)$$

becomes a non-degenerate solution of (12.3), and lies in a group of rank 1. We cannot have proportional solutions, for if

$$(\alpha_1^k, \alpha_{i_2}^k, \ldots, \alpha_{i_q}^k) \text{ and } (\alpha_1^h, \alpha_{i_2}^h, \ldots, \alpha_{i_q}^h)$$

were proportional, $(\alpha_1/\alpha_{i_2})^k = (\alpha_1/\alpha_{i_2})^h$, so that $k = h$, since α_1/α_{i_2} is not a root of 1. Thus by Theorem 12.3 we are dealing with at most $e^{2(4q)^{4q}}$ solutions of (12.3). The number of subsets $\{i_2, \ldots, i_q\} \subset \{2, \ldots, t\}$ is $\binom{t-1}{q-1}$, and summing over q we get

$$\sum_{q=2}^{t} \binom{t-1}{q-1} e^{2(4q)^{4q}} < e^{3(4t)^{4t}}.$$

We have

Corollary 12.4. *A simple, non-degenerate linear recurrence of order t has zero-multiplicity*

$$< e^{3(4t)^{4t}}.$$

Actually, the bound $e^{(6t)^{3t}}$ was given in [14], with an anticipated improvement to $\exp\exp(c_1 t)$ if the above mentioned result of [5] is used. Berstel discovered that the linear recurrence of order 3 with $u_k = 2u_{k-1} - 4u_{k-2} + 4u_{k-3}$ and $u_1 = u_2 = 0$, $u_3 = 1$ begins (for positive subscripts) with

$$0,\ 0,\ 1,\ 2,\ 0,\ -4,\ 0,\ 16,\ 16,\ -32,\ -64,\ 64,\ 256,\ 0,\ \ldots,$$

so that $u_k = 0$ for $k = 1, 2, 5, 7, 14$. In fact also $u_{53} = 0$, so that the zero-multiplicity is at least 6. On the other hand, Beukers [2] showed that a recurrence of order 3 with rational terms has zero-multiplicity at most 6, so that Berstel's sequence has zero-multiplicity 6. In Section 13 we will exhibit an example of Bavencoffe and Bézivin of a simple recurrence of order $t \geqq 3$ with zero-multiplicity $\geqq c_0 t^2 > 0$.

* * *

As indicated above, there is another instance where we can get rid of the dependency on the degree of number fields, hence can get estimates independent of being in a number field at all. This is on linear recurrences, which may be simple or not, so that we are dealing with polynomial–exponential equations

$$P_1(k)\alpha_1^k + \cdots + P_q(k)\alpha_q^k = 0$$

in one variable k. In [42] I could show that the zero-multiplicity of a non-degenerate linear recurrence of order t is below some constant $c_1(t)$, and in fact one may take

$$c_1(t) = \exp\exp\exp(3t\log t).$$

Finally in [43], making a Theorem of Mahler [20] more explicit, I established

Theorem 12.5. *Let u be a linear recurrence sequence of order t, which may be degenerate. Then the set $\mathcal{Z}$ of subscripts k with $u_k = 0$ is the union of at most $c_2(t)$ arithmetic progressions and single numbers. One may take*

$$c_2(t) = \exp\exp\exp(20t).$$

Since we already know that a non-degenerate (in fact a weakly non-degenerate) sequence has finite zero-multiplicity, this contains (except for the constant 20) the result on non-degenerate sequences enunciated above.

* * *

The remainder of these Notes will be devoted to exhibiting the Bavencoffe-Bézivin sequences, to proofs of Theorems 12.2 and 12.3, and a partial proof of Theorem 12.5. More precisely, the BB-sequences will be dealt with in Section 13, Theorem 12.2 will be proved in Section 14, and Theorem 12.3 in Sections 15-19. Finally, Theorem 12.5 will be discussed in Sections 20, 21.

13 The Bavencoffe-Bézivin Sequence

For $t \geqq 3$ set

$$\mathcal{Q}(Z) = Z^{t+1} + (-2)^{t-1}Z + (-2)^t.$$

Then

$$\mathcal{P}(Z) = \mathcal{Q}(Z)/(Z+2)$$

is in $\mathbb{Z}[Z]$. The Bavencoffe–Bézivin sequence (in short BB-sequence) of order t is the sequence $u \in V_1$ annihilated by $\mathcal{P}$ with

$$u_0 = \cdots = u_{t-2} = 0, \quad u_{t-1} = 1.$$

When $t = 3$, the BB-sequence becomes the Berstel sequence.

Theorem 13.1 (See [1]). *The BB-sequence of order t has*

(a) $u_k = 0$ *when* $k = \ell(t+1) + q$ *with* $\ell \geqq 0$, $q \geqq 0$, $\ell + q \leqq t - 2$,

(b) $u_k = 0$ *when* $k = j(2t+1) - 1$ *with* $1 \leqq j \leqq t - 1$.

The sequence is simple, and non-degenerate.

If $\ell(t+1) + q = \ell'(t+1) + q'$ with ℓ', q' satisfying the same conditions as ℓ, q, then $q \equiv q' \pmod{t+1}$, therefore $q = q'$, $\ell = \ell'$. Thus (a) gives $\frac{1}{2}t(t-1)$ distinct values of k with $u_k = 0$. Further the numbers k in (a) and in (b) are distinct, for $\ell(t+1) + q = j(2t+1) - 1$ yields $\ell + q \equiv j - 1 \pmod{t}$, hence $\ell + q = j - 1$, and we get $\ell t = 2jt$, $\ell = 2j$, which entails the impossible $2j \leqq \ell + q = j - 1$. Therefore the BB-sequence has zero-multiplicity at least

$$\frac{1}{2}t^2 + \frac{1}{2}t - 1.$$

In fact, it is shown in [1] that the sequence has a zero with $k < 0$ when t is even, and at least 2 such zeros when t is of the form $m(m+1)$.

One checks that $\mathcal{P}(Z) = Z^t - 2Z^{t-1} - \cdots - c_0$, so that the recurrence relation is of the form $u_k = 2u_{k-1} + \cdots + c_0 u_{k-t}$. Therefore $u_t = 2$. Since $\mathcal{Q}$ is of degree $t+1$, the BB-sequence is the unique sequence with $\mathcal{Q}u = 0$ and

$$u_0 = \cdots = u_{t-1} = 0, \quad u_{t-1} = 1, \quad u_t = 2. \tag{13.1}$$

The generating function of u (more precisely, of its restriction v in V_1') is by Theorem 3.1 rational with denominator $\widehat{\mathcal{Q}}(Z)$, and in fact

$$f(Z) = \sum_{k=0}^{\infty} u_k Z^k = \frac{Z^{t-1}(1+2Z)}{1 + (-2)^{t-1}Z^t + (-2)^t Z^{t+1}},$$

where the numerator is determined by (13.1). Therefore

$$\begin{aligned} f(Z) &= \sum_{n=0}^{\infty} Z^{t-1+tn}(1+2Z)(-2)^{(t-1)n}(-1+2Z)^n \\ &= (1+2Z) \sum_{0 \leqq j \leqq n} (-1)^n (-2)^{(t-1)n+j} \binom{n}{j} Z^{t(n+1)+j-1}. \end{aligned}$$

The coefficient of Z^k is certainly zero unless k is of the form

$$k = t(n+1) + j - 1, \tag{13.2}$$

or $k = t(n+1) + j$, with $0 \leqq j \leqq n$. Thus the coefficient is zero unless k is of the form (13.2) with

$$0 \leqq j \leqq n+1. \tag{13.3}$$

Now if k is as in (a), i.e., $k = \ell(t+1) + q$ with $\ell \geqq 0$, $q \geqq 0$, $\ell + q \leqq t-2$, then

$$k = \ell(t+1) + q = t(n+1) + j - 1$$

yields $\ell + q \equiv j - 1 \pmod t$, which is impossible for $j = 0$, since $0 \leqq \ell + q \leqq t-2$. Therefore $j \geqq 1$, $t(n+1) \leqq \ell(t+1) + q \leqq (t+1)(t-2) = t^2 - t - 2$, so that $n+1 \leqq t-1-2/t$, and $j \leqq n+1 \leqq t-2$, and the congruence gives $\ell + q = j-1$, so that $\ell t = t(n+1)$, and $n+1 = \ell \leqq j-1 \leqq n$, which is impossible. Part (a) of Theorem 13.1 is established.

An integer of the form (13.2) with $0 \leqq j \leqq t-1$ is uniquely written in this form: for if $t(n+1) + j - 1 = t(n_1+1) + j_1 - 1$, then $tn + t \leqq t(n_1+1) + j_1 \leqq tn_1 + 2t - 1$, so that $n+1 < n_1 + 2$, hence $n \leqq n_1$. By symmetry $n = n_1$, whence also $j = j_1$. Therefore when k is written as (13.2) with $1 \leqq j \leqq t-1$, it is also uniquely written as $k = t(n'+1) + j'$ with $0 \leqq j' \leqq t-2$ (namely with $n' = n$, $j' = j-1$).

As a consequence, when k is given by (13.2) with $1 \leqq j \leqq t-1$, the coefficient of Z^k in $f(Z)$ is

$$\begin{aligned}(-1)^n \left[(-2)^{(t-1)n+j}\binom{n}{j} + 2(-2)^{(t-1)n+j-1}\binom{n}{j-1}\right] \\ = (-1)^n(-2)^{(t-1)n+j}\frac{n!}{j!(n-j+1)!}(n+1-2j).\end{aligned}$$

This vanishes when $n = 2j-1$, and therefore $u_k = 0$ when $k = t(n+1)+j-1 = 2tj + j - 1 = (2t+1)j - 1$ with $1 \leqq j \leqq t-1$, proving (b).

To show that the sequence is simple, we have to show that the polynomial $\mathcal{P}$ has only simple roots. In fact, even $\mathcal{Q}$ has only simple roots. For if x were a multiple root of $\mathcal{Q}$, we had

$$x^{t+1} + (-2)^{t-1}x + (-2)^t = 0, \quad (t+1)x^t + (-2)^{t-1} = 0.$$

After multiplying the second equation by x, and eliminating x^{t+1}, we obtain $x = 2(t+1)/t$, and eventually

$$2(t+1)^{t+1} + (-1)^{t-1}t^t = 0,$$

which is impossible.

The proof of the non-degeneracy is surprisingly hard. Given algebraic numbers α, β with $\alpha \in \mathbb{Q}(\beta)$, we write $\mathcal{N}_\beta(\alpha)$ for the norm $\mathcal{N}_{\mathbb{Q}(\beta)/\mathbb{Q}}(\alpha)$.

Lemma 13.2. *Suppose $\zeta \neq \pm 1$ is a root of 1. Then*

$$\mathcal{N}_\zeta(\zeta - 1) = \begin{cases} p, & \text{if } \zeta \text{ has order } p^s \text{ where } p \text{ is a prime;} \\ 1, & \text{otherwise.} \end{cases}$$

This is well known; for convenience we will provide a

Proof. If ζ has order $n \geqq 3$ and F_n is the n-th cyclotomic polynomial, so that $\zeta - 1$ is a root of $F_n(X+1)$, then $\mathcal{N}_\zeta(\zeta - 1) = (-1)^{\phi(n)} F_n(1) = F_n(1)$. Now

$$\begin{aligned} F_n(X) &= \prod_{d|n} (X^d - 1)^{\mu(n/d)} = \prod_{d|n} \left((X^{d-1} + \cdots + 1)(X-1)\right)^{\mu(n/d)} \\ &= \prod_{d|n} (X^{d-1} + \cdots + 1)^{\mu(n/d)}. \end{aligned}$$

Therefore

$$F_n(1) = \prod_{d|n} d^{\mu(n/d)} = \prod_{m|n} (n/m)^{\mu(m)} = \prod_{m|n} m^{-\mu(m)} = \prod_{m|q} m^{-\mu(m)} \tag{13.4}$$

where q is the product of the prime factors of m. If $q = p_1 \cdots p_t$, the exponent of p_i in the last product in (13.4) is

$$-\sum_{r|(q/p_i)} \mu(p_i r) = \sum_{r|(q/p_i)} \mu(r) = \begin{cases} 1 & \text{if } q = p_i, \\ 0 & \text{otherwise.} \end{cases}$$

Therefore $F_n(1) = 1$ unless $t = 1$, i.e., n is a prime power p^s, in which case $F_n(1) = p$. □

In order to show that the BB-sequence is non-degenerate, it will suffice to show that no two roots of $\mathcal{P}$ have a quotient which is a root of 1. If two such roots were $x, x\zeta$ with $\zeta \neq 1$ a root of 1, then in particular $\mathcal{Q}(x) = \mathcal{Q}(x\zeta) = 0$, which yields

$$x^{t+1} + (-2)^{t-1}x + (-2)^t = 0, \quad (x\zeta)^{t+1} + (-2)^{t-1}x\zeta + (-2)^t = 0,$$

whence

$$x(\zeta^{t+1} - \zeta) = 2(\zeta^{t+1} - 1). \tag{13.5}$$

If either $\zeta^{t+1} - \zeta = 0$ or $\zeta^{t+1} - 1 = 0$, then both are 0, leading to the excluded $\zeta = 1$. Therefore $\zeta^t \neq 1$, $\zeta^{t+1} \neq 1$. Substitution of (13.5) into $\mathcal{Q}(x) = 0$ yields

$$2(\zeta^{t+1} - 1)^{t+1} + (-1)^t (1 - \zeta)(\zeta^{t+1} - \zeta)^t = 0. \tag{13.6}$$

Put

$$u = \zeta(\zeta^t - 1)/(\zeta^{t+1} - 1). \tag{13.7}$$

Then

$$u - 1 = -(\zeta - 1)/(\zeta^{t+1} - 1), \tag{13.8}$$

and therefore $1/(u-1)$ is an algebraic integer. In view of (13.6) we get

$$u^{t+1} - u^t + 2(-1)^t = 0,$$

so that also u is an algebraic integer. We may infer that $u - 1$ is a unit. We have

$$X^{t+1} - X^t + 2(-1)^t = (X+1)\left(X^t - 2\frac{X^t - (-1)^t}{X+1}\right) = (X+1)r(X),$$

say. Here r is irreducible by Eisenstein's criterion. Now $u = -1$ would lead via (13.7) to $2\zeta^{t+1} = \zeta + 1$, which is impossible for a root of unity $\zeta \neq 1$. Therefore u is a root of $r(X)$, whence $\deg u = t$. We obtain $\mathcal{N}_u(u) = \pm 2$, therefore[4] $\mathcal{N}_\zeta(u) = \pm 2^b$ with $b = d/t$, where $d = \deg \zeta$, $t = \deg u$. Our equations (13.7), (13.8) yield

$$\pm 2^b \mathcal{N}_\zeta(\zeta^{t+1} - 1) = \mathcal{N}_\zeta(\zeta^t - 1) \tag{13.9}$$

and

$$\mathcal{N}_\zeta(\zeta^{t+1} - 1) = \pm(\mathcal{N}_\zeta(u-1))^{-1}\mathcal{N}_\zeta(\zeta - 1) = \pm\mathcal{N}_\zeta(\zeta - 1). \tag{13.10}$$

As a consequence, $\mathcal{N}_\zeta(\zeta^t - 1)$ is divisible by 2, so that by virtue of Lemma 13.2, $\mathcal{N}_{\zeta^t}(\zeta^t - 1) = \pm 2$ (the minus sign only if $\zeta^t = -1$). We obtain $\mathcal{N}_\zeta(\zeta^t - 1) = \pm 2^h$ with $h = [\mathbb{Q}(\zeta) : \mathbb{Q}(\zeta^t)]$. By (13.9), (13.10), both $\mathcal{N}_\zeta(\zeta^{t+1} - 1)$ and $\mathcal{N}_\zeta(\zeta - 1)$ are powers of 2, so that by Lemma 13.2 again, the order of ζ is a power of 2, and $\mathcal{N}_\zeta(\zeta - 1) = 2 = \pm\mathcal{N}_\zeta(\zeta^{t+1} - 1)$. Say $\operatorname{ord}\zeta = 2^m$, hence $d = \deg\zeta = \phi(2^m) = 2^{m-1}$. Then also b, h are powers of 2. Now (13.9) yields $b + 1 = h$, therefore $b = 1$, $h = 2$, and $t = d = 2^{m-1}$. Since $\operatorname{ord}\zeta = 2^m$, we may observe that $(\zeta^t - 1)(\zeta^t + 1) = \zeta^{2^m} - 1 = 0$, hence in fact $\zeta^t + 1 = 0$, and $X^t + 1 = 0$ is the irreducible equation for ζ. Substituting $\zeta^t = -1$ into (13.6) we get

$$(\zeta + 1)^{t+1} + (-1)^t 2^{t-1}(1 - \zeta) = 0.$$

The irreducible polynomial $X^t + 1$ must divide $(X+1)^{t+1} + (-1)^t 2^{t-1}(1-X)$, so that

$$(X+1)^{t+1} + (-1)^t 2^{t-1}(1 - X) = (X^t + 1)(X + \gamma),$$

say. Clearly $\gamma = 1 + (-1)^t \cdot 2^{t-1}$. Substituting $X = 1$ we obtain $2^{t+1} = 4 + (-2)^t$, which is impossible for $t \geqq 3$.

Thus Theorem 13.1 has been established. □

[4] The ± signs in our equations are not related.

14 Proof of Evertse's Theorem on Roots of Unity

Our goal will be to prove Theorem 12.2. We will follow Evertse's clear exposition. We first will deal with

$$a_1\zeta_1 + \cdots + a_q\zeta_q = 0 \tag{14.1}$$

with nonzero *rational* coefficients $a_1, \dots, a_q$ and unknowns $\zeta_1, \dots, \zeta_q$ which are roots of 1. This case was first treated by Mann [22].

Lemma 14.1. *Let $\zeta_1, \dots, \zeta_q$ be a (not necessarily nondegenerate) solution of (14.1). Then for some indices $i \neq i'$, the quotient $\zeta_i/\zeta_{i'}$ is a root of unity of order $\leqq q^2$.*

Proof. When $q = 2$, then $\zeta_1/\zeta_2 = -a_2/a_1$ is rational, hence ± 1, of order $\leqq 2 < q^2$. So suppose $q > 2$, and the assertion is true for equations with fewer than q terms. As is easily seen, we may suppose that the solution is nondegenerate (for otherwise we get a nondegenerate solution of an equation with fewer terms), that $\zeta_1 = 1$, and that $(\zeta_1, \dots, \zeta_q)$ is not $(1, \dots, 1)$.

Let d be the smallest positive integer with $\zeta_1^d = \cdots = \zeta_q^d = 1$. Then $d > 1$. Let p be a prime dividing d, and suppose $p^m \,\|\, d$. Every ℓ may uniquely be written as $\ell = p\mu + (d/p^m)\nu$ with $0 \leqq \nu < p$, and when ξ is a primitive d-th root of 1,

$$\xi^\ell = \zeta^* \zeta^\nu$$

where $\zeta = \xi^{d/p^m}$ is a primitive p^m-th root of 1, and $\zeta^* = \xi^{p\mu}$ has $(\zeta^*)^{d/p} = 1$. Accordingly,

$$\zeta_i = \zeta_i^* \zeta^{\nu_i} \qquad (i = 1, \dots, q) \tag{14.2}$$

where $0 \leqq \nu_i < p$ and $(\zeta_i^*)^{d/p} = 1$. Inserting (14.2) into (14.1) we get

$$\sum_{j=0}^{p-1} a(j)\zeta^j = 0 \tag{14.3}$$

where

$$a(j) = \sum_{i:\, \nu_i = j} a_i \zeta_i^* \qquad (j = 0, \dots, p-1)$$

lie in the cyclotomic field K generated by the (d/p)-th roots of unity. We have $\nu_1 = 0$ since $\zeta_1 = 1$, and it follows from the minimality of d that at least one of the exponents $\nu_1, \dots, \nu_q$ in (14.2) is non-zero. Therefore the set of i with $\nu_i = 0$ is a proper, nonempty subset of $\{1, \dots, q\}$. Since we are dealing with a nondegenerate solution,

$$a(0) = \sum_{i:\, \nu_i = 0} a_i \zeta_i^* \neq 0. \tag{14.4}$$

In view of (14.3), (14.4), ζ has degree at most $p-1$ over K. Since otherwise ζ would have degree $\phi(d)/\phi(d/p) \geqq p$ over K, we may conclude that $p^2 \nmid d$. This holds for every prime divisor p of d, so that d is square free.

Now ζ is a primitive p-th root of 1, and since $\phi(d) = \phi(p)\phi(d/p) = (p-1)\phi(d/p)$, it has degree $p-1$ over K and minimal polynomial $X^{p-1} + \cdots + X + 1$. Because of (14.3) this implies $a(0) = \cdots = a(p-1)$, i.e.,

$$\sum_{i:\, \nu_i = j} a_i \zeta_i^* + \sum_{i:\, \nu_i = j'} (-a_i)\zeta_i^* = 0 \tag{14.5}$$

for any pair of numbers j, j' in $\{0, \ldots, p-1\}$.

In view of (14.5), the set of i with $\nu_i = j$ is nonempty for each j, $0 \leqq j \leqq p-1$. There are distinct j, j' such that the set T of indices i with $\nu_i = j$ or j' has cardinality $\leqq (2/p)q$.

Now *if* $p \geqq 3$, this cardinality is $< q$, and by induction applied to (14.5), there are distinct indices i, i' in T such that $\zeta_i^*/\zeta_{i'}^*$ is a root of unity of order $\leqq (2q/p)^2$. But by (14.2), $\zeta_i/\zeta_{i'} = \zeta^{\nu_i - \nu_{i'}} \zeta_i^*/\zeta_{i'}^*$, and therefore $\zeta_i/\zeta_{i'}$ is of order at most

$$p(2q/p)^2 = (4/p)q^2 < q^2$$

provided $p \geqq 5$. So we are home if d has a prime divisor $p \geqq 5$. But otherwise, since d is square free, $d \leqq 6$, so that each quotient $\zeta_i/\zeta_{i'}$ is a root of unity of order $\leqq 6 < q^2$. □

Corollary 14.2. *There is a set S_1 of cardinality not exceeding q^4, and depending only on q, such that for any solution $(\zeta_1, \ldots, \zeta_q)$ of* (14.1) *there are indices $i \neq i'$ with $\zeta_i/\zeta_{i'} \in S_1$.*

Proof. Let S_1 consist of all roots of unity of order $\leqq q^2$. □

We now drop the hypothesis that $a_1, \ldots, a_q$ be rational. They will be arbitrary non-zero complex numbers.

Lemma 14.3. *There is a set S_2 of cardinality not exceeding $(q!)^6$, depending on $a_1, \ldots, a_q$, such that for every solution $(\zeta_1, \ldots, \zeta_q)$ of* (14.1) *there are indices $i \neq i'$ with $\zeta_i/\zeta_{i'} \in S_2$.*

Proof. We employ a determinant argument which is occasionally used in our subject, which is often our only available argument to get rid of arbitrary coefficients such as $a_1, \ldots, a_q$, but is probably very wasteful. When $\mathbf{z}_1 = (\zeta_{11}, \ldots, \zeta_{1q})$, ..., $\mathbf{z}_q = (\zeta_{q1}, \ldots, \zeta_{qq})$ are q solutions to (14.1), then the determinant $\det(\zeta_{ij}) = 0$. Thus

$$\sum_\sigma \varepsilon(\sigma)\zeta_{1,\sigma(1)} \cdots \zeta_{q,\sigma(q)} = 0, \tag{14.6}$$

when σ runs through the permutations of $\{1, \ldots, q\}$, and $\varepsilon(\sigma)$ is the sign of σ. The sum in (14.6) is a sum of $q!$ roots of unity. By Corollary 14.2, applied with $q!$ in place of q, we see that there is a set S_3 of cardinality at most $(q!)^4$ such that for any q-tuple of solutions $\mathbf{z}_1, \ldots, \mathbf{z}_q$ of (14.1) there are distinct permutations σ, τ with

$$(\zeta_{1,\sigma(1)} \cdots \zeta_{q,\sigma(q)})/(\zeta_{1,\tau(1)} \cdots \zeta_{q,\tau(q)}) \in S_3. \tag{14.7}$$

So far, so good. Now comes Evertse's clever argument. Let $m \leqq q$ be the smallest integer with the following property: for every m-tuple of solutions $\mathbf{z}_1, \ldots, \mathbf{z}_m$ of (14.1) there are permutations $\sigma \neq \tau$ of $\{1, \ldots, q\}$ with

$$\sigma(m+1) = \tau(m+1), \ldots, \sigma(q) = \tau(q) \tag{14.8}$$

such that

$$(\zeta_{1,\sigma(1)} \cdots \zeta_{m,\sigma(m)})/(\zeta_{1,\tau(1)} \cdots \zeta_{m,\tau(m)}) \in S_3.$$

(Here the condition (14.8) is understood to be vacuous when $m = q$.) Clearly $2 \leqq m \leqq q$.

First suppose $m \geqq 3$. Since m is minimal, we may pick solutions

$$\mathbf{z}_1, \ldots, \mathbf{z}_{m-1}$$

such that for any permutation $\sigma \neq \tau$ with

$$\sigma(m) = \tau(m), \ldots, \sigma(q) = \tau(q),$$

we have

$$(\zeta_{1,\sigma(1)} \cdots \zeta_{m-1,\sigma(m-1)})/(\zeta_{1,\tau(1)} \cdots \zeta_{m-1,\tau(m-1)}) \notin S_3.$$

With $\mathbf{z}_1, \ldots, \mathbf{z}_{m-1}$ fixed, we allow $\mathbf{z}_m$ to vary. Writing $\mathbf{z}_m = \mathbf{z} = (\zeta_1, \ldots, \zeta_q)$, we see that there are permutations $\sigma \neq \tau$ depending on $\mathbf{z}$ such that we have (14.8) and

$$(\zeta_{1,\sigma(1)} \cdots \zeta_{m-1,\sigma(m-1)}\zeta_{\sigma(m)})/(\zeta_{1,\tau(1)} \cdots \zeta_{m-1,\tau(m-1)}\zeta_{\tau(m)}) \in S_3, \tag{14.9}$$

where necessarily $\sigma(m) \neq \tau(m)$.

When $m = 2$, fix a solution $\mathbf{z}_1$. For any other solution $\mathbf{z}$ there are $\sigma \neq \tau$ with (14.8) having

$$(\zeta_{1,\sigma(1)}\zeta_{\sigma(2)})/(\zeta_{1,\tau(1)}\zeta_{\tau(2)}) \in S_3.$$

Here $\sigma(2) \neq \tau(2)$, since otherwise $\sigma(2) = \tau(2), \ldots, \sigma(q) = \tau(q)$, against $\sigma \neq \tau$. Therefore in fact for every possible value of m, we can find for each solution $\mathbf{z}$ a pair of permutations with (14.9) and $\sigma(m) \neq \tau(m)$.

Writing $\sigma(m) = i$, $\tau(m) = i'$, we see from (14.9) that for every solution $\mathbf{z}$ of (14.1) there are distinct indices $i \neq i'$ such that $\zeta_i/\zeta_{i'} \in S_2$, where S_2 consists of all products of an element of S_3 and a number

$$(\zeta_{1,\tau(1)} \cdots \zeta_{m-1,\tau(m-1)})/(\zeta_{1,\sigma(1)} \cdots \zeta_{m-1,\sigma(m-1)})$$

where σ, τ are distinct permutations of $\{1, \ldots, q\}$. Since, as noted above, S_3 has cardinality $\leqq (q!)^4$, and since the number of pairs σ, τ of permutations is $(q!)^2$, our set S_2 has cardinality $\leqq (q!)^6$. □

The proof of Theorem 12.2 is now easily completed. The case $q = 2$ being trivial, we may suppose that $q \geqq 3$, and the assertion is true for $q - 1$. By Lemma 14.3, for every solution $(\zeta_1, \dots, \zeta_q)$ there are $i \neq i'$ and a $\xi \in S_2$ with $\zeta_i/\zeta_{i'} = \xi$. The number of triples (i, i', ξ) is at most

$$q^2|S_2| \leqq (q!)^6 q^2 \leqq q^{6q-4}. \tag{14.10}$$

So how many solutions are there with a fixed triple (i, i', ξ)? For simplicity, suppose $i = q$, $i' = q - 1$. Then $a_{q-1}\zeta_{q-1} + a_q\zeta_q = a'_{q-1}\zeta_{q-1}$ with $a'_{q-1} = a_{q-1} + \xi a_q$. Substitution into (14.1) gives

$$a_1\zeta_1 + \cdots + a_{q-2}\zeta_{q-2} + a'_{q-1}\zeta_{q-1} = 0. \tag{14.11}$$

As is easily seen, a nondegenerate solution of (14.1) gives rise to a nondegenerate solution of (14.11). By induction, there are, up to proportionality, fewer than

$$(q-1)^{3(q-1)^2}$$

such solutions. Multiplying this with the right hand side of (14.10) we get

$$< q^{3(q-1)^2+6q-4} < q^{3q^2}.$$

□

15 Reductions for Theorem 12.3

We will consider (12.1), i.e.,

$$a_1x_1 + \cdots + a_qx_q = 0, \tag{15.1}$$

and will show that the number of nondegenerate solutions in $\mathbb{P}(\Gamma)$, where Γ is of rank r, is at most

$$C(q,r) = \begin{cases} 1, & \text{when } q = 2; \\ e^{(4q)^{4q}(r+1)}, & \text{when } q > 2. \end{cases} \tag{15.2}$$

First Reduction

It will suffice to show that when S is a finite set of nondegenerate solutions of (15.1) in $\mathbb{P}(\Gamma)$, then $|S| \leqq C(q,r)$. In other words, given s nondegenerate solutions $\mathbf{x}_1, \dots, \mathbf{x}_s$ of (15.1) in Γ, no two of which are proportional, we have to show that $s \leqq C(q,r)$. Since Γ is of rank r, there are vectors $\mathbf{b}_1, \dots, \mathbf{b}_r$ in $(\mathbb{C}^\times)^q$ such that each $\mathbf{x}_i$ satisfies a relation

$$\mathbf{x}_i^{h_i} = \mathbf{b}_1^{h_{i1}} \cdots \mathbf{b}_r^{h_{ir}} \qquad (i = 1, \dots, s) \tag{15.3}$$

with integers $h_i > 0, h_{i1}, \ldots, h_{ir}$. (The powers and products in (15.3) are with respect to the group operation $*$.) The fact that each $\mathbf{x}_i$ is a solution to (15.1), together with the relations (15.3) (where the exponents are fixed!) means that $\mathcal{X}_0 = (\mathbf{x}_1, \ldots, \mathbf{x}_s, \mathbf{b}_1, \ldots, \mathbf{b}_r)$ lies in a certain algebraic variety $Y_0 \subset \mathbb{C}^{(r+s)q}$, hence $\mathcal{X} = (a_1, \ldots, a_q, \mathbf{x}_1, \ldots, \mathbf{x}_s, \mathbf{b}_1, \ldots, \mathbf{b}_r)$ in a variety $Y \subset \mathbb{C}^{(r+s+1)q}$.

On the other hand, (i) all the a_i and all the coordinates of $\mathbf{b}_1, \ldots, \mathbf{b}_r$ are nonzero, (ii) no two among $\mathbf{x}_1, \ldots, \mathbf{x}_s$ are proportional, so that certain (2×2)-determinants are nonzero, (iii) since $\mathbf{x}_1, \ldots, \mathbf{x}_s$ are nondegenerate solutions of (15.1), certain linear relations do not hold. All this means that $\mathcal{X} \notin F$, where F is an algebraic variety. Thus $\mathcal{X} \in Y \backslash F$. There is an algebraic specialization $\widehat{\mathcal{X}} \in Y \backslash F$ of $\mathcal{X}$. All the coordinates of $\widehat{\mathcal{X}} = (\hat{a}_1, \ldots, \hat{a}_q, \hat{\mathbf{x}}_1, \ldots, \hat{\mathbf{x}}_s, \hat{\mathbf{b}}_1, \ldots, \hat{\mathbf{b}}_r)$ are nonzero and no two of $\hat{\mathbf{x}}_1, \ldots, \hat{\mathbf{x}}_s$ are proportional. By (15.3), $\hat{\mathbf{x}}_1, \ldots, \hat{\mathbf{x}}_s$ lie in a group $\widehat{\Gamma}$ of rank $\leqq r$. Finally each $\hat{\mathbf{x}}_i$ is a nondegenerate solution of $\hat{a}_1 \hat{x}_1 + \cdots + \hat{a}_q \hat{x}_q = 0$. We therefore have reduced to the situation where each $a_i \in K^\times$ and $\Gamma \subset (K^\times)^q$, with an algebraic number field K.

Second Reduction

Proposition 15.1. *Let K be a number field, $a_1, \ldots, a_q \in K$, and $\Gamma \subset (K^\times)^q$ a group of rank r. Then the solutions $\mathbf{x} \in \Gamma$ of (15.1) will lie in the union of at most*

$$B(q, r) = e^{(4q)^{3q}(r+2)}$$

proper subspaces of the $(q-1)$-dimensional space E defined by (15.1).

We will deduce Theorem 12.3 in a way which has become routine in our subject. In fact it is similar to the way in which Theorem 12.2 was deduced from Lemma 14.3.

The case $q = 2$ being obvious, suppose $q > 2$. By the Proposition, our solutions $\mathbf{x}$ will be contained in the union of $B(q, r)$ proper subspaces. Let V be such a subspace, defined by an equation

$$\sum_{i \in I} b_i x_i = 0 \tag{15.4}$$

where I is a subset of $\{1, \ldots, q\}$ of cardinality $2 \leqq |I| \leqq q - 1$, and $b_i \neq 0$ for $i \in I$. For every nontrivial solution of (15.4) there is a nonempty subset J of I such that

$$\sum_{i \in J} b_i x_i = 0, \tag{15.5}$$

but no proper subsum of (15.5) vanishes. Clearly $2 \leqq |J| \leqq q-1$, and we may suppose that $J = \{1, \ldots, \ell\}$, say.

We are dealing with solutions in $\mathbb{P}(\Gamma)$. Denoting them by $(x_1 : \cdots : x_q)$, let $\mathbb{P}(\Gamma_1)$ be the group of elements $(x_1 : \cdots : x_\ell)$ such that

$$(x_1 : \cdots : x_\ell : \cdots : x_q) \in \mathbb{P}(\Gamma)$$

for some $x_{\ell+1}, \ldots, x_q$. The homomorphism $\varphi : \mathbb{P}(\Gamma) \longrightarrow \mathbb{P}(\Gamma_1)$ with $\varphi(x_1 : \cdots : x_q) = (x_1 : \cdots : x_\ell)$ has kernel $\mathbb{P}(\Gamma_2)$, where $\mathbb{P}(\Gamma_2)$ consists of elements $(z : \cdots : z : z_{\ell+1} : \cdots : z_q)$ of $\mathbb{P}(\Gamma)$. Therefore

$$r \geqq \operatorname{rank} \mathbb{P}(\Gamma) = \operatorname{rank} \mathbb{P}(\Gamma_1) + \operatorname{rank} \mathbb{P}(\Gamma_2) = r_1 + r_2,$$

say.

By induction, (15.5) has at most $C(\ell, r_1)$ nondegenerate solutions $(x_1 : \cdots : x_\ell) \in \mathbb{P}(\Gamma_1)$. Every solution of (15.1) with (15.5) is of the type $(x_1 z : \cdots : x_\ell z : z_{\ell+1} : \cdots : z_q)$ with $(x_1 : \cdots : x_\ell)$ as above. Moreover, $(z : z_{\ell+1} : \cdots : z_q)$ will lie in a group of rank r_2, and will have

$$bz + a_{\ell+1} z_{\ell+1} + \cdots + a_q z_q = 0 \tag{15.6}$$

where $b = a_1 x_1 + \cdots + a_\ell x_\ell$. When the solution $\mathbf{x}$ of (15.1) is nondegenerate, then so is the solution $(z, z_{\ell+1}, \ldots, z_q)$ of (15.6). By induction, and since $q - \ell + 1 < q$, (15.6) has not more than $C(q - \ell + 1, r_2)$ nondegenerate solutions. Combining this with the bound $C(\ell, r_1)$ for the number of nondegenerate solutions of (15.5), we may deduce that (15.5) leads to at most

$$C(\ell, r_1) C(q - \ell + 1, r_2) \leqq C(q - 1, r)$$

nondegenerate solutions in $\mathbb{P}(\Gamma)$ of (15.1): the last inequality is a consequence of the definition of $C(q, r)$, which entails

$$C(u, r_1) C(v, r_2) \leqq C(u + v - 2, r_1 + r_2).$$

Taking the sum over the possible subsets J of I, we see that each subspace V gives rise to at most $2^q C(q-1, r)$ solutions. We still have to multiply this with the number $B(q, r)$ of subspaces. Our bound becomes

$$2^q C(q-1, r) B(q, r) < 2^q e^{(4q)^{4q-4}(r+1)} e^{(4q)^{3q}(r+2)} < e^{(4q)^{4q}(r+1)} = C(q, r).$$

□

Third Reduction

Proposition 15.2. *Let Γ be as in Proposition 15.1. Then the solutions $\mathbf{x} \in \Gamma$ of*

$$x_1 + \cdots + x_q = 0 \tag{15.7}$$

lie in the union of at most

$$B'(q, r) = e^{(4q)^{3q}(r+1)}$$

proper subspaces of the space E' defined by (15.7).

When dealing with (15.1) we observe that $(a_1 x_1, \ldots, a_q x_q)$ lies in the group Γ' of rank $\leqq r + 1$ generated by Γ and $(a_1, \ldots, a_q)$. Thus the required number of subspaces will be $\leqq B'(q, r+1) = B(q, r)$, so that Proposition 15.1 will follow.

16 Special Solutions

Let $\mathcal{S}$ be a finite set of nondegenerate solutions of (15.7) in $\mathbb{P}(\Gamma)$. It will suffice to show that $\mathcal{S}$ is contained in the union of at most $B'(q,r)$ proper subspaces of the space defined by (15.7). We further may suppose that $\Gamma \subset (K^{\times})^q$, with K a number field. Since Γ is of rank r, $\mathbb{P}(\Gamma) = \Gamma/D$ is of rank $\leqq r$. Since the image $\bar{\mathcal{S}}$ of $\mathcal{S}$ in $\mathbb{P}(\Gamma)$ is again finite, there are $n \leqq r$ multiplicatively independent points $\bar{\mathbf{b}}_1, \ldots, \bar{\mathbf{b}}_n$ in $\mathbb{P}(\Gamma)$ such that each $\bar{\mathbf{x}} \in \bar{\mathcal{S}}$ may be written as $\bar{\boldsymbol{\xi}} \bar{\mathbf{b}}_1^{k_1} \cdots \bar{\mathbf{b}}_n^{k_n}$, with $\bar{\boldsymbol{\xi}}$ a torsion point and $\mathbf{k} = (k_1, \ldots, k_n) \in \mathbb{Z}^n$. We therefore need only consider solutions $\mathbf{x}$ of (15.7) of the form

$$\mathbf{x} = \boldsymbol{\zeta} \mathbf{b}_1^{k_1} \cdots \mathbf{b}_n^{k_n}, \tag{16.1}$$

where $\boldsymbol{\zeta}$ is torsion, i.e., its components are roots of unity. Observe that by our choice of $\bar{\mathbf{b}}_1, \ldots, \bar{\mathbf{b}}_n$, we have $\mathbf{b}_1^{k_1} \cdots \mathbf{b}_n^{k_n} \in D$ only when $\mathbf{k} = \mathbf{0}$, and in fact when $\mathbf{k} \neq \mathbf{0}$, then $\mathbf{b}_1^{k_1} \cdots \mathbf{b}_n^{k_n}$ is not proportional to a torsion point.

By homogeneity we can make a further reduction: we may suppose that the first coordinate of each point is 1. Thus we may suppose this to be true for $\mathbf{b}_1, \ldots, \mathbf{b}_n$. Write

$$\mathbf{b}_j = (a_{1j}, a_{2j}, \ldots, a_{qj}) = (1, a_{2j}, \ldots, a_{qj}) \qquad (j = 1, \ldots, n),$$

and set

$$\boldsymbol{\alpha}_i = (a_{i1}, \ldots, a_{in}) \qquad (i = 1, \ldots, q),$$

so that in particular $\boldsymbol{\alpha}_1 = (1, \ldots, 1)$. Then $\mathbf{x}$ given by (16.1) is

$$\mathbf{x} = (1, \zeta_2 \boldsymbol{\alpha}_2^{\mathbf{k}}, \ldots, \zeta_q \boldsymbol{\alpha}_q^{\mathbf{k}}).$$

Note that $\|\boldsymbol{\alpha}_i^{\mathbf{k}}\|_v = \|a_{i1}^{k_1} \cdots a_{in}^{k_n}\|_v = \|a_{i1}\|_v^{k_1} \cdots \|a_{in}\|_v^{k_n}$, so that, writing[5]

$$\mathbf{a}_{iv} = (\log \|a_{i1}\|_v, \ldots, \log \|a_{in}\|_v)$$

we get

$$\log \|\boldsymbol{\alpha}_i^{\mathbf{k}}\|_v = \mathbf{a}_{iv} \mathbf{k} \qquad (v \in M(K),\ 1 \leqq i \leqq q)$$

(where the right hand side is an inner product). Observe that $\mathbf{a}_{1v} = \mathbf{0}$.

Recall the definition of heights in Section 8, and set

$$\begin{aligned} \tilde{h}(\mathbf{x}) &= h(\mathbf{x}, \mathbf{x}^{-1}) \\ &= \sum_v \max(0, \mathbf{a}_{2v}\mathbf{k}, \ldots, \mathbf{a}_{qv}\mathbf{k}, -\mathbf{a}_{2v}\mathbf{k}, \ldots, -\mathbf{a}_{qv}\mathbf{k}). \end{aligned} \tag{16.2}$$

Since $\sum_v \max(0, -\mathbf{a}_{jv}\mathbf{k}) = \sum_v \max(\mathbf{a}_{jv}\mathbf{k}, 0)$ by the Product Formula,

[5] v always denotes a place, i.e., an element of $M(K)$. A sum $\sum_v$ is over all $v \in M(K)$.

$$
\begin{aligned}
h(\mathbf{x}) \leqq \tilde{h}(\mathbf{x}) &\leqq \sum_{v} \left(\max(0, \mathbf{a}_{2v}\mathbf{k}, \ldots, \mathbf{a}_{qv}\mathbf{k}) + \sum_{j=2}^{q} \max(0, -\mathbf{a}_{jv}\mathbf{k}) \right) \\
&= h(\mathbf{x}) + \sum_{v} \sum_{j=2}^{q} \max(0, \mathbf{a}_{jv}\mathbf{k}) \\
&\leqq q h(\mathbf{x}).
\end{aligned}
\tag{16.3}
$$

We noted above that $\mathbf{b}_1^{k_1} \cdots \mathbf{b}_n^{k_n}$ for nonzero $\mathbf{k} \in \mathbb{Z}^n$ is not proportional to a torsion point. Therefore[6] $h(\mathbf{x}) \geqq c_0(K) > 0$ when $\mathbf{k} \neq \mathbf{0}$, so that the quantity $\psi(\mathbf{k})$ given by (16.2) has

$$\psi(\mathbf{k}) \geqq c_0(K) > 0.$$

We may extend ψ to a function on $\mathbb{R}^n$ by setting

$$
\begin{aligned}
\psi(\boldsymbol{\xi}) &= \sum_{v} \max(0, \mathbf{a}_{2v}\boldsymbol{\xi}, \ldots, \mathbf{a}_{qv}\boldsymbol{\xi}, -\mathbf{a}_{2v}\boldsymbol{\xi}, \ldots, -\mathbf{a}_{qv}\boldsymbol{\xi}) \\
&= \sum_{v} \max(|\mathbf{a}_{2v}\boldsymbol{\xi}|, \ldots, |\mathbf{a}_{qv}\boldsymbol{\xi}|).
\end{aligned}
$$

Clearly

(a) $\psi(\boldsymbol{\xi}) \geqq 0$,

(b) $\psi(\alpha\boldsymbol{\xi}) = |\alpha|\, \psi(\boldsymbol{\xi})$,

(c) $\psi(\boldsymbol{\xi} + \boldsymbol{\eta}) \leqq \psi(\boldsymbol{\xi}) + \psi(\boldsymbol{\eta})$,

and the argument for Lemma 8.4 shows that

(d) $\psi(\boldsymbol{\xi}) \geqq c_1 |\boldsymbol{\xi}|$

with a positive constant c_1. By our conditions (a)-(d), the set $\Psi \subset \mathbb{R}^n$ consisting of points $\boldsymbol{\xi}$ with $\psi(\boldsymbol{\xi}) \leqq 1$ is convex, symmetric (i.e., $\boldsymbol{\xi} \in \Psi$ implies $-\boldsymbol{\xi} \in \Psi$), compact, and contains $\mathbf{0}$ in its interior. Thus Ψ is what is called a *symmetric convex body* in $\mathbb{R}^n$.

Put

$$m = 4q^2 - 4q. \tag{16.4}$$

When $\mathbf{x}$ is given by (16.1), write

$$\tilde{h} = h(\mathbf{x}, \mathbf{x}^{-1}) = \psi(\mathbf{k}), \qquad H = \exp \tilde{h}. \tag{16.5}$$

Given $\boldsymbol{\rho} \in \mathbb{R}^n$, a point given by (16.1) will be called $\boldsymbol{\rho}$-special if $\tilde{h} > 0$ and if

$$\mathbf{k} \in (\tilde{h}/m)\Psi + \tilde{h}\boldsymbol{\rho}; \tag{16.6}$$

the right hand side here signifies the set $(\tilde{h}/m)\Psi$ translated by $\tilde{h}\boldsymbol{\rho}$.

We next quote Lemma 8.1 of [27], which will however not be proved here.

[6] The numbering of constants begins anew in each Section.

Lemma 16.1. *Let Φ be a symmetric convex body in $\mathbb{R}^n$. Then $\lambda\Phi$ (where $\lambda > 0$) can be covered by a collection of not more than*

$$(2\lambda + 4)^n$$

translates of Φ.

Applying this Lemma with $\Phi = m^{-1}\Psi$ and $\lambda = m$, we see that $\Psi = m\Phi$ may be covered by

$$Z = (2m + 4)^n \tag{16.7}$$

translates of $m^{-1}\Psi$, say by $m^{-1}\Psi + \boldsymbol{\rho}_i$ $(i = 1, \dots, Z)$. Then $\tilde{h}\Psi$ is covered by the sets $(\tilde{h}/m)\Psi + \tilde{h}\boldsymbol{\rho}_i$ $(i = 1, \dots, Z)$. Since $\psi(\mathbf{k}) = \tilde{h}$ by (16.5), we obtain

Lemma 16.2. *There are points $\boldsymbol{\rho}_1, \dots, \boldsymbol{\rho}_Z$ in $\mathbb{R}^n$ such that $\mathbf{x}$ as given by* (16.1) *and with $\tilde{h}(\mathbf{x}) > 0$ is special for at least one of $\boldsymbol{\rho}_1, \dots, \boldsymbol{\rho}_Z$.*

We finally remark that in view of (16.6), $\boldsymbol{\rho}_1, \dots, \boldsymbol{\rho}_Z$ may be chosen with

$$\boldsymbol{\rho}_i \in (1 + m^{-1})\Psi. \tag{16.8}$$

17 Properties of Special Solutions

Choose a finite set $S \subset M(K)$ containing all the infinite places such that each a_{ij} is an S-unit. Set

$$g_{iv}(\boldsymbol{\xi}) = \mathbf{a}_{iv}\boldsymbol{\xi} \qquad (v \in S,\ 1 \leqq i \leqq q),$$

so that in particular $g_{1v} = 0$. By the Product Formula $\sum_v \log \|a_{ij}\|_v = 0$ for each i, j, therefore

$$\sum_v g_{iv}(\boldsymbol{\xi}) = 0 \qquad (1 \leqq i \leqq q).$$

Let $\boldsymbol{\rho} \in (1 + m^{-1})\Psi$ be fixed and set

$$m_{iv} = g_{iv}(\boldsymbol{\rho}).$$

Then

$$\sum_v m_{iv} = 0 \qquad (1 \leqq i \leqq q). \tag{17.1}$$

We have

$$\sum_v \max(|m_{1v}|, \dots, |m_{qv}|) = \psi(\boldsymbol{\rho}) \leqq 1 + 1/m. \tag{17.2}$$

Now suppose $\mathbf{x}$ is $\boldsymbol{\rho}$-special, so that (16.6) holds. Then

$$g_{iv}(\mathbf{k}) = \tilde{h} g_{iv}(\boldsymbol{\rho}) + (\tilde{h}/m) g_{iv}(\boldsymbol{\xi}) = \tilde{h} m_{iv} + (\tilde{h}/m) g_{iv}(\boldsymbol{\xi}) \tag{17.3}$$

with $\boldsymbol{\xi} \in \Psi$. Therefore

$$\begin{aligned} \sum_v \max_i |g_{iv}(\mathbf{k}) - \tilde{h} m_{iv}| &= (\tilde{h}/m) \sum_v \max_i |g_{iv}(\boldsymbol{\xi})| \\ &= (\tilde{h}/m) \psi(\boldsymbol{\xi}) \leqq (\tilde{h}/m). \end{aligned} \tag{17.4}$$

For $v \in M(K)$, let $L_1^{(v)}, \ldots, L_q^{(v)}$ be the linear forms in $\mathbf{Y} = (Y_2, \ldots, Y_q)$ given by

$$\begin{aligned} L_1^{(v)}(\mathbf{Y}) &= Y_2 + \cdots + Y_q, \\ L_j^{(v)}(\mathbf{Y}) &= Y_j \qquad (j = 2, \ldots, q). \end{aligned}$$

When $\mathbf{x} = (x_1, \ldots, x_q)$, set $\mathbf{y} = (x_2, \ldots, x_q)$.

Lemma 17.1. *Let $\boldsymbol{\rho}$ be as above. There are $(q-1)$-element subsets $\mathcal{I}(v)$ of $\{1, \ldots, q\}$ defined for $v \in M(K)$, and there are numbers ℓ_{jv} ($v \in M(K)$, $j \in \mathcal{I}(v)$) with the following properties:*

$$\mathcal{I}(v) = \{2, \ldots, q\} \quad \textit{for} \quad v \notin S, \tag{17.5}$$

$$\ell_{jv} = 0 \quad \textit{for} \quad v \notin S,\ j \in \mathcal{I}(v), \tag{17.6}$$

$$\sum_v \sum_{j \in \mathcal{I}(v)} \ell_{jv} = 0, \tag{17.7}$$

$$\sum_v \sum_{j \in \mathcal{I}(v)} |\ell_{jv}| \leqq 1. \tag{17.8}$$

Moreover, every $\boldsymbol{\rho}$-special solution of (15.7) *satisfies*

$$\prod_v \max_{j \in \mathcal{I}(v)} \left\{ \| L_j^{(v)}(\mathbf{y}) \|_v Q^{-\ell_{jv}} \right\} \leqq Q^{-1/9q^3}$$

where $Q = H^{(9/4)q}$.

Proof. For $v \in S$, let $j(v) \in \{1, \ldots, q\}$ be a subscript with

$$m_{j(v),v} = \max_i m_{iv}.$$

Define $\mathcal{I}(v)$ to be the complement of $j(v)$ in $\{1, \ldots, q\}$. Of course, for $v \notin S$, $\mathcal{I}(v)$ is already determined by (17.5). Recall

$$\tilde{h} = \psi(\mathbf{k}) = \sum_v \max(|g_{2v}(\mathbf{k})|, \ldots, |g_{qv}(\mathbf{k})|).$$

For some i_0, $2 \leqq i_0 \leqq q$,

$$\sum_v |g_{i_0 v}(\mathbf{k})| \geqq \tilde{h}/(q-1).$$

Since $\sum_v g_{i_0 v}(\mathbf{k}) = 0$, we may conclude that

$$\sum_v \max(0, g_{i_0 v}(\mathbf{k})) \geqq \tilde{h}/(2(q-1)).$$

Further (17.4) tells us that

$$\tilde{h} \sum_v \max(0, m_{i_0 v}) \geqq \sum_v \max(0, g_{i_0 v}(\mathbf{k})) - \tilde{h}/m \geqq \tilde{h}\left(\frac{1}{2(q-1)} - \frac{1}{m}\right).$$

Since $m_{j(v),v} \geqq \max(m_{0v}, m_{i_0 v}) = \max(0, m_{i_0 v})$, we obtain

$$\sum_v m_{j(v),v} \geqq \frac{1}{2(q-1)} - \frac{1}{m}. \tag{17.9}$$

Let $s = |S|$ be the cardinality of S, and set

$$\gamma = \frac{1}{(q-1)s} \sum_v m_{j(v),v}, \tag{17.10}$$

$$c_{jv} = \begin{cases} m_{jv} + \gamma, & \text{if } v \in S \text{ and } j \in \mathcal{I}(v); \\ 0, & \text{if } v \notin S \text{ and } j \in \mathcal{I}(v). \end{cases} \tag{17.11}$$

By (17.1), since $\mathcal{I}(v)$ is the complement of $j(v)$ in $\{1, \dots, q\}$, and since $|\mathcal{I}(v)| = q-1$, we get

$$\sum_{v \in S} \sum_{j \in \mathcal{I}(v)} c_{jv} = 0. \tag{17.12}$$

Also, (17.2) (with (16.4), $q \geqq 2$) yields

$$\sum_v \max_j |c_{jv}| \leqq 2(1 + 1/m) \leqq 9/4. \tag{17.13}$$

By the equation $x_1 + \cdots + x_q = 0$, and the definition of the linear forms $L_i^{(v)}$,

$$\|L_j^{(v)}(\mathbf{y})\|_v = \exp(g_{jv}(\mathbf{k})) \qquad (v \in M(K),\ 1 \leqq j \leqq q),$$

and (17.3) gives

$$\|L_j^{(v)}(\mathbf{y})\|_v \exp(-\tilde{h} m_{jv}) = \exp((\tilde{h}/m) g_{jv}(\boldsymbol{\xi})).$$

In terms of the c_{jv} introduced in (17.11), and restricting to $j \in \mathcal{I}(v)$,

$$\|L_j^{(v)}(\mathbf{y})\|_v \exp(-\tilde{h}c_{jv}) = \exp\left(-\tilde{h}\gamma + (\tilde{h}/m)g_{jv}(\boldsymbol{\xi})\right) \quad (v \in S,\ j \in \mathcal{I}(v)).$$

By the definition (16.5) of $H = \exp \tilde{h}$,

$$\|L_j^{(v)}(\mathbf{y})\|_v H^{-c_{jv}} = H^{-\gamma + m^{-1}g_{jv}(\boldsymbol{\xi})} \quad (v \in S,\ j \in \mathcal{I}(v)),$$

$$\|L_j^{(v)}(\mathbf{y})\|_v H^{-c_{jv}} = H^{m^{-1}g_{jv}(\boldsymbol{\xi})} \quad (v \notin S,\ j \in \mathcal{I}(v)),$$

where in fact both sides of the last equation are 1. We get

$$\prod_v \max_{j \in \mathcal{I}(v)} \left\{\|L_j^{(v)}(\mathbf{y})\|_v H^{-c_{jv}}\right\} \leqq H^{-s\gamma + m^{-1}}, \tag{17.14}$$

since

$$\sum_v \max_j |g_{jv}(\boldsymbol{\xi})| = \psi(\boldsymbol{\xi}) \leqq 1.$$

In view of (17.9), (17.10), (16.4),

$$-s\gamma + m^{-1} \leqq -\frac{1}{(q-1)}\left(\frac{1}{2(q-1)} - \frac{1}{m}\right) + \frac{1}{m} < -1/4q^2,$$

so that the product in (17.14) is $< H^{-1/4q^2}$. Now if we set

$$Q = H^{(9/4)q}, \qquad \ell_{jv} = (4/9q)c_{jv},$$

then (17.7), (17.8) follow from (17.12), (17.13), and the product of Lemma 17.1 is at most $Q^{-1/9q^3}$. □

18 Large Solutions

A solution $\mathbf{x}$ of (15.7) will be called *large* if

$$\tilde{h}(\mathbf{x}) = h(\mathbf{x}, \mathbf{x}^{-1}) > 8q^2 \log q. \tag{18.1}$$

A solution which is not large will be called *small.*

Proposition 18.1. *Let* ℓ_{jv} *satisfy* (17.6), (17.7), (17.8). *Suppose* $0 < \delta < 1$. *Then the points* $\mathbf{y} \in K^{q-1}$ *satisfying*

$$\prod_{v \in M(K)} \max_{j \in \mathcal{I}(v)} \left\{\|L_j^{(v)}(\mathbf{y})\|_v Q^{-\ell_{jv}}\right\} \leqq Q^{-\delta} \tag{18.2}$$

for any Q *with*

$$Q > (q-1)^{2/\delta} \tag{18.3}$$

lie in the union of at most

$$t = 4^{(q+7)^2}\delta^{-q-3}(\log 2q)\log\log 2q \tag{18.4}$$

proper subspaces of K^{q-1}.

This Proposition is an immediate consequence of a very special case of a version of the Subspace Theorem due to Evertse and Schlickewei. We may apply Theorem 2.1 of [13], with the quantities r, n, $\mathcal{H}$, K, F of [13] replaced by our quantities q, $q-1$, 1, $\mathbb{Q}$, K respectively.

In view of Lemma 17.1, we may apply the Proposition with $Q = H^{(9/4)q}$ and $\delta = 1/9q^3$. By Lemma 17.1, given $\boldsymbol{\rho}$, any $\boldsymbol{\rho}$-special solution of (15.7) has $\mathbf{y} = (x_2, \ldots, x_q)$ satisfying (18.2) with a tuple (ℓ_{jv}) depending only on $\boldsymbol{\rho}$.

With $Q = H^{(9/4)q}$, $\delta = 1/9q^3$, the condition (18.3) is satisfied if $H^{(9/4)q} > q^{18q^3}$, hence when (18.1) holds. By (18.4), a given $\boldsymbol{\rho}$ gives rise to at most

$$4^{(q+7)^2}(9q^3)^{q+3}(\log 2q)\log\log 2q < e^{4(q+7)^2}$$

subspaces (where we used that $q \geqq 3$).[7] We yet have to multiply this number by the number Z of points $\boldsymbol{\rho}_1, \ldots, \boldsymbol{\rho}_Z$. By (16.4), (16.7) we obtain

$$< (8q^2)^n e^{4(q+7)^2}.$$

Since $n \leqq r$, we may summarize our results so far as follows.

Lemma 18.2. *The set of large solutions of* (15.7) *is contained in the union of fewer than*

$$(8q^2)^r e^{4(q+7)^2}$$

proper subspaces of the $(q-1)$*-dimensional space* E' *defined by* (15.7).

19 Small Solutions, and the end of the proof of Theorem 12.3

The following is a homogeneous version of the case $d = 1$ of Theorem 5 in [41].

Proposition 19.1. *Let* Γ *be a finitely generated subgroup of* $(\overline{\mathbb{Q}}^\times)^q$ *of rank* r. *Then given* $C > 1$, *the set of nondegenerate solutions* $\mathbf{x} \in \mathbb{P}(\Gamma)$ *of* (15.7) *with*

$$\sum_{i=2}^{q} h(x_1, x_i) \leqq C \tag{19.1}$$

has cardinality at most

$$b(Cb)^r$$

with $b = \exp\left((4(q-1))^{3(q-1)}\right)$.

[7] I do not enjoy giving numerical estimates such as the one just exhibited!

Recently David and Philippon [5] proved a result of this type, with $b = \exp\exp(c_0 q)$.

Note that $\sum_{i=2}^{q} h(x_1, x_i) \leqq (q-1)h(\mathbf{x}) \leqq q\tilde{h}(\mathbf{x})$, so that the small solutions satisfy (19.1) with $C = 8q^3 \log q$. As a consequence, there are not more than

$$e^{(4q)^{3q-3}(r+1)}(8q^3 \log q)^r$$

nondegenerate small solutions. On the other hand it is clear that the degenerate solutions are contained in the union of 2^q proper subspaces of the space given by (15.7). We may summarize as follows.

Lemma 19.2. *The small solutions are contained in the union of at most*

$$e^{(4q)^{3q-1}(r+1)}$$

proper subspaces of the space given by (15.7).

Combining Lemmas 18.2 and 19.2, we see that

$$(8q^2)^r e^{4(q+7)^2} + e^{(4q)^{3q-1}(r+1)} < e^{(4q)^{3q}(r+1)}$$

subspaces suffice. Therefore Proposition 19.1, hence Theorem 12.3, has been established. □

20 Linear Recurrence Sequences Again

As was pointed out in Corollary 12.4, a *simple* non-degenerate recurrence of order t has zero-multiplicity not exceeding

$$e^{3(4t)^{4t}}.$$

On the other hand, establishing a long standing conjecture, I could prove [42] that any non-degenerate linear recurrence sequence of order t has zero-multiplicity $\leqq c(t)$. In fact I proved this with

$$c(t) = \exp\exp\exp(3t \log t). \tag{20.1}$$

Mahler [20], extending the Skolem–Mahler–Lech Theorem, dealt with completely arbitrary linear recurrences. He proved that the set $\mathcal{Z}$ of zeros, i.e., of numbers k with $u_k = 0$, is a union of finitely many arithmetic progressions and single numbers. As pointed out in Theorem 12.5, I could make this more explicit, as follows.

The set $\mathcal{Z}$ of zeros of a linear recurrence of order t is the union of at most

$$\exp\exp\exp(20t)$$

arithmetic progressions and single numbers.

Note that for large enough t, this is better than the bound (20.1). The proof of (20.1) and of Theorem 12.5 is complicated. I therefore will deal here only with a very special case, which however displays some main ideas. Even in the special case I cannot give all the details, such as, e.g., a proof of Lemma 20.5 below.

The case I will deal with here is when the companion polynomial has just one multiple root, and this root is of order 2. Thus

$$u_k = a_1\alpha_1^k + \cdots + a_{q-1}\alpha_{q-1}^k + (ak+b)\alpha_q^k \tag{20.2}$$

with nonzero $a_1, \ldots, a_{q-1}, a$. The order of u is

$$t = q + 1.$$

We will indicate a proof of

Proposition 20.1. *When u is non-degenerate and given by (20.2), its zero multiplicity is*[8]

$$< \exp((8t)^{8t}).$$

We recall an equivalence relation $\approx$ introduced in Section 11: $\alpha \approx \beta$ if α/β is a root of unity. The non-degeneracy of the sequence given by (20.2) means that $\alpha_i \not\approx \alpha_j$ for $i \neq j$.

For algebraic α it is customary to introduce its height

$$h(\alpha) = h(\alpha, 1).$$

Then $h(\alpha^k) = |k|\, h(\alpha)$, and $h(\alpha\beta) \leqq h(\alpha) + h(\beta)$. According to Dobrowolski [6], when $\alpha \not\approx 1$ is algebraic of degree n,

$$h(\alpha) \geqq (c_1/n)((\log^+ \log^+ n)/(\log^+ n))^3,$$

where $\log^+ x = \log x$ when $x \geqq e$, and $= 1$ otherwise. In fact by work of Voutier [46] we may take $c_1 = 1/4$, so that certainly

$$h(\alpha) \geqq 1/(4n(\log^+ n)^3).$$

Before going further, we have to formulate another result on purely exponential equations. Given nonzero $\alpha_1, \ldots, \alpha_q$ and $a_1, \ldots, a_q$, consider the function

$$f(k) = a_1\alpha_1^k + \cdots + a_q\alpha_q^k. \tag{20.3}$$

Say $\alpha_1, \ldots, \alpha_q$ belong to m equivalence classes (with respect to $\approx$) represented by $\beta_1, \ldots, \beta_m$. Grouping together terms whose α's are $\approx$ to each other, and changing the notation, we may write

[8] A proof is based on Lemma 20.5 below, which depends on [14]. It is likely that a version of [14] using [5] would lead to a bound $\exp(c^t)$.

$$f(k) = f_1(k)\beta_1^k + \cdots + f_m(k)\beta_m^k \tag{20.4}$$

where

$$f_i(k) = a_{i1}\alpha_{i1}^k + \cdots + a_{i,q_i}\alpha_{i,q_i}^k \qquad (i = 1, \ldots, m),$$

with $q_1 + \cdots + q_m = q$, and the α_{ij} $(1 \leqq i \leqq m,\ 1 \leqq j \leqq q_i)$ roots of unity.

Lemma 20.2 ("Second Splitting Lemma"). *All but at most*

$$F(q) = e^{4(4q)^{4q}}$$

solutions $k \in \mathbb{Z}$ of $f(k) = 0$ have

$$f_1(k) = \cdots = f_m(k) = 0. \tag{20.5}$$

Proof. The Lemma is nontrivial only when $m \geqq 2$; and then also $q = q(f) \geqq 2$. We proceed by induction on q. When $q = 2$ and $m = 2$, we have $f(k) = a\alpha_{11}^k + b\alpha_{21}^k$ with $ab \neq 0$ and $\alpha_{11} \not\approx \alpha_{21}$. There can be at most one $k \in \mathbb{Z}$ with $f(k) = 0$.

We now come to the step $q - 1 \to q$ where $q \geqq 3$. The point $\mathbf{x} = (a_1\alpha_1^k, \ldots, a_q\alpha_q^k)$ lies in the group Γ of rank $\leqq 2$ generated by $(a_1, \ldots, a_q)$ and $(\alpha_1, \ldots, \alpha_q)$. By Theorem 12.3, there are at most

$$e^{3(4q)^{4q}}$$

nondegenerate solutions of $x_1 + \cdots + x_q = 0$. Such a nondegenerate $\mathbf{x}$ determines k, for if $(a_1\alpha_1^k, \ldots, a_q\alpha_q^k)$ and $(a_1\alpha_1^{k'}, \ldots, a_q\alpha_q^{k'})$ are proportional, then $a_i\alpha_i^k/a_j\alpha_j^k = a_i\alpha_i^{k'}/a_j\alpha_j^{k'}$ $(1 \leqq i, j \leqq q)$, hence $(\alpha_i/\alpha_j)^{k-k'} = 1$, and since $\alpha_i \not\approx \alpha_j$ for $i \neq j$, we get $k = k'$.

When $\mathbf{x}$ is a degenerate solution, then

$$x_{i_1} + \cdots + x_{i_n} = x_{j_1} + \cdots + x_{j_\ell} = 0$$

where $\{i_1, \ldots, i_n\}$, $\{j_1, \ldots, j_\ell\}$ are disjoint nonempty sets whose union is $\{1, \ldots, q\}$. There are fewer than 2^q such partitions of $\{1, \ldots, q\}$. But each such partition yields nonzero f^*, f^{**} with $f = f^* + f^{**}$, and

$$f^*(k) = f^{**}(k) = 0$$

for the solutions in question. Further $q(f^*) < q$, $q(f^{**}) < q$ (where $q(f)$ signifies the number of nonzero summands of f). Write

$$f^* = f_1^*\beta_1^k + \cdots + f_m^*\beta_m^k,$$
$$f^{**} = f_1^{**}\beta_1^k + \cdots + f_m^{**}\beta_m^k$$

where f_i^*, f_i^{**} are linear combinations of some of $\alpha_{i1}^k, \ldots, \alpha_{i,q_i}^k$. By induction, all but at most $2F(q-1)$ solutions of $f^*(k) = f^{**}(k) = 0$ have

$$f_i^*(k) = 0 \quad (1 \leqq i \leqq m), \qquad f_i^{**}(k) = 0 \quad (1 \leqq i \leqq m),$$

hence (20.5) since $f_i = f_i^* + f_i^{**}$. Therefore the number of exceptions to (20.5) is

$$< e^{3(4q)^{4q}} + 2F(q-1) = e^{3(4q)^{4q}} + 2\,e^{4(4q-4)^{4q-4}} < e^{4(4q)^{4q}}.$$

□

Corollary 20.3. *Suppose f has a summand $f_i(k)$ in* (20.4) *with just one term, i.e., suppose $f(k)$ has a summand $a_{i_0}\alpha_{i_0}^k$ with $\alpha_{i_0} \not\approx \alpha_i$ for every $i \neq i_0$. Then f has at most $F(q)$ zeros $k \in \mathbb{Z}$.*

Such a term of f will be called a *singleton.*

Remark. A sequence $u_k = f(k)$ with f as in the Corollary is a weakly non-degenerate linear recurrence in the terminology introduced in Section 7.

For completeness we mention a variation on the Second Splitting Lemma. Let

$$f(k) = P_1(k)\alpha_1^k + \cdots + P_q(k)\alpha_q^k,$$

and $t = \sum_{i=1}^q (1 + \deg P_i)$. Again write $f(k)$ as in (20.4), where now

$$f_i(k) = P_{i1}(k)\alpha_{i1}^k + \cdots + P_{i,q_i}(k)\alpha_{i,q_i}^k$$

with roots of unity α_{ij}.

Lemma 20.4 ("Third Splitting Lemma"). *All but at most*

$$\exp\exp\exp(20t)$$

solutions $k \in \mathbb{Z}$ of $f(k) = 0$ have $f_1(k) = \cdots = f_m(k) = 0$.

This Lemma will not be proved here. See [43], Corollary of the main Theorem.

Our next Lemma first appeared in a paper by Schlickewei, Schmidt and Waldschmidt [40]. We will not be able to prove it here.

Lemma 20.5 ("Reduction Lemma"). *Consider an equation*

$$P_1(k)\alpha_1^k + \cdots + P_q(k)\alpha_q^k = 0 \tag{20.6}$$

where $(\alpha_1, \ldots, \alpha_q) \in \left(\overline{\mathbb{Q}}^\times\right)^q$, and the P_i are nonzero polynomials of degree δ_i $(1 \leqq i \leqq q)$ with coefficients in $\overline{\mathbb{Q}}$. Set

$$p = \sum_{i=1}^q (\delta_i + 1), \qquad \delta = \max_i \delta_i.$$

Suppose $p \geqq 3$, $\delta > 0$, $0 < \hbar \leqq 1$, and

$$\max_{i,j} h(\alpha_i, \alpha_j) \geqq \hbar.$$

Set

$$E = 16p^2\delta/\hbar, \qquad w = \exp((5p)^{5p}) + 5E\log E.$$

Then there are tuples

$$(P_1^{(\ell)}, \ldots, P_q^{(\ell)}) \neq (0, \ldots, 0) \qquad (\ell = 1, \ldots, w)$$

of polynomials where $\deg P_i^{(\ell)} \leqq \delta_i$ $(1 \leqq i < q,\ 1 \leqq \ell \leqq w)$ *and* $\deg P_q^{(\ell)} < \delta_q$ *or* $P_q^{(\ell)} = 0$ $(1 \leqq \ell \leqq w)$, *such that every solution* $k \in \mathbb{Z}$ *of (20.6) satisfies*

$$P_1^{(\ell)}(k)\alpha_1^k + \cdots + P_q^{(\ell)}(k)\alpha_q^k = 0$$

for some ℓ.

* * *

The idea of the proof of Proposition 20.1 is simple. We clearly may suppose $t \geqq 3$, $q \geqq 2$. An easy specialization argument shows that we may suppose that all our data lie in a number field K, of some degree D. Dividing the equation $u_k = 0$ (with $u_k = f(k)$ as in (20.3)) by $a_q\alpha_q^k$ and changing the notation, we need to count solutions $k \in \mathbb{Z}$ of

$$g(k) := a_1\alpha_1^k + \cdots + a_{q-1}\alpha_{q-1}^k + (k+b) = 0.$$

Here $\alpha_1, \ldots, \alpha_{q-1}, 1$ are $\not\approx$ to each other. Let $\xi \mapsto \xi^{(i)}$ $(i = 1, \ldots, D)$ signify the embeddings of K into $\mathbb{C}$, and set

$$g^{(i)}(k) = a_1^{(i)}\alpha_1^{(i)k} + \cdots + a_{q-1}^{(i)}\alpha_{q-1}^{(i)k} + (k + b^{(i)}).$$

Now $g(k) = 0$ yields $g^{(i)}(k) = 0$, hence

$$\begin{aligned} &g(k) - g^{(i)}(k) \\ &\quad = a_1\alpha_1^k - a_1^{(i)}\alpha_1^{(i)k} + \cdots + a_{q-1}\alpha_{q-1}^k - a_{q-1}^{(i)}\alpha_{q-1}^{(i)k} + b - b^{(i)} = 0 \qquad (20.7) \end{aligned}$$

for $i = 1, \ldots, D$.

Case I. $b \notin \mathbb{Q}$, so that $b^{(i)} \neq b$ for some i. We study (20.7) in the light of Corollary 20.3. Now $b - b^{(i)} = (b - b^{(i)}) \cdot 1^k$ is a singleton, since each $\alpha_j \not\approx 1$, $\alpha_j^{(i)} \not\approx 1$. By the Corollary, (20.7) for our particular i has $\leqq F(2q-1) = F(2t-3)$ solutions. Proposition 20.1 holds in this case.

Case II. $b \in \mathbb{Q}$, so that $b^{(i)} = b$ for each i, and we get the equations

$$a_1\alpha_1^k - a_1^{(i)}\alpha_1^{(i)k} + \cdots + a_n\alpha_n^k - a_n^{(i)}\alpha_n^{(i)k} = 0 \qquad (20.8)$$

$(i = 1, \ldots, D)$, where we have set

$$n = q - 1 = t - 2.$$

The function on the left of (20.8) is of the type of the Splitting Lemma 20.2, with $2n$ in place of q. Now the α_j's are pairwise $\not\approx$, and so are the $\alpha_j^{(i)}$'s. Each α_j can be $\approx$ to at most one of the $\alpha_1^{(i)}, \ldots, \alpha_n^{(i)}$. Hence the left hand side of (20.8) contains a singleton, unless there is a permutation $\sigma^{(i)}$ of $\{1, \ldots, n\}$ with

$$\alpha_j \approx \alpha^{(i)}_{\sigma^{(i)}(j)} \qquad (1 \leqq j \leqq n). \tag{20.9}$$

In view of the Corollary to the Splitting Lemma we have

Lemma 20.6. *$g(k) = 0$ has at most $F(2n) = F(2t-4)$ solutions $k \in \mathbb{Z}$, unless for each i in $1 \leqq i \leqq D$, there is a permutation $\sigma^{(i)}$ with* (20.9).[9]

We may then suppose that there are such permutations $\sigma^{(1)}, \ldots, \sigma^{(D)}$. There will be some m such that there are at least D/n among these permutations with

$$\sigma^{(i)}(m) = 1. \tag{20.10}$$

Let S be the set of $i \in \{1, \ldots, D\}$ with (20.10); its cardinality is $s = |S| \geqq D/n$. We have

$$\alpha_m \approx \alpha_1^{(i)} \qquad (i \in S). \tag{20.11}$$

$\{1, \ldots, D\}$ is the disjoint union of sets $S = S_1, S_2, \ldots, S_{D/s}$ where $\alpha_1^{(i)} \approx \alpha_1^{(j)}$ precisely when i, j lie in the same set S_r: for a relation $\alpha \approx \beta$ is preserved under isomorphisms.

Set $\Delta = \deg \alpha_1$. Then D/Δ elements among the conjugates $\alpha_1^{(i)}$ are equal. In each S_r, collections of D/Δ among the $\alpha_1^{(i)}$ are equal, so that S_r contains $s\Delta/D$ *distinct* numbers $\alpha_1^{(i)}$. Denote these by $\beta_r^{(1)}, \ldots, \beta_r^{(s\Delta/D)}$. The Galois group of the normal closure of $\mathbb{Q}(\alpha_1)$ will permute the sets $S_r' = \{\beta_r^{(1)}, \ldots, \beta_r^{(s\Delta/D)}\}$ $(1 \leqq r \leqq D/s)$. Writing

$$\gamma_r = \beta_r^{(1)} \cdots \beta_r^{(s\Delta/D)} \qquad (1 \leqq r \leqq D/s),$$

the Galois group will permute $\gamma_1, \ldots, \gamma_{D/s}$, so that each $\deg \gamma_r \leqq D/s \leqq n$. Here $\gamma_1 \not\approx 1$, for γ_1 is a product of $s\Delta/D$ elements $\beta_1^{(i)}$ which are $\approx$ to each other, so that $\gamma_1 \approx 1$ would yield $\beta_1^{(i)s\Delta/D} \approx 1$, hence $\beta_1^{(i)} \approx 1$, so that $\alpha_1^{(i)} \approx 1$ for some i, contradicting $\alpha_1 \not\approx 1$. But $\gamma_1 \not\approx 1$ yields

[9] A variation of the argument below was given by Zannier, employing the fact that by (20.9) there is a homomorphism from the Galois group of the field generated by $\alpha_1, \alpha_2, \ldots$ into the symmetric group S_n.

$$h(\gamma_1) \geqq 1/(4n(\log^+ n)^3).$$

On the other hand $h(\gamma_1) \leqq \sum_{i=1}^{s\Delta/D} h(\beta_1^{(i)}) = (s\Delta/D)h(\alpha_1)$, so that

$$h(\alpha_1) \geqq (D/s\Delta)/(4n(\log^+ n)^3). \tag{20.12}$$

The $\beta_1^{(i)}$ with $1 \leqq i \leqq s\Delta/D$, being $\approx$ to each other, all have the same absolute value, but they are distinct. There will be j_1, j_2 with

$$\beta_1^{(j_2)} = e^{2\pi i\rho}\beta_1^{(j_1)}$$

where $0 < \rho \leqq D/s\Delta$. Thus there will be $i_1, i_2 \in S$ with

$$\alpha_1^{(i_2)} = e^{2\pi i\rho}\alpha_1^{(i_1)}. \tag{20.13}$$

We now consider (20.8) for $i = i_1$ and $i = i_2$. By the Splitting Lemma, all but at most $F(2n)$ solutions of (20.8) with $i = i_1$ will have

$$a_1^{(i_1)}\alpha_1^{(i_1)k} = a_m\alpha_m^k,$$

and all but $F(2n)$ solutions of (20.8) with $i = i_2$ will have

$$a_1^{(i_2)}\alpha_1^{(i_2)k} = a_m\alpha_m^k.$$

Then with at most $2F(2n)$ exceptions,

$$e^{2\pi i\rho k} = (\alpha_1^{(i_2)}/\alpha_1^{(i_1)})^k = a_1^{(i_1)}/a_1^{(i_2)}.$$

When k_0 is such a solution, and k another, then

$$e^{2\pi i\rho(k-k_0)} = 1,$$

hence $\rho(k - k_0) \in \mathbb{Z}$. Say $\rho = u/r$ (in lowest terms)[10]; then $u(k - k_0) \equiv 0 \pmod r$, so that

$$k = k_0 + rx \tag{20.14}$$

with $x \in \mathbb{Z}$. We have shown: All but at most $2F(2n)$ solutions k are of the form (20.14). We substitute this into the original equation $g(k) = 0$, which becomes

$$b_1\eta_1^x + \cdots + b_{q-1}\eta_{q-1}^x + (rx + k_0 + b) = 0 \tag{20.15}$$

with $b_i = a_i\alpha_i^{k_0}$, $\eta_i = \alpha_i^r$ $(1 \leqq i \leqq q-1)$. So what have we gained?
We had $\rho \leqq D/s\Delta$, therefore $r \geqq s\Delta/D$, and by (20.12),

[10] This r has nothing to do with the r above.

$$h(\eta_1) = h(\alpha_1^r) = rh(\alpha_1) \geqq 1/(4n(\log^+ n)^3).$$

We now apply the Reduction Lemma 20.5 to (20.15) with

$$\hbar = 1/(4n(\log^+ n)^3),$$

with $P_i = b_i$ $(1 \leqq i < q)$, $P_q(x) = rx + k_0 + b$. Now the $P_q^{(\ell)}$ of Lemma 20.5 will be constants (some perhaps zero), so that we are reduced to w equations of the type

$$b_1^{(\ell)}\eta_1^x + \cdots + b_q^{(\ell)}\eta_q^x = 0$$

with $\eta_q = 1$. The left hand side here is the value at x of a simple, non-degenerate linear recurrence of order $\leqq q$, so that there are at most

$$e^{3(4q)^{4q}} < e^{(4t)^{4t}}$$

solutions. Thus dealing with $\ell = 1, \dots, w$ we get not more than $w \cdot e^{(4t)^{4t}}$ solutions. In our situation $p = q + 1 = t$, $\delta = 1$, so that

$$E = 16t^2 \cdot 4n(\log^+ n)^3 < 64t^3(\log t)^3,$$
$$w < \exp((5t)^{5t}) + 320t^3(\log t)^3(\log 64 + 6\log t) < \exp((6t)^{5t}),$$

and

$$w \cdot e^{(4t)^{4t}} < \exp\left((6t)^{5t} + (4t)^{4t}\right) < \exp\left((7t)^{5t}\right).$$

To this we have to add the $2F(2n) = 2\,e^{4(8n)^{8n}} < e^{(8t)^{8t-1}}$ "exceptional" solutions noted above. Thus altogether there will be fewer than $\exp((8t)^{8t})$ solutions. □

21 Final Remarks

The Reduction Lemma 20.5 suggests an approach to Theorem 12.5: induction on $p = \sum_i(1+\deg P_i)$. One finally obtains purely exponential equations, which can be dealt with. Consider the case when the sequence is non-degenerate. Then $\alpha_i \not\approx \alpha_j$ for $i \neq j$ yields $h(\alpha_i/\alpha_j) > 0$. But using Dobrowolski's estimate, one only gets a lower bound $\hbar$ depending on the degree of α_i/α_j. The whole point of Theorem 12.5 is a bound independent of degrees.

An n-tuple of linear forms $M_1, \dots, M_n$ in $\mathbf{X} = (X_1, \dots, X_q)$ will be called *linearly independent over* $\mathbb{Q}$ if there is no identity

$$y_1M_1 + \cdots + y_nM_n = 0$$

with $\mathbf{y} = (y_1, \dots, y_n) \in \mathbb{Q}^n \backslash \{\mathbf{0}\}$. The crux of the argument is encapsuled in the following "Main Lemma":

Lemma 21.1. *Let $M_1, \dots, M_n$ with algebraic coefficients be linearly independent over $\mathbb{Q}$. Let $\alpha_1, \dots, \alpha_q$ be algebraic with $\alpha_i \not\approx \alpha_j$ for $i \neq j$. Then the numbers $k \in \mathbb{Z}$ for which the numbers*

$$M_1(\alpha_1^k, \dots, \alpha_q^k), \dots, M_n(\alpha_1^k, \dots, \alpha_q^k) \tag{21.1}$$

are linearly dependent over $\mathbb{Q}$ fall into

$$H(q, n)$$

classes as follows. For each class C there is an $r \in \mathbb{N}$ such that

(i) $k \equiv k' \pmod{r}$ *for* $k, k' \in C$,
(ii) $\max_{i,j} h(\alpha_i^r / \alpha_j^r) \geqq c(q, n) > 0$.

The non-degenerate case of Theorem 12.5 follows easily. Given the equation

$$\sum_{i=1}^{q} P_i(k) \alpha_i^k = 0, \tag{21.2}$$

write $n = 1 + \max_i \deg P_i$,

$$P_i = \sum_{j=1}^{n} a_{ij} X^{j-1} \qquad (i = 1, \dots, q),$$

$$M_j = \sum_{i=1}^{q} a_{ij} X_i \qquad (j = 1, \dots, n).$$

Then (21.2) becomes

$$\sum_{j=1}^{n} M_j(\alpha_1^k, \dots, \alpha_q^k) k^{j-1} = 0. \tag{21.3}$$

In order to apply the Main Lemma, we need $M_1, \dots, M_n$ to be linearly independent over $\mathbb{Q}$. But let us ignore this easily circumvented hindrance. The solutions fall into $H(q, n)$ classes. In a fixed class C, when $k_0 \in C$, every other k is of the form

$$k = rx + k_0$$

with $x \in \mathbb{Z}$. Substitution into (21.2) gives

$$\sum_{i=1}^{q} \widehat{P}_i(x) \widehat{\alpha}_i^x = 0$$

with $\widehat{P}_i(X) = \alpha_i^{k_0} P_i(rX + k_0)$, $\widehat{\alpha}_i = \alpha_i^r$. This looks just like (21.2), so what have we gained?

$$\max_{i,j} h(\widehat{\alpha}_i/\widehat{\alpha}_j) = \max_{i,j} h(\alpha_i^r/\alpha_j^r) \geqq c(q,n) > 0.$$

So now we may apply the Reduction Lemma with $\hbar = c(q,n)$, which is independent of degrees!

And what is the idea behind a proof of the Main Lemma? The linear dependence of the n quantities (21.1) means that

$$y_1 M_1(\alpha_1^k, \ldots, \alpha_q^k) + \cdots + y_n M_n(\alpha_1^k, \ldots, \alpha_q^k) = 0$$

with $(y_1, \ldots, y_n) \in \mathbb{Q}^n \backslash \{\mathbf{0}\}$. Let us suppose the α_i and the coefficients of the forms M_j lie in a number field K of degree D. Let $\xi \mapsto \xi^{(i)}$ $(i = 1, \ldots, D)$ be the embeddings $K \hookrightarrow \mathbb{C}$. Then

$$y_1 M_1^{(i)}(\alpha_1^{(i)k}, \ldots, \alpha_q^{(i)k}) + \cdots + y_n M_n^{(i)}(\alpha_1^{(i)k}, \ldots, \alpha_q^{(i)k}) = 0$$

for $i = 1, \ldots, D$. The matrix

$$M_j^{(i)}(\alpha_1^{(i)k}, \ldots, \alpha_q^{(i)k}) \qquad (1 \leqq j \leqq n,\ 1 \leqq i \leqq D)$$

has rank $< n$. In other words, certain determinants vanish. Such a determinant is of the form

$$\det\left(M_j^{(i_\ell)}(\alpha_1^{(i_\ell)k}, \ldots, \alpha_q^{(i_\ell)k})_{1 \leqq j,\ell \leqq n}\right) = 0 \tag{21.4}$$

where $i_1, \ldots, i_n$ is an n-tuple of numbers in $1 \leqq i \leqq D$. Now (21.4) is an equation in k of purely exponential type! However we don't have exponentials α_i^k, but

$$\left(\alpha_{u_1}^{(i_1)} \cdots \alpha_{u_n}^{(i_n)}\right)^k.$$

So the situation is complicated, but eventually manageable.

As has already been pointed out, it would be desirable to have a version of Theorem 12.1 not involving degrees. One would have to generalize the approach discussed in this Section to equations in n variables $k_1, \ldots, k_n$.

References

1. E. Bavencoffe et J.-P. Bézivin. Une Famille Remarquable de Suites Recurrentes Linéaires. Monatsh. f. Math. 120 (1995), 189–203.
2. F. Beukers. The zero multiplicity of ternary sequences. Compositio Math. 77 (1991), 165–177.
3. F. Beukers and H. P. Schlickewei. The equation $x + y = 1$ in finitely generated groups. Acta Arith. 78 (1996), 189–199.
4. E. Bombieri, J. Mueller and U. Zannier. Equations in one variable over function fields. Acta Arith. 99 (2001), 27–39.
5. S. David et P. Philippon. Minorations des hauteurs normalisées des sous-variétés des tores. Ann. Scuola Norm. Sup. Pisa Cl. Sci. (4) 28 (1999), 489–543. Errata, ibid. 29 (2000), 729–731.

6. E. Dobrowolski. On a question of Lehmer and the number of irreducible factors of a polynomial. Acta Arith. 34 (1979), 391–401.
7. P. Erdős, C. L. Stewart and R. Tijdeman. Some diophantine equations with many solutions. Compositio Math. 66 (1988), 37–56.
8. G. R. Everest and A. J. van der Poorten. Factorisation in the ring of exponential polynomials. Proc. Amer. Math. Soc. 125 (1997), 1293–1298.
9. J.-H. Evertse. On equations in S-units and the Thue–Mahler equation. Invent. Math. 75 (1984), 561–584.
10. J.-H. Evertse. On sums of S-units and linear recurrences. Compositio Math. 53 (1984), 225–244.
11. J.-H. Evertse. The number of solutions of decomposable form equations. Invent. Math. 122 (1995), 559–601.
12. J.-H. Evertse. The number of solutions of linear equations in roots of unity. Acta Arith. 89 (1999), 45–51.
13. J.-H. Evertse and H. P. Schlickewei. A quantitative version of the Absolute Subspace Theorem. J. reine angew. Math. 548 (2002), 21–127.
14. J.-H. Evertse, H. P. Schlickewei and W. M. Schmidt. Linear equations in variables which lie in a multiplicative group. Ann. of Math. (2) 155 (2002), 807–836.
15. E. Gourin. On Irreducible Polynomials in Several Variables Which Become Reducible When the Variables Are Replaced by Powers of Themselves. Trans. Amer. Math. Soc. 32 (1930), 485–501.
16. S. Lang. Integral points on curves. Publ. Math. I.H.E.S. 6 (1960), 27–43.
17. M. Laurent. Équations exponentielles polynômes et suites récurrentes linéaires. Journées arithmétiques de Besançon. Astérisque 147–148 (1987), 121–139, 343–344. Équations exponentielles polynômes et suites récurrentes linéaires, II. J. Number Theory 31 (1989), 24–53.
18. C. Lech. A note on recurring series. Ark. Math. 2 (1953), 417–421.
19. K. Mahler. Zur Approximation algebraischer Zahlen I. Über den grössten Primteiler binärer Formen. Math. Ann. 107 (1933), 691–730.
20. K. Mahler. Eine arithmetische Eigenschaft der Taylor–Koeffizienten rationaler Funktionen. Proc. Akad. Wetensch. Amsterdam 38 (1935), 50–60.
21. K. Mahler. On the Taylor coefficients of rational functions. Proc. Camb. Phil. Soc. 52 (1956) 39–48. Addendum 53, 544.
22. H. B. Mann. On linear relations between roots of unity. Mathematika 12 (1965), 107–117.
23. A. J. van der Poorten. Solution de la conjecture de Pisot sur le quotient de Hadamard de deux fractions rationnelles. C. R. Acad. Sci. Paris 306 (1988), Série I, 97–102.
24. A. J. van der Poorten. Some facts that should be better known, especially about rational functions. Macquarie Mathematics Reports No. 88-0022 (1988).
25. A. J. van der Poorten and H. P. Schlickewei. Additive relations in fields. J. Austral. Math. Soc. (Ser. A) 51 (1991), 154–170.
26. J. F. Ritt. A factorization theory for functions $\sum_{i=1}^{n} a_i e^{\alpha_i x}$. Trans. Amer. Math. Soc. 29 (1929), 584–596.
27. C. A. Rogers. A note on coverings and packings. J. London Math. Soc. 25 (1950), 327–331.
28. R. S. Rumely. Notes on van der Poorten's proof of the Hadamard quotient theorem. I, II. Séminaire de Théorie des Nombres, Paris 1986-87, 349–382, 383–409. Progr. Math. 75, Birkhäuser, 1988.

29. R. S. Rumely and A. J. van der Poorten. A note on the Hadamard kth root of a rational function. J. Austral. Math. Soc. (Ser. A) 43 (1987), 314–327.
30. A. Schinzel. Selected Topics on Polynomials. The University of Michigan Press, 1982.
31. H. P. Schlickewei. S-unit equations over number fields. Invent. Math. 102 (1990), 95–107.
32. H. P. Schlickewei. Equations $ax + by = 1$. Manuscript (1996).
33. H. P. Schlickewei. Equations in roots of unity. Acta Arith. 76 (1996), 99–108.
34. H. P. Schlickewei. Lower Bounds for Heights in Finitely Generated Groups. Monatsh. f. Math. 123 (1997), 171–178.
35. H. P. Schlickewei. The multiplicity of binary recurrences. Invent. Math. 129 (1997), 11–36.
36. H. P. Schlickewei and W. M. Schmidt. On polynomial–exponential equations. Math. Ann. 296 (1993), 339–361.
37. H. P. Schlickewei and W. M. Schmidt. Linear equations in members of recurrence sequences. Ann. Scuola Norm. Sup. Pisa Cl. Sci. (4) 20 (1993), 219–246.
38. H. P. Schlickewei and W. M. Schmidt. Equations $au_n^\ell = bu_m^k$ satisfied by members of recurrence sequences. Proc. Amer. Math. Soc. 118 (1993), 1043–1051.
39. H. P. Schlickewei and W. M. Schmidt. The number of solutions of polynomial–exponential equations. Compositio Math. 120 (2000), 193–225.
40. H. P. Schlickewei, W. M. Schmidt and M. Waldschmidt. Zeros of linear recurrence sequences. Manuscripta Math. 98 (1999), 225–241.
41. W. M. Schmidt. Heights of points on subvarieties of $\mathbb{G}_m^n$. London Math. Soc. Lecture Note Series 235 (1996), 157–187.
42. W. M. Schmidt. The zero multiplicity of linear recurrence sequences. Acta Math. 182 (1999), 243–282.
43. W. M. Schmidt. Zeros of linear recurrences. Publicationes Math. Debrecen 56 (2000), 609–630.
44. T. N. Shorey and R. Tijdeman. Exponential Diophantine Equations. Cambridge Tracts in Mathematics 87, Cambridge Univ. Press, 1986.
45. Th. Skolem. Ein Verfahren zur Behandlung gewisser exponentialer Gleichungen und diophantischer Gleichungen. (8. Skand. Mat. Kongr. Stockholm 1934) (1935), 163–188.
46. P. Voutier. An effective lower bound for the height of algebraic numbers. Acta Arith. 74 (1996), 81–95.
47. U. Zannier. (Unpublished 1998 manuscript on polynomial–exponential equations where some α_i/α_j is transcendental).
48. U. Zannier. A proof of Pisot's dth root conjecture. Ann. of Math. (2) 151 (2000), 375–383.

Linear Independence Measures for Logarithms of Algebraic Numbers

Michel Waldschmidt

Institut de Mathématiques, Université Pierre et Marie Curie (Paris VI)

Let $\alpha_1, \dots, \alpha_n$ be nonzero algebraic numbers and $b_1, \dots, b_n$ rational integers. Assume $\alpha_1^{b_1} \cdots \alpha_n^{b_n} \neq 1$. According to Liouville's inequality (Proposition 1.13), the lower bound

$$\left|\alpha_1^{b_1} \cdots \alpha_n^{b_n} - 1\right| \geq e^{-cB}$$

holds with $B = \max\{|b_1|, \dots, |b_n|\}$ and with a positive number c depending only on $\alpha_1, \dots, \alpha_n$. A fundamental problem is to prove a sharper estimate.

Transcendence methods lead to linear independence measures, over the field of algebraic numbers, for logarithms of algebraic numbers. Such measures are nothing else than lower bounds for numbers of the form

$$\Lambda = \beta_0 + \beta_1 \log \alpha_1 + \cdots + \beta_n \log \alpha_n,$$

where $\beta_0, \dots, \beta_n$ are algebraic numbers, $\alpha_1, \dots, \alpha_n$ are nonzero algebraic numbers, while $\log \alpha_1, \dots, \log \alpha_n$ are logarithms of $\alpha_1, \dots, \alpha_n$ respectively.

In the special case where $\beta_0 = 0$ and $\beta_1, \dots, \beta_n$ are rational integers, writing b_i for β_i, we have

$$\Lambda = b_1 \log \alpha_1 + \cdots + b_n \log \alpha_n,$$

which is the so-called *homogeneous rational case.* The importance of this special case is due to the fact that for $|\Lambda| \leq 1/2$ we have

$$\frac{1}{2}|\Lambda| \leq \left|e^{\Lambda} - 1\right| \leq 2|\Lambda|$$

with

$$e^{\Lambda} - 1 = \alpha_1^{b_1} \cdots \alpha_n^{b_n} - 1.$$

Hence we are back to the problem of estimating from below the distance between 1 and a number of the form $\alpha_1^{b_1} \cdots \alpha_n^{b_n}$.

The first three lectures are devoted to the qualitative theory of transcendental numbers, the last three ones to the quantitative theory of Diophantine approximation.

According to Hermite-Lindemann's Theorem, a number $\Lambda = \beta - \log\alpha$, with algebraic α and β, is zero only in the trivial case $\beta = \log\alpha = 0$. We start by assuming further that α and β are positive integers. It is a nontrivial fact that a positive integer b cannot be the logarithm of another positive integer a. In the first lecture we give two proofs of this result: the first one uses an auxiliary function, the second one uses an interpolation determinant together with a zero estimate. In the second lecture we complete the proof of Hermite-Lindemann's Theorem in the general case (with algebraic α and β), by means of the interpolation determinant method, but without a zero estimate: this is achieved thanks to an extrapolation argument.

In the third lecture we introduce Baker's Theorem on the linear independence of logarithms of algebraic numbers. After a brief survey of the available methods, we produce a proof by means of an interpolation determinant involving an extrapolation.

An introduction to Diophantine approximation is given in Section 4, where we address the question of estimating from below the distance between b and $\log a$, for a and b positive integers. A conjecture attributed to K. Mahler states that this distance should be at least a negative power of a:

$$|b - \log a| \overset{?}{\geq} a^{-c} \quad \text{for} \quad a \geq 2.$$

So far one does not know how to prove this result with a constant exponent, but only with exponent a constant times $\log\log a$ (K. Mahler; see (4.3) below):

$$|b - \log a| \geq a^{-c\log\log a} \quad \text{for} \quad a \geq 3.$$

We discuss a proof of this result by means of a method which is inspired by a recent work of M. Laurent and D. Roy [15].

The last two sections are devoted to Baker's method and to the question of measures of linear independence for an arbitrary number of logarithms of algebraic numbers. In the fifth lecture we survey available methods and in the last one we explain how to replace Matveev's auxiliary function by an interpolation determinant.

Notation. As a general rule we use the notation of [34]. In particular the absolute logarithmic height is denoted by h. The *length* of a polynomial $P \in \mathbb{C}[X_1, \ldots, X_n]$ (which is nothing else than the sum of the absolute values of its coefficients) is denoted by $\mathrm{L}(P)$.

For $\underline{z} = (z_1, \ldots, z_n) \in \mathbb{C}^n$, we set

$$|\underline{z}| = \max_{1\leq i\leq n} |z_i| \quad \text{and} \quad \|\underline{z}\| = |z_1| + \cdots + |z_n|.$$

In Section 4.1, $\|\cdot\|_{\mathbb{Z}}$ denotes the distance of a real number to the nearest integer:

$$\|x\|_{\mathbb{Z}} = \min_{k\in\mathbb{Z}} |x - k|.$$

For $x \in \mathbb{R}$ we set

$$\log_+ x = \log\max\{1, x\} \quad \text{and} \quad |x|_+ = \max\{1, |x|\}.$$

The integral part of x is denoted by $[x]$:

$$[x] \in \mathbb{Z}, \quad 0 \le x - [x] < 1$$

while $\lceil x \rceil$ denotes the least integer $\ge x$:

$$\lceil x \rceil \in \mathbb{Z}, \quad 0 \le \lceil x \rceil - x < 1.$$

For $r \ge 0$ we denote by

$$B_n(0, r) = \{\underline{z} \in \mathbb{C}^n \ ; \ |\underline{z}| \le r\}$$

the closed polydisk of $\mathbb{C}^n$ of center 0 and radius r, by $|f|_r$ the supremum norm of a continuous function $f\colon B_n(0,r) \to \mathbb{C}$ and by $\mathcal{H}_n(r)$ the set of continuous functions $f\colon B_n(0,r) \to \mathbb{C}$ which are holomorphic in the interior of $B_n(0,r)$.

Several differential operators will be used. For $\underline{k} = (k_1, \dots, k_n) \in \mathbb{N}^n$ and F a function of n variables $z_1, \dots, z_n$, $\mathcal{D}^{\underline{k}}F$ is the derivative

$$\left(\frac{\partial}{\partial z_1}\right)^{k_1} \cdots \left(\frac{\partial}{\partial z_n}\right)^{k_n} F.$$

For a function F of a single variable z we write $F^{(k)}$ in place of $(d/dz)^k F$.

The notation $\underline{k}!$ stands for $k_1! \cdots k_n!$ and $\underline{z}^{\underline{k}}$ for $z_1^{k_1} \cdots z_n^{k_n}$. In Section 3 for $\underline{x} = (x_0, \dots, x_n) \in \mathbb{C}^{n+1}$ we shall introduce also the notation

$$\mathcal{D}_{\underline{x}} = x_0 \frac{\partial}{\partial z_0} + \cdots + x_n \frac{\partial}{\partial z_n}.$$

In Section 4 we shall denote by D_b the following derivation, attached to a complex number b,

$$\frac{\partial}{\partial X} + bY\frac{\partial}{\partial Y},$$

on the ring $\mathbb{C}[X, Y]$.

For n and k rational integers, the binomial coefficient

$$\binom{n}{k} = \frac{n!}{k!(n-k)!}$$

is considered to be 0 unless $0 \le k \le n$.

The symmetric group on $\{1, \dots, L\}$ will be denoted by $\mathfrak{S}_L$.

Acknowledgements. Many thanks to Stéphane Fischler who carefully checked a preliminary version of the first draft and made a lot of useful comments. It is also a great pleasure to thank Francesco Amoroso, Umberto Zannier and the Fondazione C.I.M.E. (Centro Internazionale Matematico Estivo) for their invitation to deliver these lectures in the superb surroundings provided by Grand Hotel San Michele in Cetraro (Cosenza).

1 First Lecture. Introduction to Transcendence Proofs

We shall provide two proofs of the following result.

Theorem 1.1. *Let a and b be two positive integers. Then $e^b \neq a$.*

This statement is a special case of Hermite-Lindemann's Theorem:

Theorem 1.2. *Let α and β be two nonzero algebraic numbers. Then $e^\beta \neq \alpha$.*

A proof of Theorem 1.2 will be given in Section 2.

1.1 Sketch of Proof

Here are the basic ideas of both proofs of Theorem 1.1. The guest star of these proofs is the exponential function e^z. It is a transcendental function: this means that the exponential monomials $z^\tau e^{tz}$ ($\tau \geq 0$, $t \geq 0$) are linearly independent. Consider the values at a point $z = sb$ with $s \in \mathbb{N}$:

$$(z^\tau e^{tz})(sb) = (sb)^\tau (e^b)^{ts}.$$

If both b and e^b are integers, then this number is also a rational integer.

We need to use a special property for the number e: Theorem 1.1 would not be true if e were replaced by 2 for instance! We take derivatives of our exponential monomials. For $\sigma = 0, 1, \ldots,$

$$\left(\frac{d}{dz}\right)^\sigma (z^\tau e^{tz}) = \sum_{\kappa=0}^{\min\{\tau,\sigma\}} \frac{\sigma!\tau!}{\kappa!(\sigma-\kappa)!(\tau-\kappa)!} z^{\tau-\kappa} t^{\sigma-\kappa} e^{tz}, \tag{1.1}$$

and

$$\left(\frac{d}{dz}\right)^\sigma (z^\tau e^{tz})(sb) = \sum_{\kappa=0}^{\min\{\tau,\sigma\}} \frac{\sigma!\tau!}{\kappa!(\sigma-\kappa)!(\tau-\kappa)!} (sb)^{\tau-\kappa} t^{\sigma-\kappa} (e^b)^{ts}.$$

These numbers again belong to the ring $\mathbb{Z}[e^b, b]$.

Starting with these numbers, there are several ways of performing the proof. We indicate two of them.

The first one rests on the construction of an auxiliary polynomial (AP)[1]. Since the functions z and e^z are algebraically independent, if $P \in \mathbb{Z}[X, Y]$ is a nonzero polynomial, the exponential polynomial $F(z) = P(z, e^z)$ is not the zero function. Together with its derivatives, it takes values in $\mathbb{Z}[e^b, b]$ at all points sb, $s \in \mathbb{N}$. Assuming b and e^b are in $\mathbb{Z}$, one wants to construct such a nonzero polynomial P for which F is the zero function, which will be a contradiction. Now the numbers $F^{(\sigma)}(sb)$ are rational integers, hence

[1] With Masser's notation in [19].

have absolute value either 0 or at least 1; this is the lower bound (LB). If P is constructed so that many numbers $F^{(\sigma)}(sb)$ have absolute value < 1, then F will have a lot of zeroes, hence (by a rigidity principle for analytic functions, called *Schwarz' Lemma*) $|F|$ will be small on a rather large disk: this is the upper bound (UB). This will enable us to deduce that $F^{(\sigma)}(sb)$ vanishes for further values of (σ, s). Once we succeed in increasing the number of known zeroes of F, there is an alternative: either we proceed by induction and extrapolate until we get so many zeroes that F has to be the zero function (for instance if all derivatives of F vanish at one point), or else we prove an auxiliary result (the *zero estimate*, or non-vanishing condition (NV)) which yields the desired conclusion.

One should say a little bit more about the initial construction of P, for which many numbers $F^{(\sigma)}(sb)$ have absolute value < 1. One solution is to select P so that the first coefficients in the Taylor expansion at the origin of F have small absolute values (see [34], Section 4.5 and [26]). Another (more classical) way is to require that many numbers $F^{(\sigma)}(sb)$ vanish. Then the existence of $P \neq 0$ is clear as soon as the number of equations we consider is smaller than the number of unknowns (the unknowns are the coefficients of P), because the conditions are linear and homogeneous. In the special case we consider here with b a real positive number, it can be proved that when the number of unknowns is the same as the number of equations, then the determinant of the system is not zero (*Pólya's Lemma* 1.6). Therefore the extrapolation can be reduced to the minimum. On the other hand for the proof of Theorem 1.1 such an argument is not required.

The second method was suggested by M. Laurent [13]: instead of solving a system of homogeneous linear equations $F^{(\sigma)}(sb) = 0$ for several values of (σ, s), consider the matrix of this system. To be more precise the matrix one considers is the one which arises from the zero estimate: if the zero estimate shows, for a given set of pairs (σ, s), that no nonzero polynomial P (with suitable bounds for its degree) can satisfy all equations $F^{(\sigma)}(sb) = 0$, then the matrix of the associated linear system has maximal rank. Consider a maximal nonsingular submatrix and its nonzero determinant Δ. The main observation of M. Laurent is that a sharp upper bound for $|\Delta|$ can be reached by means of Schwarz' Lemma. Since Δ lies in the ring $\mathbb{Z}[e^b, b]$, as soon as the estimate $0 < |\Delta| < 1$ is established one deduces that one at least of the two numbers b, e^b is not a rational integer.

1.2 Tools for the Auxiliary Function

We introduce four main tools for the proof of Theorem 1.1 by means of an auxiliary function: Liouville's inequality (LB), Schwarz' Lemma (UB), the Zero Estimate (NV) and Thue-Siegel's Lemma (AP).

We shall use here only a trivial case of Liouville's inequality (a more general statement is Proposition 1.13 below):

$$\textit{For any } n \in \mathbb{Z} \textit{ with } n \neq 0 \textit{ we have } |n| \geq 1. \tag{1.2}$$

The analytic upper bound for our auxiliary function is a consequence of the following Schwarz' Lemma (see [34] Exercise 4.3, [9] and also Lemma 1.12 for a quantitative refinement):

Lemma 1.3. *Let m, $\sigma_1, \ldots, \sigma_m$ be positive integers and r, R positive real numbers with $r \leq R$. Let $\zeta_1, \ldots, \zeta_m$ be distinct elements in the disk $|\zeta| \leq R$ and $F \in \mathcal{H}_1(R)$ an analytic function which vanishes at each ζ_i with multiplicity $\geq \sigma_i$ $(1 \leq i \leq m)$. Then*

$$|F|_r \leq |F|_R \prod_{i=1}^{m} \left(\frac{R^2 + r|\zeta_i|}{R(r + |\zeta_i|)} \right)^{-\sigma_i}.$$

We recall the definition: a function F vanishes at a point ζ with multiplicity $\geq \sigma$ if $F^{(k)}(\zeta) = 0$ for $0 \leq k < \sigma$.

In our applications, we shall introduce a parameter $E > 1$ such that $R \geq Er$ and $R \geq E|\zeta_i|$ for $1 \leq i \leq m$. The conclusion yields

$$|F|_r \leq |F|_R \left(\frac{E^2 + 1}{2E} \right)^{-N}$$

where $N = \sigma_1 + \cdots + \sigma_m$ is a lower bound for the number of zeroes of F in the disk $|z| \leq R/E$. In practice N will be large, $E \geq e$, and E^{-N} will be the main term in the right hand side. In particular $|F|_R$ will not be too large, and from the conclusion of Lemma 1.3 we shall infer that $|F|_r$ is quite small.

For our first transcendence proof the zero estimate is a very simple one:

- *If F is a nonzero analytic function near z_0, there exists $\sigma \in \mathbb{N}$ such that*

$$F^{(\sigma)}(z_0) \neq 0.$$

Our last tool is Thue-Siegel's Lemma:

Lemma 1.4. *Let m and n be positive integers with $n > m$ and a_{ij} $(1 \leq i \leq n$, $1 \leq j \leq m)$ rational integers. Define*

$$A = \max\{1, \max_{\substack{1 \leq i \leq n \\ 1 \leq j \leq m}} |a_{ij}|\}.$$

There exist rational integers $x_1, \ldots, x_n$ which satisfy

$$0 < \max\{|x_1|, \ldots, |x_n|\} \leq (nA)^{m/(n-m)}$$

and

$$\sum_{i=1}^{n} a_{ij} x_i = 0 \quad \textit{for} \quad 1 \leq j \leq m.$$

For a proof of this result, we refer for instance to [2] Lemma 1, Chap. 2, [6], Theorem 6.1, Chap. 1, Section 6.1, [7], Theorem 1.10, Chap. 1, Section 4.1 or [11], Lemma 1, Chap. VII, Section 2.

1.3 Proof with an Auxiliary Function and without Zero Estimate

Here is a first proof of Theorem 1.1. Let b be a positive integer such that e^b is also a positive integer. We denote by T_0, T_1, S_0 and S_1 positive integers which we shall choose later: during the proof we shall introduce conditions on these parameters and at the end of the proof we shall check that it is possible to select the parameters so that these conditions are satisfied. Right now let us just say that these integers will be sufficiently large.

We want to deduce from Lemma 1.4 that there exists a nonzero polynomial $P \in \mathbb{Z}[X,Y]$, of degree $< T_0$ in X and degree $< T_1$ in Y, such that the exponential polynomial $F(z) = P(z, e^z)$ has a zero of multiplicity $\geq S_0$ at each point $0, b, 2b, \dots, (S_1 - 1)b$. If this unknown polynomial P is

$$P(X,Y) = \sum_{\tau=0}^{T_0-1} \sum_{t=0}^{T_1-1} c_{\tau t} X^\tau Y^t,$$

then the conditions

$$F^{(\sigma)}(sb) = 0 \quad (0 \leq \sigma < S_0,\ 0 \leq s < S_1)$$

can be written

$$\sum_{\tau=0}^{T_0-1} \sum_{t=0}^{T_1-1} c_{\tau t} \left(\frac{d}{dz}\right)^\sigma (z^\tau e^{tz})(sb) = 0 \quad (0 \leq \sigma < S_0,\ 0 \leq s < S_1).$$

Finding P amounts to solving a system of S_0S_1 linear equations, with rational integers coefficients, in T_0T_1 unknowns $c_{\tau t}$ $(0 \leq \tau < T_0,\ 0 \leq t < T_1)$. We are going to apply Lemma 1.4 with $n = T_0T_1$ and $m = S_0S_1$. In place of the condition $n > m$ we shall require $n \geq 2m$, so that the so-called "Dirichlet's exponent" $m/(n-m)$ is at most 1. This yields the first main condition on our parameters:

$$\boxed{T_0T_1 \geq 2S_0S_1.}$$

We need an upper bound for the number A occurring in Lemma 1.4. Consider (1.1). We wish to estimate from above the modulus of the complex number

$$\sum_{\kappa=0}^{\min\{\tau,\sigma\}} \frac{\sigma!\tau!}{\kappa!(\sigma-\kappa)!(\tau-\kappa)!} z^{\tau-\kappa} t^{\sigma-\kappa}$$

for $|z| \leq R$ with $R > 0$, and for $0 \leq t < T_1$. A first upper bound is given by

$$\sum_{\kappa=0}^{\tau} \frac{\tau!}{\kappa!(\tau-\kappa)!} \sigma^\kappa R^{\tau-\kappa} T_1^{\sigma-\kappa} = T_1^\sigma \left(\frac{\sigma}{T_1} + R\right)^\tau$$

and another one is

$$\sum_{\kappa=0}^{\sigma} \frac{\sigma!}{\kappa!(\sigma-\kappa)!} \tau^\kappa R^{\tau-\kappa} T_1^{\sigma-\kappa} = R^\tau \left(\frac{\tau}{R} + T_1\right)^\sigma.$$

Putting these estimates together yields

$$\sup_{|z|\le R} \left| \sum_{\kappa=0}^{\min\{\tau,\sigma\}} \frac{\sigma!\tau!}{\kappa!(\sigma-\kappa)!(\tau-\kappa)!} z^{\tau-\kappa} t^{\sigma-\kappa} \right| \le R^\tau T_1^\sigma \min\left\{ \left(1+\frac{\sigma}{T_1 R}\right)^\tau ; \left(1+\frac{\tau}{T_1 R}\right)^\sigma \right\}. \quad (1.3)$$

Similar estimates are known for more general exponential polynomials, also in several variables: see for instance [34] Lemmas 4.9 and 13.6.

Here we shall not use the full force of this estimate. Taking $z = bs$, $R = bS_1$ we deduce

$$A \le T_1^{S_0}(bS_1 + S_0)^{T_0} e^{bT_1(S_1-1)}.$$

Hence a nonzero polynomial P exists, satisfying the required conditions and with

$$\max_{\substack{0\le\tau<T_0\\ 0\le t<T_1}} |c_{\tau t}| \le T_0 T_1 A.$$

We need an upper bound for the length

$$\mathrm{L}(P) = \sum_{\tau=0}^{T_0-1} \sum_{t=0}^{T_1-1} |c_{\tau t}|$$

of P. As soon as T_0, T_1, S_0, S_1 are sufficiently large, we have

$$bS_1 \le \frac{1}{4} bS_0 S_1, \quad S_0 \le \frac{1}{4} bS_0 S_1, \quad T_0^2 \le 2^{T_0}, \quad T_1^2 \le e^{bT_1},$$

hence

$$bS_1 + S_0 \le \frac{1}{2} bS_0 S_1, \quad T_0^2 (bS_1 + S_0)^{T_0} \le (bS_0 S_1)^{T_0}$$

and therefore

$$\mathrm{L}(P) \le T_0^2 T_1^2 A \le T_1^{S_0} (bS_0 S_1)^{T_0} e^{bT_1 S_1}.$$

Since the two functions z and e^z are algebraically independent, the function F is not the zero function: at each point sb with $0 \le s < S_1$ its vanishing order is finite (and $\ge S_0$, by construction). We denote by S_0' the minimum of these orders. In other terms S_0' is the largest integer such that the conditions

$$F^{(\sigma)}(sb) = 0 \quad (0 \le \sigma < S_0',\ 0 \le s < S_1)$$

hold and therefore there is an integer s' in the range $0 \le s' < S_1$ with

$$F^{(S_0')}(s'b) \ne 0.$$

An upper bound for S_0' follows from Lemma 1.6 below, namely:

$$S_0'S_1 \le T_0T_1,$$

but this estimate will not be used in the present proof: we need only the lower bound $S_0' \ge S_0$.

By assumption the number $F^{(S_0')}(s'b)$ is a nonzero rational integer, hence has absolute value ≥ 1. We shall deduce from Cauchy's inequalities and Schwarz' Lemma 1.3 an upper bound for this number in terms of the parameters T_0, T_1, S_0 and S_1. It will then suffice to check that the parameters can be selected so that this upper bound is less than 1 and the contradiction will follow.

Cauchy's inequalities yield

$$\left|F^{(S_0')}(s'b)\right| \le S_0'!|F|_r$$

for any $r \ge s'b+1$. We take $r = 2bS_1$. Next we apply Lemma 1.3 with $R = Er$, where $E > 1$ is a new parameter which we are free to choose. As we shall see a suitable choice is $E = S_0'/b$; notice that E, which is selected at this stage of the proof, is allowed to depend on S_0', while S_0' in turn depends on T_0, T_1, S_0 and S_1.

Define $m = S_1$, $\sigma_1 = \cdots = \sigma_m = S_0'$ and $\zeta_i = (i-1)b \ \ (1 \le i \le m)$. Since $\max_{1\le i\le m}|\zeta_i| \le r$, by Lemma 1.3 we have

$$|F|_r \le \left(\frac{E^2+1}{2E}\right)^{-S_0'S_1} |F|_R.$$

It remains to bound $|F|_R$ from above:

$$\begin{aligned}|F|_R &\le \sum_{\tau=0}^{T_0-1}\sum_{t=0}^{T_1-1} |c_{\tau t}| \sup_{|z|=R} \left|z^\tau e^{tz}\right| \\ &\le (T/S)T_1^{S_0}(bS_0S_1)^{T_0} e^{bT_1S_1} R^{T_0} e^{T_1R} \\ &\le T_1^{S_0}(bES_0S_1)^{2T_0} e^{3bET_1S_1}.\end{aligned}$$

Hence

$$|F|_r \le (E/2)^{-S_0'S_1} T_1^{S_0}(bES_0S_1)^{2T_0} e^{3bET_1S_1}.$$

This explains our second main condition on the parameters: taking into account the inequality $S_0'!|F|_r \ge 1$, we shall deduce the desired contradiction as soon as we are able to check

$$\boxed{S_0'!T_1^{S_0}(bES_0S_1)^{2T_0} e^{3bET_1S_1} < (E/2)^{S_0'S_1}.}$$

Here is an admissible choice for these parameters. Recall that b is a fixed positive integer. We start by selecting a sufficiently large, but fixed, positive integer S_1 and we set $T_1 = S_1$. Next let S_0 be an integer, which is much larger than S_1; the required estimates below are easy to check by letting $S_0 \to \infty$. Now define $T_0 = 3S_0$ and $E = S'_0/b$. With this choice we have

$$T_0T_1 > 2S_0S_1,$$

$$S'_0! < (S'_0)^{S'_0} < (E/2)^{S'_0S_1/4}$$

because $S'_0 < E^2 < (E/2)^{S_1/4}$,

$$T_1^{S_0} < (E/2)^{S'_0S_1/4}$$

because $T_1 < E/2 < (E/2)^{S_1/4}$ and $S_0 \leq S'_0$,

$$bES_0S_1 < (S'_0)^3 \quad \text{and} \quad (S'_0)^{6T_0} < (E/2)^{S'_0S_1/4},$$

and finally

$$3bET_1S_1 = 3S'_0T_1S_1 < \frac{1}{4}S'_0S_1\log(E/2).$$

This completes the proof of Theorem 1.1. □

Remark. To a certain extent this proof of Theorem 1.1 involves an extrapolation: we get more and more derivatives of F vanishing at all points $0, b, \dots, (S_1 - 1)b$. It is only a matter of presentation: instead of defining S'_0 as we did, it amounts to the same to check by induction on $S'_0 \geq S_0$ that F has a zero of multiplicity at least S'_0 at each point sb with $0 \leq s < S_1$. At the end of the induction we get a contradiction.

We could also extrapolate on the points at the same time as on the derivatives. Here is a variant of the proof.

We may assume[2] $b \geq 3$. Fix a large[3] positive integer N and set

$$T_0(N) = 2N^2b[\log b], \qquad T_1(N) = N^2[\log b],$$
$$S_0(N) = N^3b[\log b], \qquad S_1(N) = N[\log b]$$

and $L(N) = T_0(N)T_1(N)$, so that

$$L(N) = 2N^4b[\log b]^2 = 2S_0(N)S_1(N).$$

The first step in the preceding proof yields a nonzero polynomial $P \in \mathbb{Z}[X, Y]$ of degree $< T_0(N)$ in X and $< T_1(N)$ in Y, of length bounded by

---

[2] For the proof of Theorem 1.1, this involves no loss of generality and the only reason for this assumption is that we prefer to write $\log b$ in place of $\log_+ b$. For the same reason when we shall need to introduce $\log\log b$ later we shall assume $b \geq 16$.

[3] We assume that N is larger than some absolute constant (independent of b); here one could assume as well than N is larger than some function of b, but it turns out not to be necessary. The relevance of this fact will appear in Section 4 only.

$$\begin{aligned}\mathrm{L}(P) &\le T_1^{S_0}(bS_0S_1)^{T_0}e^{bT_1S_1}\\ &\le \exp\{3L(N)(\log N)/N\}\\ &\le \exp\{L(N)/\sqrt{N}\},\end{aligned}$$

such that the function $F(z) = P(z, e^z)$ satisfies

$$F^{(\sigma)}(sb) = 0 \quad \text{for } 0 \le \sigma < S_0(N) \text{ and } 0 \le s < S_1(N).$$

The second step is an inductive argument: we prove that for any $M \ge N$ we have

$$F^{(\sigma)}(sb) = 0 \quad \text{for } 0 \le \sigma < S_0(M) \text{ and } 0 \le s < S_1(M). \tag{1.4}$$

This is true by construction for $M = N$. Assuming (1.4) is true for M, we deduce it for $M+1$ as follows. Let $(\sigma', s') \in \mathbb{N}^2$ satisfy $0 \le \sigma' < S_0(M+1)$ and $0 \le s' < S_1(M+1)$. Combining the induction hypothesis with Lemma 1.3 where we choose

$$m = S_1(M), \quad \zeta_i = (i-1)b \quad (1 \le i \le m), \quad \sigma_1 = \cdots = \sigma_m = S_0(M),$$
$$r = 2bS_1(M+1) \text{ and } R = 2er,$$

we deduce

$$\begin{aligned}|F|_r &\le e^{-S_0(M)S_1(M)}|F|_R\\ &\le e^{-S_0(M)S_1(M)}\mathrm{L}(P)R^{T_0(N)}e^{T_1(N)R}\\ &\le e^{-S_0(M)S_1(M)/2},\end{aligned}$$

because

$$\log \mathrm{L}(P) + T_0(N)\log R + T_1(N)R \le \frac{1}{2}S_0(M)S_1(M).$$

Since

$$\log(S_0(M+1)!) \le \frac{1}{4}S_0(M)S_1(M),$$

Cauchy's inequalities yield

$$\left|F^{(\sigma')}(s'b)\right| < 1$$

for $0 \le \sigma' < S_0(M+1)$ and $0 \le s' < S_1(M+1)$. Since the left hand side is a rational integer, we deduce $F^{(\sigma')}(s'b) = 0$ and the inductive argument follows.

Plainly we conclude $F = 0$, which completes this new proof of Theorem 1.1.

Remark. In this inductive argument from M to $M+1$, the first step (with $M = N$) is the hardest one: as soon as M is large with respect to N, the required estimates are easier to check.

1.4 Tools for the Interpolation Determinant Method

Some tools which have already been introduced above will be needed for the proof involving interpolation determinants. For instance Liouville's inequality is just the same (1.2). On the other hand in place of Schwarz' Lemma 1.3 we shall use M. Laurent's fundamental observation that interpolation determinants have a small absolute value (see [13], Section 6.3, Lemma 3 and [14], Section 6, Lemma 6). The following estimate ([34] Lemma 2.8) is a consequence of the case $m = 1$, $\zeta_1 = 0$ of Lemma 1.3.

Lemma 1.5. *Let $\varphi_1, \dots, \varphi_L$ be entire functions in $\mathbb{C}$, $\zeta_1, \dots, \zeta_L$ elements of $\mathbb{C}$, $\sigma_1, \dots, \sigma_L$ nonnegative integers and $0 < r \leq R$ real numbers, with $|\zeta_\mu| \leq r$ $(1 \leq \mu \leq L)$. Then the absolute value of the determinant*

$$\Delta = \det\left(\varphi_\lambda^{(\sigma_\mu)}(\zeta_\mu)\right)_{1\leq\lambda,\mu\leq L}$$

is bounded from above by

$$|\Delta| \leq \left(\frac{R}{r}\right)^{-(L(L-1)/2)+\sigma_1+\cdots+\sigma_L} L! \prod_{\lambda=1}^{L} \max_{1\leq\mu\leq L} \sup_{|z|=R} \left|\varphi_\lambda^{(\sigma_\mu)}(z)\right|.$$

The zero estimate we need is the following result due to G. Pólya ([34] Corollary 2.3):

Lemma 1.6. *Let $w_1, \dots, w_n$ be pairwise distinct real numbers, $x_1, \dots, x_m$ also pairwise distinct real numbers and $\tau_1, \dots, \tau_n$, $\sigma_1, \dots, \sigma_m$ nonnegative integers, with*

$$\tau_1 + \cdots + \tau_n = \sigma_1 + \cdots + \sigma_m.$$

Choose any ordering for the pairs (τ, i) with $0 \leq \tau < \tau_i$ and $1 \leq i \leq n$ and any ordering for the pairs (σ, j) with $0 \leq \sigma < \sigma_j$ and $1 \leq j \leq m$. Then the square matrix

$$\left(\left(\frac{d}{dz}\right)^{\sigma} (z^{\tau} e^{w_i z})(x_j)\right)_{\substack{(\tau,i)\\(\sigma,j)}}$$

is nonsingular.

We call this result a *zero estimate* because it can be stated as follows: if $c_{\tau i}$ are complex numbers $(0 \leq \tau < \tau_i,\ 1 \leq i \leq n)$, not all of which are zero, then the exponential polynomial

$$f(z) = \sum_{i=1}^{n} \sum_{\tau=0}^{\tau_i - 1} c_{\tau i} z^{\tau} e^{w_i z}$$

cannot vanish at each x_j with multiplicity $\geq \sigma_j$ $(1 \leq j \leq m)$.

One main characteristic of Laurent's interpolation determinant method is that there is no need of Thue-Siegel's Lemma 1.4.

1.5 Proof with an Interpolation Determinant and a Zero Estimate

Here is another proof of Theorem 1.1.

We start with a positive real number $b \geq 1$, without any other assumption. We introduce auxiliary parameters T_0, T_1, S_0 and S_1, which are positive integers and $E > 1$ a real number. These parameters will be specified later, but it is convenient to assume T_0, T_1, S_0 and S_1 are all ≥ 2.

Consider the matrix

$$\mathtt{M} = \left(\left(\frac{d}{dz} \right)^{\sigma} \left(z^{\tau} e^{tz} \right)(sb) \right)_{\substack{0 \leq \tau < T_0,\ 0 \leq t < T_1 \\ 0 \leq \sigma < S_0,\ 0 \leq s < S_1}},$$

with $T_0 T_1$ rows labeled with (τ, t) and $S_0 S_1$ columns labeled with (σ, s). Here we shall work with a square matrix, which means that we require

$$\boxed{T_0 T_1 = S_0 S_1.}$$

We denote by L this number, so that $\mathtt{M}$ is a square $L \times L$ matrix. By Lemma 1.6 with $n = T_1$,

$$w_i = i - 1, \quad \tau_i = T_0 \qquad (1 \leq i \leq n),$$

$m = S_1$ and

$$x_j = (j-1)b, \quad \sigma_j = S_0 \qquad (1 \leq j \leq m),$$

it follows that $\mathtt{M}$ is nonsingular. Let Δ be the determinant of $\mathtt{M}$. By Lemma 1.5 with

$$\{\varphi_1, \ldots, \varphi_L\} = \{z^{\tau} e^{tz}\ ;\ 0 \leq \tau < T_0,\ 0 \leq t < T_1\},$$
$$\{(\sigma_1, \zeta_1), \ldots, (\sigma_L, \zeta_L)\} = \{(\sigma, sb)\ ;\ 0 \leq \sigma < S_0,\ 0 \leq s < S_1\}$$

and $r = bS_1$, $R = Er$, we have

$$|\Delta| \leq E^{-L(L-1-S_0)/2} L! \prod_{\tau=0}^{T_0-1} \prod_{t=0}^{T_1-1} \max_{0 \leq \sigma < S_0} \sup_{|z|=R} \left| \left(\frac{d}{dz} \right)^{\sigma} \left(z^{\tau} e^{tz} \right) \right|.$$

Since, for $0 \leq \sigma < S_0$ and $|z| = R$, we have by (1.3) and (1.1)

$$\begin{aligned} \left| \left(\frac{d}{dz} \right)^{\sigma} \left(z^{\tau} e^{tz} \right) \right| &\leq T_1^{S_0} (R + S_0)^{T_0} e^{T_1 R} \\ &\leq T_1^{S_0} (bES_0S_1)^{T_0} e^{bET_1S_1}, \end{aligned}$$

we deduce

$$\begin{aligned} |\Delta| &\leq E^{-L(L-1-S_0)/2} L! T_1^{LS_0} (bES_0S_1)^{LT_0} e^{bELT_1S_1} \\ &\leq E^{-L^2/2} L^L (ET_1)^{LS_0} (bES_0S_1)^{LT_0} e^{bELT_1S_1}. \end{aligned}$$

This estimate holds unconditionally. If we can select our parameters so that $|\Delta| < 1$, then this will prove that the nonzero number Δ cannot be a rational integer, hence one at least of b and e^b is not a rational integer. Therefore the proof of Theorem 1.1 will be completed if we show that our parameters may be selected so that

$$\boxed{E^{L/2} > L(ET_1)^{S_0}(bES_0S_1)^{T_0}e^{bET_1S_1}.}$$

Here is an admissible choice: let N be a sufficiently large positive integer (independent of b). Assuming $b \geq 3$, define $E = e$,

$$T_0 = N^2 b[\log b], \qquad T_1 = N^2[\log b],$$
$$S_0 = N^3 b[\log b], \qquad S_1 = N[\log b],$$

so that $L = N^4 b[\log b]^2$.

This completes the proof of Theorem 1.1. □

1.6 Remarks

In this last proof of Theorem 1.1, we did not need to assume that b and e^b are integers: Liouville's inequality (1.2) is used at the very end and provides the conclusion. More precisely it is plain that the interpolation determinant method of Section 1.5 yields the following explicit result.

Proposition 1.7. *Let b be a positive real number. Let T_0, T_1, S_0, S_1 and L be positive integers satisfying*

$$L = T_0 T_1 = S_0 S_1.$$

Let E be a positive number, $E > 1$. Then there exists a polynomial $f \in \mathbb{Z}[Z_1, Z_2]$, of degree $< LT_1S_1$ in Z_1 and $< LT_0$ in Z_2, of length bounded by

$$\mathrm{L}(f) \leq L! T_1^{LS_0}(S_0S_1)^{LT_0},$$

such that

$$0 < |f(e^b, b)| \leq E^{-L^2/2} L! (ET_1)^{LS_0}(|b|_+ ES_0S_1)^{LT_0} e^{|b|ELT_1S_1}.$$

We now explain how to modify the first proof (in Section 1.3) involving an auxiliary function and deduce the following variant of Proposition 1.7.

Proposition 1.8. *Let b be a positive real number. Let T_0, T_1, S_0, S_1 and L be positive integers such that*

$$L = T_0 T_1 = S_0 S_1.$$

Let E, U, V, W be positive real numbers satisfying

$$E \geq e, \quad W \geq 12 \log E,$$

$$U \geq \log L + T_0 \log(|b|_+ E S_1) + |b|_+ E T_1 S_1 \tag{1.5}$$

and

$$4(U + V + W)^2 \leq LW \log E. \tag{1.6}$$

There exists a nonzero polynomial $f \in \mathbb{Z}[Z_1, Z_2]$ *of degree* $< T_1 S_1$ *in* Z_1 *and* $< T_0$ *in* Z_2*, of length*

$$\mathrm{L}(f) \leq L e^W T_1^{S_0} (S_0 + S_1)^{T_0},$$

such that

$$0 < |f(e^b, b)| \leq S_0! e^{-V}.$$

Remark. It is interesting to compare the two estimates provided by Propositions 1.7 and 1.8. Condition (1.6) is satisfied with[4]

$$U = V = W = \frac{1}{36} L \log E.$$

Up to terms of smaller order (when T_0, T_1, S_0 and S_1 are all sufficiently large), the estimates one deduces from Proposition 1.7 for the degrees, the logarithm of the length and the logarithm of the absolute value are L times the corresponding ones in Proposition 1.8.

For all practical purposes, Proposition 1.8, which is obtained by the auxiliary function method, is much sharper that Proposition 1.7. This fact has been an obstacle during a while to develop the interpolation determinant method. For instance it took several years before proofs of algebraic independence results could be achieved by means of Laurent's interpolation determinant method. A nice solution has been provided by M. Laurent and D. Roy in [15], who point out that the polynomial f given by the proof of Proposition 1.7 has a further quite interesting property: its first derivatives

$$\left(\frac{\partial}{\partial Z_1}\right)^{k_1} \left(\frac{\partial}{\partial Z_2}\right)^{k_2} f$$

with $(k_1, k_2) \in \mathbb{N}^2$ satisfying, say, $k_1 + k_2 < L/2$, also have a small absolute value at the point (e^b, b). We shall develop this argument later (see Theorem 4.5).

Our proof of Proposition 1.8 uses an auxiliary function. Since we do not assume b and e^b are integers, we cannot apply Lemma 1.4 as we did in Section 1.3. Instead of solving linear equations, we select (again by means of Dirichlet's box principle) the coefficients $c_{\tau t}$ of the auxiliary polynomial P so that a set of inequalities is satisfied. There are several possibilities. Here we shall use the following auxiliary function.

---

[4] This choice yields a weak upper bound for the length of f. From this point of view a better choice is for instance $U = V = (1/20N) L \log E$ and $W = U/N$ with $N \geq 5$.

Lemma 1.9. *Let L be a positive integer, U, V, W, R, r positive real numbers and $\varphi_1, \ldots, \varphi_L$ functions in $\mathcal{H}_1(R)$. Assume*

$$U+V+W \geq 12, \quad e \leq \frac{R}{r} \leq e^{(U+V+W)/6}, \quad \sum_{\lambda=1}^{L} |\varphi_\lambda|_R \leq e^U$$

and

$$4(U+V+W)^2 \leq LW \log(R/r).$$

Then there exist rational integers $p_1, \ldots, p_L$, with

$$0 < \max_{1 \leq \lambda \leq L} |p_\lambda| \leq e^W,$$

such that the function $F = p_1\varphi_1 + \cdots + p_L\varphi_L$ satisfies

$$|F|_r \leq e^{-V}.$$

We do not give the proof of Lemma 1.9 (see [34] Proposition 4.10, which provides a similar statement in several variables). It suffices to say that it combines Dirichlet's box principle (Lemma 1.10) with an interpolation formula (Lemma 1.11).

Here is Lemma 4.12 of [34].

Lemma 1.10. *Let ν, μ, X be positive integers, U, V positive real numbers and u_{ij} $(1 \leq i \leq \nu,\ 1 \leq j \leq \mu)$ complex numbers. Assume*

$$\sum_{i=1}^{\nu} |u_{ij}| \leq e^U, \qquad (1 \leq j \leq \mu)$$

and

$$\left(\sqrt{2} X e^{U+V} + 1\right)^{2\mu} \leq (X+1)^\nu.$$

Then there exists $(\xi_1, \ldots, \xi_\nu) \in \mathbb{Z}^\nu$ satisfying

$$0 < \max_{1 \leq i \leq \nu} |\xi_i| \leq X$$

and

$$\max_{1 \leq j \leq \mu} \left| \sum_{i=1}^{\nu} u_{ij}\xi_i \right| \leq e^{-V}.$$

The next result is Lemma 4.13 of [34] (in case $r = 0$ we agree that $r^{\|\underline{k}\|} = 1$ for $\underline{k} = \underline{0}$). For the proof of Lemma 1.9, the case $n = 1$ suffices, but we shall need the general case in Section 4.4.

Lemma 1.11. *Let n, K be positive integers, r and R real numbers satisfying $0 \leq r < R$ and F an entire function in $\mathbb{C}^n$. Then*

$$|F|_r \leq (1+\sqrt{K}) \left(\frac{r}{R}\right)^K |F|_R + \sum_{\|\underline{k}\| < K} |\mathcal{D}^{\underline{k}} F(0)| \frac{r^{\|\underline{k}\|}}{\underline{k}!}.$$

Proof of Proposition 1.8. We apply Lemma 1.9 to the functions

$$\{\varphi_1, \dots, \varphi_L\} = \{z^\tau e^{tz} \; ; \; 0 \le \tau < T_0, \; 0 \le t < T_1\}$$

with

$$r = |b|_+ S_1, \quad R = Er.$$

From hypothesis (1.5) we derive

$$\sum_{\lambda=1}^{L} |\varphi_\lambda|_R = \sum_{\tau=0}^{T_0-1} \sum_{t=0}^{T_1-1} \sup_{|z|=R} \left| z^\tau e^{tz} \right| \le L R^{T_0} e^{T_1 R}$$
$$\le L(|b|_+ E S_1)^{T_0} e^{|b|_+ E T_1 S_1} \le e^U.$$

We deduce the existence of a nonzero polynomial

$$P(X,Y) = \sum_{\tau=0}^{T_0-1} \sum_{t=0}^{T_1-1} c_{\tau t} X^\tau Y^t \in \mathbb{Z}[X,Y],$$

of degree $< T_0$ in X and $< T_1$ in Y, with integer coefficients bounded in absolute value by e^W, such that the function $F(z) = P(z, e^z)$ satisfies

$$|F|_r \le e^{-V}.$$

By Lemma 1.6 there is a nonzero element γ in the set

$$\left\{ F^{(\sigma)}(sb) \; ; \; 0 \le \sigma < S_0, \; 0 \le s < S_1 \right\}.$$

From Cauchy's inequality, and since $r \ge s|b| + 1$, we deduce the upper bound

$$|\gamma| \le S_0! |F|_r \le S_0! e^{-V}.$$

Writing

$$\gamma = F^{(\sigma)}(sb),$$

define $f \in \mathbb{Z}[Z_1, Z_2]$ by

$$\sum_{\tau=0}^{T_0-1} \sum_{t=0}^{T_1-1} c_{\tau t} \sum_{\kappa=0}^{\min\{\tau,\sigma\}} \frac{\sigma! \tau!}{\kappa!(\sigma-\kappa)!(\tau-\kappa)!} s^{\tau-\kappa} t^{\sigma-\kappa} Z_1^{ts} Z_2^{\tau-\kappa}$$

so that, using (1.1), we can write

$$\gamma = f(e^b, b).$$

The degrees of f plainly satisfy the required conditions in the conclusion of Proposition 1.8, and finally the length of f is bounded thanks to (1.3):

$$\mathrm{L}(f) \le \sum_{\tau=0}^{T_0-1} \sum_{t=0}^{T_1-1} |c_{\tau t}| \sum_{\kappa=0}^{\min\{\tau,\sigma\}} \frac{\sigma! \tau!}{\kappa!(\sigma-\kappa)!(\tau-\kappa)!} s^{\tau-\kappa} t^{\sigma-\kappa}$$
$$\le L e^W T_1^{S_0} (S_0 + S_1)^{T_0}.$$

□

One can prove a variant of Proposition 1.8 by constructing the auxiliary function $F(z) = P(z, e^z)$ in a slightly different way. One applies Lemma 1.10 again, but now we require that many values at points sb of the function F and of its first derivatives have a small absolute value[5]. A rigidity principle for analytic functions (Lemma 1.12) enables us to deduce that $|F|_r$ is rather small for a suitable parameter r. We are back to the situation of our first proof: we invoke Pólya's Lemma 1.6 and produce a nonzero value of F (or of one of its derivatives). This nonzero number is the value at the point (e^b, b) of a polynomial $f \in \mathbb{Z}[Z_1, Z_2]$ which satisfies the desired conclusion.

Lemma 1.11 is quite simple, since only one point $z = 0$ is involved. For functions of a single variable one can consider an arbitrary finite set of points[6]. Here is Lemma 5.1 of [26].

Lemma 1.12. *Let ℓ be a positive integer, $w_1, \ldots, w_\ell$ pairwise distinct complex numbers and $m_1, \ldots, m_\ell$ positive integers. Put*

$$L = \sum_{j=1}^{\ell} m_j, \qquad \varrho = \max_{1 \le j \le \ell} \max\{1, |w_j|\},$$

and

$$\delta_1 = \min_{1 \le j \le \ell} \prod_{\substack{1 \le j' \le \ell \\ j' \ne j}} |w_j - w_{j'}|^{m_{j'}/L},$$

$$\delta_2 = \min\Big\{1, \min_{\substack{1 \le j, j' \le \ell \\ j \ne j'}} |w_j - w_{j'}|^{m_{j'}/L}\Big\},$$

with the convention that $\delta_1 = \delta_2 = 1$ when $\ell = 1$. Then, for any pair of real numbers r and R with $R \ge 2r$ and $r \ge 2\varrho$ and for any function $F \in \mathcal{H}_1(R)$, we have

$$|F|_r \le \left(\frac{6r}{\delta_1 \delta_2}\right)^L \max_{\substack{1 \le j \le \ell \\ 0 \le \kappa < m_j}} \frac{1}{\kappa!} \left|F^{(\kappa)}(w_j)\right| + \left(\frac{6r}{R}\right)^L |F|_R.$$

Each of the two Propositions 1.7 and 1.8 yields the real case of Hermite-Lindemann's Theorem 1.2: in place of the trivial Liouville's inequality (1.2) we have used so far, it suffices to invoke the next result, which is Proposition 3.14 of [34]:

Proposition 1.13. (Liouville's Inequality). *Let $\mathbb{K}$ be a number field of degree D, v an Archimedean absolute value of $\mathbb{K}$ and $\nu_1, \ldots, \nu_\ell$ positive integers. For*

---

[5] Lemma 1.9 is proved in [34] by constructing P so that the first Taylor coefficients of F at the origin have a small absolute value; hence it may be considered as a variant of this approach, which consists in taking only $s = 0$ at this stage of the proof - the number b does not occur in this case.

[6] Such a statement is called an "approximate Schwarz' Principle" in [23], Section 3.a.

$1 \le i \le \ell$, let $\gamma_{i1}, \dots, \gamma_{i\nu_i}$ be elements of $\mathbb{K}$. Further, let f be a polynomial in $\nu_1 + \cdots + \nu_\ell$ variables, with coefficients in $\mathbb{Z}$, which does not vanish at the point $\underline{\gamma} = \left(\gamma_{ij}\right)_{1 \le j \le \nu_i, 1 \le i \le \ell}$. Assume f has total degree at most N_i with respect to the ν_i variables corresponding to $\gamma_{i1}, \dots, \gamma_{i\nu_i}$. Then

$$\log |f(\underline{\gamma})|_v \ge -(D-1) \log \mathrm{L}(f) - D \sum_{i=1}^{\ell} N_i \mathrm{h}(1 \colon \gamma_{i1} \colon \cdots \colon \gamma_{i\nu_i}).$$

Finally the complex case of Hermite-Lindemann's Theorem 1.2 can also be proved easily by the same arguments, either with an auxiliary function or with an interpolation determinant. The only new feature is to replace Pólya's Lemma 1.6 by another zero estimate, for instance Lemma 4.3. We refer to [2], Chap. 1, Section 3, [6], Chap. 1, Section 9, [7], Chap. 2, Section 2 and [30], Section 3.1 for proofs of the Hermite-Lindemann's Theorem by means of an auxiliary function and to [34] Chap. 2 for the interpolation determinant method with a zero estimate.

In the next section we provide a new proof of Hermite-Lindemann's Theorem 1.2 by means of an interpolation determinant but without any zero estimate: we shall extrapolate like in Section 1.3.

2 Second Lecture. Extrapolation with Interpolation Determinants

The proof given in Section 1.3 rests on an auxiliary function and involves an extrapolation; this extrapolation enabled us to conclude without using the zero estimate Lemma 1.6. We explain here how to perform an extrapolation by means of the interpolation determinant method of Section 1.5.

2.1 Upper Bound for a Determinant in a Single Variable

We are looking for an upper bound for an interpolation determinant. Lemma 1.5 is proved by M. Laurent in [14], Section 6 (also in [34] Lemma 2.8) by means of Schwarz' Lemma 1.3 for the function

$$\Phi : z \longmapsto \det\left(\varphi_\lambda^{(\sigma_\mu)}(z\zeta_\mu)\right)_{1 \le \lambda, \mu \le L}$$

which has a zero at the origin of multiplicity at least

$$0 + 1 + \cdots + (L-1) - (\sigma_1 + \cdots + \sigma_L) = \frac{L(L-1)}{2} - \sum_{\mu=1}^{L} \sigma_\mu.$$

Example 2.1. (See Masser's Lecture 1 in [19]). Define

$$\varphi(z) = z + z^2 + z^4 + z^8 + \cdots = \sum_{m=0}^{\infty} z^{2^m}.$$

Set $L = 6$ and take for $\varphi_1, \ldots, \varphi_6$ the functions

$$1,\ z,\ \varphi(z),\ z\varphi(z),\ \varphi^2(z),\ z\varphi^2(z).$$

Further set

$$\sigma_1 = 0,\ \sigma_2 = 1,\ \sigma_3 = 2,\ \sigma_4 = 3,\ \sigma_5 = 4,\ \sigma_6 = 0$$

and

$$\zeta_1 = \zeta_2 = \zeta_3 = \zeta_4 = \zeta_5 = 0,\ \zeta_6 = 1.$$

The function

$$\begin{aligned}\Phi(z)\prod_{\mu=1}^{6}\frac{1}{\sigma_\mu!} &= \det\left(\frac{1}{\sigma_\mu!}\varphi_\lambda^{(\sigma_\mu)}(z\zeta_\mu)\right)_{1\le\lambda,\mu\le L}\\ &= \det\begin{pmatrix} 1\,0\,0\,0\,0 & 1\\ 0\,1\,0\,0\,0 & z\\ 0\,1\,1\,0\,1 & \varphi(z)\\ 0\,0\,1\,1\,0 & z\varphi(z)\\ 0\,0\,1\,2\,1 & \varphi^2(z)\\ 0\,0\,0\,1\,2 & z\varphi^2(z)\end{pmatrix}\\ &= 2z\varphi^2(z) + 4z\varphi(z) - 3\varphi^2(z) + z - \varphi(z)\end{aligned}$$

has a zero of multiplicity ≥ 5 at the origin; as pointed out by D.W. Masser [19], actually the multiplicity is 6.

Back to the general case, we need to take into account further zeroes. Such an upper bound is given in Corollary 2.4 of [33]; the proof relies on a Schwarz Lemma for Cartesian products (see [33] Proposition 2.3; see also [9] for a general discussion of this issue). Philippon ([23] Lemme 4) also gave upper estimates for interpolation determinants and he does not need to deal with Cartesian products: he uses a much more simple inductive argument which suffices for interpolation determinants (but does not seem to extend to Cartesian products). Here we follow his approach.

We first combine Schwarz' Lemma 1.3 with Cauchy's inequalities.

Lemma 2.2. *Let R, r, ϱ and E be positive real numbers, F an element of $\mathcal{H}_1(R)$, m a positive integer, $\zeta_1, \ldots, \zeta_m$ pairwise distinct complex numbers, ξ a complex number and κ, $\sigma_1, \ldots, \sigma_m$ nonnegative integers. Set*

$$N = \sigma_1 + \cdots + \sigma_m.$$

Assume

$$R \geq \max\{r,\, |\xi| + \varrho\}, \quad r \geq \max_{1 \leq i \leq m} |\zeta_i| \quad \text{and} \quad 1 \leq E \leq \frac{R^2 + r(|\xi| + \varrho)}{R(r + |\xi| + \varrho)}.$$

Assume also that F *satisfies*

$$F^{(\sigma)}(\zeta_i) = 0 \quad \text{for} \quad 0 \leq \sigma < \sigma_i \quad \text{and} \quad 1 \leq i \leq m.$$

Then

$$\left| F^{(\kappa)}(\xi) \right| \leq \kappa! \varrho^{-\kappa} E^{-N} |F|_R.$$

Proof. By Cauchy's inequalities

$$\left| F^{(\kappa)}(\xi) \right| \leq \kappa! \varrho^{-\kappa} |F|_{|\xi| + \varrho}.$$

Since F has at least N zeroes (counting multiplicities) in the disk $B_1(0, r)$, we deduce from Schwarz' Lemma 1.3:

$$|F|_{|\xi|+\varrho} \leq E^{-N} |F|_R.$$

Hence the result. □

The next result is an extension of Corollary 2.4 of [33] where we include multiplicities.

Proposition 2.3. *Let* R *be a positive real number,* $\varphi_1, \ldots, \varphi_L$ *elements of* $\mathcal{H}_1(R)$ *with* $L \geq 1$, $\xi_1, \ldots, \xi_L$ *complex numbers in* $B_1(0, R)$ *and* $\kappa_1, \ldots, \kappa_L$ *nonnegative integers. Consider the determinant*

$$\Delta = \det \left(\varphi_\lambda^{(\kappa_\mu)}(\xi_\mu) \right)_{1 \leq \lambda, \mu \leq L}.$$

Further let $m_1, \ldots, m_L$ *be nonnegative integers and, for* $1 \leq \mu \leq L$ *and* $1 \leq i \leq m_\mu$, *let* $\zeta_{\mu i}$ *be a complex number and* $\sigma_{\mu i}$ *a nonnegative integer. We assume that for each* $\mu = 1, \ldots, L$, *the* m_μ *numbers* $\zeta_{\mu 1}, \ldots, \zeta_{\mu m_\mu}$ *are pairwise distinct. Set*

$$N_\mu = \sum_{i=1}^{m_\mu} \sigma_{\mu i} \quad (1 \leq \mu \leq L).$$

For $1 \leq \mu \leq L$, *let* r_μ, R_μ, ϱ_μ *and* E_μ *be positive real numbers satisfying*

$$R \geq R_\mu \geq \max\{r_\mu,\, |\xi_\mu| + \varrho_\mu\}, \quad r_\mu \geq \max_{1 \leq i \leq m_\mu} |\zeta_{\mu i}|$$

and

$$1 \leq E_\mu \leq \frac{R_\mu^2 + r_\mu(|\xi_\mu| + \varrho_\mu)}{R_\mu(r_\mu + |\xi_\mu| + \varrho_\mu)}.$$

Denote by Φ the analytic mapping

$$(\varphi_1, \ldots, \varphi_L) \colon B_1(0,R) \to \mathbb{C}^L .$$

Assume that for any $(\mu, i, \kappa) \in \mathbb{N}^3$ satisfying $1 \le \mu \le L$, $1 \le i \le m_\mu$ and $0 \le \kappa < \sigma_{\mu i}$, the μ vectors

$$\Phi^{(\kappa_1)}(\xi_1), \ldots, \Phi^{(\kappa_{\mu-1})}(\xi_{\mu-1}),\ \Phi^{(\kappa)}(\zeta_{\mu i}) \tag{2.1}$$

in $\mathbb{C}^L$ are linearly dependent. Then

$$|\Delta| \le L! \max_{\tau \in \mathfrak{S}_L} \prod_{\mu=1}^{L} \left(\kappa_\mu ! \varrho_\mu^{-\kappa_\mu} E_\mu^{-N_\mu} |\varphi_{\tau(\mu)}|_{R_\mu} \right).$$

Proof. We prove Proposition 2.3 by induction on L. For $L = 1$ we have $\Phi = \varphi_1$,

$$\Delta = \Phi^{(\kappa_1)}(\xi_1), \qquad 1 \le E_1 \le \frac{R_1^2 + r_1(|\xi_1| + \varrho_1)}{R_1(r_1 + |\xi_1| + \varrho_1)}$$

and hypothesis (2.1) reads

$$\Phi^{(\kappa)}(\zeta_{1i}) = 0 \quad \text{for} \quad 0 \le \kappa < \sigma_{1i} \quad \text{and} \quad 1 \le i \le m_1.$$

From Lemma 2.2 we deduce

$$|\Delta| \le \kappa_1 ! \varrho_1^{-\kappa_1} E_1^{-N_1} |\varphi_1|_{R_1}.$$

Hence Proposition 2.3 is true in case $L = 1$.

Assume now that the conclusion is true for L replaced by $L - 1$. Define $F \in B_1(0,R)$ by

$$F(z) = \det\left(\Phi^{(\kappa_1)}(\xi_1), \ldots, \Phi^{(\kappa_{L-1})}(\xi_{L-1}),\ \Phi(z) \right).$$

By assumption (2.1) with $\mu = L$, for $1 \le i \le m_L$ and $0 \le \kappa < \sigma_{Li}$, we have

$$F^{(\kappa)}(\zeta_{Li}) = 0.$$

Since

$$\Delta = F^{(\kappa_L)}(\xi_L),$$

we deduce from Lemma 2.2

$$|\Delta| \le \kappa_L ! \varrho_L^{-\kappa_L} E_L^{-N_L} |F|_{R_L}.$$

We expand the determinant F with respect to the last column: define, for $1 \le \lambda \le L$,

$$\Phi_\lambda = (\varphi_1, \ldots, \varphi_{\lambda-1}, \varphi_{\lambda+1}, \ldots, \varphi_L) \colon B_1(0,R) \to \mathbb{C}^{L-1}$$

and let Δ_λ denote the determinant of the $(L-1)\times(L-1)$ matrix

$$\left(\Phi_\lambda^{(\kappa_1)}(\xi_1),\dots,\Phi_\lambda^{(\kappa_{L-1})}(\xi_{L-1})\right),$$

so that

$$F(z)=\sum_{\lambda=1}^{L}(-1)^{L-\lambda}\varphi_\lambda(z)\Delta_\lambda.$$

Hence

$$|F|_{R_L}\le L\max_{1\le\lambda\le L}|\varphi_\lambda|_{R_L}|\Delta_\lambda|$$

and therefore

$$|\Delta|\le\kappa_L!L\varrho_L^{-\kappa_L}E_L^{-N_L}\max_{1\le\lambda\le L}|\varphi_\lambda|_{R_L}|\Delta_\lambda|.$$

We fix an index $\lambda^0\in\{1,\dots,L\}$ such that

$$|\Delta|\le\kappa_L!L\varrho_L^{-\kappa_L}E_L^{-N_L}|\varphi_{\lambda^0}|_{R_L}|\Delta_{\lambda^0}|.$$

Using the induction hypothesis, we deduce that there exists a bijective map t from $\{1,\dots,L-1\}$ to $\{1,\dots,\lambda^0-1,\lambda^0+1,\dots,L\}$ such that

$$|\Delta_{\lambda^0}|\le(L-1)!\prod_{\mu=1}^{L-1}\left(\kappa_\mu!\varrho_\mu^{-\kappa_\mu}E_\mu^{-N_\mu}|\varphi_{t(\mu)}|_{R_\mu}\right).$$

Define $\tau\in\mathfrak{S}_L$ by $\tau(\mu)=t(\mu)$ for $1\le\mu<L$ and $\tau(L)=\lambda^0$. Proposition 2.3 follows. □

We shall use a special case of Proposition 2.3.

We consider a finite sequence $(\zeta_0,\dots,\zeta_N)$ of complex numbers, which are not supposed to be pairwise distinct. We define the associated *multiplicity sequence* $(\sigma_0,\dots,\sigma_N)$ as follows:

$$\sigma_\nu=\mathrm{Card}\{i;\ 0\le i<\nu\,,\zeta_i=\zeta_\nu\}\quad(0\le\nu\le N).$$

If $\zeta_0,\dots,\zeta_N$ are pairwise distinct then $\sigma_0=\cdots=\sigma_N=0$. In general, $(\zeta_0,\dots,\zeta_N)$ consists of ℓ distinct complex numbers $w_1,\dots,w_\ell$, where w_j is repeated m_j times $(1\le j\le\ell)$, so that

$$\prod_{\nu=0}^{N}(z-\zeta_\nu)=\prod_{j=1}^{\ell}(z-w_j)^{m_j}\quad\text{and}\quad m_1+\cdots+m_\ell=N+1.$$

Then for an analytic function F the $N+1$ equations

$$F^{(\sigma_\nu)}(\zeta_\nu)=0\quad\text{for}\quad 0\le\nu\le N$$

are nothing else than

$$F^{(\kappa)}(w_j) = 0 \text{ for } 0 \leq \kappa < m_j \text{ and } 1 \leq j \leq \ell.$$

The pairs (σ_ν, ζ_ν) $(0 \leq \nu \leq N)$ are pairwise distinct and for each $\nu = 0, \ldots, N$ and each $\sigma = 0, \ldots, \sigma_\nu$ there is an index μ in the range $0 \leq \mu \leq \nu$ with $(\sigma_\mu, \zeta_\mu) = (\sigma, \zeta_\nu)$.

Corollary 2.4. *Let L, N be integers with $1 \leq L \leq N+1$, R a positive real number and $(\zeta_0, \ldots, \zeta_N)$ a sequence of $N+1$ elements in $B_1(0,R)$. Denote by $(\sigma_0, \ldots, \sigma_N)$ the associated multiplicity sequence. Let $0 = \nu_0 \leq \nu_1 < \cdots < \nu_L \leq N$ be integers and $\varphi_1, \ldots, \varphi_L$ elements of $\mathcal{H}_1(R)$. Consider the determinant*

$$\Delta = \det\left(\varphi_\lambda^{(\sigma_{\nu_\mu})}(\zeta_{\nu_\mu})\right)_{1 \leq \lambda, \mu \leq L}.$$

For $1 \leq \mu \leq L$, let r_μ, R_μ, ϱ_μ and E_μ be positive real numbers satisfying

$$R \geq R_\mu \geq \max\{r_\mu, |\zeta_{\nu_\mu}| + \varrho_\mu\}, \quad r_\mu \geq \max_{0 \leq \nu < \nu_\mu} |\zeta_\nu|$$

and

$$1 \leq E_\mu \leq \frac{R_\mu^2 + r_\mu(|\zeta_{\nu_\mu}| + \varrho_\mu)}{R_\mu(r_\mu + |\zeta_{\nu_\mu}| + \varrho_\mu)}.$$

Denote by Φ the analytic mapping

$$(\varphi_1, \ldots, \varphi_L) \colon B_1(0,R) \to \mathbb{C}^L.$$

Assume that for any $(\mu, \nu) \in \mathbb{N}^2$ satisfying $1 \leq \mu \leq L$ and $\nu_{\mu-1} \leq \nu < \nu_\mu$, the μ vectors

$$\Phi^{(\sigma_{\nu_1})}(\zeta_{\nu_1}), \ldots, \Phi^{(\sigma_{\nu_{\mu-1}})}(\zeta_{\nu_{\mu-1}}),\ \Phi^{(\sigma_\nu)}(\zeta_\nu)$$

in $\mathbb{C}^L$ are linearly dependent. Then

$$|\Delta| \leq L! \max_{\tau \in \mathfrak{S}_L} \prod_{\mu=1}^{L} \left(\sigma_{\nu_\mu}! \varrho_\mu^{-\sigma_{\nu_\mu}} E_\mu^{-\nu_\mu} |\varphi_{\tau(\mu)}|_{R_\mu}\right).$$

Proof. We apply Proposition 2.3 with

$$\xi_\mu = \zeta_{\nu_\mu} \text{ and } \kappa_\mu = \sigma_{\nu_\mu} \quad (1 \leq \mu \leq L).$$

We define m_μ, $\zeta_{\mu i}$ and $\sigma_{\mu i}$ as follows: for $1 \leq \mu \leq L$, we denote by m_μ the number of distinct elements in the sequence $(\zeta_0, \ldots, \zeta_{\nu_\mu - 1})$, by $\zeta_{\mu i}$ these distinct elements and by $\sigma_{\mu i}$ the number of ν in the range $0 \leq \nu < \nu_\mu$ such that $\zeta_\nu = \zeta_{\mu i}$. Therefore

$$N_\mu = \sigma_{\mu 1} + \cdots + \sigma_{\mu m_\mu} = \nu_\mu.$$

□

Remark. Corollary 2.4 includes Corollary 2.4 of [33]: if some σ_{ν_μ} is zero, then $\varrho_\mu^{-\sigma_{\nu_\mu}} = 1$ even if we replace ϱ_μ by 0 in the definitions of R_μ and E_μ. Another special case of Corollary 2.4 is related to Lemma 2.5 of [34], which is nothing else than the case $\sigma_1 = \cdots = \sigma_L = 0$ of Lemma 1.5. Indeed we can take in Corollary 2.4

$$\zeta_0, \ldots, \zeta_{L-1} \text{ pairwise distinct}, \qquad N = L-1, \qquad \sigma_0 = \cdots = \sigma_{L-1} = 0,$$
$$\nu_\mu = \mu - 1, \quad R_\mu = R, \quad r_\mu = r, \quad E_\mu = (R^2 + r^2)/2rR \quad (1 \leq \mu \leq L),$$

with

$$R \geq r \geq \max\{|\zeta_0|, \ldots, |\zeta_{L-1}|\}.$$

Since $\nu_1 + \cdots + \nu_L = L(L-1)/2$, we deduce from Corollary 2.4

$$\left| \det\Big(\varphi_\lambda(\zeta_\mu)\Big)_{1 \leq \lambda, \mu \leq L} \right| \leq \left(\frac{R^2 + r^2}{2rR}\right)^{-L(L-1)/2} L! \prod_{\lambda=1}^{L} \sup_{|z|=R} |\varphi_\lambda(z)|.$$

Apart from the quantity $(R^2+r^2)/2rR$ which replaces R/r, this is the estimate of Lemma 2.5 of [34].

It is not clear to me whether Proposition 2.3 contains the general case of Lemma 1.5 (without the restriction $\sigma_1 = \cdots = \sigma_L = 0$).

2.2 Proof of Hermite-Lindemann's Theorem with an Interpolation Determinant and without Zero Estimate

Thanks to Corollary 2.4, one can modify the proof of Section 1.5 involving an interpolation determinant so that Lemma 1.6 (zero estimate) is not required any more. In this section we explain how to extrapolate and to increase either the number of derivatives, or the number of points, or both.

Proof of Hermite-Lindemann's Theorem 1.2. Let α and β be two complex numbers with $\beta \neq 0$ and $\alpha = e^\beta$.

`Step 1. Introducing the Parameters`

Consider two nondecreasing sequences $(S_0(N))_{N\geq 0}$ and $(S_1(N))_{N\geq 0}$ of nonnegative integers with $S_0(0) = S_1(0) = 0$ and such that the sequence $(S_0(N)S_1(N))_{N\geq 0}$ is increasing.

We construct a sequence $\zeta_0, \zeta_1, \ldots$ as follows. For each $N \geq 0$, the sequence

$$\big(\zeta_{S_0(N)S_1(N)}, \ldots, \zeta_{S_0(N+1)S_1(N+1)-1}\big)$$

consists of

- each element $s\beta$ with $0 \leq s < S_1(N)$ repeated $S_0(N+1) - S_0(N)$ times,

and

- each element $s\beta$ with $S_1(N) \leq s < S_1(N+1)$ repeated $S_0(N+1)$ times.

Denote by $(\sigma_0, \sigma_1, \ldots, \sigma_\nu, \ldots)$ the associated multiplicity sequence.

For each $\nu \geq 0$ denote by N_ν the least integer $N \geq 1$ for which $\nu < S_0(N)S_1(N)$. Hence we have $N_0 = 1$ and for $\nu \geq 0$

$$\begin{gathered} S_0(N_\nu - 1)S_1(N_\nu - 1) \leq \nu < S_0(N_\nu)S_1(N_\nu), \\ \sigma_\nu < S_0(N_\nu) \text{ and } |\zeta_\nu| < S_1(N_\nu)|\beta|. \end{gathered} \tag{2.2}$$

We also introduce two sufficiently large integers T_0 and T_1 and we set $L = T_0T_1$.

Step 2. The Matrix M and the Determinant Δ

Consider the matrix with L rows and infinitely many columns

$$\mathtt{M} = \Big(\mathcal{C}_0, \mathcal{C}_1, \ldots, \mathcal{C}_\nu, \ldots\Big),$$

where $\mathcal{C}_\nu$ is the column vector (with $L = T_0T_1$ rows)

$$\left(\left(\frac{d}{dz}\right)^{\sigma_\nu}\left(z^\tau e^{tz}\right)(\zeta_\nu)\right)_{0\leq\tau<T_0,\, 0\leq t<T_1}.$$

We claim that the rank of M is L. Indeed a linear relation between the rows

$$\sum_{\tau_0=0}^{T_0-1}\sum_{t=0}^{T_1-1} c_{\tau t}\left(\frac{d}{dz}\right)^{\sigma_\nu}\left(z^\tau e^{tz}\right)(\zeta_\nu) = 0 \text{ for } \nu \geq 0$$

would mean that the exponential polynomial

$$F(z) = \sum_{\tau_0=0}^{T_0-1}\sum_{t=0}^{T_1-1} c_{\tau t} z^\tau e^{tz}$$

satisfies

$$F^{(\sigma_\nu)}(\zeta_\nu) = 0 \text{ for } \nu \geq 0.$$

These relations can also be written

$$F^{(\sigma)}(sb) = 0 \text{ for } \sigma \geq 0 \text{ and } s \geq 0,$$

and they plainly imply $c_{\tau t} = 0$ for $0 \leq \tau < T_0$ and $0 \leq t < T_1$.

We select L columns of M as the minimal ones in the lexicographic ordering such that we obtain a nonsingular matrix. Concretely we define $\nu_1, \ldots, \nu_L$ as follows:

$$\nu_1 = \min\{\nu \geq 0,\ \mathcal{C}_\nu \neq 0\},$$

and for $2 \leq \mu \leq L$

$$\nu_\mu = \min\{\nu > \nu_{\mu-1}\,;\ \mathcal{C}_{\nu_1}, \ldots, \mathcal{C}_{\nu_{\mu-1}}, \mathcal{C}_\nu \text{ are linearly independent}\}.$$

Hence we have $0 = \nu_1 < \nu_2 < \cdots < \nu_L$ and the matrix

$$\Big(\mathcal{C}_{\nu_1}, \ldots, \mathcal{C}_{\nu_L}\Big)$$

is nonsingular. We denote its determinant by Δ.

Remark. We may assume $\zeta_0 = 0$ and $\sigma_0 = 0$; in this case the first column has T_1 components 1 (those with index $(0,t)$ such that $0 \le t < T_1$) and $(T_0-1)T_1$ components 0 (the other ones, with index (τ,t) such that $1 \le \tau < T_0$ and $0 \le t < T_1$).

Step 3. Upper Bound for $|\Delta|$

We apply Corollary 2.4 with $\varrho_\mu = 1$, $r_\mu = |\beta|_+ S_1(N_{\nu_\mu})$, $R_\mu = 3er_\mu$, $E_\mu = e$ and

$$\{\varphi_1, \dots, \varphi_L\} = \{z^\tau e^{tz} \; ; \; 0 \le \tau < T_0, \, 0 \le t < T_1\}.$$

We deduce

$$|\Delta| \le L! \prod_{\mu=1}^{L} \left(e^{-\nu_\mu} \sigma_{\nu_\mu}! R_\mu^{T_0} e^{T_1 R_\mu}\right) \le L! \exp\left(\sum_{\mu=1}^{L} \rho_\mu\right).$$

where

$$\rho_\mu = -\nu_\mu + \log\big(S_0(N_{\nu_\mu})!\big) + T_0 \log\big(3e|\beta|_+ S_1(N_{\nu_\mu})\big) + 3e|\beta|_+ T_1 S_1(N_{\nu_\mu}).$$

Step 4. Lower Bound for $|\Delta|$

The number Δ is not zero and lies in the ring $\mathbb{Z}[\alpha, \beta]$: there is a polynomial $f \in \mathbb{Z}[Z_1, Z_2]$ such that

$$\Delta = f(\alpha, \beta),$$

the degree of f in Z_1 and Z_2 respectively is at most

$$\sum_{\mu=1}^{L} T_1 S_1(N_{\nu_\mu}) \quad \text{and} \quad LT_0,$$

while the length of f is bounded by

$$\mathrm{L}(f) \le L! \prod_{\mu=1}^{L} T_1^{S_0(N_{\nu_\mu})} \big(S_0(N_{\nu_\mu}) + S_1(N_{\nu_\mu})\big)^{T_0}.$$

Assume now that α and β are both algebraic. Then we may use Liouville's inequality Proposition 1.13:

$$|\Delta| \ge L!^{-D+1} e^{-LDT_0 \mathrm{h}(\beta)}$$
$$\times \prod_{\mu=1}^{L} \left(T_1^{-(D-1)S_0(N_{\nu_\mu})} \big(S_0(N_{\nu_\mu}) + S_1(N_{\nu_\mu})\big)^{-(D-1)T_0} e^{-DT_1 S_1(N_{\nu_\mu}) \mathrm{h}(\alpha)}\right).$$

Step 5. Choice of parameters

Define, for $N \ge 0$,

$$S_0(N) = N^3 \quad \text{and} \quad S_1(N) = N.$$

By (2.2) we have for $\nu \geq 1$

$$(N_\nu - 1)^4 \leq \nu < N_\nu^4,$$

hence

$$S_0(N_\nu) \leq (\nu^{1/4}+1)^3 \leq 8\nu^{3/4}, \quad S_1(N_\nu) \leq \nu^{1/4}+1 \leq 2\nu^{1/4}.$$

Fix a sufficiently large integer T_0 (larger than some constant depending only on α and β) and define $T_1 = T_0$, so that $L = T_0^2$.

For $1 \leq \mu \leq L$ the following estimates are plain:

$$\begin{aligned} D\log L + DT_0\mathrm{h}(\beta) + (D-1)T_0\log\bigl(S_0(N_{\nu_\mu})S_1(N_{\nu_\mu})\bigr) \\ + T_0\log\bigl(3e|\beta|_+S_1(N_{\nu_\mu})\bigr) \leq (c_1 + c_2\log\nu_\mu)T_0 \end{aligned}$$

and

$$\begin{aligned} (D-1)S_0(N_{\nu_\mu})\log\bigl(T_1S_0(N_{\nu_\mu})\bigr) &\leq c_3\nu_\mu^{3/4}\log(T_0\nu_\mu), \\ T_1S_1(N_{\nu_\mu})\bigl(D\mathrm{h}(\alpha) + 3e|\beta|_+\bigr) &\leq c_4T_0\nu_\mu^{1/4}, \end{aligned}$$

where $c_1, \ldots, c_4$ are positive real numbers which depend only on α and β. Therefore we have

$$\begin{aligned} D\log L + DT_0\mathrm{h}(\beta) + (D-1)T_0\log\bigl(S_0(N_{\nu_\mu})S_1(N_{\nu_\mu})\bigr) \\ + T_0\log\bigl(3e|\beta|_+S_1(N_{\nu_\mu})\bigr) + (D-1)S_0(N_{\nu_\mu})\log T_1 \\ + S_0(N_{\nu_\mu})\log\bigl(S_0(N_{\nu_\mu})\bigr) + T_1S_1(N_{\nu_\mu})\bigl(D\mathrm{h}(\alpha) + 3e|\beta|_+\bigr) \leq Q_\mu, \end{aligned}$$

where

$$Q_\mu = c_5T_0\nu_\mu^{1/4} + c_6\nu_\mu^{3/4}\log(T_0\nu_\mu)$$

and again c_5, c_6 depend only on α and β.

`Step 6. Conclusion`

We claim

$$\sum_{\mu=1}^{L}(\nu_\mu - Q_\mu) > 0. \tag{2.3}$$

Indeed, since $\mu \mapsto Q_\mu$ is increasing, we have

$$\sum_{1\leq\mu\leq L/2} Q_\mu \leq \sum_{L/2<\mu\leq L} Q_\mu.$$

The estimate

$$\nu_\mu \geq 0 \text{ for } 1 \leq \mu \leq L/2$$

is trivial, while for $L/2 < \mu \leq L$ the lower bound

$$\nu_\mu \geq \mu - 1 \geq (L-1)/2 = (T_0^2 - 1)/2$$

implies

$$\nu_\mu^{3/4} > 4c_5 T_0 \quad \text{and} \quad \nu_\mu^{1/4} > 4c_6 \log(T_0 \nu_\mu),$$

hence

$$\nu_\mu > 2Q_\mu \quad \text{for} \quad L/2 < \mu \leq L.$$

Therefore our claim (2.3) is vindicated.

According to steps 5 and 6, the conclusions of steps 3 and 4 are not compatible, hence one at least of the two numbers α, β is transcendental.

This completes the proof of Hermite-Lindemann's Theorem 1.2. □

3 Third Lecture. Linear Independence of Logarithms of Algebraic Numbers

The main result, due to A. Baker ([1] and [2] Th. 1.2), is the following:

Theorem 3.1. *Let $\alpha_1, \ldots, \alpha_n$ be nonzero algebraic numbers. For each $i = 1, \ldots, n$, let $\lambda_i \in \mathbb{C}$ satisfy $e^{\lambda_i} = \alpha_i$. Assume the n numbers $\lambda_1, \ldots, \lambda_n$ are linearly independent over $\mathbb{Q}$. Then the $n+1$ numbers $1, \lambda_1, \ldots, \lambda_n$ are linearly independent over the field $\overline{\mathbb{Q}}$ of algebraic numbers.*

We shall use the notation $\log \alpha_i$ in place of λ_i. One should keep in mind that this notation may be troublesome: for instance Theorem 3.1 can be applied with

$$\alpha_1 = \alpha_2 = 2, \quad \lambda_1 = \log 2, \quad \lambda_2 = \log 2 + 2i\pi,$$

and the conclusion shows that the three numbers $1, \log 2, \pi$ are linearly independent over $\overline{\mathbb{Q}}$. However the same conclusion can be obtained by taking $\alpha_1 = 2$ and $\alpha_2 = -1$ for instance.

By the way, when $\alpha_1, \ldots, \alpha_n$ are nonzero complex numbers, for any choice $(\lambda_1, \ldots, \lambda_n) \in \mathbb{C}^n$ with $e^{\lambda_i} = \alpha_i \quad (1 \leq i \leq n)$, the following conditions are clearly equivalent:

(i) The numbers $\alpha_1, \ldots, \alpha_n$ are multiplicatively independent, which means that any relation
$$\alpha_1^{a_1} \cdots \alpha_n^{a_n} = 1$$
with $(a_1, \ldots, a_n) \in \mathbb{Z}^n$ implies $a_1 = \cdots = a_n = 0$.

(ii) The $n+1$ complex numbers $2\pi i, \lambda_1, \ldots, \lambda_n$ are linearly independent over $\mathbb{Q}$.

Hence, given complex numbers $\lambda_1, \ldots, \lambda_n$, the multiplicative subgroup of $\mathbb{C}^\times$ generated by $e^{\lambda_1}, \ldots, e^{\lambda_n}$ has rank (as a $\mathbb{Z}$-module) equal to $r-1$, where r is the dimension of the $\mathbb{Q}$-vector space spanned by $2\pi i, \lambda_1, \ldots, \lambda_n$. In particular, if $\lambda_1, \ldots, \lambda_n$ are $\mathbb{Q}$-linearly independent, then this rank is

$$\begin{cases} n & \text{if } 2\pi i, \lambda_1, \ldots, \lambda_n \text{ are linearly independent over } \mathbb{Q}, \\ n-1 & \text{if } 2\pi i, \lambda_1, \ldots, \lambda_n \text{ are } \mathbb{Q}-\text{linearly dependent.} \end{cases}$$

Proofs of Baker's Theorem 3.1 on linear independence of logarithms are given in [2] Chap. 2, [6] Chap. 10, Section 1, [7], Chap. 4, Section 1.3, [30], Chap. 8 and [34], Section 10.1. A "dual" argument (extension of Schneider's method, while Baker's method is an extension of Gel'fond's method) is worked out in [34] Chap. 6 (for the homogeneous case) and Section 9.1 (for the non-homogeneous case). See also [34], Section 4.2 for a proof, following Bertrand and Masser, which rests on Schneider-Lang's Criterion for Cartesian products (involving again Gel'fond's method).

Here we consider only Baker's method. In Section 3.1 we explain why Baker's method can be introduced by means of functions of either one or several variables. A proof of Theorem 3.1 by means of Baker's method, involving an auxiliary function with an extrapolation argument, is given in Section 3.2, which includes also a sketch of proof with an interpolation determinant but without any extrapolation. For both proofs the zero estimate which is used there is due to Philippon [22]. In Section 3.3 we show how to replace this zero estimate by a much simpler one, due to R. Tijdeman, by means of a further extrapolation with the auxiliary function. Our ultimate goal in this third lecture is to extrapolate with an interpolation determinant (Section 3.5), in order to complete the proof of Theorem 3.1 without any auxiliary function, and with Tijdeman's zero estimate in place of Philippon's one. This is achieved thanks to a generalization (in Section 3.4) in several variables of the results of Section 2.1 giving upper bounds for interpolation determinants.

3.1 Introduction to Baker's Method

We explain the basic ideas of the proof of Theorem 3.1 by means of Baker's method with an auxiliary function involving an extrapolation.

Assume $\alpha_1, \dots, \alpha_n$ are nonzero algebraic numbers, $\log \alpha_1, \dots, \log \alpha_n$ are linearly independent over $\mathbb{Q}$, $\beta_0, \dots, \beta_{n-1}$ are algebraic numbers, and

$$\log \alpha_n = \beta_0 + \beta_1 \log \alpha_1 + \cdots + \beta_{n-1} \log \alpha_{n-1}. \tag{3.1}$$

We shall eventually reach a contradiction.

From now on α^z stands for $\exp(z \log \alpha)$, which has a meaning as soon as a complex number $\lambda = \log \alpha$ has been selected with $e^\lambda = \alpha$.

Hence relation (3.1) implies

$$\alpha_n = e^{\beta_0} \alpha_1^{\beta_1} \cdots \alpha_{n-1}^{\beta_{n-1}},$$

and more generally for $z \in \mathbb{C}$

$$\alpha_n^z = e^{\beta_0 z} \alpha_1^{\beta_1 z} \cdots \alpha_{n-1}^{\beta_{n-1} z}.$$

To each polynomial $P \in \mathbb{Z}[Y_0, Y_1, \dots, Y_n]$ we associate analytic functions of 1, n and $n+1$ complex variables. The proof of Baker's qualitative Theorem 3.1 on linear independence of logarithms of algebraic numbers requires a Schwarz'

Lemma, while the quantitative refinements (measures of linear independence; see Section 5) will require an approximate Schwarz' Lemma. A fundamental fact is that one needs such auxiliary results only for functions of a single variable: even if we introduce functions of several variables, we shall consider only the values of our functions at multiples of a single point (but derivatives are taken in several directions); therefore it would be possible to avoid completely the introduction of several variables, but we use them only to explain the role of certain differential operators.

a) Using a Single Variable

Consider the entire functions

$$z, \alpha_1^z, \dots, \alpha_n^z.$$

To the auxiliary polynomial $P \in \mathbb{Z}[\underline{Y}]$ is attached the exponential polynomial

$$G(z) = P(z, \alpha_1^z, \dots, \alpha_n^z),$$

which can be written also

$$G(z) = P\bigl(z, \alpha_1^z, \dots, \alpha_{n-1}^z, e^{\beta_0 z}\alpha_1^{\beta_1 z} \cdots \alpha_{n-1}^{\beta_{n-1} z}\bigr).$$

In order to take (3.1) into account, we consider derivatives of G. We avoid difficulties (related with Liouville's inequality) arising from the unwanted transcendental numbers $\log \alpha_i$ $(1 \le i < n)$ by writing the derivatives of G as polynomials in $\log \alpha_1, \dots, \log \alpha_{n-1}$, and the coefficients of these polynomials are themselves exponential polynomials with algebraic coefficients.

We start with the first derivative $G'(z) = (d/dz)G(z)$ of G: this is the value, at $(z, \alpha_1^z, \dots, \alpha_n^z)$, of the polynomial

$$\left(\frac{\partial}{\partial Y_0} + \sum_{k=1}^{n} (\log \alpha_k) \frac{\partial}{\partial Y_k}\right) P = (\partial_0 + \partial_1 \log \alpha_1 + \cdots + \partial_{n-1} \log \alpha_{n-1}) P,$$

where $\partial_0, \dots, \partial_{n-1}$ are the differential operators

$$\partial_0 = \frac{\partial}{\partial Y_0} + \beta_0 Y_n \frac{\partial}{\partial Y_n}, \qquad \partial_k = Y_k \frac{\partial}{\partial Y_k} + \beta_k Y_n \frac{\partial}{\partial Y_n} \quad (1 \le k \le n-1)$$

on the ring $\mathbb{C}[\underline{Y}]$

We now take higher derivatives. For $\underline{\sigma} = (\sigma_0, \dots, \sigma_{n-1}) \in \mathbb{N}^n$ we write $\partial^{\underline{\sigma}}$ in place of

$$\partial_0^{\sigma_0} \cdots \partial_{n-1}^{\sigma_{n-1}}.$$

Since

$$\left(\partial_0 + \partial_1 \log \alpha_1 + \cdots + \partial_{n-1} \log \alpha_{n-1}\right)^k = \sum_{\|\underline{\sigma}\|=k} \frac{k!}{\underline{\sigma}!} (\log \alpha_1)^{\sigma_1} \cdots (\log \alpha_{n-1})^{\sigma_{n-1}} \partial^{\underline{\sigma}},$$

we have

$$G^{(k)}(z) = \sum_{\|\underline{\sigma}\|=k} \frac{k!}{\underline{\sigma}!} (\log \alpha_1)^{\sigma_1} \cdots (\log \alpha_{n-1})^{\sigma_{n-1}} G_{\underline{\sigma}}(z),$$

where

$$G_{\underline{\sigma}}(z) = P_{\underline{\sigma}}(z, \alpha_1^z, \ldots, \alpha_n^z) \text{ and } P_{\underline{\sigma}} = \partial^{\underline{\sigma}} P \in \mathbb{C}[\underline{Y}].$$

Now if we compose derivations we easily deduce for $\underline{\sigma}$ and $\underline{\lambda}$ in $\mathbb{N}^n$

$$\partial^{\underline{\sigma}+\underline{\lambda}} = \partial^{\underline{\sigma}} \circ \partial^{\underline{\lambda}},$$

which yields the fundamental relation for $\underline{\sigma} \in \mathbb{N}^n$ and $\ell \in \mathbb{N}$

$$G_{\underline{\sigma}}^{(\ell)}(z) = \sum_{\|\underline{\lambda}\|=\ell} \frac{\ell!}{\underline{\lambda}!} (\log \alpha_1)^{\lambda_1} \cdots (\log \alpha_{n-1})^{\lambda_{n-1}} G_{\underline{\sigma}+\underline{\lambda}}(z). \tag{3.2}$$

As a consequence, if S_0 and S_1 are positive integers for which

$$G_{\underline{\sigma}}(s) = 0 \text{ for } \|\underline{\sigma}\| < S_0 \text{ and } 0 \leq s < S_1, \tag{3.3}$$

then for any $\underline{\sigma} \in \mathbb{N}^n$ with $\|\underline{\sigma}\| < S_0$ the function $G_{\underline{\sigma}}$ has a zero at $s = 0, \ldots, S_1 - 1$ of multiplicity $\geq S_0 - \|\underline{\sigma}\|$.

This is really the main point in Baker's method [1], I, p.212, which has no counterpart in the dual method of [34] Chap. 6 (there is no efficient extrapolation so far when one deals with functions of several variables). By means of the one dimensional approximate Schwarz' Lemma 1.12, one deduces from (3.3) a sharp upper bound for $|G_{\underline{\sigma}}|_r$ and gets more equations like (3.3): this is the extrapolation.

Remark. If we were to replace (3.1) by an algebraic relation between logarithms of algebraic numbers, for instance

$$\log \alpha_n = A(\log \alpha_1, \ldots, \log \alpha_{n-1})$$

where A is a polynomial of total degree > 1, then one could also write the derivatives of G as polynomials in $(\log \alpha_1, \ldots, \log \alpha_{n-1})$, but there are no nice relations like (3.2) between the corresponding exponential polynomials replacing the $G_{\underline{\sigma}}$.

b) Introducing n Variables

As we said, from a strict logical point of view, introducing functions of several variables is not required. But it may help to understand better the meaning of the differential operators ∂_k $(0 \le k < n)$.

To the auxiliary polynomial P is also associated an analytic function of n complex variables

$$\Phi(z_0, z_1 \dots, z_{n-1}) = P\bigl(z_0, e^{z_1}, \dots, e^{z_{n-1}}, e^{\beta_0 z_0 + \beta_1 z_1 + \cdots + \beta_{n-1} z_{n-1}}\bigr).$$

We take derivatives of Φ with respect to the n variables, and consider the values of these derivatives at the point

$$\underline{v} = (1, \log \alpha_1, \dots, \log \alpha_{n-1}) \in \mathbb{C}^n.$$

Obviously we have

$$\Phi(z\underline{v}) = G(z)$$

for $z \in \mathbb{C}$, but what is more interesting is the connection between the derivatives. From the definition of the differential operators $\partial_0, \dots, \partial_{n-1}$, it is plain that for $0 \le k \le n-1$ we have

$$\frac{\partial}{\partial z_k}\Phi(z_0, z_1 \dots, z_{n-1}) = (\partial_k P)\bigl(z_0, e^{z_1}, \dots, e^{z_{n-1}}, e^{\beta_0 z_0 + \beta_1 z_1 + \cdots + \beta_{n-1} z_{n-1}}\bigr).$$

Hence for $\underline{\sigma} = (\sigma_0, \dots, \sigma_{n-1}) \in \mathbb{N}^n$ and $z \in \mathbb{C}$ we have

$$G_{\underline{\sigma}}(z) = (\mathcal{D}^{\underline{\sigma}}\Phi)(z\underline{v})$$

where

$$\mathcal{D}^{\underline{\sigma}} = \left(\frac{\partial}{\partial z_0}\right)^{\sigma_0} \cdots \left(\frac{\partial}{\partial z_{n-1}}\right)^{\sigma_{n-1}}.$$

c) Introducing $n+1$ Variables

Instead of working with n variables it is sometimes convenient (for instance for the zero estimate) to consider $n+1$ variables: define

$$F(z_0, z_1 \dots, z_n) = P\bigl(z_0, e^{z_1}, \dots, e^{z_n}\bigr).$$

The point

$$\underline{u} = (1, \log \alpha_1, \dots, \log \alpha_n) \in \mathbb{C}^{n+1}$$

lies in the hyperplane W of equation

$$z_n = \beta_0 z_0 + \beta_1 z_1 + \cdots + \beta_{n-1} z_{n-1}.$$

A basis of W is $\mathbf{w} = (\underline{w}_0, \dots, \underline{w}_{n-1})$ where

$$\underline{w}_k = (\delta_{k0}, \ldots, \delta_{k,n-1}, \beta_k) \quad (0 \le k \le n-1).$$

These elements $\underline{w}_0, \ldots, \underline{w}_{n-1}$ are the n column vectors of the matrix

$$\begin{pmatrix} 1 & 0 & \cdots & 0 \\ 0 & 1 & \cdots & 0 \\ \vdots & \vdots & \ddots & \vdots \\ 0 & 0 & \cdots & 1 \\ \beta_0 & \beta_1 & \cdots & \beta_{n-1} \end{pmatrix} = \begin{pmatrix} & \mathrm{I}_n & \\ \beta_0 & \cdots & \beta_{n-1} \end{pmatrix}.$$

Since

$$z_0\underline{w}_0 + \cdots + z_{n-1}\underline{w}_{n-1} = (z_0, \ldots, z_{n-1}, \beta_0 z_0 + \cdots + \beta_{n-1} z_{n-1}),$$

we have

$$\Phi(z_0, \ldots, z_{n-1}) = F(z_0\underline{w}_0 + \cdots + z_{n-1}\underline{w}_{n-1}),$$

and one may view Φ as the restriction of F to the hyperplane W, equipped with the basis $\mathbf{w}$, by means of the isomorphism

$$\begin{array}{ccc} \mathbb{C}^n & \longrightarrow & W \\ (z_0, \ldots, z_{n-1}) & \longmapsto & z_0\underline{w}_0 + \cdots + z_{n-1}\underline{w}_{n-1}. \end{array}$$

To take the derivatives of Φ in all n directions amounts to taking the derivatives of F in the directions of W. More precisely, for $\underline{x} = (x_0, \ldots, x_n) \in \mathbb{C}^{n+1}$, define

$$\mathcal{D}_{\underline{x}} = x_0 \frac{\partial}{\partial z_0} + \cdots + x_n \frac{\partial}{\partial z_n}.$$

For instance

$$\mathcal{D}_{\underline{w}_k} = \frac{\partial}{\partial z_k} + \beta_k \frac{\partial}{\partial z_n} \quad \text{for } 0 \le k \le n-1,$$

hence for $0 \le k \le n-1$

$$\mathcal{D}_{\underline{w}_k} F(z_0, \ldots, z_n) = (\partial_k P)(z_0, e^{z_1}, \ldots, e^{z_n}).$$

Define also, for $\underline{\sigma} = (\sigma_0, \ldots, \sigma_{n-1}) \in \mathbb{N}^n$,

$$\mathcal{D}_{\mathbf{w}}^{\underline{\sigma}} = \mathcal{D}_{\underline{w}_0}^{\sigma_0} \cdots \mathcal{D}_{\underline{w}_{n-1}}^{\sigma_{n-1}}.$$

Then

$$\mathcal{D}_{\mathbf{w}}^{\underline{\sigma}} F(z_0, \ldots, z_n) = (\partial^{\underline{\sigma}} P)(z_0, e^{z_1}, \ldots, e^{z_n})$$

and

$$\mathcal{D}_{\mathbf{w}}^{\underline{\sigma}} F(z_0, \ldots, z_{n-1}, \beta_0 z_0 + \cdots + \beta_{n-1} z_{n-1}) = \left(\frac{\partial}{\partial z_0}\right)^{\sigma_0} \cdots \left(\frac{\partial}{\partial z_{n-1}}\right)^{\sigma_{n-1}} \Phi(z_0, \ldots, z_{n-1}).$$

In particular

$$G_{\underline{\sigma}}(z) = \mathcal{D}_{\mathbf{w}}^{\underline{\sigma}} F(z\underline{u}).$$

3.2 Proof of Baker's Theorem

We prove Theorem 3.1 following basically [30] Chap. 8. One main difference is that we shall use a sharper zero estimate than Theorem 6.1.1 of [30], and therefore we do not need a long extrapolation like in [30]: here a single step will be sufficient. This explains why the interpolation determinant method could easily be used in [34], Section 10.1. In Section 3.3 we shall explain how a longer extrapolation enables one to use a weaker zero estimate.

Denote by $\mathbb{K}$ the number field

$$\mathbb{K} = \mathbb{Q}(\alpha_1, \ldots, \alpha_n, \beta_0, \ldots, \beta_{n-1}).$$

Let $T_0, T_1, \ldots, T_n, S_0$ and S_1 be sufficiently large positive integers. Explicit conditions on these parameters will occur along the proof, and we shall discuss them later, but it may help the reader to know that a suitable choice is

$$\begin{aligned} T_0 &= 2[\mathbb{K}:\mathbb{Q}]N^{2n+1}, & T_1 &= \cdots = T_n = N^{2n-1}, \\ S_0 &= N^{2n+1}, & S_1 &= N. \end{aligned} \tag{3.4}$$

We also set $L = T_0T_1\cdots T_n$ and $T = T_1 + \cdots + T_n$.

a) Construction of the Auxiliary Polynomial

By means of Thue-Siegel's Lemma 1.4, we show the existence of a nonzero auxiliary polynomial $P \in \mathbb{Z}[Y_0, \ldots, Y_n]$, with degree $< T_i$ in Y_i $(0 \le i \le n)$, such that the equations (3.3) hold.

These conditions amount to a homogeneous linear system of equations with coefficients in $\mathbb{K}$, where the unknowns are the coefficients of P.

Our first condition on the parameters will be

$$\boxed{2[\mathbb{K}:\mathbb{Q}]\binom{S_0+n-1}{n}S_1 \le L,}$$

so that the number of equations is at most half the number of unknowns[7].

The coefficients of the linear system are the numbers

$$\gamma_{\tau\underline{t}}^{(\underline{\sigma},s)} = \partial^{\underline{\sigma}}\left(Y_0^{\tau}Y_1^{t_1}\cdots Y_n^{t_n}\right)\left(s, \alpha_1^s, \ldots, \alpha_n^s\right) \tag{3.5}$$

with $0 \le \tau < T_0$, $0 \le t_i < T_i$ $(1 \le i \le n)$ and $\underline{\sigma} \in \mathbb{N}^n$, $\|\underline{\sigma}\| < S_0$, $0 \le s < S_1$.

We write them explicitly by computing the derivatives of

$$z_0^{\tau}e^{t_1z_1}\cdots e^{t_{n-1}z_{n-1}}e^{t_n(\beta_0z_0+\cdots+\beta_{n-1}z_{n-1})}.$$

One obtains easily

[7] One could construct P with coefficients in the field $\mathbb{K}$, and then omit the factor $[\mathbb{K}:\mathbb{Q}]$.

$$\gamma_{\tau\underline{t}}^{(\underline{\sigma},s)} = \sum_{\kappa=0}^{\min\{\tau,\sigma_0\}} \frac{\sigma_0!\tau!}{\kappa!(\sigma_0-\kappa)!(\tau-\kappa)!} s^{\tau-\kappa}(\beta_0 t_n)^{\sigma_0-\kappa} \prod_{i=1}^{n-1}(t_i+t_n\beta_i)^{\sigma_i} \prod_{j=1}^{n} \alpha_j^{t_j s}.$$

One should use a variant of Lemma 1.4 taking into account the fact that the coefficients of our linear system are in $\mathbb{K}$ rather than in $\mathbb{Z}$, but anyway a rough estimate shows that we end up with a nonzero polynomial P of length at most

$$\mathrm{L}(P) \le \exp\{c_1(T_0+S_0)\log L + c_2 T S_1\}.$$

Here and below, $c_1, \dots, c_{16}$ denote positive numbers which do not depend on $T_0, T_1, \dots, T_n, S_0, S_1$. For instance with our choice (3.4) we get

$$\mathrm{L}(P) \le \exp\{c_3 N^{2n+1}\log N\}.$$

Remark. The whole point in this argument is that (3.1) allows us to consider values of polynomials in $n-1$ variables at the point

$$(\log\alpha_1, \dots, \log\alpha_{n-1})$$

in place of values of polynomials in n variables at the point

$$(\log\alpha_1, \dots, \log\alpha_n).$$

Without (3.1) it would be necessary to replace

$$\binom{S_0+n-1}{n} \quad \text{with} \quad \binom{S_0+n}{n+1}$$

and the only difference with the present proof is that no choice of parameters would be admissible!

b) Extrapolation on Integral Points

We introduce further parameters S_0' and S_1' which are positive integers with $S_0' < S_0$ and $S_1' > S_1$, and we are going to prove

$$G_{\underline{\sigma}}(s) = 0 \quad \text{for} \quad (\underline{\sigma}, s) \in \mathbb{N}^n \times \mathbb{N} \quad \text{with} \quad \|\underline{\sigma}\| < S_0' \quad \text{and} \quad 0 \le s < S_1'. \tag{3.6}$$

With the choice of parameters (3.4) we shall take

$$S_0' = [S_0/2], \quad S_1' = N^2.$$

Fix $\underline{\sigma} \in \mathbb{N}^n$ with $\|\underline{\sigma}\| < S_0'$. The function

$$G_{\underline{\sigma}}(z) = \partial^{\underline{\sigma}} P(z, \alpha_1^z, \dots, \alpha_n^z)$$

has a zero at each point $s = 0, \dots, S_1 - 1$ of multiplicity $\geq S_0 - S_0'$. The one variable Schwarz' Lemma 1.3 with $r = S_1'$, $E = e$, $R = 2er$ provides the following upper bound:

$$|G_{\underline{\sigma}}|_r \leq e^{-(S_0 - S_0')S_1} |G_{\underline{\sigma}}|_R.$$

This yields an upper bound for $|G_{\underline{\sigma}}(s)|$ with $s \in \mathbb{Z}$ in the range $0 \leq s < S_1'$ which is not compatible with Liouville's lower bound provided that

$$\boxed{c_4(T_0 + S_0)\log L + c_5 T S_1' < (S_0 - S_0')S_1.}$$

Hence (3.6).

c) Using Philippon's Zero Estimate

The next auxiliary Lemma is a very special case of Philippon's zero estimate [22] (see also Chap. 8 of [34] by D. Roy).

Proposition 3.2. *Let $\alpha_1, \dots, \alpha_n$ be nonzero complex numbers which generate a multiplicative subgroup of $\mathbb{C}^\times$ of rank $\geq n-1$ and let $\beta_0, \dots, \beta_{n-1}$ be complex numbers. Assume that $1, \beta_1, \dots, \beta_{n-1}$ are linearly independent over $\mathbb{Q}$. Let $T_0, T_1, \dots, T_n$, S_0, S_1 and L be positive integers satisfying the following conditions:*

$$L = T_0 T_1 \cdots T_n, \qquad T_1 \leq T_2 \leq \cdots \leq T_n,$$

$$S_0 > (n+1)T_n, \qquad S_0 S_1 > \frac{1}{2} n!(n+1)! \max\{T_0,\ 2T_n\},$$

and

$$\binom{S_0 + n - 1}{n} S_1 > (n+1)! T_0 T_1 \cdots T_n. \tag{3.7}$$

Assume also

- *either $\beta_0 \neq 0$*
- *or else $n \geq 2$ and*

$$\binom{S_0 + n - 2}{n - 1} S_1 > (n+1)! T_1 \cdots T_n. \tag{3.8}$$

For $\tau \in \mathbb{N}$, $\underline{t} \in \mathbb{N}^n$, $\underline{\sigma} \in \mathbb{N}^n$ and $s \in \mathbb{N}$, consider the number $\gamma_{\tau \underline{t}}^{(\underline{\sigma} s)}$ given by (3.5) and build up the matrix:

$$\mathtt{M} = \left(\gamma_{\tau \underline{t}}^{(\underline{\sigma} s)}\right)_{\substack{(\tau, \underline{t}) \\ (\underline{\sigma}, s)}}$$

where the index of rows $(\tau, \underline{t})$ runs over the elements in $\mathbb{N} \times \mathbb{N}^n$ with $0 \leq \tau < T_0$, $0 \leq t_i < T_i$ $(1 \leq i \leq n)$, while the index of columns $(\underline{\sigma}, s)$ runs over the elements in $\mathbb{N}^n \times \mathbb{N}$ with $\|\underline{\sigma}\| < (n+1)S_0$ and $0 \leq s < (n+1)S_1$. Then $\mathtt{M}$ has rank L.

Remark. Using (3.7) it is easily checked that (3.8) can be replaced by

$$nT_0 > S_0 + n - 1.$$

Remark. Given $\mathbb{Q}$-linearly independent complex numbers $\lambda_1, \ldots, \lambda_n$, the n numbers $\alpha_i = e^{\lambda_i}$ $(1 \le i \le n)$ generate a multiplicative subgroup of $\mathbb{C}^\times$ of rank $\ge n-1$. Conversely, if $\alpha_1, \ldots, \alpha_n$ are nonzero elements of $\mathbb{C}$ which generate a multiplicative subgroup of $\mathbb{C}^\times$ of rank $\ge n-1$, then there exist $\mathbb{Q}$-linearly independent complex numbers $\lambda_1, \ldots, \lambda_n$ such that $\alpha_i = e^{\lambda_i}$ for $1 \le i \le n$.

Proposition 3.2 is essentially Proposition 10.2 of [34], with a few differences:

- We do not assume $\beta_0 \neq 0$ here. At the same time our points are $(s, \alpha_1^s, \ldots, \alpha_n^s)$ in place[8] of $(s\beta_0, \alpha_1^s, \ldots, \alpha_n^s)$.
- We work with polynomials in $Y_0, \ldots, Y_n$ of degree $< T_i$ in Y_i $(0 \le i \le n)$, while in [34] we considered polynomials in $X_0, X_1^{\pm 1}, \ldots, X_n^{\pm 1}$ of degree $\le T_0$ in X_0 and degree $\le T_1$ in each of the variables $X_i^{\pm 1}$. Also here we consider nonnegative integers s with $0 \le s < (n+1)S_1$, while in [34] we had $s \in \mathbb{Z}$ with $|s| \le (n+1)S_1$. Also here we use strict inequalities for $\|\underline{\sigma}\|$.

These changes introduce few modifications in the proof of Proposition 10.2 of [34], but for the convenience of the reader we provide the details.

Proof. Consider the algebraic groups $G_0 = \mathbb{G}_{\mathrm{a}}$, $G_1 = \mathbb{G}_{\mathrm{m}}^n$, $G = G_0 \times G_1$, of dimensions $d_0 = 1$, $d_1 = n$ and $d = n+1$ respectively. Let W be the hyperplane in $\mathbb{C}^{n+1}$ of equation

$$\beta_0 z_0 + \beta_1 z_1 + \cdots + \beta_{n-1} z_{n-1} = z_n.$$

Introduce also the set

$$\Sigma = \left\{(s, \alpha_1^s, \ldots, \alpha_n^s)\,;\ s \in \mathbb{N},\ 0 \le s < S_1\right\} \subset G(\mathbb{C}) = \mathbb{C} \times (\mathbb{C}^\times)^n.$$

If the rank of the matrix $\mathtt{M}$ is less than L, then there exists a nonzero polynomial P in $\mathbb{C}[Y_0, Y_1, \ldots, Y_n]$, of degree $< T_i$ in Y_i $(0 \le i \le n)$, for which the functions $G_{\underline{\sigma}}(z) = \partial^{\underline{\sigma}} P(z, \alpha_1^z, \ldots, \alpha_n^z)$ satisfy

$$G_{\underline{\sigma}}(s) = 0 \quad \text{for} \quad (\underline{\sigma}, s) \in \mathbb{N}^n \times \mathbb{N}$$
$$\text{with} \quad \|\underline{\sigma}\| < (n+1)S_0 \quad \text{and} \quad 0 \le s < (n+1)S_1.$$

According to Philippon's zero estimate there exists an algebraic subgroup G^* of G of dimension $d^* \le n$ and codimension $d' = n+1-d^*$ such that

$$\binom{S_0 + \ell_0 - 1}{\ell_0} \mathrm{Card}\left(\frac{\Sigma + G^*}{G^*}\right) \mathcal{H}(G^*; \underline{T}) \le \mathcal{H}(G; \underline{T}), \tag{3.9}$$

[8] This was an oversight in Proposition 10.2 of [34]!

where

$$\ell_0 = \begin{cases} d'-1 & \text{if } T_e(G^*) \subset W, \\ d' & \text{otherwise.} \end{cases}$$

The notation $\mathcal{H}(G^*; \underline{T})$ stands for a multihomogeneous Hilbert-Samuel polynomial (see [34], Section 5.2.3); for instance $\mathcal{H}(G; \underline{T}) = (n+1)!L$.

We first check that this inequality (3.9) is not satisfied with $G^* = \{e\}$: indeed when $G^* = \{e\}$ we have

$$d^* = 0, \quad \ell_0 = n, \quad \mathrm{Card}\left(\frac{\Sigma + G^*}{G^*}\right) = \mathrm{Card}(\Sigma) = S_1, \quad \mathcal{H}(G^*; \underline{T}) = 1,$$

so that by (3.7)

$$\binom{S_0 + \ell_0 - 1}{\ell_0} \mathrm{Card}\left(\frac{\Sigma + G^*}{G^*}\right) \mathcal{H}(G^*; \underline{T}) > (n+1)!L.$$

Therefore $d^* \geq 1$.

Write $G^* = G_0^* \times G_1^*$ where G_0^* is an algebraic subgroup of G_0 and G_1^* an algebraic subgroup of G_1. Denote by d_0^* and d_1^* the dimensions of G_0^* and G_1^* respectively, and by d_0' and d_1' their codimensions:

$$d_0^* + d_0' = d_0 = 1, \quad d_1^* + d_1' = d_1 = n.$$

Assume first $T_e(G^*) \subset W$. Since $1, \beta_1, \ldots, \beta_{n-1}$ are linearly independent over $\mathbb{Q}$, the hyperplane of $\mathbb{C}^n$ of equation

$$\beta_1 z_1 + \cdots + \beta_{n-1} z_{n-1} = z_n$$

does not contain any nonzero element of $\mathbb{Q}^n$. Since $T_e(G_1^*)$ is a subspace of $\mathbb{C}^n$ which is rational over $\mathbb{Q}$, we deduce $G_1^* = \{e\}$, hence $G^* = \mathbb{G}_a \times \{e\}$, $d^* = 1$, $d' = n$, $\ell_0 = n-1$ and $\mathcal{H}(G^*; \underline{T}) = T_0$. Now the condition $T_e(G^*) \subset W$ implies $\beta_0 = 0$, hence (3.8) gives $n \geq 2$ and

$$\binom{S_0 + n - 2}{n-1} S_1 > (n+1)! T_1 \cdots T_n.$$

Since $n \geq 2$ and since $\alpha_1, \ldots, \alpha_n$ generate a subgroup of $\mathbb{C}^\times$ of rank $\geq n-1$, we have

$$\mathrm{Card}\left(\frac{\Sigma + G^*}{G^*}\right) = \mathrm{Card}\{(\alpha_1^s, \ldots, \alpha_n^s)\,;\, s \in \mathbb{N},\, 0 \leq s < S_1\} = S_1.$$

Hence (3.9) does not hold and we get a contradiction.

Therefore $T_e(G^*) \not\subset W$ and $\ell_0 = d'$.

Consider the case $G_0^* = \{0\}$. We have $d_0^* = 0$, $d^* = d_1^*$, $d' = n + 1 - d_1^*$,

$$\mathcal{H}(G^*; \underline{T}) \geq (d_1^* + 1)! T_1 \cdots T_{d_1^*} \quad \text{and} \quad \mathrm{Card}\left(\frac{\Sigma + G^*}{G^*}\right) = S_1.$$

Therefore (3.9) implies

$$\binom{S_0+d'-1}{d'}S_1 \le \frac{(n+1)!}{(n+2-d')!}T_0T_{d_1^*+1}\cdots T_n,$$

hence

$$S_0^{d'}S_1 \le \frac{(n+1)!d'!}{(n+2-d')!}T_0T_{d_1^*+1}\cdots T_n.$$

However we have $S_0 > T_n$, $1 \le d' \le n$ and

$$\frac{d'!}{(n+2-d')!} \le \frac{1}{2}n!,$$

hence we get a contradiction with the inequality

$$S_0S_1 > \frac{1}{2}n!(n+1)!T_0.$$

So we have $d_0^* = 1$, $d^* = d_1^*+1$, $d' = n - d_1^*$ and

$$\mathcal{H}(G^*;\underline{T}) \ge (d_1^*+1)!T_0T_1\cdots T_{d_1^*}.$$

Now (3.9) gives

$$\binom{S_0+d'-1}{d'}\mathrm{Card}\left(\frac{\Sigma+G^*}{G^*}\right) \le \frac{(n+1)!}{(n+1-d')!}T_{d_1^*+1}\cdots T_n$$

from which we deduce

$$S_0^{d'}\,\mathrm{Card}\left(\frac{\Sigma+G^*}{G^*}\right) \le \frac{(n+1)!d'!}{(n+1-d')!}T_{d_1^*+1}\cdots T_n.$$

Using the estimates

$$S_0 > T_n, \quad 1 \le d' \le n, \quad \frac{d'!}{(n+1-d')!} \le n!$$

and

$$S_0S_1 > n!(n+1)!T_n$$

we obtain

$$\mathrm{Card}\left(\frac{\Sigma+G^*}{G^*}\right) < S_1,$$

which means $\Sigma \cap G^* \ne \{e\}$. The assumption on the rank of the subgroup of $\mathbb{C}^*$ generated by $\alpha_1,\dots,\alpha_n$ then implies $d_1^* = n-1$, $d' = 1$ and we get the estimate

$$S_0 \le (n+1)T_n$$

which is not compatible with our assumptions. □

d) End of the Proof of Baker's Theorem 3.1

We apply Proposition 3.2 with S_0 and S_1 replaced respectively by

$$\lceil S_0'/(n+1)\rceil \quad\text{and}\quad \lceil S_1'/(n+1)\rceil.$$

From (3.4) and (3.6), using the estimates

$$S_0' > (n+1)^2T, \quad S_0'S_1' > n!(n+1)!(n+1)^2 \max\left\{\frac{1}{2}T_0,\, T\right\},$$

$$nT_0 > S_0 + n - 1 \quad\text{and}\quad {S_0'}^n S_1' > n!(n+1)!L,$$

we deduce a contradiction, which completes the proof of Baker's Theorem 3.1. □

Remark. The basic ideas for a proof of Baker's Theorem with an interpolation determinant in [34], Section 10.1.4 are essentially the same. Instead of using Dirichlet's pigeonhole principle to solve a system of linear equations, we only consider the matrix of this linear system. More precisely the relevant matrix $\mathtt{M}$ is the one occurring in the zero estimate (Proposition 3.2): it has maximal rank, and enables one to start with a nonzero determinant Δ. As usual the required lower bound for $|\Delta|$ is given by Liouville's estimate (Proposition 1.13). On the other hand the argument occurring above turns out to be perfectly adaptable to yield an upper bound for $|\Delta|$ which gives just what we need.

The difference between the proof with an auxiliary function and the proof with an interpolation determinant is that in the latter Dirichlet's box principle is not required. However there is a substitute to the auxiliary function, which is the (explicit) exponential polynomial given by a determinant

$$\det\Big(\mathcal{C}_1,\, \dots,\, \mathcal{C}_{L-1},\, \Phi(z)\Big),$$

where $\mathcal{C}_1, \dots, \mathcal{C}_{L-1}$ are $L-1$ vector columns of $\mathtt{M}$, while the last column vector $\Phi : \mathbb{C} \to \mathbb{C}^L$ is given by

$$\left(z^\tau \alpha_1^{t_1 z} \cdots \alpha_n^{t_n z}\right)_{0\le\tau<T_0,\ 0\le t_i<T_i\ (1\le i\le n)}.$$

3.3 Further Extrapolation with the Auxiliary Function

In this section we shall explain how to replace, in the previous proof, Philippon's zero estimate (Proposition 3.2) by a simpler one. The idea is to extrapolate further and to prove by induction on $j = 0, 1, \dots,$

$$G_{\underline{\sigma}}(s) = 0 \quad\text{for}\quad (\underline{\sigma}, s) \in \mathbb{N}^n \times \mathbb{N}$$
$$\text{with}\quad \|\underline{\sigma}\| < S_0/2^j \quad\text{and}\quad 0 \le s < S_1^{(j)},$$

with $S_1^{(0)} = S_1$ and $S_1^{(1)} = S_1'$. One cannot continue such an induction forever (in any case one needs $2^j \le S_0$). On the other hand, obviously, when the number of equations increases, it is easier to derive a contradiction by means of a zero estimate. This is not only of historical interest: our motivation is related with the problem of linear independence measures, where a short extrapolation yields weaker estimates than a longer one (see Section 5).

Baker used a variety of arguments for concluding his proofs, including clever non-vanishing results for certain determinants. In the real case one could just appeal to Pólya's Lemma 1.6. Dealing with the general case of complex algebraic numbers α_i and β_j, we shall use Tijdeman's zero estimate for exponential polynomials in one variable ([29]; see also [6], Chap. 9, Section 4, Lemma 8.9, [2], Chap. 12, Section 2, Lemma 6 and [30], Chap. 6).

Lemma 3.3. *Let $a_1, \dots, a_n$ be polynomials in $\mathbb{C}[z]$, not all of which are zero, of degrees $d_1, \dots, d_n$. Let $w_1, \dots, w_n$ be pairwise distinct complex numbers. Define*

$$\Omega = \max\{|w_1|, \dots, |w_n|\}.$$

Then for $R > 0$ the number of zeroes (counting multiplicities) of the function

$$F(z) = \sum_{i=1}^{n} a_i(z) e^{w_i z}$$

in the disk $B_1(0, R)$ is at most $2(d_1 + \cdots + d_n + n - 1) + 5R\Omega$.

We complete the proof of Baker's Theorem 3.1 as follows. We repeat the argument of Section 3.2 b) and perform an induction on j with $1 \le j \le J$. We introduce further parameters $S_0^{(j)}$ and $S_1^{(j)}$ which are positive integers with

$$S_0^{(0)} = S_0, \quad S_1^{(0)} = S_1, \qquad S_0^{(1)} = S_0', \quad S_1^{(1)} = S_1',$$
$$S_0^{(j)} < S_0^{(j-1)} \text{ and } S_1^{(j)} > S_1^{(j-1)} \quad (1 \le j \le J).$$

One may keep in mind the following picture:

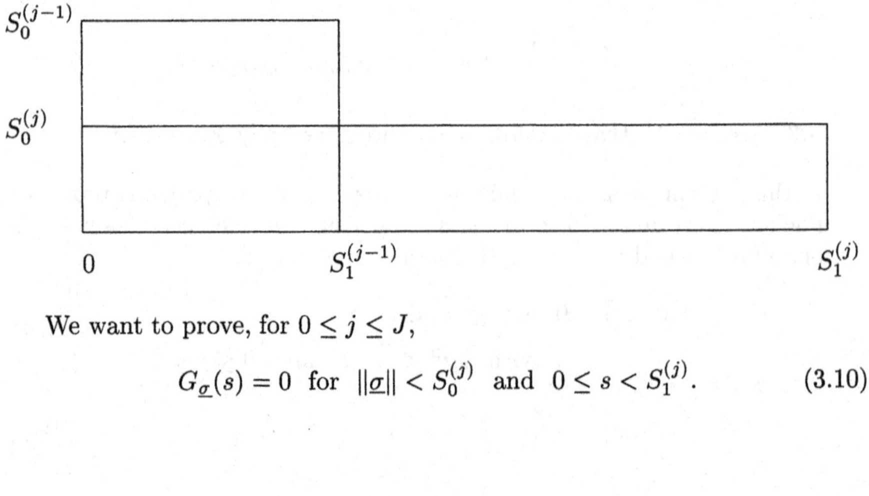

We want to prove, for $0 \le j \le J$,

$$G_{\underline{\sigma}}(s) = 0 \text{ for } \|\underline{\sigma}\| < S_0^{(j)} \text{ and } 0 \le s < S_1^{(j)}. \tag{3.10}$$

This is true by construction (Section 3.2, a) for $j = 0$, and by the first extrapolation (Section 3.2, b) for $j = 1$. Relations (3.10) for $j+1$ follow from those for j provided that

$$\boxed{c_6(T_0+S_0)\log\bigl(LS_1^{(j+1)}\bigr) + c_7 T S_1^{(j+1)} < \bigl(S_0^{(j)} - S_0^{(j+1)}\bigr)S_1^{(j)}.}$$

Assuming (3.4), let us choose

$$J = 2n^2, \quad S_0^{(j)} = \bigl[S_0/2^j\bigr], \quad S_1^{(j)} = N^{j+1} \quad (1 \le j \le J). \tag{3.11}$$

At the end of the induction ($j = J$) we get an exponential polynomial G with a zero at each point s with $0 \le s < S_1^{(J)}$ of multiplicity $\ge S_0^{(J)}$. Since

$$\boxed{S_0^{(J)} S_1^{(J)} > 2L + c_8 T S_1^{(J)},}$$

we get a contradiction with Lemma 3.3.

3.4 Upper Bound for a Determinant in Several Variables

Let $F \in \mathcal{H}_n(R)$ be a function with a zero of multiplicity $\ge S_0$ at S_1 distinct points of $B_n(0,r)\cap\mathbb{C}\underline{v}$ with $r \le R$. Then $G(z) = F(\underline{v}z)$ (which may be viewed as the restriction of F to $\mathbb{C}\underline{v}$) is a function of a single variable with S_0S_1 zeroes in $B_1(0, r/|\underline{v}|)$[9]. From Schwarz' Lemma 1.3 we deduce

$$\begin{aligned}\sup_{\underline{z}\in B_n(0,r)\cap\mathbb{C}\underline{v}} |F(\underline{z})| = |G|_{r/|\underline{v}|} &\le \left(\frac{R^2+r^2}{2rR}\right)^{-S_0S_1} |G|_{R/|\underline{v}|}\\ &\le \left(\frac{R^2+r^2}{2rR}\right)^{-S_0S_1} |F|_R.\end{aligned}$$

Unfortunately there is no similar upper bound for $|F|_r$ and we cannot use Cauchy's inequalities to bound derivatives of F in other directions than $\mathbb{C}\underline{v}$. For instance taking $n = 2$, $\underline{v} = (1,0)$, $F(z_1,z_2) = z_2^L$, we have $G(z) = F(z,0) = 0$, but

$$\left(\frac{\partial}{\partial z_2}\right)^L F(z,0) = L!.$$

According to Baker's remark (see Section 3.1, a)), for $\|\underline{\sigma}\| < S_0$, the one variable function $G_{\underline{\sigma}}(z) = \mathcal{D}^{\underline{\sigma}}F(\underline{v}z)$ has at least $(S_0 - \|\underline{\sigma}\|)S_1$ zeroes in $B_1(0, r/|\underline{v}|)$. Hence

[9] Here we do not use all the information: a zero of multiplicity $\ge S_0$ for F involves $\binom{S_0+n-1}{n}$ conditions, while for G it involves only S_0 conditions.

$$\sup_{\underline{z}\in B_n(0,r)\cap\mathbb{C}\underline{v}} |\mathcal{D}^{\underline{\sigma}}F(\underline{z})| \le \left(\frac{R^2+r^2}{2rR}\right)^{-(S_0-\|\underline{\sigma}\|)S_1} |\mathcal{D}^{\underline{\sigma}}F|_R.$$

This is the key point which explains why, in Baker's extrapolation argument, the order of derivation needs to decrease; one compensates by increasing the set of points.

Assuming for simplicity $\|\underline{\sigma}\| \le S_0/2$, the exponent of $2rR/(R^2+r^2)$ is $\ge S_0S_1/2$. For a function of n variables with S_1 zeroes of multiplicity $\ge S_0$, one should expect only an exponent $S_0S_1^{1/n}$ (up to a small absolute multiplicative constant), but the point here is that these zeroes lie on a complex line, and this explains why the exponent can be as large as a constant multiple of S_0S_1. It is interesting to compare with the interpolation determinant method: the exponent which arises more naturally (see [34], Section 10.1.4) is $S_0S_1^{1/n}$ and this is sufficient to achieve nontrivial estimates, but a refinement can be included (see [34] Proposition 10.5), so that one reaches the same exponent (namely S_0S_1, up to a constant) as with the auxiliary function.

We extend Corollary 2.4 to the situation arising in Baker's method. We deal with derivatives of functions of several variables, but the points we consider lie on a complex line $V = \mathbb{C}\underline{v} \subset \mathbb{C}^n$.

We first give a variant of Lemma 2.2 for functions of several variables.

Lemma 3.4. *Let $n \ge 1$ be an integer, V a complex subspace of dimension 1 of $\mathbb{C}^n$, R, r, ϱ and E be positive real numbers, F an element of $\mathcal{H}_n(R)$, $\underline{\zeta}_1,\dots,\underline{\zeta}_m$ pairwise distinct elements of V, $\underline{v}$ an element of V, $\sigma_1,\dots,\sigma_m$ nonnegative integers and $\underline{\kappa}$ an element of $\mathbb{N}^n$. Set*

$$M = \sum_{i=1}^{m} \max\{0, \sigma_i - \|\underline{\kappa}\|\}.$$

Assume

$$R \ge \varrho + \max\{r, |\underline{v}|\}, \quad r \ge \max_{1\le i\le m} |\underline{\zeta}_i| \quad \text{and} \quad 1 \le E \le \frac{(R-\varrho)^2 + r|\underline{v}|}{(R-\varrho)(r+|\underline{v}|)}.$$

Assume also that F satisfies

$$\mathcal{D}^{\underline{\sigma}}F(\underline{\zeta}_i) = 0 \quad \text{for } \underline{\sigma}\in\mathbb{N}^n \text{ with } \|\underline{\sigma}\| < \sigma_i \text{ and } 1 \le i \le m.$$

Then

$$|\mathcal{D}^{\underline{\kappa}}F(\underline{v})| \le \underline{\kappa}!\varrho^{-\|\underline{\kappa}\|}E^{-M}|F|_R.$$

Proof. Let $\underline{v}_0 \in V$ satisfy $|\underline{v}_0| = 1$. Define a function $G \in \mathcal{H}_1(R)$ of a single variable z by

$$G(z) = \mathcal{D}^{\underline{\kappa}}F(\underline{v}_0 z).$$

Define also $\zeta_1,\dots,\zeta_m$ in $B_1(0,r)$ by $\underline{\zeta}_i = \underline{v}_0\zeta_i$ $(1 \le i \le m)$. The formula

$$G^{(k)}(z) = \sum_{\|\underline{\tau}\|=k} \frac{k!}{\underline{\tau}!} \underline{v}_0^{\underline{\tau}} \mathcal{D}^{\underline{\kappa}+\underline{\tau}} F(\underline{v}_0 z) \qquad (k \geq 0)$$

shows that G has a zero of multiplicity $\geq \max\{0, \sigma_i - \|\underline{\kappa}\|\}$ at ζ_i for $1 \leq i \leq m$. According to Lemma 1.3 (with r replaced by $|\underline{v}|$ and R by $R - \varrho$) we have

$$|G|_{|\underline{v}|} \leq E^{-M} |G|_{R-\varrho}.$$

We derive the conclusion from Cauchy's inequalities:

$$|G|_{R-\varrho} = \sup_{|z|=R-\varrho} |\mathcal{D}^{\underline{\kappa}} F(\underline{v}_0 z)| \leq \underline{\kappa}! \varrho^{-\|\underline{\kappa}\|} |F|_R.$$

□

Remark. For $n = 1$ Lemma 3.4 contains Lemma 1.3, but not Lemma 2.2 when $\kappa > 0$.

The next result is a variant of Proposition 2.3 for functions of several variables.

Proposition 3.5. *Let $n \geq 1$ be an integer, V a complex line in $\mathbb{C}^n$, R a positive real number, $\varphi_1, \dots, \varphi_L$ elements of $\mathcal{H}_n(R)$, $\underline{\xi}_1, \dots, \underline{\xi}_L$ elements of $V \cap B_n(0, R)$ and $\underline{\kappa}_1, \dots, \underline{\kappa}_L$ elements of $\mathbb{N}^n$. Consider the determinant*

$$\Delta = \det \left(\mathcal{D}^{\underline{\kappa}_\mu} \varphi_\lambda (\underline{\xi}_\mu) \right)_{1 \leq \lambda, \mu \leq L}.$$

Let $m_1, \dots, m_L$ be nonnegative integers and, for $1 \leq \mu \leq L$ and $1 \leq i \leq m_\mu$, let $\underline{\zeta}_{\mu i}$ be an element of V and $\sigma_{\mu i}$ a nonnegative integer. We assume that for each $\mu = 1, \dots, L$, the m_μ elements $\underline{\zeta}_{\mu 1}, \dots, \underline{\zeta}_{\mu m_\mu}$ are pairwise distinct. Set

$$M_\mu = \sum_{i=1}^{m_\mu} \max\{0, \sigma_{\mu i} - \|\underline{\kappa}_\mu\|\} \quad (1 \leq \mu \leq L).$$

For $1 \leq \mu \leq L$, let r_μ, R_μ, ϱ_μ and E_μ be positive real numbers satisfying

$$R \geq R_\mu \geq \varrho_\mu + \max\{r_\mu, |\underline{\xi}_\mu|\}, \quad r_\mu \geq \max_{1 \leq i \leq m_\mu} |\underline{\zeta}_{\mu i}|$$

and

$$1 \leq E_\mu \leq \frac{(R_\mu - \varrho_\mu)^2 + r_\mu |\underline{\xi}_\mu|}{(R_\mu - \varrho_\mu)(r_\mu + |\underline{\xi}_\mu|)}.$$

Denote by Φ the analytic mapping

$$(\varphi_1, \dots, \varphi_L) \colon B_n(0, R) \to \mathbb{C}^L.$$

Assume that for any $(\mu, i, \underline{\kappa}) \in \mathbb{N}^{2+n}$ satisfying $1 \leq \mu \leq L$, $1 \leq i \leq m_\mu$ and $\|\underline{\kappa}\| < \sigma_{\mu i}$, the μ vectors

$$\mathcal{D}^{\underline{\kappa}_1}\Phi(\underline{\xi}_1),\ldots,\mathcal{D}^{\underline{\kappa}_{\mu-1}}\Phi(\underline{\xi}_{\mu-1}),\ \mathcal{D}^{\underline{\kappa}}\Phi(\underline{\zeta}_{\mu i}) \tag{3.12}$$

in $\mathbb{C}^L$ *are linearly dependent. Then*

$$|\Delta| \le L! \max_{\tau\in\mathfrak{S}_L} \prod_{\mu=1}^{L} \left(\underline{\kappa}_\mu!\varrho_\mu^{-\|\underline{\kappa}_\mu\|} E_\mu^{-M_\mu} |\varphi_{\tau(\mu)}|_{R_\mu}\right).$$

Proof. We prove Proposition 3.5 by induction on L. For $L = 1$ we have $\Phi = \varphi_1$,

$$\Delta = \mathcal{D}^{\underline{\kappa}_1}\Phi(\underline{\xi}_1), \qquad 1 \le E_1 \le \frac{(R_1-\varrho_1)^2 + r_1|\underline{\xi}_1|}{(R_1-\varrho_1)(r_1+|\underline{\xi}_1|)}$$

and hypothesis (3.12) reads

$$\mathcal{D}^{\underline{\kappa}}\Phi(\underline{\zeta}_{1i}) = 0 \quad\text{for}\quad \|\underline{\kappa}\| < \sigma_{1i} \quad\text{and}\quad 1 \le i \le m_1.$$

From Lemma 3.4 we deduce

$$|\Delta| \le \underline{\kappa}_1!\varrho_1^{-\|\underline{\kappa}_1\|} E_1^{-M_1} |\varphi_1|_{R_1}.$$

Hence Proposition 3.5 is true in case $L = 1$.

Assume now that the conclusion is true for L replaced by $L-1$. Define $F \in B_n(0,R)$ by

$$F(\underline{z}) = \det\left(\mathcal{D}^{\underline{\kappa}_1}\Phi(\underline{\xi}_1),\ldots,\mathcal{D}^{\underline{\kappa}_{L-1}}\Phi(\underline{\xi}_{L-1}),\ \Phi(\underline{z})\right). \tag{3.13}$$

By assumption (3.12) with $\mu = L$, for $1 \le i \le m_L$ and $\|\underline{\kappa}\| < \sigma_{Li}$, we have

$$\mathcal{D}^{\underline{\kappa}}F(\underline{\zeta}_{Li}) = 0.$$

Since

$$\Delta = \mathcal{D}^{\underline{\kappa}_L}F(\underline{\xi}_L),$$

we deduce from Lemma 3.4

$$|\Delta| \le \underline{\kappa}_L!\varrho_L^{-\|\underline{\kappa}_L\|} E_L^{-M_L} |F|_{R_L}.$$

We expand the determinant in the right hand side of (3.13) with respect to the last column: define, for $1 \le \lambda \le L$,

$$\Phi_\lambda = (\varphi_1,\ldots,\varphi_{\lambda-1},\varphi_{\lambda+1},\ldots,\varphi_L) : B_n(0,R) \to \mathbb{C}^{L-1}$$

and let Δ_λ denote the determinant of the $(L-1)\times(L-1)$ matrix

$$\left(\mathcal{D}^{\underline{\kappa}_1}\Phi_\lambda(\underline{\xi}_1),\ldots,\mathcal{D}^{\underline{\kappa}_{L-1}}\Phi_\lambda(\underline{\xi}_{L-1})\right).$$

We have

$$F(\underline{z}) = \sum_{\lambda=1}^{L} (-1)^{L-\lambda} \varphi_\lambda(\underline{z}) \Delta_\lambda,$$

hence

$$|F|_{R_L} \le L \max_{1\le\lambda\le L} |\varphi_\lambda|_{R_L} |\Delta_\lambda|$$

and therefore

$$|\Delta| \le \underline{\kappa}_L! L \varrho_L^{-\|\underline{\kappa}_L\|} E_L^{-M_L} \max_{1\le\lambda\le L} |\varphi_\lambda|_{R_L} |\Delta_\lambda|.$$

We fix an index $\lambda^0 \in \{1, \ldots, L\}$ such that

$$|\Delta| \le \underline{\kappa}_L! L \varrho_L^{-\|\underline{\kappa}_L\|} E_L^{-M_L} |\varphi_{\lambda^0}|_{R_L} |\Delta_{\lambda^0}|.$$

Using the induction hypothesis, we deduce that there exists a bijective map t from $\{1, \ldots, L-1\}$ to $\{1, \ldots, \lambda^0 - 1, \lambda^0 + 1, \ldots, L\}$ such that

$$|\Delta_{\lambda^0}| \le (L-1)! \prod_{\mu=1}^{L-1} \left(\underline{\kappa}_\mu! \varrho_\mu^{-\|\underline{\kappa}_\mu\|} E_\mu^{-M_\mu} |\varphi_{t(\mu)}|_{R_\mu} \right).$$

Define $\tau \in \mathfrak{S}_L$ by $\tau(\mu) = t(\mu)$ for $1 \le \lambda < L$ and $\tau(L) = \lambda^0$. Proposition 3.5 follows. □

We shall use a special case of Proposition 3.5.

Given a sequence $(\underline{\sigma}_\nu, \zeta_\nu)_{0\le\nu\le N}$ of elements in $\mathbb{N}^n \times \mathbb{C}$, an index ν in the range $0 \le \nu \le N$ and a complex number ζ, we define the *weight* $w_\nu(\zeta)$ *of index* ν *of* ζ *in this sequence* as follows: $w_\nu(\zeta) = 0$ if $\zeta_i \ne \zeta$ for $0 \le i \le \nu$ and otherwise

$$w_\nu(\zeta) = \max\bigl\{ \|\underline{\sigma}\| \; ; \; \exists i \text{ with } 0 \le i \le \nu \text{ such that } (\underline{\sigma}_i, \zeta_i) = (\underline{\sigma}, \zeta) \bigr\}. \quad (3.14)$$

We consider a sequence $(\underline{\sigma}_\nu, \zeta_\nu)_{0\le\nu\le N}$ which satisfies the following property:

$$\text{For } 0 \le \nu \le N \text{ and for any } \underline{\sigma} \in \mathbb{N}^n \text{ satisfying } \|\underline{\sigma}\| < \|\underline{\sigma}_\nu\|,$$
$$\text{there exists } i \text{ with } 0 \le i < \nu \text{ such that } (\underline{\sigma}_i, \zeta_i) = (\underline{\sigma}, \zeta_\nu). \quad (3.15)$$

Such a sequence will be called *admissible.*

In the case $n = 1$ the sequence $(\sigma_\nu, \zeta_\nu)_{0\le\nu\le N}$ which occurred in the hypotheses of Corollary 2.4 is admissible and satisfies

$$w_\nu(\zeta_\nu) = \sigma_\nu \quad \text{for} \quad 0 \le \nu \le N.$$

For an admissible sequence $(\underline{\sigma}_\nu, \zeta_\nu)_{0\le\nu\le N}$, for an analytic function F of n variables and for $\underline{v} \in \mathbb{C}^n$, the $N+1$ conditions

$$\mathcal{D}^{\underline{\sigma}_\nu} F(\underline{v}\zeta_\nu) = 0 \quad \text{for} \quad 0 \le \nu \le N \quad (3.16)$$

imply that the one-variable function $G(z) = F(\underline{v}z)$ has a zero of multiplicity $\geq w_N(\zeta)$ at each point $\zeta \in \mathbb{C}$. Moreover, for any $\underline{\tau} \in \mathbb{N}^n$, the same conditions (3.16) imply that the function $G_{\underline{\tau}}(z) = \mathcal{D}^{\underline{\tau}} F(\underline{v}z)$ has a zero at ζ of multiplicity

$$\geq \max\{0, w_N(\zeta) - \|\underline{\tau}\|\}.$$

Corollary 3.6. *Let n and L be positive integers, $\varphi_1, \dots, \varphi_L$ elements of $\mathcal{H}_n(R)$, $0 = \nu_0 \leq \nu_1 < \cdots < \nu_L \leq N$ nonnegative integers, $\underline{\sigma}_0, \dots, \underline{\sigma}_N$ elements of $\mathbb{N}^n$, $\zeta_0, \dots, \zeta_N$ complex numbers and $\underline{v} \in \mathbb{C}^n \setminus \{0\}$. Consider the determinant*

$$\Delta = \det\left(\mathcal{D}^{\underline{\sigma}_{\nu_\mu}} \varphi_\lambda(\underline{v}\zeta_{\nu_\mu})\right)_{1 \leq \lambda, \mu \leq L}.$$

Assume $(\underline{\sigma}_\nu, \zeta_\nu)_{0 \leq \nu \leq N}$ is an admissible sequence. For $0 \leq \nu \leq N$ define

$$M_\nu = \sum_{\zeta \in B_1(0, R/|\underline{v}|)} \max\{0, w_\nu(\zeta) - \|\underline{\sigma}_\nu\|\}.$$

For $1 \leq \mu \leq L$, let r_μ, R_μ and ϱ_μ be positive real numbers satisfying

$$R \geq R_\mu \geq \varrho_\mu + \max\{r_\mu, |\underline{v}\zeta_{\nu_\mu}|\} \quad \textit{and} \quad r_\mu \geq \max_{0 \leq \nu < \nu_\mu} |\underline{v}\zeta_\nu|.$$

Let E_μ satisfy

$$1 \leq E_\mu \leq \frac{(R_\mu - \varrho_\mu)^2 + r_\mu |\underline{v}\zeta_{\nu_\mu}|}{(R_\mu - \varrho_\mu)(r_\mu + |\underline{v}\zeta_{\nu_\mu}|)}.$$

Denote by Φ the analytic mapping

$$(\varphi_1, \dots, \varphi_L) \colon B_n(0, R) \to \mathbb{C}^L.$$

Assume that for $0 \leq \mu \leq L-1$ and for $\nu_\mu \leq \nu < \nu_{\mu+1}$ the system of $\mu + 1$ vectors

$$\mathcal{D}^{\underline{\sigma}_{\nu_1}} \Phi(\underline{v}\zeta_{\nu_1}), \dots, \mathcal{D}^{\underline{\sigma}_{\nu_\mu}} \Phi(\underline{v}\zeta_{\nu_\mu}),\ \mathcal{D}^{\underline{\sigma}_\nu} \Phi(\underline{v}\zeta_\nu)$$

in $\mathbb{C}^L$ is linearly dependent. Then

$$|\Delta| \leq L! \max_{\tau \in \mathfrak{S}_L} \prod_{\mu=1}^{L} \left(\underline{\sigma}_{\nu_\mu}! \varrho_\mu^{-\|\underline{\sigma}_{\nu_\mu}\|} E_\mu^{-M_{\nu_\mu}} |\varphi_{\tau(\mu)}|_{R_\mu}\right).$$

Remark. This statement does not contain Corollary 2.4, because the exponent M_ν is usually smaller than

$$\sum_{\zeta \in B_1(0, R/|\underline{v}|)} w_\nu(\zeta).$$

It does not seem clear how to get a result containing both the one variable Proposition 2.3 and Corollary 3.6. It might be easier to combine Corollary 3.6 with Proposition 10.5 of [34].

Proof. We apply Proposition 3.5 with $V = \mathbb{C}\underline{v}$,

$$\underline{\xi}_\mu = \underline{v}\zeta_{\nu_\mu} \quad \text{and} \quad \underline{\kappa}_\mu = \underline{\sigma}_{\nu_\mu} \quad (1 \le \mu \le L).$$

We define m_μ, $\underline{\zeta}_{\mu i}$ and $\sigma_{\mu i}$ as follows: for $1 \le \mu \le L$, we consider the sequence $(\zeta_0, \dots, \zeta_{\nu_\mu})$ given by the hypotheses of Corollary 3.6, we denote by m_μ the number of distinct elements in this sequence, by $\zeta_{\mu i}$ these distinct elements, by $\sigma_{\mu i}$ the weight of index ν_μ of $\zeta_{\mu i}$ in this sequence, and we set $\underline{\zeta}_{\mu i} = \underline{v}\zeta_{\mu i}$. Therefore

$$\sum_{\zeta \in B_1(0, R/|\underline{v}|)} \max\{0, w_{\nu_\mu}(\zeta) - \|\underline{\sigma}_{\nu_\mu}\|\} \ge \sum_{i=1}^{m_\mu} \max\{0, \sigma_{\mu i} - \|\underline{\kappa}_\mu\|\} \quad (1 \le \mu \le L).$$

Corollary 3.6 follows. □

3.5 Extrapolation with an Interpolation Determinant

We give a proof of Baker's Theorem 3.1 by means of an interpolation determinant and an extrapolation; the zero estimate which will enable us to get the conclusion is Tijdeman's Lemma 3.3.

Let $T_0, T_1, \dots, T_n, J, S_0^{(j)}$ and $S_1^{(j)}$ $(0 \le j \le J)$ be positive integers. Assume that the sequence $(S_0^{(j)})_{0 \le j \le J}$ is decreasing and that the sequence $(S_1^{(j)})_{0 \le j \le J}$ increases. We write S_0 and S_1 for $S_0^{(0)}$ and $S_1^{(0)}$ respectively and we set $L = T_0 \cdots T_n$, $T = T_1 + \cdots + T_n$.

We denote by $\{\varphi_1, \dots, \varphi_L\}$ the L exponential monomials

$$z^\tau \alpha_1^{t_1 z} \cdots \alpha_n^{t_n z} \quad \text{for} \quad 0 \le \tau < T_0 \quad \text{and} \quad 0 \le t_i < T_i \ (1 \le i \le n).$$

For $0 \le j \le J$, define $\mathcal{S}_j$ as the set of $(\underline{\sigma}, s)$ in $\mathbb{N}^n \times \mathbb{N}$ satisfying $\|\underline{\sigma}\| < S_0^{(j)}$ and $0 \le s < S_1^{(j)}$. We need to choose an ordering $(\underline{\sigma}_\nu, s_\nu)_{0 \le \nu < N_J}$ on the union $\mathcal{S}$ of these sets. Define

$$N_{-1} = 0, \quad N_0 = \binom{S_0 + n - 1}{n} S_1$$

and, for $1 \le j \le J$,

$$N_j = N_{j-1} + \binom{S_0^{(j)} + n - 1}{n} (S_1^{(j)} - S_1^{(j-1)}).$$

We choose an ordering so that

$$\{(\underline{\sigma}_\nu, s_\nu) \,;\, 0 \le \nu < N_0\} = \{(\underline{\sigma}, s) \in \mathbb{N}^n \times \mathbb{N} \,;\, \|\underline{\sigma}\| < S_0 \ \text{ and } \ 0 \le s < S_1\}$$

and, for $1 \le j \le J$,

$$\{(\underline{\sigma}_\nu, s_\nu)\,;\; N_{j-1} \le \nu < N_j\}$$
$$= \{(\underline{\sigma}, s) \in \mathbb{N}^n \times \mathbb{N}\,;\; \|\underline{\sigma}\| < S_0^{(j)} \text{ and } S_1^{(j-1)} \le s < S_1^{(j)}\}.$$

It remains to specify the order inside each interval $N_{j-1} \le \nu < N_j$ $(0 \le j \le J)$. There is no complete rigidity, but it suffices to say that we choose an order for which the resulting sequence is admissible (see (3.15)).

For each τ, $\underline{t}$, $\underline{\sigma}$ and s, consider the algebraic number $\gamma_{\tau\underline{t}}^{(\underline{\sigma}s)}$ given by (3.5) and build up the matrix

$$\mathtt{M} = \left(\gamma_{\tau\underline{t}}^{(\underline{\sigma}_\nu s_\nu)}\right)_{\substack{(\tau,\underline{t})\\ 0\le\nu<N_J}}$$

with L rows indexed by $(\tau, \underline{t})$ with $0 \le \tau < T_0$ and $0 \le t_i < T_i$ $(1 \le i \le n)$, and N_J columns indexed by ν with $0 \le \nu < N_J$.

We first deduce from Lemma 3.3 that $\mathtt{M}$ has rank L. Indeed otherwise there exists a nonzero polynomial $P \in \mathbb{C}[Y_0, \ldots, Y_n]$ for which

$$\partial^{\underline{\sigma}_\nu} P\big(s_\nu, \alpha_1^{s_\nu}, \ldots, \alpha_n^{s_\nu}\big) = 0 \quad \text{for} \quad 0 \le \nu < N_J,$$

which means

$$\partial^{\underline{\sigma}} P\big(s, \alpha_1^{s}, \ldots, \alpha_n^{s}\big) = 0 \quad \text{for} \quad (\underline{\sigma}, s) \in \mathcal{S}.$$

In particular the function $G(z) = P(z, \alpha_1^z, \ldots, \alpha_n^z)$ satisfies

$$G^{(k)}(s) = 0 \quad \text{for} \quad 0 \le k < S_0^{(J)} \quad \text{and} \quad 0 \le s < S_1^{(J)}.$$

In order to apply Lemma 3.3, we assume

$$\boxed{S_0^{(J)} S_1^{(J)} > 2L + c_8 T S_1^{(J)}.}$$

Since $\mathtt{M}$ has maximal rank L, one can select L columns which produce a nonzero determinant Δ. We select them minimal as in the proof of Section 2.2 and we write

$$\Delta = \det\left(\gamma_{\tau\underline{t}}^{(\underline{\sigma}_{\nu_\mu} s_{\nu_\mu})}\right)_{\substack{(\tau,\underline{t})\\ 1\le\mu\le L}}.$$

We want to derive from Corollary 3.6 an upper bound for $|\Delta|$. Set

$$\underline{v} = (1, \log\alpha_1, \ldots, \log\alpha_{n-1}).$$

We need first to estimate the weights (3.14) related to our sequence

$$(\underline{\sigma}_\nu, s_\nu)_{0\le\nu<N_J}.$$

Let ν be an index with $N_0 \le \nu < N_J$; define j in the range $1 \le j \le J$ by $N_{j-1} \le \nu < N_j$. Thanks to the construction of the sequence $(\underline{\sigma}_\nu, s_\nu)_{0\le\nu<N_J}$, we have

$$w_\nu(s) = S_0^{(j-1)} - 1 \quad \text{for} \quad 0 \le s < S_1^{(j-1)}.$$

Since $\|\underline{\sigma}_\nu\| < S_0^{(j)}$, it follows that the number

$$M_\nu = \sum_{\zeta \in B_1(0, R/|\underline{v}|)} \max\{0, w_\nu(\zeta) - \|\underline{\sigma}_\nu\|\}$$

satisfies

$$M_\nu \ge \left(S_0^{(j-1)} - S_0^{(j)}\right) S_1^{(j-1)}. \tag{3.17}$$

For each μ with $1 \le \mu \le L$ we define $j_\mu \in \{0, \ldots, J\}$ by

$$N_{j_\mu - 1} \le \nu_\mu < N_{j_\mu}.$$

Since $N_{-1} = 0$, for $\nu_\mu < N_0$ we have $j_\mu = 0$. Hence for any $\mu = 1, \ldots, L$, we have

$$\|\underline{\sigma}_{\nu_\mu}\| < S_0^{(j_\mu)} \quad \text{and} \quad 0 \le s_{\nu_\mu} < S_1^{(j_\mu)}.$$

We want to use (3.17) for each ν_μ with μ in the range $L/2 < \mu \le L$, so we need to check $\nu_\mu \ge N_0$ for these μ. For this reason we require our parameters to satisfy

$$\boxed{2\binom{S_0 + n - 1}{n} S_1 \le L.} \tag{3.18}$$

Since

$$\Delta = \det\left(\mathcal{D}^{\underline{\sigma}_{\nu_\mu}} \varphi_\lambda(\underline{v} s_{\nu_\mu})\right)_{1 \le \lambda, \mu \le L},$$

we may apply Corollary 3.6 with

$$r_\mu = c_9 S_1^{(j_\mu)}, \quad \varrho_\mu = 1, \quad R_\mu = 3 e r_\mu, \quad E_\mu = e.$$

We deduce

$$\begin{aligned} |\Delta| &\le L! \prod_{\mu=1}^{L} \left(\underline{\sigma}_{\nu_\mu}! e^{-M_{\nu_\mu}} R_\mu^{T_0} e^{c_{10} T R_\mu}\right) \\ &\le L! \exp\left\{\sum_{\mu=1}^{L} \left(-M_{\nu_\mu} + c_{11}(T_0 + S_0)\log\left(L S_1^{(j_\mu)}\right) + c_{12} T S_1^{(j_\mu)}\right)\right\} \end{aligned}$$

(recall that $S_0^{(j)} \le S_0$ for $0 \le j \le J$).

Since Δ is a nonzero number in the field $\mathbb{K}$, Liouville's Proposition 1.13 produces a lower bound for $|\Delta|$:

$$|\Delta| \ge L!^{-D+1} \exp\left\{\sum_{\mu=1}^{L} \left(-c_{13}(T_0 + S_0)\log\left(L S_1^{(j_\mu)}\right) - c_{14} T S_1^{(j_\mu)}\right)\right\}.$$

Define, for $1 \le \mu \le L$,

$$Q_\mu = c_{15}(T_0 + S_0)\log\bigl(LS_1^{(j_\mu)}\bigr) + c_{16}TS_1^{(j_\mu)}.$$

We assume our parameters are selected so that

$$\boxed{2Q_\mu < \left(S_0^{(j_\mu-1)} - S_0^{(j_\mu)}\right) S_1^{(j_\mu-1)}.}$$

Then we have

$$M_{\nu_\mu} \ge 0 \quad\text{for}\quad 1 \le \mu < \frac{L}{2} - 1,$$

$$\sum_{1\le\mu\le L/2} Q_\mu \le \sum_{L/2<\mu\le L} Q_\mu$$

and by (3.17)

$$M_{\nu_\mu} > 2Q_\mu \quad\text{for}\quad \frac{L}{2} - 1 \le \mu \le L.$$

Therefore

$$\sum_{\mu=1}^{L}(M_{\nu_\mu} - Q_\mu) > 0,$$

and the contradiction follows.

It remains to select our parameters. We take the same values as in Section 3.3, namely (3.4) and (3.11):

$$T_0 = 2[\mathbb{K}:\mathbb{Q}]N^{2n+1}, \quad T_1 = \cdots = T_n = N^{2n-1}, \quad S_0 = N^{2n+1}, \quad S_1 = N$$

and

$$J = 2n^2, \qquad S_0^{(j)} = [S_0/2^j], \qquad S_1^{(j)} = N^{j+1} \quad (1 \le j \le J).$$

4 Fourth Lecture. Introduction to Diophantine Approximation

4.1 On a Conjecture of Mahler

Consider the successive powers of e, namely

$$e,\ e^2,\ e^3,\ e^4,\ \dots,$$

and their distances to the nearest integer

$$\|e\|_{\mathbb{Z}},\ \|e^2\|_{\mathbb{Z}},\ \|e^3\|_{\mathbb{Z}},\ \|e^4\|_{\mathbb{Z}},\ \dots$$

It is an open problem to prove that these numbers are equidistributed[10] in the interval $(0, 1/2)$. We are interested in estimating $\Psi(B) = \min_{1 \le b \le B} \|e^b\|_{\mathbb{Z}}$ as $B \to \infty$ from below, but let us first say one word on upper bounds. Essentially nothing is known: it is not yet proved that $\Psi(B)$ tends to 0 when $B \to \infty$:

(?) *For any $\epsilon > 0$ there exists $b \in \mathbb{N}$ such that $\|e^b\|_{\mathbb{Z}} < \epsilon$,*

but it is expected that $B\Psi(B)$ tends to 0 when $B \to \infty$:

(?) *For any $\epsilon > 0$ there exists $b \in \mathbb{N}$ such that $\|e^b\|_{\mathbb{Z}} < \dfrac{\epsilon}{b}\cdot$*

We come back to lower bounds for $\Psi(B)$. In [17] p. 397, K. Mahler says:

- *... one can easily show that*

$$|\log a - b| < \frac{1}{a}$$

for an infinite increasing sequence of positive integers a and suitable integers b.

Indeed, this inequality holds for each pair (a, b) of positive integers for which $\log a < b < \log(a + 1)$.

For a given positive integer B, if the B numbers

$$\|e^b\|_{\mathbb{Z}} \quad (1 \le b \le B)$$

were evenly distributed in the interval $(0, 1/2)$, the smallest of them would not be less than a constant times $1/B$. This is likely to be too optimistic for a conjecture, but a more reasonable question is:

- *Is there an absolute constant $\kappa \ge 1$ such that*

$$|e^b - a| \ge \frac{1}{b^\kappa} \tag{4.1}$$

for any positive integers a and b with $b \ge 2$?

A straightforward computation shows that this inequality holds with $\kappa = 2.25$ for $2 \le b \le 100$.

If $|e^b - a|$ is small, then b is close to $\log a$, hence

$$e^b - a = a\left(e^{b - \log a} - 1\right)$$

is close to $a(b - \log a)$. One easily deduces that (4.1) is equivalent to:

---

[10] A result of Koksma ([10], Chap. VIII, Section 3, N° 12 Satz 16) states that for almost all real $\theta \ge 1$ (in the sense of Lebesgue measure) the sequence of fractional parts of θ^k is uniformly distributed modulo 1. On the other hand there there is no known example of a transcendental number which satisfies this property.

- *Is there an absolute constant $\kappa \geq 1$ such that*

$$|b - \log a| \geq \frac{1}{a(\log a)^{\kappa}} \tag{4.2}$$

for any positive integers a and b with $a \geq 3$?

A symmetric refinement of both (4.1) and (4.2) could be formulated by restricting to large values of a and b, say $a \geq a_0$ and $b \geq b_0$, and by introducing functions $x(\log x)(\log_2 x) \cdots (\log_{k-1} x)(\log_k x)^{\kappa}$, where $\log_2 = \log\log$ and $\log_k$ is the k-th iterated logarithm; but since (4.1) and (4.2) are not yet known, it seems more reasonable to consider weaker statements rather than stronger ones!

A weaker estimate than (4.1) and (4.2), which is still an open question, is usually attributed to Mahler (for instance in [34] Chap. 14, open problems and [35]) but is not explicit in [17] nor in [18].

Conjecture 4.1. Does there exist a positive absolute constant c such that, for any positive rational integers a and b with $a \geq 2$,

$$|e^b - a| \geq a^{-c} \ ?$$

The same argument as above shows that Conjecture 4.1 amounts to the existence of $c' > 1$ with

$$|b - \log a| \overset{?}{\geq} e^{-c'b}.$$

If we restrict to sufficiently large a and b, then c' is essentially $c + 1$.

The best known estimate in the direction of Conjecture 4.1 is due to Mahler himself [17]:

$$|e^b - a| \geq a^{-c \log\log a} \quad \text{and} \quad |b - \log a| \geq b^{-cb} \tag{4.3}$$

for $a \geq 3$ and $b \geq 2$, with an absolute constant $c > 0$. Assuming a is sufficiently large, K. Mahler in [17] gave a sharp explicit numerical value for c, namely $c = 40$ (for both estimates), which he refined in [18], getting $c = 33$. A further refinement is due to F. Wielonsky [36]: for sufficiently large a, Mahler's estimates (4.3) hold with $c = 20$.

Fel'dman proved several lower bounds for $|e^{\beta} - \alpha|$ and $|\beta - \log \alpha|$ when α and β are algebraic numbers. References are given in [21], where a refinement including most previously known results was established.

The following result is a slight refinement of the main estimate of [21] (apart from the fact that our constant c is not explicit).

Theorem 4.2. *There is an absolute constant $c > 0$ with the following property. Let α, β be two nonzero algebraic numbers and λ a logarithm of α. Define $\mathbb{K} = \mathbb{Q}(\alpha, \beta)$ and $D = [\mathbb{K} : \mathbb{Q}]$. Let A, B and E be positive real numbers satisfying $E \geq e$,*

$$\log A \geq \max\{\mathrm{h}(\alpha)\,,\, D^{-1}\log E\,,\, E|\lambda|/D\}$$

and[11]

$$\log B \geq \mathrm{h}(\beta) + \log(D\log A) + \log D + \frac{1}{D}\log E. \tag{4.4}$$

Then

$$|\beta - \lambda| \geq \exp\Big\{-cD^2(\log A)(\log B)\big(D\log D + \log E\big)(\log E)^{-2}\Big\}.$$

According to Theorem 5 of [21], if we replace the condition (4.4) on B by

$$\log B \geq \mathrm{h}(\beta) + \log_+ \log A + \log D + \log E,$$

then $c = 105\,500$ is an admissible value. It is not a big challenge to produce a smaller numerical value; for instance, taking $D = 1$ and $E = e$, one deduces from [21] that for any positive rational numbers a and b,

$$|e^b - a| \geq \exp\Big\{-1000\big(\mathrm{h}(b) + \log\log A + 12\big)\big(\log A + 12\big)\Big\},$$

where $\log A = \max\{1, \mathrm{h}(a)\}$. In particular we recover Mahler's results (4.3) on $|e^b - a|$ and $|b - \log a|$ as a consequence of Theorem 4.2.

On the other hand, any further improvement of Theorem 4.2 in terms of either A, B, D or E seems to require a new idea.

For the proof of Theorem 4.2, a refinement of Propositions 1.7 and 1.8 is needed. To begin with, assume for simplicity that $\alpha = a$ and $\beta = b$ are positive integers. As we shall see, the condition $f(e^b, b) \neq 0$, which occurs in both Propositions 1.7 and 1.8, does not suffice any more, but we need $f(a, b) \neq 0$. This is achieved by the following zero estimate due to Yu.V. Nesterenko.
Let b be a complex number. The derivation $\mathrm{D}_b = (\partial/\partial X) + bY(\partial/\partial Y)$, on the ring of polynomials in two variables X and Y, has the following remarkable property: for a polynomial $P \in \mathbb{C}[X, Y]$, the derivative of the complex function $P(z, e^{bz})$ is the exponential polynomial $(\mathrm{D}_b P)(z, e^{bz})$.

Here is Lemma 2 of [21] (see also [34] Prop. 2.14).

Lemma 4.3. *Assume $b \neq 0$. Let T_0, T_1, S_0 and S_1 be positive integers satisfying*

$$S_0 S_1 > (T_0 + S_1 - 1)T_1.$$

Let $(\xi_0, \eta_0), \ldots, (\xi_{S_1-1}, \eta_{S_1-1})$ be elements in $\mathbb{C} \times \mathbb{C}^\times$ with $\xi_0, \ldots, \xi_{S_1-1}$ pairwise distinct. Then there is no nonzero polynomial $P \in \mathbb{C}[X, Y]$, of degree $< T_0$ in X and of degree $< T_1$ in Y, which satisfies

$$\mathrm{D}_b^\sigma P(\xi_s, \eta_s) = 0 \quad \textit{for} \quad 0 \leq s < S_1 \quad \textit{and} \quad 0 \leq \sigma < S_0.$$

[11] It does not make any difference if one omits the summand $\log(D\log A)$ in (4.4), provided that one replaces the factor $(\log B)$ in the conclusion by $(\log B + \log(D\log A))$.

Remark. An obvious necessary condition for the nonexistence of P is

$$S_0S_1 \geq T_0T_1.$$

Indeed if $S_0S_1 < T_0T_1$, then the homogeneous linear system of equations given by the conditions $\mathrm{D}_b^\sigma P(\xi_\mu, \eta_\mu) = 0$ has a nontrivial solution, because the number of unknowns (which are the coefficients of P) is larger than the number of equations. On the other hand, if $\eta_0 = \cdots = \eta_{S_1-1}$ and $S_0 < T_1$, then

$$(Y - \eta_0)^{S_0}$$

is a nontrivial solution.

Hence one cannot replace $T_0 + S_1 - 1$ in the hypothesis of Lemma 4.3 by a smaller number than $\max\{T_0\,,\, S_1\}$.

Sketch of proof of Theorem 4.2 . We first consider the special case where $\alpha = a$ and $\beta = b$ are positive integers with, say, $b \geq 3$. Our goal is to check

$$|b - \log a| \geq \exp\{-cb(\log b)\} \tag{4.5}$$

for some absolute constant $c > 0$ (which we shall not compute explicitly). Recall that when b and $\log a$ are close together, then $b(\log b)$ is close to $(\log a)(\log b)$.

Here the parameter E of Theorem 4.2 is not helpful and we shall just take $E = e$, but we shall keep it for a while to show why we cannot get a better result by taking a large quotient R/r in Schwarz' Lemma.

If we were following the proof of Proposition 1.8 in Section 1.6, we would get $b(\log b)^2$ instead of $b \log b$ in (4.5). Let us explain why.

As we shall see in (4.10), the conclusion will have the shape

$$|b - \log a| \geq E^{-L}.$$

Set $U = V = W = (1/36)L \log E$, so that (1.6) holds (for simplicity we forget the upper bound for the length of f). In order to check (1.5) we need the number $L \log E$ to be larger than bET_1S_1 and than $T_0 \log(bE)$. Since $L = T_0T_1$, we get the conditions

$$T_0 \log E > bES_1 \quad \text{and} \quad T_1 \log E > \log b.$$

Writing now $S_0 = T_0T_1/S_1$, we obtain

$$S_0 > b(\log b)E(\log E)^{-2}.$$

In particular this explains our choice $E = e$.

The conclusion of Proposition 1.8 involves $S_0!$, so we shall require $L > \log(S_0!)$, hence $S_1 > \log S_0$ (we discard lower order terms), and therefore $L = S_0S_1$ should satisfy $L > b(\log b)^2$.

From this point of view the choice of parameters in Section 1.5 yields the smallest possible value for L in terms of b.

The main term which is responsible for $(\log b)^2$ is $S_0!$. If we could forget it, we would get only a constant times $b(\log b)(\log\log b)$ for L by taking

$$T_0 = N^2 b, \quad T_1 = N^2[\log b], \quad S_0 = N^3 b[\log b], \quad \text{and} \quad S_1 = N$$

with a suitable (sufficiently large) integer N. The quantity $b(\log b)(\log\log b)$ arises from the term $S_0 \log T_1$ which appears in the estimate for the length of f in Proposition 1.8.

We perform a very simple change of variables[12] which will allow us to take a smaller value for S_0. In place of evaluating the functions $z^\tau e^{tz}$ at the points $z = sb$, we consider rather $z^\tau e^{btz}$ at the points $z = s$. This means that we replace

$$\left(\frac{d}{dz}\right)^\sigma (z^\tau e^{tz})(sb) = \sum_{\kappa=0}^{\min\{\tau,\sigma\}} \frac{\sigma!\tau!}{\kappa!(\sigma-\kappa)!(\tau-\kappa)!}(sb)^{\tau-\kappa} t^{\sigma-\kappa} e^{tsb}$$

by

$$\left(\frac{d}{dz}\right)^\sigma (z^\tau e^{btz})(s) = \sum_{\kappa=0}^{\min\{\tau,\sigma\}} \frac{\sigma!\tau!}{\kappa!(\sigma-\kappa)!(\tau-\kappa)!} s^{\tau-\kappa}(tb)^{\sigma-\kappa} e^{tsb}.$$

While the former number is a positive real number bounded from above by

$$(bS_1)^\tau T_1^\sigma \min\left\{\left(1+\frac{\sigma}{bT_1S_1}\right)^\tau ; \left(1+\frac{\tau}{bT_1S_1}\right)^\sigma\right\} e^{bT_1S_1},$$

(see (1.3)), the latter is at most

$$S_1^\tau (bT_1)^\sigma \min\left\{\left(1+\frac{\sigma}{bT_1S_1}\right)^\tau ; \left(1+\frac{\tau}{bT_1S_1}\right)^\sigma\right\} e^{bT_1S_1}.$$

Thanks to this change of variables a variant of Proposition 1.8 can be deduced, where the upper bound for the length of f is

$$Le^W S_1^{T_0}(T_0+T_1)^{S_0},$$

the degree of f in Z_2 is $< S_0$ and (1.5) is replaced by

$$U \geq \log L + T_0 \log(ES_1) + |b|_+ ET_1S_1.$$

A suitable choice for the parameters is then

---

[12] Compare with the "duality" of [34], Section 13.7.

$$T_0 = N^2 b[\log b], \qquad T_1 = N^2[\log\log b],$$
$$S_0 = N^3 b[\log\log b], \qquad S_1 = N[\log b],$$

which may be used to prove (4.5) with

$$\exp\{-cb(\log b)\} \text{ replaced by } \exp\{-cb(\log b)(\log\log b)\}$$

(for $b \geq 16$, say).

In order to remove the extra factor $\log\log b$ and to reach the desired estimate (4.5) involving only $\exp\{-cb(\log b)\}$, we need to get rid of the term $S_1^{T_0}$ in the upper bound for $\mathrm{L}(f)$ in Proposition 1.8, so that one could take

$$T_0 = N^2 b[\log b], \quad T_1 = N^2, \quad S_0 = N^3 b, \text{ and } S_1 = N[\log b].$$

This is achieved by means of the so-called Fel'dman's polynomials. We use here a variant due to E.M. Matveev (see [34] Lemma 9.8): we replace z^τ by the polynomials[13] $\Delta(z;\tau,T_0^\sharp)$ which we now define.

4.2 Fel'dman's Polynomials

For r a nonnegative integer and $z \in \mathbb{C}$, consider the binomial (or Fel'dman's) polynomial

$$\Delta(z;r) = \frac{z(z+1)\cdots(z+r-1)}{r!}$$

if $r > 0$ and $\Delta(z;0) = 1$.

Let $\tau \geq 0$ and $T_0^\sharp > 0$ be two integers. Following Matveev [20], I, Section 7, we define a polynomial $\Delta(z;\tau,T_0^\sharp) \in \mathbb{Q}[z]$ of degree τ by

$$\Delta(z;\tau,T_0^\sharp) = \left(\Delta(z;T_0^\sharp)\right)^q \cdot \Delta(z;r),$$

where q and r are the quotient and remainder of the division of τ by $T_0^\sharp$:

$$\tau = T_0^\sharp q + r, \quad 0 \leq r < T_0^\sharp.$$

For $\sigma \geq 0$, define

$$\Delta(z;\tau,T_0^\sharp,\sigma) = \left(\frac{d}{dz}\right)^\sigma \Delta\left(z;\tau,T_0^\sharp\right).$$

For any positive integer n, denote by $\nu(n)$ the least common multiple of $1, 2, \ldots, n$. The estimate

$$\nu(n) \leq e^{107n/103}$$

---

[13] The polynomial denoted here by $\Delta(z;\tau,T_0^\sharp)$ is denoted $\delta_{T_0^\sharp}(z;\tau)$ in [34], Section 9.2.1, while our notation $\Delta(z;\tau,T_0^\sharp,\sigma)$ corresponds to $\delta_{T_0^\sharp}(z;\tau,\sigma)$ in [34], Section 9.2.1. Here we follow Matveev's notation in [20], I, Section 7.

can be deduced from the prime number Theorem (see for instance [34], Section 9.2.1).

The next result is due to Matveev [20], I, Lemma 7.1 (see also [34] Lemma 9.8).

Lemma 4.4. *Let $T_0^\sharp > 0$, $\tau > 0$ and $\sigma \geq 0$ be rational integers. For any integer κ in the interval $0 \leq \kappa \leq \sigma$ and any rational integer $m \in \mathbb{N}$, the number*

$$\nu(T_0^\sharp)^\sigma \cdot \frac{1}{\kappa!} \Delta\left(m; \tau, T_0^\sharp, \kappa\right)$$

is a nonnegative rational integer. Moreover, for any complex number z, we have

$$\sum_{\kappa=0}^{\sigma} \binom{\sigma}{\kappa} \left|\Delta(z; \tau, T_0^\sharp, \kappa)\right| \leq \sigma! e^{\tau + T_0^\sharp} \left(\frac{|z|}{T_0^\sharp} + 1\right)^\tau.$$

4.3 Output of the Transcendence Argument

Here is the basic estimate which follows from the transcendence proof (by means of the interpolation determinant method) and will enable us to deduce Theorem 4.2. It is worth to notice that Theorem 4.5 does not involve any arithmetic assumption (the proof does not use Liouville's inequality).

Theorem 4.5. *Let b be a nonzero complex number. Let T_0, $T_0^\sharp$, T_1, S_0, S_1, K be positive integers and E a real number with $E \geq e$. Set $L = T_0 T_1$ and assume $0 \leq K < L$ and*

$$S_0 S_1 > (T_0 + S_1 - 1) T_1.$$

Then there exists a set $\{f_1, \ldots, f_M\}$ of polynomials in the ring $\mathbb{Z}[Z_1, Z_2]$, each of degree $< L T_1 S_1$ in Z_1 and degree $< L S_0$ in Z_2, of length at most

$$L! \nu(T_0^\sharp)^{L S_0} (S_0!)^L e^{(T_0^\sharp + T_0) L} \left(\frac{S_1}{T_0^\sharp} + 1\right)^{T_0 L} T_1^{S_0 L},$$

such that the polynomials $\{f_1(Z_1, b), \ldots, f_M(Z_1, b)\}$ in $\mathbb{C}[Z_1]$ have no common zero in $\mathbb{C}^\times$ and such that

$$\frac{1}{L} \max_{\substack{\underline{k} \in \mathbb{N}^2 \\ \|\underline{k}\| < K}} \log \left| \frac{1}{\underline{k}!} \mathcal{D}^{\underline{k}} f_i(e^b, b) \right| \leq -\frac{1}{2L}(L - K)(L - K - 1) \log E + \log(8L)$$
$$+ S_0 \log\left(\nu(T_0^\sharp) S_0 E\right) + T_0^\sharp + T_0 + T_0 \log\left(\frac{S_1 E}{T_0^\sharp} + 1\right)$$
$$+ S_0 \log(2 T_1 |b|_+) + T_1 S_1 E |b|$$

for $1 \leq i \leq M$.

In the conclusion $\mathcal{D}^{\underline{k}}$ stands for $(\partial/\partial Z_1)^{k_1}(\partial/\partial Z_2)^{k_2}$.

For the proof of Theorem 4.5 we shall need the following refinement of Lemma 1.5:

Lemma 4.6. *Let L and L' be positive integers with $L' \le L$ and $\varphi_1, \dots, \varphi_{L'}$ entire functions in $\mathbb{C}$. Let also $\zeta_1, \dots, \zeta_L$ be complex numbers and $\sigma_1, \dots, \sigma_L$ nonnegative integers. Let S_0 satisfy $S_0 \ge \max_{1\le\mu\le L} \sigma_\mu$. Furthermore, for $L'+1 \le \lambda \le L$ and $1 \le \mu \le L$ let $\delta_{\lambda\mu}$ be a complex number. For $1 \le \lambda \le L'$ and $1 \le \mu \le L$ we define*

$$\delta_{\lambda\mu} = \varphi_\lambda^{(\sigma_\mu)}(\zeta_\mu).$$

Finally, let $E > 1$ and $Q_1, \dots, Q_L$ be positive real numbers satisfying

$$Q_\lambda \ge \log \sup_{|z|=E} \max_{1\le\mu\le L} |\varphi_\lambda^{(\sigma_\mu)}(z\zeta_\mu)| \quad (1 \le \lambda \le L'),$$

$$Q_\lambda \ge \log \max_{1\le\mu\le L} |\delta_{\lambda,\mu}| \quad (L'+1 \le \lambda \le L).$$

We consider the determinant

$$\Delta = \det\Bigl(\delta_{\lambda\mu}\Bigr)_{1\le\lambda,\mu\le L}.$$

Then we have

$$\log|\Delta| \le -\frac{1}{2}L'(L'-1)\log E + L'S_0 \log E + \log(L!) + Q_1 + \cdots + Q_L.$$

Proof. Lemma 4.6 is essentially the case $n = 1$ of Lemma 7.5 in [34] but it includes derivatives like in Proposition 9.13 of [34].

For $1 \le \mu \le L$, we define functions $d_{1\mu}(z), \dots, d_{L\mu}(z)$ by

$$d_{\lambda\mu}(z) = \begin{cases} \varphi_\lambda^{(\sigma_\mu)}(\zeta_\mu z) & \text{for } 1 \le \lambda \le L', \\ \delta_{\lambda\mu} & \text{for } L' < \lambda \le L. \end{cases}$$

From Lemma 9.2 of [34] we deduce that the function

$$D(z) = \det\Bigl(d_{\lambda\mu}(z)\Bigr)_{1\le\lambda,\mu\le L}$$

has a zero at the origin of multiplicity $\ge (1/2)L'(L'-1) - L'S_0$. As $D(1) = \Delta$, we conclude the proof of Lemma 4.6 by using Schwarz' Lemma 1.3. □

Proof of Theorem 4.5. For each $a \in \mathbb{C}^\times$ we shall construct a polynomial $f \in \mathbb{Z}[Z_1, Z_2]$ (depending on a) satisfying the required conditions and such that

$$f(a, b) \ne 0.$$

Since the degrees and lengths of these polynomials will be bounded (independently of a), this will mean that we end up with a finite set of polynomials

$\{f_1, \ldots, f_M\}$ in $\mathbb{Z}[Z_1, Z_2]$, such that the only possible common zero in $\mathbb{C}$ of the polynomials $f_i(Z_1, b)$ $(i = 1, \ldots, M)$ is $Z_1 = 0$.

Define, for $0 \le \tau < T_0$ and $0 \le t < T_1$,

$$\varphi_{\tau t}(z) = \Delta(z; \tau, T_0^\sharp) e^{tbz}.$$

For $\sigma = 0, 1, \ldots, S_0 - 1$, we have

$$\varphi_{\tau t}^{(\sigma)}(z) = \sum_{\kappa=0}^{\sigma} \binom{\sigma}{\kappa} \Delta\left(z; \tau, T_0^\sharp, \kappa\right)(tb)^{\sigma-\kappa} e^{tbz}.$$

Define $\widetilde{f}_{\tau t}^{(\sigma s)} \in \mathbb{Q}[Z_1, Z_2]$ by

$$\widetilde{f}_{\tau t}^{(\sigma s)}(Z_1, Z_2) = \sum_{\kappa=0}^{\sigma} \binom{\sigma}{\kappa} \Delta\left(s; \tau, T_0^\sharp, \kappa\right)(tZ_2)^{\sigma-\kappa} Z_1^{ts},$$

so that[14]

$$\widetilde{f}_{\tau t}^{(\sigma s)}(e^b, b) = \varphi_{\tau t}^{(\sigma)}(s) \tag{4.6}$$

and

$$\widetilde{f}_{\tau t}^{(\sigma s)}(a, b) = \mathrm{D}_b^{\sigma}\left(\Delta(X; \tau, T_0^\sharp) Y^t\right)(s, a^s).$$

This polynomial $\widetilde{f}_{\tau t}^{(\sigma s)}$ has rational coefficients. In order to get a polynomial with integer coefficients, we multiply it by a denominator: using Lemma 4.4, define $f_{\tau t}^{(\sigma s)} \in \mathbb{Z}[Z_1, Z_2]$ by

$$f_{\tau t}^{(\sigma s)}(Z_1, Z_2) = \nu(T_0^\sharp)^{\sigma} \widetilde{f}_{\tau t}^{(\sigma s)}(Z_1, Z_2).$$

In place of the exponent σ for $\nu(T_0^\sharp)$ we could as well put the exponent $\min\{\sigma, \tau\}$, because $\Delta(z; \tau, T_0^\sharp, \kappa) = 0$ for $\kappa > \tau$. Notice also that we use only the fact that

$$\nu(T_0^\sharp)^{\sigma} \Delta\left(s; \tau, T_0^\sharp, \kappa\right)$$

is an integer; indeed, according to Lemma 4.4, it is a multiple of $\kappa!$, but we do not use this fact[15].

The polynomial $f_{\tau t}^{(\sigma s)}$ has degree $\le ts$ in Z_1, degree $\le \sigma$ in Z_2 and length at most

$$\nu(T_0^\sharp)^{\sigma} \sum_{\kappa=0}^{\sigma} \binom{\sigma}{\kappa} \Delta\left(s; \tau, T_0^\sharp, \kappa\right) t^{\sigma-\kappa} \le \nu(T_0^\sharp)^{S_0} S_0! e^{T_0^\sharp + T_0} \left(\frac{S_1}{T_0^\sharp} + 1\right)^{T_0} T_1^{S_0}.$$

---

[14] The upper index (σs) in the notation $\widetilde{f}_{\tau t}^{(\sigma s)}$ is not a derivation, while $\varphi_{\tau t}^{(\sigma)} = (d/dz)^{\sigma} \varphi_{\tau t}$.

[15] Maybe one should, in order to remove the extra $\log \log A$ which arises in $\log B + \log \log A \ldots$

Consider the matrix

$$\mathtt{M} = \left(f_{\tau t}^{(\sigma s)}(a,b) \right)_{\substack{(\tau,t) \\ (\sigma,s)}}$$

with L rows indexed by (τ, t), $(0 \leq \tau < T_0,\ 0 \leq t < T_1)$ and $S_0 S_1$ columns indexed by (σ, s) with $0 \leq \sigma < S_0$, $0 \leq s < S_1$.

We claim that $\mathtt{M}$ has rank $L = T_0 T_1$. Indeed consider a linear dependence relation between the rows:

$$\sum_{\tau=0}^{T_0-1} \sum_{t=0}^{T_1-1} c_{\tau t} f_{\tau t}^{(\sigma s)}(a,b) = 0 \quad \text{for} \quad 0 \leq \sigma < S_0 \quad \text{and} \quad 0 \leq s < S_1.$$

Define

$$P(X,Y) = \sum_{\tau=0}^{T_0-1} \sum_{t=0}^{T_1-1} c_{\tau t} \, \Delta\left(X; \tau, T_0^\sharp\right) Y^t.$$

Then we have

$$\mathrm{D}_b^\sigma P(s, a^s) = \sum_{\tau=0}^{T_0-1} \sum_{t=0}^{T_1-1} c_{\tau t} \widetilde{f}_{\tau t}^{(\sigma s)}(a,b),$$

hence

$$\mathrm{D}_b^\sigma P(s, a^s) = 0 \quad \text{for} \quad 0 \leq \sigma < S_0 \quad \text{and} \quad 0 \leq s < S_1.$$

According to Lemma 4.3 with

$$\xi_s = s, \quad \eta_s = a^s \quad (0 \leq s < S_1)$$

and thanks to the assumption

$$S_0 S_1 > (T_0 + S_1 - 1) T_1,$$

we deduce $c_{\tau t} = 0$ for $0 \leq \tau < T_0$ and $0 \leq t < T_1$.

Hence $\mathtt{M}$ has rank $L = T_0 T_1$. Select L columns of $\mathtt{M}$ with indices (σ_μ, s_μ) $(1 \leq \mu \leq L)$, which are linearly independent. This means that the polynomial

$$f(Z_1, Z_2) = \det\left(f_{\tau t}^{(\sigma_\mu s_\mu)}(Z_1, Z_2) \right)_{\substack{(\tau,t) \\ 1 \leq \mu \leq L}}$$

does not vanish at the point (a, b). The degree of f in Z_1 is $< L T_1 S_1$ and in Z_2 is $< L S_0$. We bound from above the length of f as follows: write

$$f_{\tau t}^{(\sigma_\mu s_\mu)}(Z_1, Z_2) = Z_1^{t s_\mu} \sum_{j=0}^{\sigma_\mu} q_{j \tau t \mu} Z_2^j$$

with

$$q_{j \tau t \mu} = \nu(T_0^\sharp)^{\sigma_\mu} \binom{\sigma_\mu}{j} \Delta\left(s_\mu; \tau, T_0^\sharp, \sigma_\mu - j\right) t^j.$$

By Lemma 4.4 we have

$$\begin{aligned}
\mathrm{L}(f_{\tau t}^{(\sigma_\mu s_\mu)}) &= \sum_{j=0}^{\sigma_\mu} |q_{j\tau t\mu}| \\
&\le \nu(T_0^\sharp)^{S_0} \sum_{j=0}^{\sigma_\mu} \binom{\sigma_\mu}{j} \Delta\left(s_\mu; \tau, T_0^\sharp, \sigma_\mu - j\right) t^j \\
&\le \nu(T_0^\sharp)^{S_0} S_0! e^{T_0^\sharp + T_0} \left(\frac{S_1}{T_0^\sharp} + 1\right)^{T_0} T_1^{S_0}.
\end{aligned}$$

Therefore

$$\mathrm{L}(f) \le L!\nu(T_0^\sharp)^{LS_0} S_0!^L e^{L(T_0^\sharp + T_0)} \left(\frac{S_1}{T_0^\sharp} + 1\right)^{LT_0} T_1^{LS_0}.$$

These crude estimates could be slightly improved, but this is irrelevant for our purpose, since we do not pay too much attention to the absolute constants.

We want to estimate from above the number

$$\max_{\substack{\underline{k} \in \mathbb{N}^2 \\ \|\underline{k}\| < K}} \log \left| \frac{1}{\underline{k}!} \mathcal{D}^{\underline{k}} f(e^b, b) \right|.$$

Fix $\underline{k} \in \mathbb{N}^2$ with $\|\underline{k}\| < K$. We have

$$\frac{1}{\underline{k}!} \mathcal{D}^{\underline{k}} f(e^b, b) = \sum_{\underline{\underline{\kappa}}} \Delta_{\underline{\underline{\kappa}}} \quad \text{with} \quad \Delta_{\underline{\underline{\kappa}}} = \det\left(\frac{1}{\underline{\kappa}_{\tau t}!} \mathcal{D}^{\underline{\kappa}_{\tau t}} f_{\tau t}^{(\sigma_\mu s_\mu)}(e^b, b) \right)_{\substack{(\tau, t) \\ 1 \le \mu \le L}},$$

where $\underline{\underline{\kappa}}$ ranges over the set of elements

$$(\underline{\kappa}_{\tau t})_{\substack{0 \le \tau < T_0 \\ 0 \le t < T_1}} \in (\mathbb{N}^2)^L \quad \text{with} \quad \sum_{\tau=0}^{T_0 - 1} \sum_{t=0}^{T_1 - 1} \underline{\kappa}_{\tau t} = \underline{k}. \tag{4.7}$$

For $\underline{k} = (k_1, k_2)$ the number of such $\underline{\underline{\kappa}}$ is

$$\binom{k_1 + L - 1}{L - 1} \binom{k_2 + L - 1}{L - 1} < 2^{K + 2L} < 2^{3L}.$$

Fix $\underline{\underline{\kappa}}$ satisfying (4.7) and define

$$\widetilde{\Delta}_{\underline{\underline{\kappa}}} = \det\left(\frac{1}{\underline{\kappa}_{\tau t}!} \mathcal{D}^{\underline{\kappa}_{\tau t}} \widetilde{f}_{\tau t}^{(\sigma_\mu s_\mu)}(e^b, b) \right)_{\substack{(\tau, t) \\ 1 \le \mu \le L}},$$

so that

$$\Delta_{\underline{\underline{\kappa}}} = \widetilde{\Delta}_{\underline{\underline{\kappa}}} \nu(T_0^\sharp)^{\sigma_1 + \cdots + \sigma_L}.$$

It remains to check the hypotheses of Lemma 4.6 with $\zeta_\mu = s_\mu$, $L' = L - K$ and

$$Q_{\tau t} = \log(S_0!) + T_0 + T_0^\sharp + T_0 \log\left(\frac{S_1 E}{T_0^\sharp} + 1\right) + S_0 \log(2T_1|b|_+) + T_1 S_1 E|b|$$

for $0 \leq \tau < T_0$ and $0 \leq t < T_1$.

From (4.7) it follows that there are at least $L - K$ indices (τ, t) with $0 \leq \tau < T_0$ and $0 \leq t < T_1$ for which $\underline{\kappa}_{\tau t} = (0,0)$; for these indices we use (4.6). We conclude by means of the estimates

$$\max_{1 \leq \mu \leq L} \sup_{|z|=E} \left|\varphi_{\tau t}^{(\sigma_\mu)}(s_\mu z)\right| \leq e^{Q_{\tau t}}$$

and, for $\underline{\kappa}_{\tau t} = (\kappa_1, \kappa_2)$,

$$\left|\frac{1}{\underline{\kappa}_{\tau t}!} \mathcal{D}^{\underline{\kappa}_{\tau t}} \widetilde{f}_{\tau t}^{(\sigma_\mu s_\mu)}(e^b, b)\right|$$
$$\leq \sum_{\kappa=0}^{\sigma_\mu} \binom{\sigma_\mu}{\kappa} \Delta\left(s_\mu; \tau, T_0^\sharp, \kappa\right) t^{\sigma_\mu - \kappa} \binom{\sigma_\mu - \kappa}{\kappa_2} |b|_+^{\sigma_\mu - \kappa - \kappa_2} \binom{t s_\mu}{\kappa_1} e^{|b|(t s_\mu - \kappa_1)},$$

which implies

$$\max_{1 \leq \mu \leq L} \left|\frac{1}{\underline{\kappa}_{\tau t}!} \mathcal{D}^{\underline{\kappa}_{\tau t}} \widetilde{f}_{\tau t}^{(\sigma_\mu s_\mu)}(e^b, b)\right|$$
$$\leq S_0! e^{T_0 + T_0^\sharp} \left(\frac{S_1}{T_0^\sharp} + 1\right)^{T_0} (2|b|_+ T_1)^{S_0} (2e^{|b|})^{T_1 S_1} \leq e^{Q_{\tau t}}.$$

□

4.4 From Polynomial Approximation to Algebraic Approximation

Let $\underline{\theta} \in \mathbb{C}^n$ and $\underline{\gamma} \in \overline{\mathbb{Q}}^n$. We want to estimate from below $|\underline{\theta} - \underline{\gamma}|$. Our strategy is as follows (see [34] Proposition 15.3 and Exercise 15.3). Assume there is a polynomial $f \in \mathbb{Z}[\underline{X}]$ such that $f(\underline{\gamma}) \neq 0$ while $|f(\underline{\theta})|$ is sufficiently small. Liouville's inequality (Proposition 1.13) yields a lower bound for $|f(\underline{\gamma})|$ (depending on the degrees and length of f, as well as the degrees and heights of the components of γ). Since $|f(\underline{\theta}) - f(\underline{\gamma})|$ can be estimated from above in terms of $|\underline{\theta} - \underline{\gamma}|$, we deduce the desired lower bound for $|\underline{\theta} - \underline{\gamma}|$.

A refined estimate can be obtained when not only $f(\underline{\theta})$, but also the values of several derivatives of f at $\underline{\theta}$, have a small absolute value. The required upper bound for $|f(\underline{\theta}) - f(\underline{\gamma})|$ in terms of $|\underline{\theta} - \underline{\gamma}|$ will be a consequence of the interpolation formula (Lemma 1.11).

Lemma 4.7. *Let ℓ, K, $\nu_1, \dots, \nu_\ell$, $N_1, \dots, N_\ell$ be positive integers. Define*

$$m = \nu_1 + \cdots + \nu_\ell.$$

Let $\mathbb{K}$ be a number field of degree $D = [\mathbb{K}\colon \mathbb{Q}]$. For $1 \le i \le \ell$, write

$$\underline{\theta}_i = (\theta_{i1}, \dots, \theta_{i\nu_i}) \in \mathbb{C}^{\nu_i} \quad \text{and} \quad \underline{\gamma}_i = (\gamma_{i1}, \dots, \gamma_{i\nu_i}) \in \mathbb{K}^{\nu_i}.$$

Set

$$\underline{\theta} = (\underline{\theta}_1, \dots, \underline{\theta}_\ell) \in \mathbb{C}^m \quad \text{and} \quad \underline{\gamma} = (\underline{\gamma}_1, \dots, \underline{\gamma}_\ell) \in \mathbb{K}^m$$

and assume

$$|\underline{\theta} - \underline{\gamma}| \le \frac{1}{2m}.$$

Let f be a polynomial with integer coefficients in m variables X_{ij} $(1 \le j \le \nu_i$, $1 \le i \le \ell)$, of total degree $\le N_i$ with respect to the variables $X_{i1}, \dots, X_{i\nu_i}$, $(1 \le i \le \ell)$, such that

$$f(\underline{\gamma}) \ne 0.$$

Define

$$\epsilon = \frac{1}{2}\mathrm{L}(f)^{-D} \exp\left\{ -D \sum_{i=1}^{\ell} N_i \mathrm{h}(1\colon \gamma_{i1}\colon \cdots \colon \gamma_{i\nu_i}) \right\}$$

and assume

$$\max_{\substack{\underline{k} \in \mathbb{N}^m \\ \|\underline{k}\| < K}} \frac{1}{\underline{k}!} |\mathcal{D}^{\underline{k}} f(\underline{\theta})| \le \frac{1}{2}\epsilon \mathrm{L}(f). \tag{4.8}$$

Then

$$|\underline{\theta} - \underline{\gamma}|^K \ge \epsilon (1 + \sqrt{K})^{-1} \prod_{i=1}^{\ell} (1 + |\underline{\theta}_i|)^{-N_i}.$$

Proof. We use the interpolation formula of Lemma 1.11 with $F(\underline{z}) = f(\underline{\theta} + \underline{z})$, $r = |\underline{\theta} - \underline{\gamma}|$, $R = 1$. We estimate $|F|_R$ from above as follows:

$$|F|_R \le \mathrm{L}(f) \prod_{i=1}^{\ell} (1 + |\underline{\theta}_i|)^{N_i}.$$

Since

$$\sum_{\substack{\underline{k} \in \mathbb{N}^m \\ \|\underline{k}\| < K}} r^{\|\underline{k}\|} \le \sum_{\underline{k} \in \mathbb{N}^m} r^{\|\underline{k}\|} = (1 - r)^{-m}$$

and since

$$\left(1 - \frac{1}{2m}\right)^{-m} \le 2,$$

by the assumption $r \le 1/(2m)$ we have:

$$\sum_{\substack{\underline{k}\in\mathbb{N}^m\\ \|\underline{k}\|<K}} r^{\|\underline{k}\|} \le 2.$$

Therefore

$$|f(\underline{\gamma})| \le |\underline{\theta}-\underline{\gamma}|^K(1+\sqrt{K})\mathrm{L}(f)\prod_{i=1}^{\ell}(1+|\underline{\theta}_i|)^{N_i} + 2\max_{\|\underline{k}\|<K}\frac{1}{\underline{k}!}|\mathcal{D}^{\underline{k}}f(\underline{\theta})|.$$

By Liouville's inequality (Proposition 1.13), we deduce from the assumption $f(\underline{\gamma})\neq 0$:

$$|f(\underline{\gamma})| \ge \mathrm{L}(f)^{1-D}\prod_{i=1}^{\ell} e^{-DN_i\mathrm{h}(1:\,\gamma_{i1}:\,\cdots:\,\gamma_{i\nu_i})}.$$

Hence

$$\begin{aligned}\mathrm{L}(f)^{1-D}&\prod_{i=1}^{\ell} e^{-DN_i\mathrm{h}(1:\,\gamma_{i1}:\,\cdots:\,\gamma_{i\nu_i})}\\ &\le |\underline{\theta}-\underline{\gamma}|^K(1+\sqrt{K})\mathrm{L}(f)\prod_{i=1}^{\ell}(1+|\underline{\theta}_i|)^{N_i} + 2\max_{\|\underline{k}\|<K}\frac{1}{\underline{k}!}|\mathcal{D}^{\underline{k}}f(\underline{\theta})|. \qquad (4.9)\end{aligned}$$

We now use the definition of ϵ and the assumption (4.8): the left hand side of (4.9) is $\ge 2\epsilon\mathrm{L}(f)$ and the second term on the right hand side is at most $\epsilon\mathrm{L}(f)$, hence

$$|\underline{\theta}-\underline{\gamma}|^K(1+\sqrt{K})\prod_{i=1}^{\ell}(1+|\underline{\theta}_i|)^{N_i} \ge \epsilon.$$

□

Remark. In the special case $m=1$ (hence $\ell=\nu_1=1$, $\theta\in\mathbb{C}$, and, say, $N=N_1$), here is a slightly different estimate, suggested by Exercise 15.3.a of [34]:

$$|\underline{\theta}-\underline{\gamma}|^K \ge \frac{\epsilon}{K\binom{N}{K}(1+|\theta|)^{N-K}}.$$

The proof is quite similar, but one replaces Lemma 1.11 by Taylor's expansion

$$f(\gamma)=\sum_{k=0}^{K-1}\frac{1}{k!}f^{(k)}(\theta)(\gamma-\theta)^k+\frac{1}{(K-1)!}\int_\theta^\gamma f^{(K)}(t)(\gamma-t)^{K-1}dt.$$

The conclusion follows from the estimates

$$\sum_{k=0}^{K-1}\frac{1}{k!}\left|f^{(k)}(\theta)\right||\gamma-\theta|^k \le \epsilon\mathrm{L}(f), \qquad |f(\gamma)|\ge 2\epsilon\mathrm{L}(f)$$

and

$$\frac{1}{(K-1)!}\sup\Big\{\big|f^{(K)}(t)\big| \; ; \; t=\theta(1-u)+\gamma u,\ 0\leq u\leq 1\Big\}$$
$$\leq K\binom{N}{K}\mathrm{L}(f)(1+|\theta|)^{N-K}.$$

4.5 Proof of Theorem 4.2

We start by selecting a sufficiently large absolute constant N. Next we assume the hypotheses of Theorem 4.2 are satisfied. From the assumptions we deduce $\log B \geq (1/D)+\log D \geq 1$, hence

$$\log\max\{B,N\} \leq (\log N)(\log B).$$

Replacing if necessary c by $c\log N$, we may assume $B\geq N$.

Next from Liouville's inequality (see [34] Exercise 3.7) we deduce

$$|\beta| \geq e^{-D\mathrm{h}(\beta)},$$

hence there is no loss of generality to assume $|\beta|\leq 2|\lambda|$ (and consequently $\lambda\neq 0$).

Define first T_1, S_1 and $T_0^\sharp$ by

$$T_1=\left[N^2\frac{D\log D+\log E}{\log E}\right], \quad S_1=\left[N\frac{D\log B}{\log E}\right], \quad T_0^\sharp=[\log B],$$

and then T_0 and S_0 by

$$T_0=\left[N\frac{D\log A}{\log E}\right]S_1, \quad S_0=2\left[N\frac{D\log A}{\log E}\right]T_1.$$

Plainly the number $L=T_0T_1$ is also equal to $(1/2)S_0S_1$. Up to terms of lower order[16], $L\log E$ is

$$N^4D^2(\log A)(\log B)\big(D\log D+\log E\big)(\log E)^{-2},$$

which is the main term in the final estimate of the conclusion of Theorem 4.2.

Denoting by $c_1,\ldots,c_{12}$ positive absolute constants, the following estimates are plain:

$$T_1S_1\leq\frac{c_1L\log E}{ND\log A}, \quad S_0\leq\frac{c_2L\log E}{ND\log B},$$
$$T_0^\sharp S_0\leq\frac{c_3}{ND}L\log E, \quad \log S_0\leq\frac{c_4}{ND}S_1\log E,$$
$$\log T_1\leq\frac{c_5}{ND}S_1\log E \ \text{ and } \ T_1S_1E|\beta|\leq\frac{c_6}{N}L\log E.$$

[16] Beware of integral parts!

We apply Theorem 4.5 with $b = \beta$ and $K = [L/2]$. Since the polynomials $f_i(Z_1, \beta)$ $(1 \le i \le n)$ have no common zero in $\mathbb{C}^\times$, one of $f_1, \dots, f_M$, say f, has $f(\alpha, \beta) \neq 0$. This polynomial has degree at most

$$N_1 = \frac{c_7 L^2 \log E}{ND \log A} \quad \text{and} \quad N_2 = \frac{c_8 L^2 \log E}{ND \log B}$$

in Z_1 and Z_2 respectively, and length at most

$$\exp\left(\frac{c_9 L^2 \log E}{ND}\right).$$

Moreover, if we set $\ell = m = 2$, $\nu_1 = \nu_2 = 1$, $(\theta_1, \theta_2) = (e^\beta, \beta)$ and $(\gamma_1, \gamma_2) = (\alpha, \beta)$, then the left hand side of (4.8) is at most

$$E^{-c_{10} L^2}.$$

Since

$$\epsilon \ge E^{-c_{11} L^2 / N}$$

we deduce from Lemma 4.7

$$|e^\beta - \alpha|^{L/2} \ge E^{-c_{12} L^2 / N}.$$

Therefore

$$|\beta - \log \alpha| \ge E^{-L} \tag{4.10}$$

and the conclusion of Theorem 4.2 follows. □

5 Fifth Lecture. Measures of Linear Independence of Logarithms of Algebraic Numbers

5.1 Introduction

The last two lectures are devoted to the question of measures of linear independence for logarithms of algebraic numbers. Here are a few references on this topic.

A simple proof for a homogeneous measure of linear independence of an arbitrary number of logarithms is given in Chap. 7 of [32], using an extension of Schneider's method. A refined estimate (relying on the same ideas) is given in Chap. 7 of [34]. The sharpest known estimate (in the general case, homogeneous or not) arising from this method is established in Chap. 9 of [34].

Baker's method is explained in [2] Chap. 3, [11] Chap. 8, 10 and 11, [6] Chap. 10, [28] Chap. 3, [7] Chap. 4 and [34], Section 10.1 (see also the introduction of [20], I, for a historical survey). While these proofs involve an auxiliary function, a measure of linear independence for logarithms of algebraic numbers is obtained in Section 10.2 of [34] by means of an interpolation determinant (without extrapolation).

A comparative discussion of these methods can be found in [34] (see Section 14.4), where a more general estimate is established (the so-called *quantitative version of the linear subgroup Theorem*).

The state of the art including references to the sharpest known measures of $\mathbb{Q}$-linear independence for logarithms of algebraic numbers is given in [34], Section 10.4.6. Now one should add to this picture Matveev's recent result in [20], II, (see Theorem 6.1; we refer also to Nesterenko's lectures).

Here is the main result for the rational case.

Theorem 5.1. *For each positive integer n there exists a positive constant $C(n)$ with the following property. Let $\alpha_1, \ldots, \alpha_n$ be nonzero algebraic numbers and $\log\alpha_1, \ldots, \log\alpha_n$ logarithms of $\alpha_1, \ldots, \alpha_n$ respectively. Assume that the numbers $\log\alpha_1, \ldots, \log\alpha_n$ are $\mathbb{Q}$-linearly independent. Let $b_1, \ldots, b_n$ be rational integers. Denote by D the degree of the number field $\mathbb{Q}(\alpha_1, \ldots, \alpha_n)$ over $\mathbb{Q}$. Further, let W, E, E^* be positive real numbers, each $\geq e$ and let $V_1, \ldots, V_n$ be positive real numbers. Assume*

$$V_j \geq \max\left\{\mathrm{h}(\alpha_j),\ \frac{E|\log\alpha_j|}{D},\ \frac{\log E}{D}\right\} \quad (1 \leq j \leq n),$$

$$\log E^* \geq \max\left\{\frac{1}{D}\log E,\ \log\left(\frac{D}{\log E}\right)\right\}$$

and $W \geq \log E^*$. Further, assume $b_n \neq 0$ and

$$e^W \geq \max_{1 \leq j \leq n-1}\left\{\frac{|b_n|}{V_j} + \frac{|b_j|}{V_n}\right\} \cdot \frac{\log E}{D}.$$

Then the number

$$\Lambda = b_1\log\alpha_1 + \cdots + b_n\log\alpha_n$$

has absolute value bounded from below by

$$|\Lambda| > \exp\{-C(n)D^{n+2}WV_1\cdots V_n(\log E^*)(\log E)^{-n-1}\}.$$

For several applications (especially for solving explicitly Diophantine equations) it is quite important to produce an estimate with a small numerical constant $C(n)$, even to the cost of relaxing the dependence in some parameters. A striking example arises with the case $n = 2$ where a very small value of $C(2)$ can be reached, provided that W is replaced by W^2 (see Theorem 5.10 below). We shall explain in an appendix to this Section 5 how such an estimate can be used also for the case $n \geq 2$ in order to get a lower bound for $|\Lambda|$ which involves a small constant – we need to replace W by $e^{\kappa W}$ for some $\kappa > 0$, but the point is that an admissible value for κ is < 1, hence the result is not trivial.

Apart from the explicit value for C, Theorem 5.1 includes essentially all known estimates on this number $|\Lambda|$.

Admissible numerical values for C are given in [34], Section 10.4.6. Assuming $E = e$, Matveev proved recently in [20], II, that $C(n)$ can be replaced by C^n where now C is a positive absolute constant (see Section 6).

Our goal is to explain some of the main ideas of the proof of Theorem 5.1 by means of Baker's method. In this lecture we discuss the classical approach following Baker's method with an auxiliary function and in the last one we shall show how to replace the auxiliary function by an interpolation determinant.

5.2 Baker's Method with an Auxiliary Function

Our main goal is to introduce the strategy of the proof, with an emphasis on the ideas and tools. We do not produce exact estimates, but we consider only the dominating terms which will explain the choice of parameters and at the same time provide some explanation for the limitation of the present method.

a) Main Conditions on the Parameters

Assume the hypotheses of Theorem 5.1 are satisfied: $\alpha_1, \dots, \alpha_n$ are nonzero algebraic numbers, $\log \alpha_1, \dots, \log \alpha_n$ are $\mathbb{Q}$-linearly independent, $b_1, \dots, b_n$ are rational integers, $b_n \neq 0$ and

$$\Lambda = b_1 \log \alpha_1 + \cdots + b_n \log \alpha_n.$$

As usual we need to introduce parameters: let $T_0, T_1, \dots, T_n, T_0^\sharp, S_0, S_1$ and L be positive integers with $L = T_0 \cdots T_n$. We also introduce another important parameter, U, which is a positive real number and will play the main role: assuming

$$0 < |\Lambda| \leq e^{-U},$$

we plan to derive a contradiction as soon as U is sufficiently large. Of course all the point is to be explicit on this condition that U is large enough.

The proof will have a lot in common with the transcendence proof of Theorem 3.1 in Section 3 (where the assumption was $\Lambda = 0$). Instead of setting $\beta_k = -b_k/b_n$ and $\beta_0 = 0$, it is slightly more convenient to change the definitions of $\partial_0, \dots, \partial_{n-1}$ and to set now

$$\partial_0 = \frac{\partial}{\partial Y_0}, \qquad \partial_k = b_n Y_k \frac{\partial}{\partial Y_k} - b_k Y_n \frac{\partial}{\partial Y_n} \qquad (1 \leq k \leq n-1).$$

The auxiliary polynomial will have the shape

$$P(Y_0, \dots, Y_n) = \sum_{(\tau, \underline{t})} c_{\tau, \underline{t}} \, \Delta\left(Y_0; \tau, T_0^\sharp\right) Y_1^{t_1} \cdots Y_n^{t_n}.$$

In the sum, $(\tau,\underline{t})$ ranges over the set of all elements in $\mathbb{N}\times\mathbb{N}^n$ with $0\le\tau<T_0$ and $0\le t_j<T_j$ $(1\le j\le n)$.

We are interested in the algebraic numbers

$$\partial^{\underline{\sigma}}P(s,\alpha_1^s,\dots,\alpha_n^s)$$

for $(\underline{\sigma},s)\in\mathbb{N}^n\times\mathbb{N}$ satisfying $\|\underline{\sigma}\|<S_0$ and $0\le s<S_1$. They can be explicitly written down as

$$\sum_{(\tau,\underline{t})} c_{\tau,\underline{t}}\gamma_{\tau\underline{t}}^{(\underline{\sigma}s)},$$

where

$$\gamma_{\tau\underline{t}}^{(\underline{\sigma}s)}=\partial^{\underline{\sigma}}\bigl(\Delta(Y_0;\tau,T_0^\sharp)Y_1^{t_1}\cdots Y_n^{t_n}\bigr)(s,\alpha_1^s,\dots,\alpha_n^s)$$

and

$$\partial^{\underline{\sigma}}\bigl(\Delta(Y_0;\tau,T_0^\sharp)Y_1^{t_1}\cdots Y_n^{t_n}\bigr)=\Delta(Y_0;\tau,T_0^\sharp,\sigma_0)\prod_{k=1}^{n-1}(b_nt_k-b_kt_n)^{\sigma_k}\cdot Y_1^{t_1}\cdots Y_n^{t_n}.$$

The auxiliary polynomial P is selected so that equations (3.3) hold.

Our first condition on the parameters will be written

$$\boxed{\frac{1}{n!}S_0^nS_1<L.}\tag{5.1}$$

This is a lousy way of requiring that the number of equations is less than the number of unknowns (namely the coefficients $c_{\tau\underline{t}}$). Here we do not pay attention to the exact value of the absolute constants which come into the picture, but we are only interested with the main constraints. For the same reason we do not look at the estimate for the coefficients $c_{\tau\underline{t}}$.

For $\underline{\sigma}\in\mathbb{N}^n$ define

$$G_{\underline{\sigma}}(z)=\partial^{\underline{\sigma}}P(z,\alpha_1^z,\dots,\alpha_n^z).$$

According to their definitions, the operators $\partial^{\underline{\sigma}}$ are related to derivatives along the hyperplane W of equation $b_1z_1+\cdots+b_nz_n=0$ in $\mathbb{C}^{n+1}$. The point $\underline{v}=(1,\log\alpha_1,\dots,\log\alpha_n)$ does not lie in this hyperplane (because $\Lambda\neq0$), but it is not far from it: to a certain extent $|\Lambda|$ measures the "distance" between $\underline{v}$ and W. If the point $\underline{v}$ were on the hyperplane W, the conditions (3.3) would imply that the one variable entire function $G_{\underline{\sigma}}$ has a zero of multiplicity $\ge S_0-\|\underline{\sigma}\|$ as each integer s with $0\le s<S_1$. When $|\Lambda|$ is small, the first $S_0-\|\underline{\sigma}\|$ derivatives of $G_{\underline{\sigma}}$ at these points have a small absolute value. More precisely define

$$g_{\underline{\sigma}}(z)=\partial^{\underline{\sigma}}P\bigl(z,\alpha_1^z,\dots,\alpha_{n-1}^z,(\alpha_1^{b_1}\cdots\alpha_{n-1}^{b_{n-1}})^{-z/b_n}\bigr).$$

Then

$$G_{\underline{\sigma}}(z) - g_{\underline{\sigma}}(z)$$

$$= \sum_{(\tau,\underline{t})} c_{\tau,\underline{t}} \,\Delta\left(z; \tau, T_0^\sharp, \sigma_0\right) \prod_{k=1}^{n-1} (b_n t_k - b_k t_n)^{\sigma_k} \cdot \prod_{i=1}^{n} \alpha_i^{t_i z} \cdot \left(1 - e^{-t_n z \Lambda / b_n}\right).$$

From the inequality $|e^w - 1| \le |w| e^{|w|}$ which is valid for any $w \in \mathbb{C}$ one deduces

$$\left|1 - e^{-t_n z \Lambda / b_n}\right| \le |t_n z \Lambda / b_n| e^{|t_n z \Lambda / b_n|},$$

hence $|G_{\underline{\sigma}}(z) - g_{\underline{\sigma}}(z)|$ is quite small for $|z|$ not too large:

$$|G_{\underline{\sigma}}(z) - g_{\underline{\sigma}}(z)| \le |\Lambda|^{1/2} = e^{-U/2}$$

for all relevant $\underline{\sigma}$ and z. In particular from the relations $G_{\underline{\sigma}}(s) = 0$ for $\|\underline{\sigma}\| < S_0$ and $0 \le s < S_1$ one deduces

$$\left|g_{\underline{\sigma}}^{(\ell)}(s)\right| \le e^{-U/3}$$

for $\|\underline{\sigma}\| < S_0 - S_0'$ and $0 \le s < S_1$. Using an approximate Schwarz' Lemma like Lemma 1.3 one deduces

$$|g_{\underline{\sigma}}(s)| \le e^{-2U'}$$

for the same $\underline{\sigma}$, s, with

$$U' = \min\{U, (S_0 - S_0')S_1 \log E\}.$$

Hence

$$|G_{\underline{\sigma}}(s)| \le e^{-U'}$$

for the same $\underline{\sigma}$ and s. Combining this estimate with Liouville's inequality (Proposition 1.13) we deduce that $G_{\underline{\sigma}}$ satisfies (3.6). This conclusion can be reached only if the parameters satisfy certain conditions. In the estimate arising from Liouville's inequality (Proposition 1.13), the dominating terms are the following ones:

$$e^{-DWS_0}(E^*)^{-DT_0} \quad \text{arising from} \quad \Delta\left(s; T_0^\sharp, \tau, \sigma_0\right)$$

and

$$\prod_{j=1}^{n} e^{-DT_j V_j S_1'} \quad \text{arising from} \quad \prod_{j=1}^{n} \alpha_j^{-t_j s}.$$

The exact conditions which motivate the definitions of $V_1, \ldots, V_n$, W and E^* are explained in [34] Chap. 9 and 10. In particular the difference between the "homogeneous rational case" considered here and the "general case" (where the rational coefficients b_i's are replaced by algebraic numbers β_i, and there may be also an extra β_0) occurs here: the factor

$$(b_n t_k - b_k t_n)^{\sigma_k}$$

like it stands would require stronger assumptions on W, namely

$$W \geq \max_{1\leq k\leq n} \log V_k \quad \text{and} \quad W \geq \max_{1\leq k\leq n} \log|b_k|,$$

so that $DS_0 \log T_i$ is bounded by U' for $1 \leq i \leq n$. In order to deal with our weaker condition on W, the idea is to replace this factor by

$$\Delta(b_n t_k - b_k t_n, \sigma_k).$$

See [34] Lemma 9.11.

Therefore it is reasonable to require

$$T_0 \leq \frac{U'}{D\log E^*}, \quad S_0 \leq \frac{U'}{DW} \quad \text{and} \quad T_j \leq \frac{U'}{nDS_1'V_j} \quad (1 \leq j \leq n).$$

Let us take

$$T_0 = \left[c_1\frac{U}{D\log E^*}\right], \quad S_0 = \left[c_2\frac{U}{DW}\right], \quad T_j = \left[c_3\frac{U}{nDS_1'V_j}\right] \quad (1 \leq j \leq n), \tag{5.2}$$

where c_1, c_2, c_3 (as well as $c_4, \ldots, c_{11}$ below) denote absolute positive constants, which we are not interested in.

From condition (5.1) we deduce

$$U > c_4 S_1 (S_1'/W)^n D V_1 \cdots V_n (\log E^*). \tag{5.3}$$

At the end of this first extrapolation step we derive (3.6), provided that

$$(S_0 - S_0')S_1 \log E \geq U.$$

As a first try, let us take $S_0' = [S_0/2]$ and then

$$S_1 = \left[c_5\frac{DW}{\log E}\right]. \tag{5.4}$$

We started with $\binom{S_0+n}{n}S_1$ equations and we end up with $\binom{S_0'+n}{n}S_1'$ new equations. It is conceivable that no real progress has been achieved unless the number of new equations exceeds the number of old equations, which means essentially that ${S_0'}^n S_1'$ should not be smaller than $S_0^n S_1$. Hence we need at least $S_1' \geq 2^n S_1$.

There are several possibilities now. The easiest one is to apply immediately a zero estimate. This is possible only if $({S_0'}^n/n!)S_1'$ is somewhat larger than $L = T_0 T_1 \cdots T_n$ (which is the opposite of (5.1) when S_0 and S_1 are replaced by S_0' and S_1'). So if we wish to conclude immediately by means of the zero estimate we need to require

$$\binom{S_0' + n - 1}{n} S_1' > c_{ZE} L, \tag{5.5}$$

where $c_{ZE} = (n+1)!$ is the loss arising from Proposition 3.2 (this c_{ZE} is a notation of [34], Section 13.5). This gives rise to the condition

$$\boxed{S_0'^{\,n} S_1' > n!^2 L.}$$

Comparing with (5.1) gives $S_1' > n! S_1$, hence (5.3) yields the condition

$$U > c_6^n (n!)^n D^{n+2} W V_1 \cdots V_n (\log E^*)(\log E)^{-n-1}. \tag{5.6}$$

The estimate (5.6) corresponds to the sharpest available result with respect to $V_1, \dots, V_n$, W, E and E^* (but not with respect to n), and the choice of parameters is given by (5.2) and (5.4).

b) End of the Proof

What we have shown so far is only that the above scheme of proof cannot reach a better estimate than $|\Lambda| \geq e^{-U}$ with U satisfying (5.6). It is a different issue to prove that one can indeed reach such an estimate by means of these arguments. However let us now say that there is mainly a single serious difficulty in doing so: (5.5) is only a necessary condition for applying the zero estimate (see (3.7)). More precisely the zero estimate enables one to conclude the proof, unless there is a linear dependence relation between the rational integers $b_1, \dots, b_n$ with "small" coefficients. Indeed, starting from $0 < |\Lambda| \leq e^{-U}$, the transcendence machinery produces a tuple $(t_1, \dots, t_n) \in \mathbb{Z}^n \setminus \{0\}$ such that $t_1 b_1 + \cdots + t_n b_n = 0$ together with a sharp upper bound for $\max_{1 \leq i \leq n} |t_i|$. This small linear dependence relation is the explanation for the fact that the system of equations (3.6) is somehow degenerate.

At this stage, the idea is to eliminate one b_i thanks to this relation and to work with a linear combination of $n-1$ logarithms instead of n. This is not the most efficient way; one may proceed by induction indeed, but it is better to repeat the transcendence argument and to take these relations into account. In [24] (and also [25] for a more general situation dealing with commutative algebraic groups), the induction is done as follows. The zero estimate produces an algebraic subgroup of $\mathbb{G}_a \times \mathbb{G}_m^n$ and the strategy is to start the proof from scratch (construction of the auxiliary polynomial) with such an obstructing subgroup. Thanks to this obstructing subgroup one has a better control on the rank of the system of linear equations to which we apply Thue-Siegel's Lemma. After the extrapolation, one produces a new set of relations (3.6) to which one applies again the zero estimate; the extremality of the initial obstructing subgroup enables one to conclude (by contradiction).

A different way of performing this induction is used by [5], but the underlying ideas are basically the same. This induction is somewhat technical but it is well under control now, and we shall not tell more about it. We assume implicitly that there is no "small" linear dependence relation between the b_i's, in which case the zero estimate shows that there is no nonzero polynomial $P \in \mathbb{C}[Y_0, \ldots, Y_n]$ for which a set of equalities (3.6) holds for $(\underline{\sigma}, s)$ ranging over a set with slightly more than $T_0 T_1 \cdots T_n$ elements (compare with condition (5.1)).

So our goal is to get more than $T_0 T_1 \cdots T_n$ equations (3.6) for P.

c) Baker's Method with an Interpolation Determinant and without Extrapolation

This Section 5.2 is devoted to Baker's method with an auxiliary function, but we make a small digression to point out that the arguments described in Section 5.2, a) and Section 5.2, b) involving an auxiliary function work out perfectly well for the interpolation determinant method. This is the topic of Chap. 10 of [34].

The basic scheme of proof is the following: Philippon's zero estimate enables us to produce a nonzero determinant ([34], Proposition 10.9) which is an algebraic number. Liouville's estimate provides a lower bound for its absolute value. The upper bound is obtained by analytic means ([34], Proposition 10.5). In this analytic argument, one takes into account the order of vanishing of an interpolation determinant at the origin only: extrapolation like in Section 3.5 is not necessary.

The role of the obstructing subgroup in connection with interpolation determinants is explained in [34], Section 10.2.3.

d) Dependence on n

We consider the dependence on n now.

The factor $(n!)^n$ in (5.6) is very large (it is comparable with the estimate which occurs in the interpolation determinant method of [34] Chap. 10 — see Section 14.4.3).

Even if one were to replace the condition $S_1' > n!S_1$ by the weaker $S_1' > 2^n S_1$, one would get a large "constant" in terms of n, involving 2^{n^2}.

In order to refine condition (5.6), an obvious solution would be to improve the zero estimate. However if one could replace $(n+1)!$ in (3.7) by, say, n^{c_7}, then one would still end up with $n^{c_8 n}$ in place of $c_6^n (n!)^n$ in the right hand side of (5.6)[17].

As soon as we wish to increase S_1'/S_1 by a factor $\lambda > 1$, we get a factor λ^n in U. Hence we should not take S_1'/S_1 larger than an absolute constant if we do not wish to introduce some n^n. We shall take $S_1' = 2S_1$. Now, since

[17] This is a significative difference with the "dual" method in Chap. 9 of [34], dealing with the algebraic group $\mathbb{G}_a^n \times \mathbb{G}_m$, where any improvement of the constant in the zero estimate immediately applies to the final estimate of Theorem 5.1.

we want to end up with at least as many equations as we started with, we need that ${S'_0}^n S'_1$ is not less than $S_0^n S_1$. It is natural to require that S'_0/S_0 is comparable with $2^{-1/n}$, which is not far from $1-(c_9/n)$. Let us take

$$S'_0 = S_0\left(1-\frac{1}{2n}\right).$$

We need to replace (5.4) by

$$S_1 = \left[c_{10}n\frac{DW}{\log E}\right].$$

From (5.3) we obtain the condition

$$U > c_{11}^n n^n D^{n+2} W V_1 \cdots V_n (\log E^*)(\log E)^{-n-1}. \tag{5.7}$$

We shall see later (Section 5.2, h)) how to remove the coefficient n^n in the right hand side of (5.7), but let us continue.

It seems we have not earned much: we started with roughly $(S_0^n/n!)S_1$ equations and we got $({S'_0}^n/n!)S'_1$ new ones, which is about the same. The only improvement concerns the term $S_0S_1\log E$, that we can replace at the end of this extrapolation by $S'_0S'_1\log E$, which is essentially twice as large.

This procedure may be repeated: we do it n times (this seems to be an optimal choice) with

$$S_0^{(j)} = S_0\left(1-\frac{j}{2n}\right) \quad\text{and}\quad S_1^{(j)} = 2^j S_1.$$

There is a small cost: at the end we shall get $|\Lambda| \geq e^{-2^n U}$ in place of $|\Lambda| \geq e^{-U}$. If one wishes to keep the conclusion $|\Lambda| \geq e^{-U}$ then one should replace everywhere else U by $U/2^n$. Indeed this is just the choice of parameters in [31].

At the end of the n extrapolation steps one has $S_0^{(n)} = S_0/2$ and $S_1^{(n)} = 2^n S_1$. The number of equations is not large enough to apply immediately the zero estimate 3.2.

Since we have been unable to increase the number of equations enough, we shall follow another strategy which originates in works by Baker and Stark and relies on arguments arising from Kummer's theory. The idea is to introduce division points and prove

$$\partial^{\underline{\sigma}} P(s/p, \alpha_1^{s/p}, \ldots, \alpha_n^{s/p}) = 0 \tag{5.8}$$

for various $\underline{\sigma} \in \mathbb{N}^n$, $s \in \mathbb{Z}$ and p. At an early stage of the theory, p was selected as a sufficiently large prime number. Later, another argument, which we explain in Section 5.2, f), enabled us to work with a smaller value for p, until it was realized that $p = 2$ also works!

We refer to [11] for several proofs involving (5.8):

- In Chap. 8, Section 2 and Section 6, for many values (s,p) with $\underline{\sigma} = \underline{0}$.
- In Chap. 10, Section 2 and Section 5, for a single sufficiently large prime number p and several values $(\underline{\sigma}, s)$.
- In Chap. 11, Section 2 and Section 4, (5.8) is applied with $p = 2$ only.

In recent works involving Kummer's theory only $p = 2$ is used; however it may be instructive to recall briefly what was done earlier with a large prime p.

e) Kummer's Theory

In this section we denote by n a positive integer, $\mathbb{K}$ a number field and $\alpha_1, \ldots, \alpha_n$ nonzero elements of $\mathbb{K}$.

We first quote Lemma 3 of [4].

Lemma 5.2. *Let p be a prime. For $1 \le j \le n$ denote by $\alpha_j^{1/p}$ any p-th root of α_j. For $1 \le r \le n$ set*

$$\mathbb{K}_r = \mathbb{K}\bigl(\alpha_1^{1/p}, \ldots, \alpha_r^{1/p}\bigr).$$

Assume

$$\bigl[\mathbb{K}_n : \mathbb{K}_{n-1}\bigr] < p.$$

Then there exist an element $\gamma \in \mathbb{K}^\times$ and integers $j_1, \ldots, j_{n-1}$ satisfying $0 \le j_\ell < p \quad (1 \le \ell \le n-1)$, such that

$$\alpha_n = \alpha_1^{j_1} \cdots \alpha_{n-1}^{j_{n-1}} \gamma^p. \tag{5.9}$$

Remark. Lemma 5.2 is proved by induction in [4]. When $\mathbb{K}$ contains the p-th roots of unity it can also be proved by using arguments from Kummer's theory as follows. Define

$$G_0 = \bigl\{x^p \; ; \; x \in \mathbb{K}^\times\bigr\}$$

and, for $1 \le r \le n$ let G_r denote the multiplicative subgroup of $\mathbb{K}^\times$ spanned by $G_0, \alpha_1, \ldots, \alpha_r$. According to [12] Chap. 6, Section 8 Th. 8.1, the field $\mathbb{K}_r$ is an abelian extension of $\mathbb{K}$ and the degree of this extension is the index of G_0 in G_r. Assume p is such that $\bigl[\mathbb{K}_n : \mathbb{K}_{n-1}\bigr] < p$. Then $(G_n : G_{n-1}) < p$, hence there is a relation

$$\alpha_n^{a_n} = \alpha_1^{a_1} \cdots \alpha_{n-1}^{a_{n-1}} \beta^p,$$

with some $\beta \in \mathbb{K}^\times$ and integers $a_1, \ldots, a_n$ satisfying $0 \le a_\ell < p \quad (1 \le \ell \le n-1)$ and $1 \le a_n < p$. Writing Bézout's relation $a_n u + pv = 1$ with rational integers u, v, one deduces the desired relation (5.9).

Lemma 5.3. *For $1 \le i \le n$ select a complex logarithm $\log \alpha_j$ of α_j. Assume that the numbers $\log \alpha_1, \ldots, \log \alpha_n$ are linearly independent over $\mathbb{Q}$ and denote by $\mathcal{G}$ the set of $\lambda \in \mathbb{C}$ such that*

$$e^\lambda \in \mathbb{K}^\times \quad \textit{and} \quad \lambda, \log\alpha_1, \dots, \log\alpha_n \textit{ are linearly dependent over } \mathbb{Q}.$$

Then $\mathcal{G}$ is a free $\mathbb{Z}$-module of rank n, containing $\mathbb{Z}\log\alpha_1 + \cdots + \mathbb{Z}\log\alpha_n$ as a subgroup of finite index.

Further, if $\log\theta_1, \dots, \log\theta_n$ is a basis of $\mathcal{G}$ over $\mathbb{Z}$, then for any prime p for which $\mathbb{K}$ contains a primitive p-th root of unity, we have

$$\left[\mathbb{K}(\theta_1^{1/p}, \dots, \theta_n^{1/p}) : \mathbb{K}\right] = p^n,$$

where $\theta_i^{1/p}$ stands for $\exp\{(1/p)\log\theta_i\}$ $(1 \le i \le n)$.

Remark. The assumption in the last sentence that $\mathbb{K}$ contains a primitive p-th root of unity cannot be omitted. Here is an example: take $\mathbb{K} = \mathbb{Q}$, $n = 1$, $\alpha_1 = 1$ and $\log\alpha_1 = 2i\pi$. In this case $\log\theta_1 = \pm 2i\pi$ and for each odd prime p we have

$$\left[\mathbb{K}(\theta_1^{1/p}) : \mathbb{K}\right] = p - 1.$$

Proof. It will be more convenient to work with

$$\Lambda = \left\{(r_1, \dots, r_n) \in \mathbb{Q}^n \; ; \; r_1\log\alpha_1 + \cdots + r_n\log\alpha_n \in \mathcal{G}\right\},$$

which is a subgroup of $\mathbb{R}^n$ isomorphic to $\mathcal{G}$ under the mapping

$$\begin{aligned} \Lambda &\longrightarrow \quad G \\ \underline{r} &\longmapsto r_1\log\alpha_1 + \cdots + r_n\log\alpha_n. \end{aligned}$$

We have $\mathbb{Z}^n \subset \Lambda \subset \mathbb{Q}^n$. We first want to check that Λ is discrete in $\mathbb{R}^n$. For $X > 0$ and for $\underline{r} \in \Lambda$ satisfying $|\underline{r}| \le X$ we have

$$\mathrm{h}(\alpha_1^{r_1} \cdots \alpha_n^{r_n}) \le X \sum_{i=1}^n \mathrm{h}(\alpha_i),$$

hence $\alpha_1^{r_1} \cdots \alpha_n^{r_n}$ belongs to a finite subset of $\mathbb{K}^\times$ as $\underline{r}$ ranges over the elements of $\Lambda \cap B_n(0, X)$. This means that the image of $\Lambda \cap B_n(0, X)$ under the mapping

$$\begin{aligned} \Lambda &\longrightarrow \quad K^\times \\ \underline{r} &\longmapsto \alpha_1^{r_1} \cdots \alpha_n^{r_n} \end{aligned}$$

is finite. Hence the proof that $\Lambda \cap B_n(0, X)$ is finite will be completed if we check that for each $\underline{r}^0 \in \mathbb{Q}^n \cap B_n(0, X)$, the set of $\underline{r} \in \mathbb{Q}^n \cap B_n(0, X)$ such that

$$\alpha_1^{r_1} \cdots \alpha_n^{r_n} = \alpha_1^{r_1^0} \cdots \alpha_n^{r_n^0}$$

is finite. Indeed, since $\log\alpha_1, \dots, \log\alpha_n$ are linearly independent over $\mathbb{Q}$, there exists $\underline{s} \in \mathbb{Q}^n$ such that

$$\left\{\underline{r} \in \mathbb{Q}^n \; ; \; \alpha_1^{r_1} \cdots \alpha_n^{r_n} = 1\right\} = \mathbb{Z}\underline{s};$$

we deduce that the set

$$\{\underline{r} \in \mathbb{Q}^n \cap B_n(0,2X)\,;\, \alpha_1^{r_1}\cdots\alpha_n^{r_n} = 1\}$$

is finite.

This shows that Λ is a lattice (discrete subgroup of rank n) in $\mathbb{R}^n$, hence a free $\mathbb{Z}$-module, and therefore $\mathcal{G}$ is also a free $\mathbb{Z}$-module of rank n.

Let $\log\theta_1,\dots,\log\theta_n$ be a basis of $\mathcal{G}$ over $\mathbb{Z}$. Assume

$$\left[\mathbb{K}(\theta_1^{1/p},\dots,\theta_n^{1/p}) : \mathbb{K}\right] < p^n$$

for some prime p. Let m be minimal with $1 \le m \le n$ such that

$$\left[\mathbb{K}(\theta_1^{1/p},\dots,\theta_m^{1/p}) : \mathbb{K}\right] < p^m.$$

We use Lemma 5.2:

$$\theta_m = \theta_1^{j_1}\cdots\theta_{m-1}^{j_{m-1}}\gamma^p,$$

for some $\gamma \in \mathbb{K}^\times$ and $0 \le j_\ell < p$, $(1 \le \ell \le m-1)$. Define

$$\lambda = \frac{1}{p}\log\theta_m - \sum_{\ell=1}^{m-1}\frac{j_\ell}{p}\log\theta_\ell.$$

Since $\mathbb{K}$ contains the p-th roots of unity, from $(e^\lambda/\gamma)^p \in \mathbb{K}^\times$ we deduce $e^\lambda \in \mathbb{K}^\times$, hence $\lambda \in \mathcal{G}$, and therefore $\lambda \in \mathbb{Z}\log\theta_1 + \cdots + \mathbb{Z}\log\theta_n$, which is clearly a contradiction. □

In the transcendence proof following Baker's method, Lemma 5.3 is applied as follows: when we need estimates for the height, we use the algebraic numbers α_i, while when we want to apply Kummer's condition, we use the numbers θ_i. This does not make a difference when using interpolation determinants; if we use an auxiliary function, the systems of equations are equivalent, but in order to investigate small values we need estimates for the transition matrix. The index N of $\mathbb{Z}^n$ in Λ plays an important role in Matveev's paper [20], II.

f) Kummer's Theory with a Large Prime p

The results in this subsection are no more used in recent papers dealing with linear independence measures for logarithms of algebraic numbers, but they keep their independent interest.

We denote by n a positive integer and by $\alpha_1,\dots,\alpha_n$ nonzero algebraic numbers.

Lemma 5.4. *Let $\mathbb{K}$ be a number field containing $\alpha_1,\dots,\alpha_n$ and let p be a sufficiently large prime.*

a) Assume $\alpha_1,\dots,\alpha_n$ are multiplicatively independent. Denote by $\alpha_i^{1/p}$ any p-th root of α_i $(1 \le i \le n)$. Then

$$[\mathbb{K}(\alpha_1^{1/p},\dots,\alpha_n^{1/p}):\mathbb{K}] = p^n.$$

b) Assume α_1 is a primitive q-th root of unity for some positive integer q, while $\alpha_2,\dots,\alpha_n$ are multiplicatively independent. Denote by $\alpha_1^{1/p}$ a primitive pq-th root of unity and by $\alpha_i^{1/p}$ any p-th root of α_i $(2 \le i \le n)$. Then

$$[\mathbb{K}(\alpha_1^{1/p},\dots,\alpha_n^{1/p}):\mathbb{K}] = (p-1)p^{n-1}.$$

Proof. a) To start with, consider the case $n = 1$. The assumption is that $\alpha = \alpha_1$ is not a root of unity, and the conclusion amounts to say that if α is a p-th power in $\mathbb{K}$, then p is bounded (depending on α and $\mathbb{K}$). We prove this claim by using heights: if $\alpha = \beta^p$ for some $\beta \in \mathbb{K}$, then $\mathrm{h}(\beta) = (1/p)\mathrm{h}(\alpha)$. Moreover β has degree $\le [\mathbb{K}:\mathbb{Q}]$ and is not a root of unity. From a result of Kronecker's (see for instance [34], Section 3.6) $\mathrm{h}(\beta)$ is bounded from below by a positive constant c depending only on $[\mathbb{K}:\mathbb{Q}]$. Hence $p \le c^{-1}\mathrm{h}(\alpha)$.

More generally, assume $\alpha_1,\dots,\alpha_n$ are multiplicatively independent. Define

$$\mathbb{K}_r = \mathbb{K}(\alpha_1^{1/p},\dots,\alpha_r^{1/p}) \quad (1 \le r \le n)$$

and $\mathbb{K}_0 = \mathbb{K}$. Let p be a prime number such that $[\mathbb{K}_n:\mathbb{K}] < p^n$. Denote by r the least integer such that $[\mathbb{K}_r:\mathbb{K}] < p^r$. We have $1 \le r \le n$, $[\mathbb{K}_{r-1}:\mathbb{K}] = p^{r-1}$ and $[\mathbb{K}_r:\mathbb{K}_{r-1}] < p$. By Lemma 5.2 there exists a nontrivial relation

$$\alpha_r = \alpha_1^{j_1}\cdots\alpha_{r-1}^{j_{r-1}}\gamma^p \tag{5.10}$$

with $j_\ell \in \mathbb{Z}, 0 \le j_\ell < p$ $(1 \le \ell < r)$ and $\gamma \in \mathbb{K}^\times$. From the properties of the height (see for instance [34] Chap. 3) we deduce

$$\mathrm{h}(\gamma) \le \mathrm{h}(\alpha_1) + \cdots + \mathrm{h}(\alpha_r),$$

hence γ belongs to a finite subset of $\mathbb{K}^\times$ which depends only on $\alpha_1,\dots,\alpha_n$. For each fixed γ there is a unique multiplicative dependence relation like (5.10), hence p is bounded. Moreover explicit upper bounds for the exponents in such a multiplicative dependence relation are known, depending only on $\alpha_1,\dots,\alpha_r$ and $[\mathbb{K}:\mathbb{Q}]$ (see for instance [34] Lemma 7.19).

b) Assume now α_1 is a primitive q-th root of unity, while $\alpha_1^{1/p}$ is a primitive pq-th root of unity. For any prime p which does not divide q the number α_1 is a p-th power of an element in $\mathbb{K}$: indeed writing $up+vq = 1$ we get $\alpha_1 = (\alpha_1^u)^p$. Hence the field $\mathbb{K}_1 = \mathbb{K}(\alpha_1^{1/p})$ has degree $< p$ over $\mathbb{K}$. Moreover, since α_1^q is a primitive p-th root of unity, the field $\mathbb{Q}(\alpha_1^q)$ is the cyclotomic field of degree $p-1$ and discriminant $\pm p^{p-2}$. Hence, as soon as p does not divide the discriminant of $\mathbb{K}$, the field $\mathbb{K}_1$ has degree $p-1$ over $\mathbb{K}$.

Next we apply part a) of this Lemma 5.4 to the numbers $\alpha_2,\dots,\alpha_n$ and the field $\mathbb{K}$: we deduce

$$[\mathbb{K}(\alpha_2^{1/p},\dots,\alpha_n^{1/p}):\mathbb{K}] = p^{n-1}.$$

Finally, $\mathbb{K}(\alpha_1^{1/p},\dots,\alpha_n^{1/p})$ is the compositum of $\mathbb{K}_1$ and $\mathbb{K}(\alpha_2^{1/p},\dots,\alpha_n^{1/p})$, hence has degree $(p-1)p^{n-1}$ over $\mathbb{K}$. □

Lemma 5.5. *Assume $\alpha_1, \ldots, \alpha_n$ are multiplicatively independent. Then there exists a positive integer D with the following property: if*

$$Q \in \mathbb{K}[X_1, \ldots, X_n]$$

is a nonzero polynomial of degree $< D$ in each variable X_i $(1 \le i \le n)$ and p a prime with $p \ge D$, then

$$Q\big(\alpha_1^{1/p}, \ldots, \alpha_n^{1/p}\big) \neq 0.$$

Proof. Using Lemma 5.4, take D sufficiently large so that for $p \ge D$,

$$\big[\mathbb{K}\big(\alpha_1^{1/p}, \ldots, \alpha_n^{1/p}\big) : \mathbb{K}\big] = p^n.$$

For such a p we prove the result by induction on n. For $n = 1$, since $\alpha_1^{1/p}$ has degree $p \ge D > \deg Q$ over $\mathbb{K}$, and since Q has coefficients in $\mathbb{K}$, it follows that $\alpha_1^{1/p}$ is not a root of Q.

If Lemma 5.5 holds for $n - 1$, then the polynomial

$$P(X) = Q\big(\alpha_1^{1/p}, \ldots, \alpha_{n-1}^{1/p}, X\big) \in \mathbb{K}_{n-1}[X]$$

is nonzero, has degree $< D$ and coefficients in the field

$$\mathbb{K}_{n-1} = \mathbb{K}\big(\alpha_1^{1/p}, \ldots, \alpha_{n-1}^{1/p}\big),$$

while $\alpha_n^{1/p}$ is algebraic of degree $p \ge D$ over this field $\mathbb{K}_{n-1}$. Hence $\alpha_n^{1/p}$ is not a root of P. □

Lemma 5.6. *Assume $\alpha_1, \ldots, \alpha_n$ are multiplicatively independent. Then there exists a positive integer D with the following property. Let*

$$Q \in \mathbb{K}[X_0, X_1, \ldots, X_n]$$

be a nonzero polynomial of degree $< D_0$ in X_0 and $< D$ in each variable X_i $(1 \le i \le n)$. Let p be a prime with $p > D$ and $\mathcal{S}$ a subset of $\mathbb{Z}$ with at least D_0 elements such that $(s, p) = 1$ for any $s \in \mathcal{S}$. Then one at least of the numbers

$$Q\big(s/p, \alpha_1^{s/p}, \ldots, \alpha_n^{s/p}\big) \quad (s \in \mathcal{S})$$

is not 0.

Proof. For $(s, p) = 1$ Bézout's relations imply

$$\mathbb{Q}(\alpha, \alpha^{1/p}) = \mathbb{Q}(\alpha, \alpha^{s/p}).$$

Therefore the same arguments as for Lemma 5.5 yield Lemma 5.6. □

Lemmas 5.5 and 5.6 occurred in Baker's method at an earlier stage of the theory (see [11] pp. 176–177 and 185). Later, similar results taking into account several values of $\underline{\sigma}$ were required (see [11], pp. 223–226 and 237–238).

g) Kummer's Theory with $p = 2$

Let p be a prime for which the field generated over $\mathbb{K} = \mathbb{Q}(\alpha_1, \dots, \alpha_n)$ by the p-th roots $\alpha_i^{1/p} = \exp((1/p)\log\alpha_i)$ with $1 \le i \le n$ has maximal degree p^n:

$$[\mathbb{K}(\alpha_1^{1/p}, \dots, \alpha_n^{1/p}) : \mathbb{K}] = p^n. \tag{5.11}$$

This condition (5.11) yields a decomposition of each relation (5.8)

$$\sum_{(\tau,\underline{t})} c_{\tau,\underline{t}} \, \Delta\,(s/p; \tau, T_0^\sharp, \sigma_0) \prod_{k=1}^{n-1} (b_n t_k - b_k t_n)^{\sigma_k} \cdot \alpha_1^{t_1 s/p} \cdots \alpha_n^{t_n s/p} = 0$$

for which s is prime to p into p^n equations, where the sum over $(\tau, \underline{t})$ is restricted to the elements with $t_i \equiv t_i^0 \pmod p$. For one at least of these equations, say corresponding to some $\underline{t}^0 \in \mathbb{N}^n$ with $0 \le t_i^0 < p$, the coefficients $c_{\tau,\underline{t}}$ with

$$t_i \equiv t_i^0 \pmod p \quad (1 \le i \le n)$$

do not all vanish. We reduce the number of coefficients of P as follows: fix such a $\underline{t}^0$, write $t_i = t_i^0 + pt_i'$ $(1 \le i \le n)$ and consider now the new auxiliary polynomial obtained as a chunk of P

$$\sum_{(\tau,\underline{t}')} c_{\tau,\underline{t}^0+p\underline{t}'} \, \Delta\,(Y_0; \tau, T_0^\sharp) Y_1^{t_1'} \cdots Y_n^{t_n'}.$$

In this process the number of coefficients[18] of the auxiliary polynomial P decreases (the upper bound T_i for the degree in Y_i with $1 \le i \le n$ is divided by p). It turns out that the number of relations (3.6) will be essentially fixed along the inductive process, but at the end of the extrapolation the number of relations will be higher than the number of coefficients of the last auxiliary polynomial P, which is what we were looking for.

Each time Kummer's condition (5.11) is used with, say, $p = 2$, we replace T_i by $T_i/2$. If one performs this extrapolation sufficiently far, one ends up (after the double induction) with some T_j, say T_n, replaced by 0 and the system of equations one gets in this case is easily shown (see Lemma 5.7) to have no nontrivial solution: there is no need to appeal to the zero estimate 3.2.

If one wishes to get $T_n/2^J < 1$ (after J steps), we need $2^J > T_n$. It turns out that such a long extrapolation procedure has a cost: there is another factor $\log V_{n-1}$ (assuming $V_1 \le \cdots \le V_{n-1} \le V_n$) in the final estimate. This is exactly the main result in [31], which improves earlier results by Baker in [3]. See also [20], I, as well as [37] for the p-adic case.

Now Proposition 3.2 enables us to perform a shorter extrapolation, which avoids this cost of $\log V_{n-1}$. This is done in [24] as well as in [5] and [20], II. See also [38], I, for the p-adic case.

---

[18] By "the number of coefficients" of P we mean the number $T_0 \cdots T_n$ where T_i is the known upper bound for the degree of P with respect to Y_i $(1 \le i \le n)$. This is a loose way of speaking, but the values of $T_0, \dots, T_n$ should be clear from the context.

h) A Simple Zero Estimate

As we saw in Section 5.2, f), combining the extrapolation with (5.11), one reduces the set of coefficients of the auxiliary polynomial. The next result shows that we reach a contradiction as soon as all coefficients $c_{\tau \underline{t}}$ with $1 \leq t_n < T_n$ vanish.

Lemma 5.7. *Let $T_0, \ldots, T_{n-1}$, S_0, S_1 be positive integers and $\alpha_1, \ldots, \alpha_{n-1}$ nonzero complex numbers. Assume*

$$S_0 S_1 \geq T_0 \quad \text{and} \quad S_0 \geq T_j \quad \text{for} \quad 1 \leq j < n.$$

Then the matrix

$$\mathtt{M} = \left(\partial^{\underline{\sigma}}\left(Y_0^{\tau} Y_1^{t_1} \cdots Y_{n-1}^{t_{n-1}}\right)(s, \alpha_1^s, \ldots, \alpha_{n-1}^s)\right)_{\substack{(\tau,\underline{t})\\(\underline{\sigma},s)}},$$

where the rows are indexed by $(\tau, \underline{t})$ and the columns by $(\underline{\sigma}, s)$ with

$$0 \leq \tau < T_0, \quad 0 \leq t_i < T_i \quad (1 \leq i \leq n-1) \quad \text{and} \quad \|\underline{\sigma}\| < S_0, \quad 0 \leq s < S_1,$$

has rank $T_0 T_1 \cdots T_{n-1}$.

Remark. Because of (5.2) and (5.4), the conditions $S_0 S_1 \geq T_0$ and $S_0 \geq T_j$ are responsible for the requirements $D \log E^* \geq \log E$ and $V_j \geq \log E$ in the hypotheses of Theorem 5.1.

Proof. On the subring $\mathbb{C}[Y_0, Y_1, \ldots, Y_{n-1}]$ of $\mathbb{C}[Y_0, Y_1, \ldots, Y_n]$, we have

$$\partial^{\underline{\sigma}} = \left(\frac{\partial}{\partial Y_0}\right)^{\sigma_0} \cdots \left(\frac{\partial}{\partial Y_{n-1}}\right)^{\sigma_{n-1}}.$$

Hence the problem is reduced to a Cartesian product situation which one deals with by induction as follows.

Consider a relation between the rows of $\mathtt{M}$:

$$\sum_{\tau=0}^{T_0-1} \sum_{t_1=0}^{T_1-1} \cdots \sum_{t_{n-1}=0}^{T_{n-1}-1} c_{\tau \underline{t}} \partial^{\underline{\sigma}}\left(Y_0^{\tau} Y_1^{t_1} \cdots Y_{n-1}^{t_{n-1}}\right)(s, \alpha_1^s, \ldots, \alpha_{n-1}^s) = 0$$

for any $(\underline{\sigma}, s)$ satisfying $\|\underline{\sigma}\| < S_0$ and $0 \leq s < S_1$. This means that the polynomial

$$P(\underline{Y}) = \sum_{\tau=0}^{T_0-1} \sum_{t_1=0}^{T_1-1} \cdots \sum_{t_{n-1}=0}^{T_{n-1}-1} c_{\tau \underline{t}} Y_0^{\tau} Y_1^{t_1} \cdots Y_{n-1}^{t_{n-1}}$$

satisfies

$$\partial^{\underline{\sigma}} P(s, \alpha_1^s, \ldots, \alpha_{n-1}^s) = 0 \quad \text{for} \quad \|\underline{\sigma}\| < S_0 \quad \text{and} \quad 0 \leq s < S_1.$$

For $(\sigma_0,\ldots,\sigma_{n-2})\in\mathbb{N}^{n-1}$ with $\sigma_0+\cdots+\sigma_{n-2}<S_0$ and for $0\le s<S_1$, the polynomial

$$\left(\frac{\partial}{\partial Y_0}\right)^{\sigma_0}\cdots\left(\frac{\partial}{\partial Y_{n-2}}\right)^{\sigma_{n-2}}P(s,\alpha_1^s,\ldots,\alpha_{n-2}^s,Y_{n-1})\in\mathbb{C}[Y_{n-1}]$$

has degree $<T_{n-1}\le S_0$ and a zero at the point α_{n-1}^s of multiplicity $\ge S_0$, hence this polynomial is 0. By induction one shows in the same way that for $1\le k<n$, $(\sigma_0,\ldots,\sigma_{k-1})\in\mathbb{N}^k$ with $\sigma_0+\cdots+\sigma_{k-1}<S_0$ and $0\le s<S_1$, the polynomial

$$\left(\frac{\partial}{\partial Y_0}\right)^{\sigma_0}\cdots\left(\frac{\partial}{\partial Y_{k-1}}\right)^{\sigma_{k-1}}P(s,\alpha_1^s,\ldots,\alpha_{k-1}^s,Y_k,\ldots,Y_{n-1})\in\mathbb{C}[Y_k,\ldots,Y_{n-1}]$$

is zero. Hence for $k=1$ we get

$$\left(\frac{\partial}{\partial Y_0}\right)^{\sigma_0}P(s,Y_1,\ldots,Y_{n-1})=0\quad\text{for}\quad 0\le\sigma_0<S_0\quad\text{and}\quad 0\le s<S_1.$$

Since $S_0S_1>T_0$ we deduce $P=0$, and therefore $\mathtt{M}$ has rank $T_0T_1\cdots T_{n-1}$. □

Remark. There is nothing special with the field $\mathbb{C}$: the result is valid for any field of zero characteristic. Also the same result holds if we select another basis than $Y_0^\tau Y_1^{t_1}\cdots Y_{n-1}^{t_{n-1}}$ for the space $\mathbb{C}[Y_0,\ldots,Y_{n-1}]$, for instance $\Delta(Y_0;\tau,T_0^\sharp)Y_1^{t_1}\cdots Y_{n-1}^{t_{n-1}}$.

i) Removing n^n under Kummer's Condition

In [20], I, E.M. Matveev succeeds to remove the factor n^n in (5.7) under Kummer's condition (5.11) for $p=2$. The idea is the following. He restricts the set of exponents $(\tau,\underline{t})$ of his auxiliary polynomial to a subset satisfying

$$\left|t_1\log\alpha_1+\cdots+t_n\log\alpha_n\right|\le\frac{U}{MES_1},\tag{5.12}$$

where $M\ge1$ is a new parameter. This allows him to take a larger radius R for the disk where he will apply a Schwarz Lemma (or an approximate Schwarz Lemma): in place of $R=Er$, he may take (almost without extra cost at this place) $R=EMr$. In the conclusion, in place of $(\log E)^{-n-1}$ appears now $\bigl(\log(ME)\bigr)^{-n-1}$. On the other hand, for application of Thue-Siegel's Lemma in the construction of P, it is necessary to estimate from below the number of such elements $(\tau,\underline{t})$. Matveev in [20], I, uses arguments from geometry of numbers. Yu Kunrui [38], II, worked out simpler arguments, using only Dirichlet's box principle, which suffice (the final numerical estimate may be not as sharp, but the argument works as well for the p-adic case; see also [34], Section 9.3 for another adaptation of this argument). Essentially, requiring (5.12) divides the number of tuples $(\tau, \underline{t})$ by M. So in the final result the new parameter U is the old one multiplied by $M\big(\log(ME)\big)^{-n-1}$. Assuming $E = e$ and taking $M = c^n$, one gets rid of the unwanted term n^n in (5.7).

Appendix. From 2 to n Logarithms

In this appendix we develop a remark which originates in Gel'fond's work and has been also used by Bombieri, Bilu and Bugeaud (see [34], Theorem 1.9 and Corollary 10.18). The goal is to deduce, from a nontrivial irrationality measure for the quotient of two logarithms of algebraic numbers, a nontrivial measure of linear independence for n logarithms. The idea is to write the coefficients b_i as $\widetilde{b}q_i + r_i$ with integers $\widetilde{b}$, $q_1, \ldots, q_n$, $r_1, \ldots, r_n$, so that the linear form

$$b_1X_1 + \cdots + b_nX_n$$

has the same value at the point $(\lambda_1, \ldots, \lambda_n)$ than the binary form

$$\widetilde{b}Y_1 + Y_2$$

at the point

$$\big(q_1\lambda_1 + \cdots + q_n\lambda_n, r_1\lambda_1 + \cdots + r_n\lambda_n\big).$$

We need to select $\widetilde{b}$ so that the "remainders" $r_i \in \mathbb{Z}$ have comparatively small absolute values.

We expand this argument in the following lemma.

Lemma 5.8. *Let K be a number field of degree D, $\lambda_1, \ldots, \lambda_n$ elements of $\mathcal{L}$ such that the algebraic numbers $\alpha_i = e^{\lambda_i}$ are in K. Define $V_1, \ldots, V_n$ and V by*

$$V_i = \max\left\{h(\alpha_i), \frac{|\lambda_i|}{D}, \frac{1}{D}\right\} \quad (i = 1, \ldots, n) \ \ \textit{and} \ \ V = \max\{V_1, \ldots, V_n\}.$$

Further, let $b_1, \ldots, b_n$ be rational integers and let B be a positive integer satisfying $B \geq \max\{|b_1|, \ldots, |b_n|\}$.

Then there exist $\widetilde{\lambda}_1, \widetilde{\lambda}_2$ in $\mathcal{L}$ and $\widetilde{b}$ in $\mathbb{Z}$ such that $\widetilde{\alpha}_1 = e^{\widetilde{\lambda}_1}$ and $\widetilde{\alpha}_2 = e^{\widetilde{\lambda}_2}$ are in K,

$$b_1\lambda_1 + \cdots + b_n\lambda_n = \widetilde{b}\,\widetilde{\lambda}_1 + \widetilde{\lambda}_2,$$

$1 \leq \widetilde{b} \leq B$, and such that the numbers

$$\widetilde{V}_i = \max\left\{h(\widetilde{\alpha}_i), \frac{|\widetilde{\lambda}_i|}{D}, \frac{i}{D}\right\} \qquad (i = 1, 2)$$

satisfy

$$\widetilde{V}_1\widetilde{V}_2 \leq 2n^2B^{1-\kappa_n}V^2 \ \ \textit{with} \ \ \kappa_n = \frac{1}{2n-1}.$$

For the proof of Lemma 5.8, we shall use Minkowski's Linear Forms Theorem (see for instance [27], Chap. II, S 1, Theorem 2C)[19].

Lemma 5.9. *Let $(x_{ij})_{1\le i,j\le n}$ be a $n\times n$ matrix with determinant ± 1. Let $A_1,\dots,A_n$ be positive real numbers with product $A_1\cdots A_n=1$. Then there exists $(q_1,\dots,q_n)\in\mathbb{Z}^n\setminus\{0\}$ such that*

$$|q_1x_{i1}+\cdots+q_nx_{in}|<A_i\quad(1\le i<n)$$

and

$$|q_1x_{n1}+\cdots+q_nx_{nn}|\le A_n.$$

Proof of Lemma 5.8. In the trivial case $b_1=\cdots=b_n=0$ we set $\widetilde{b}=1$, $\widetilde{\lambda}_1=\widetilde{\lambda}_2=0$ and the conclusion is satisfied. So we may assume $(b_1,\dots,b_n)\ne 0$. By symmetry, we may assume $b_1\ge|b_i|$ for $1\le i\le n$.

Further, if $n=1$, we set

$$\widetilde{b}=b_1,\quad \widetilde{\lambda}_1=\lambda_1,\quad \widetilde{\lambda}_2=0,$$

so that $\widetilde{V}_1=V_1$, $\widetilde{V}_2=2/D$ and again the conclusion is satisfied. Hence we shall assume $n\ge 2$.

Define

$$Q=B^{(n-1)\kappa_n}.$$

Notice that the exponent is

$$(n-1)\kappa_n=\frac{1}{2}(1-\kappa_n).$$

Using Lemma 5.9, we deduce that there exist rational integers $q_1,\dots,q_n$, not all of which are zero, satisfying

$$\left|\frac{b_i}{b_1}q_1-q_i\right|<Q^{-1/(n-1)}\ \text{ for }\ 2\le i\le n\ \text{ and }\ |q_1|\le Q.$$

Since

$$|q_i|<\left|\frac{b_i}{b_1}\cdot q_1\right|+1\le|q_1|+1,$$

we have $|q_i|\le|q_1|$. In particular $q_1\ne 0$. Replacing, if necessary, all q_i by $-q_i$, we may assume $q_1>0$.

We define

$$\widetilde{b}=[b_1/q_1].$$

Clearly the inequalities $1\le\widetilde{b}\le B$ are satisfied. Further, set

[19] If one applies Dirichlet's box principle in place of Minkowski's Theorem, one deduces a weaker estimate for $\widetilde{V}_1\widetilde{V}_2$, where $2n^2B^{1-\kappa_n}V^2$ is replaced by $4n^2B^{1-\kappa_n}V^2$.

$$r_i = b_i - \widetilde{b}q_i \quad (1 \le i \le n),$$

$$\widetilde{\lambda}_1 = \sum_{i=1}^{n} q_i\lambda_i \quad \text{and} \quad \widetilde{\lambda}_2 = \sum_{i=1}^{n} r_i\lambda_i,$$

so that

$$\widetilde{b}\widetilde{\lambda}_1 + \widetilde{\lambda}_2 = \sum_{i=1}^{n}(\widetilde{b}q_i + r_i)\lambda_i = \sum_{i=1}^{n} b_i\lambda_i.$$

It remains to estimate $\widetilde{V}_1$ and $\widetilde{V}_2$. Define $R = \max_{1\le i\le n} |r_i|$. We have

$$\widetilde{V}_1 \le nq_1V, \quad \widetilde{V}_2 \le nRV,$$

hence

$$\widetilde{V}_1\widetilde{V}_2 \le n^2q_1RV^2.$$

Finally from the inequalities

$$|b_i - \widetilde{b}q_i| \le \frac{1}{q_1}|b_iq_1 - b_1q_i| + |q_i|\left|\frac{b_1}{q_1} - \widetilde{b}\right| \le q_1^{-1}B^{1-\kappa_n} + q_1$$

and

$$q_1^2 \le Q^2 \le B^{1-\kappa_n}$$

we deduce

$$q_1R \le B^{1-\kappa_n} + q_1^2 \le 2B^{1-\kappa_n}.$$

□

We shall combine Lemma 5.8 with the following sharp estimate for two logarithms (Corollary 1 in [16]).

Theorem 5.10. *Let λ_1, λ_2 be two elements in $\mathcal{L}$ and b_1, b_2 two nonzero rational integers. Define*

$$\alpha_1 = e^{\lambda_1}, \quad \alpha_2 = e^{\lambda_2} \quad \text{and} \quad D = [\mathbb{Q}(\alpha_1, \alpha_2) : \mathbb{Q}].$$

Assume α_1 and α_2 are multiplicatively independent. Let V_1, V_2 and W be positive real numbers satisfying

$$V_i = \max\left\{h(\alpha_i), \frac{|\lambda_i|}{D}, \frac{1}{D}\right\} \qquad (i = 1, 2)$$

and

$$W \ge 1, \quad W \ge \frac{21}{D}, \quad e^W \ge \frac{|b_2|}{DV_1} + \frac{|b_1|}{DV_2}.$$

Then

$$|b_1\lambda_1 + b_2\lambda_2| \ge \exp\{-31D^4V_1V_2W^2\}.$$

Corollary 5.11. *Let $\alpha_1, \dots, \alpha_n$ be nonzero multiplicatively independent algebraic numbers. Under the assumptions of Theorem 5.1, if we set* $V = \max\{V_1, \dots, V_n\}$, $E = e$,

$$B = \max\{e,\ e^{21/D},\ |b_1|, \dots, |b_n|\}$$

and $\kappa_n = 1/(2n-1)$, *then we have*

$$|\Lambda| > \exp\{-62n^2 D^4 B^{1-\kappa_n} (\log B)^2 V^2\}.$$

Further similar estimates are easy to produce.

Remark. Such an argument is used in [8] in order to reduce a linear form in logarithms from 5 to 2 terms and to show that all integer solutions to the Diophantine equation in 3 variables

$$x^5 + (z-1)^2 x^4 y - (2z^3 + 4z + 4)x^3 y^2 + (z^4 + z^3 + 2z^2 + 4z - 3)x^2 y^3 \\ + (z^3 + z^2 + 5z + 3)xy^4 + y^5 = \pm 1$$

are the trivial ones given by

$$\pm(x,y) = \begin{cases} (1,0),\ (0,1), & z \neq -1,\ 0, \\ (1,0),\ (0,1),\ (\pm 1, 1),\ (-2,1), & z \in \{-1,\ 0\}. \end{cases}$$

6 Sixth Lecture. Matveev's Theorem with Interpolation Determinants

We refer to Nesterenko's lectures for an introduction to Matveev's proof of the following Theorem:

Theorem 6.1. *There exists an absolute positive constant C with the following property. Let n be a positive integer, $\alpha_1, \dots, \alpha_n$ nonzero algebraic numbers and $\log \alpha_1, \dots, \log \alpha_n$ logarithms of $\alpha_1, \dots, \alpha_n$ respectively. Assume that the numbers $\log \alpha_1, \dots, \log \alpha_n$ are $\mathbb{Q}$-linearly independent. Let $b_1, \dots, b_n$ be rational integers, not all of which are zero. Denote by D the degree of the number field $\mathbb{Q}(\alpha_1, \dots, \alpha_n)$ over $\mathbb{Q}$. Further, let $W, V_1, \dots, V_n$ be positive real numbers. Assume*

$$V_j \geq \max\left\{\mathrm{h}(\alpha_j),\ \frac{\mathrm{e}|\log \alpha_j|}{D}\right\} \quad (1 \leq j \leq n) \quad \text{and} \quad W \geq 1 + \log D.$$

Further, assume $b_n \neq 0$ and

$$e^W \geq \frac{1}{D} \max_{1 \leq j \leq n-1} \left\{\frac{|b_n|}{V_j} + \frac{|b_j|}{V_n}\right\}.$$

Then the number

$$\Lambda = b_1 \log \alpha_1 + \cdots + b_n \log \alpha_n$$

has absolute value bounded from below by

$$|\Lambda| > \exp\{-C^n D^{n+2}(1 + \log D) W V_1 \cdots V_n\}.$$

Our goal is to explain how to prove Matveev's Theorem 6.1 by means of interpolation determinants and to avoid the construction of an auxiliary function involving Thue-Siegel-Bombieri-Vaaler's Lemma in [20], II. We aim at obtaining the conclusion with an unspecified (but effectively computable) value for the absolute constant C.

6.1 First Extrapolation

Matveev uses several auxiliary polynomials which are of the form

$$P(\underline{Y}) = \sum_{(\tau,\underline{t})\in\mathcal{L}} c_{\tau,\underline{t}} \bigtriangleup (Y_0; \tau, T_0^\sharp) Y_1^{t_1} \cdots Y_n^{t_n} \in \mathbb{C}[Y_0, \ldots, Y_n],$$

where $\mathcal{L}$ is a suitable set of $(\tau, \underline{t}) \in \mathbb{N} \times \mathbb{N}^n$ with $0 \leq \tau < T_0$ and $0 \leq t_j < T_j$ $(1 \leq j \leq n)$. He starts the construction by means of Bombieri-Vaaler's version of Thue-Siegel's Lemma and solves a system of equations (3.3). Next a first extrapolation enables him to get more relations like (3.6), say

$$\begin{aligned} \partial^{\underline{\sigma}} P(s, \alpha_1^s, \ldots, \alpha_n^s) = 0 \quad &\text{for} \quad \|\underline{\sigma}\| < S_0^{(j)} \\ &\text{and} \quad 0 \leq s < S_1^{(j)} \quad (0 \leq j \leq J). \end{aligned} \tag{6.1}$$

At this stage of his proof we consider the matrix $\mathtt{M}$ of the linear system given by equations (6.1). We write this matrix as follows. Consider the column vector with $|\mathcal{L}|$ rows and entries in $\mathbb{Q}[\underline{Y}]$:

$$\mathcal{P}(\underline{Y}) = \Big(\bigtriangleup (Y_0; \tau, T_0^\sharp) Y_1^{t_1} \cdots Y_n^{t_n} \Big)_{(\tau,\underline{t})\in\mathcal{L}}.$$

For each $\underline{\sigma} \in \mathbb{N}^n$, define

$$\mathcal{G}_{\underline{\sigma}}(z) = \partial^{\underline{\sigma}} \mathcal{P}(z, \alpha_1^z, \ldots, \alpha_n^z)$$

and set

$$\mathtt{M} = \Big(\mathcal{G}_{\underline{\sigma}}(s) \Big)_{(\underline{\sigma},s)},$$

where $(\underline{\sigma}, s)$ ranges over the set of elements in $\mathbb{N}^n \times \mathbb{N}$ such that $\|\underline{\sigma}\| < S_0^{(j)}$ and $0 \leq s < S_1^{(j)}$ for at least one j in the range $0 \leq j \leq J$.

We consider two cases. The first one is when the rank of $\mathtt{M}$ is maximal, equal to $|\mathcal{L}|$. In this case the proof is very short: we select a maximal nonvanishing determinant

$$\Delta = \det\Big(\mathcal{G}_{\underline{\sigma}_\mu}(s_\mu)\Big)_{1\le\mu\le|\mathcal{L}|}$$

by taking the first $|\mathcal{L}|$ columns in minimal lexicographic ordering like in Section 2.2 and Section 3.5, we bound its absolute value from below by means of Liouville's inequality (Proposition 1.13), and from above thanks to analytic arguments. Corollary 3.6 is not quite sufficient for this purpose: it corresponds to a Schwarz' Lemma like Lemma 1.3 for a function with many zeroes, while we need a statement corresponding to an approximated Schwarz' Principle like Lemma 1.12 for a function with many small values. These estimates provide the required conclusion.

Now we assume $\mathtt{M}$ has rank $< |\mathcal{L}|$. Denote by $L_1 - 1$ this rank. We select a subset $\mathcal{L}_1$ of $\mathcal{L}$ with L_1 elements such that, if we set

$$\mathcal{P}_1(\underline{Y}) = \Big(\Delta(Y_0;\tau,T_0^\sharp)Y_1^{t_1}\cdots Y_n^{t_n}\Big)_{(\tau,\underline{t})\in\mathcal{L}_1}$$

and

$$\mathcal{G}^{(1)}_{\underline{\sigma}}(z) = \partial^{\underline{\sigma}}\mathcal{P}_1(z,\alpha_1^z,\dots,\alpha_n^z),$$

then the associated truncated matrix with L_1 rows only

$$\mathtt{M}_1 = \Big(\mathcal{G}^{(1)}_{\underline{\sigma}}(s)\Big)_{(\underline{\sigma},s)}$$

has rank $L_1 - 1$. Thanks to Lemma 5.7 we may assume that all $(\tau,\underline{t})\in\mathcal{L}$ with $t_n = 0$ belong to $\mathcal{L}_1$.

Again we select the first $L_1 - 1$ columns of $\mathtt{M}_1$ which are linearly independent like in Section 2.2 and Section 3.5, say

$$\mathcal{G}^{(1)}_{\underline{\sigma}_1}(s_1),\dots,\mathcal{G}^{(1)}_{\underline{\sigma}_{L_1-1}}(s_{L_1-1}),$$

and we consider the polynomial

$$P_1(\underline{Y}) = \det\Big(\mathcal{G}^{(1)}_{\underline{\sigma}_1}(s_1),\dots,\mathcal{G}^{(1)}_{\underline{\sigma}_{L_1-1}}(s_{L_1-1}),\mathcal{P}_1(\underline{Y})\Big).$$

From the construction it follows that P_1 is not zero. We extrapolate on the division points as follows: for $\|\underline{\sigma}\| < S_0^{(J+1)}$ and $0\le s < S_1^{(J+1)}$ we prove

$$\partial^{\underline{\sigma}}P_1(s/2,\alpha_1^{s/2},\dots,\alpha_n^{s/2}) = 0.$$

Thanks to Lemma 5.3, for odd s each such equation decomposes into 2^n equations. We explain now how to use this fact.

6.2 Using Kummer's Condition

Lemma 6.2. *Let ℓ, $L_1,\dots,L_\ell$, M, N be positive integers, $\mathbb{L}$ a field, $\mathbb{K}$ a subfield of $\mathbb{L}$, $t_1,\dots,t_\ell$ elements in $\mathbb{L}$ which are linearly independent over $\mathbb{K}$,*

$A_1,\dots,A_\ell$, $B_1,\dots,B_\ell$ matrices with entries in $\mathbb{K}$, where A_ν has size $L_\nu \times M$ and B_ν size $L_\nu \times N$ $(1 \le \nu \le \ell)$. Define

$$A = \begin{pmatrix} A_1 \\ \vdots \\ A_\ell \end{pmatrix} \quad \textit{and} \quad C = \begin{pmatrix} A_1 & t_1 B_1 \\ \vdots & \vdots \\ A_\ell & t_\ell B_\ell \end{pmatrix}$$

and assume

$$\operatorname{rank}(C) = \operatorname{rank}(A) < L_1 + \cdots + L_\ell.$$

Then for at least one index ν in the range $1 \le \nu \le \ell$ we have

$$\operatorname{rank}(B_\nu) < L_\nu.$$

Proof. For $1 \le \nu \le \ell$, write

$$B_\nu = \left(b_{\lambda j}^{(\nu)}\right)_{\substack{1 \le \lambda \le L_\nu \\ 1 \le j \le N}} = \left(\underline{b}_1^{(\nu)}, \dots, \underline{b}_N^{(\nu)}\right),$$

where $\underline{b}_1^{(\nu)}, \dots, \underline{b}_N^{(\nu)}$ are the column vectors in $\mathbb{K}^{L_\nu}$. Let r be the rank of the matrix A. Since C has also rank r, each of the column vectors

$$\begin{pmatrix} t_1 \underline{b}_j^{(1)} \\ \vdots \\ t_\ell \underline{b}_j^{(\ell)} \end{pmatrix} \quad \text{of} \quad \begin{pmatrix} t_1 B_1 \\ \vdots \\ t_\ell B_\ell \end{pmatrix}$$

$(1 \le j \le N)$ belongs to the space spanned by the column vectors of A. We write the N relations

$$\operatorname{rank} \begin{pmatrix} A_1 & t_1 \underline{b}_j^{(1)} \\ \vdots & \vdots \\ A_\ell & t_\ell \underline{b}_j^{(\ell)} \end{pmatrix} = r \qquad (1 \le j \le N)$$

expanding by minors:

$$\sum_{\nu=1}^{\ell} \sum_{\lambda=1}^{L_\nu} \alpha_{\nu\lambda}^{(\varrho)} t_\nu b_{\lambda j}^{(\nu)} = 0 \quad \text{for} \quad 1 \le \varrho \le R \quad \text{and} \quad 1 \le j \le N,$$

where the coefficients $\alpha_{\nu\lambda}^{(\varrho)}$ are in $\mathbb{K}$ (independent of j) and are not all zero (because A has rank r). Since $t_1,\dots,t_\ell$ are linearly independent over $\mathbb{K}$, we deduce

$$\sum_{\lambda=1}^{L_\nu} \alpha_{\nu\lambda}^{(\varrho)} b_{\lambda j}^{(\nu)} = 0 \quad \text{for} \quad 1 \le \varrho \le R, \quad 1 \le j \le N \quad \text{and} \quad 1 \le \nu \le \ell. \tag{6.2}$$

There is at least one ν in the range $1 \le \nu \le \ell$ such that the coefficients $\alpha_{\nu\lambda}^{(\varrho)}$ with $1 \le \varrho \le R$ and $1 \le j \le N$ are not all zero. For such a ν the relations (6.2) yield a nontrivial dependence relation between the row vectors of B_ν, hence B_ν has rank $< L_\nu$. □

6.3 Second Extrapolation

At this stage of the proof, after the first extrapolation including division points, we get a new matrix

$$\mathtt{M}_2 = \left(\mathcal{G}^{(2)}_{\underline{\sigma}}(s)\right)_{(\underline{\sigma},s)}$$

built with

$$\mathcal{G}^{(2)}_{\underline{\sigma}}(s) = \partial^{\underline{\sigma}}\mathcal{P}_2(s, \alpha_1^s, \ldots, \alpha_n^s)$$

and

$$\mathcal{P}_2(\underline{Y}) = \left(\triangle(Y_0; \tau, T_0^\sharp) Y_1^{t_1} \cdots Y_n^{t_n}\right)_{(\tau,\underline{t})\in\mathcal{L}_2}.$$

Now the set $\mathcal{L}_1$ has been reduced to $\mathcal{L}_2$, with

$$|t_i| < T_i/2 \text{ for } (\tau, \underline{t}) \in \mathcal{L}_2 \text{ and } 1 \le i \le n,$$

while the set of $(\underline{\sigma}, s)$ is bigger than for $\mathtt{M}_1$, say

$$\|\underline{\sigma}\| < S_0^{(2,j)} \text{ and } 0 \le s < S_1^{(2,j)} \quad (0 \le j \le J_2). \tag{6.3}$$

The upper index $(2, j)$ corresponds to the j-th step in the second extrapolation (hence we may set $S_0^{(1,j)} = S_0^{(j)}$ and $S_1^{(1,j)} = S_1^{(j)}$).

Before starting our second induction let us have a look at Matveev's arguments in [20], II: his second auxiliary polynomial P_2 is a linear combination of the components of $\mathcal{P}_2(\underline{Y})$; he shows by induction on j the relations

$$\partial^{\underline{\sigma}} P_2(s, \alpha_1^s, \ldots, \alpha_n^s) = 0$$

for all $(\underline{\sigma}, s) \in \mathbb{N}^n \times \mathbb{N}$ in (6.3).

As far as we are concerned we repeat the argument of Section 5.1. However it is necessary to be slightly careful with the construction of the next matrix: for an efficient application of the analytic argument we need to introduce a condition like (3.18). This difficulty did not arise in the first extrapolation (Section 6.1), because the condition

$$|\mathcal{L}| > 2\binom{S_0 + n - 1}{n} S_1$$

was in force: Matveev needed it to solve his system of linear equations and produce his first auxiliary polynomial. For the second extrapolation this condition is not there and we proceed as follows.

Let $L_2 - 1$ be the rank of $\mathtt{M}_2$. Define $\mu_0 = L_2/2^{n+2}$. We select the first μ_0 columns which are linearly independent. Next we consider the matrix $\widetilde{\mathtt{M}}_2$ which consists of these μ_0 columns together with all other columns of $\mathtt{M}_2$ for which

$$\|\underline{\sigma}\| < \widetilde{S}_0^{(2,j)} \text{ and } 0 \le s < S_1^{(2,j)} \quad (0 \le j \le J_2),$$

where

$$\begin{cases} \widetilde{S}_0^{(2,0)} = \frac{1}{2}\left(S_0^{(2,0)} + S_0^{(2,1)}\right) & \\ \widetilde{S}_0^{(2,j)} = S_0^{(2,j)} & \text{for } 1 \le j \le J_2. \end{cases}$$

When μ_0 is large, there is no point to distinguish between $\mathtt{M}_2$ and $\widetilde{\mathtt{M}}_2$, but otherwise the picture is the following:

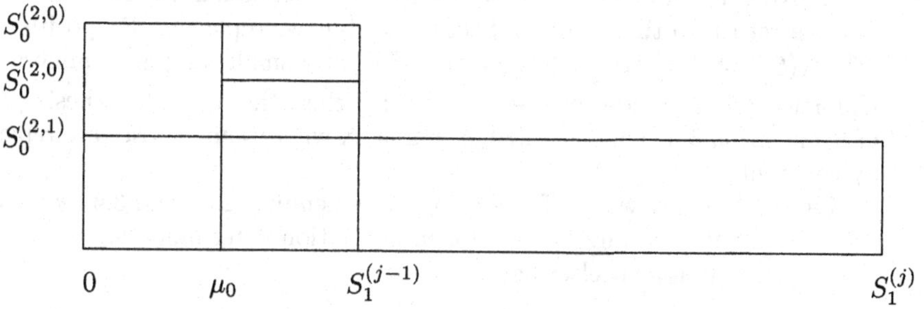

Denote by $\widetilde{L}_2$ the rank of $\widetilde{\mathtt{M}}_2$. We already selected μ_0 columns, we select $\widetilde{L}_2 - \mu_0$ further columns in the usual way in order to get a maximal number of independent columns.

If $\widetilde{L}_2 - 1 = |\mathcal{L}_2|$, we easily conclude the proof by usual means: $\widetilde{\mathtt{M}}_2$ is a nonsingular square matrix with nonzero determinant, say Δ_2. We bound the absolute value of Δ_2 from below by arithmetic means and from above by analytic means.

If $\widetilde{L}_2 < \mu_0 = L_2/2^{n+2}$, then the second extrapolation is complete: our goal is precisely to produce a matrix with such a low rank.

Assume now $L_2/2^{n+2} \le \widetilde{L}_2 \le |\mathcal{L}_2|$. We select $\widetilde{L}_2$ rows of $\widetilde{\mathtt{M}}_2$, corresponding to a subset $\widetilde{\mathcal{L}}_2$ of $\mathcal{L}_2$ with $\widetilde{L}_2$ elements. Our new auxiliary polynomial is the determinant $\widetilde{P}_2(\underline{Y})$ of the square matrix

$$\left(\widetilde{\mathcal{G}}_{\underline{\sigma}_1}^{(2)}(s_1), \ldots, \widetilde{\mathcal{G}}_{\underline{\sigma}_{\widetilde{L}_2-1}}^{(2)}(s_{\widetilde{L}_2-1}), \widetilde{\mathcal{P}}_2(\underline{Y})\right).$$

We extrapolate on division points and complete the second induction as we did for the first one.

6.4 An Approximate Schwarz Lemma for Interpolation Determinants

The results of Section 3.4 deal with zeroes of interpolation determinants and involve Schwarz' Lemma. They are sufficient for the proof of qualitative statements like Baker's Theorem 3.1.

In Section 4, we were able to prove a quantitative result by means of Lemma 4.6, which is also a Schwarz' Lemma; however only the zero at the origin is used, and there is no extrapolation; this explains why the required approximate Schwarz' Lemma in the proof of Lemma 4.7 is just the interpolation formula of Lemma 1.11 (or just Taylor's expansion as in the remark in Section 4.4).

Such an argument does not seem to suffice when we want to extrapolate. Indeed the determinant

$$\left(\partial^{\underline{\kappa}}\mathcal{P}\big(s,\alpha_1^s,\ldots,\alpha_{n-1}^s,(\alpha_1^{b_1}\cdots\alpha_{n-1}^{b_{n-1}})^{-s/b_n}\big)(s_\mu)\right)$$

does not satisfy the assumption (3.12) of Proposition 3.5.

Another attempt is to follow the method of Section 7.3 in [34]: in this case we return to the results of Section 3.4 but we replace in Proposition 3.5 $\mathcal{D}^{\underline{\kappa}_\mu}\varphi_\lambda(\underline{\xi}_\mu)$ by $\mathcal{D}^{\underline{\kappa}_\mu}\varphi_\lambda(\underline{\xi}_\mu)+\epsilon_{\lambda\mu}$ with sufficiently small complex numbers $\epsilon_{\lambda\mu}$. Unfortunately this does not seem to work either: the main hypothesis (3.12) of Proposition 3.5 is not satisfied for a matrix where some entries are replaced by constants.

Finally the only way so far is to combine Lemmas 1.12 and 3.4 by proving an approximate Schwarz Lemma for interpolation determinants.

Details will appear elsewhere.

References

1. Baker, A. – Linear forms in the logarithms of algebraic numbers. I, II, III, IV. Mathematika **13** (1966), 204–216; ibid. **14** (1967), 102–107; ibid. **14** (1967), 220–228; ibid. **15** (1968), 204–216.
2. Baker, A. – *Transcendental number theory.* Cambridge Mathematical Library. Cambridge University Press, Cambridge, 1975. Second edition, 1990.
3. Baker, A. – The theory of linear forms in logarithms. *Transcendence theory: advances and applications* (Proc. Conf., Univ. Cambridge, Cambridge, 1976), pp. 1–27. Academic Press, London, 1977.
4. Baker, A.; Stark, H. M. – On a fundamental inequality in number theory. Ann. of Math. (2) **94** (1971), 190–199.
5. Baker, A.; Wüstholz, G. – Logarithmic forms and group varieties. J. reine angew. Math. **442** (1993), 19–62.
6. Fel'dman, N. I. – *Hilbert's seventh problem.* Moskov. Gos. Univ., Moscow, 1982.
7. Fel'dman, N. I.; Nesterenko, Y. V. – *Number theory. IV. Transcendental Numbers.* Encyclopaedia of Mathematical Sciences **44**. Springer-Verlag, Berlin, 1998.
8. Gaál, I.; Lettl, G. – A parametric family of quintic Thue equations II. Monatsh. Math. **131**, No. 1 (2000), 29–35.
9. Gramain, F. – Lemme de Schwarz pour des produits cartésiens. Ann. Math. Blaise Pascal **8**, No. 2 (2001), 67-75.
10. Koksma, J. F.– *Diophantische Approximationen.* Reprint. Springer-Verlag, Berlin-New York, 1974.
11. Lang, S. – *Elliptic curves: Diophantine analysis.* Grundlehren der Mathematischen Wissenschaften **231**. Springer-Verlag, Berlin-New York, 1978.
12. Lang, S. – *Algebra.* Third edition. Addison-Wesley Publishing Co., Reading, Mass., 1993.
13. Laurent, M. – Sur quelques résultats récents de transcendance. *Journées Arithmétiques, 1989* (Luminy, 1989). Astérisque No. **198–200** (1991), 209–230.
14. Laurent, M. – Linear forms in two logarithms and interpolation determinants. Acta Arith. **66**, No. 2 (1994), 181–199.

15. Laurent, M.; Roy, D. – Criteria of algebraic independence with multiplicities and approximation by hypersurfaces; J. reine angew. Math. **536** (2001), 65–114.
16. Laurent, M.; Mignotte, M.; Nesterenko, Y. V. – Formes linéaires en deux logarithmes et déterminants d'interpolation. J. Number Theory **55**, No. 2 (1995), 285–321.
17. Mahler, K. – On the approximation of logarithms of algebraic numbers. Philos. Trans. Roy. Soc. London. Ser. A. **245** (1953), 371–398.
18. Mahler, K. – Applications of some formulae by Hermite to the approximation of exponentials and logarithms. Math. Ann. **168** (1967), 200–227.
19. Masser, D.W. – Heights, transcendence, and linear independence on commutative group varieties. This volume.
20. Matveev, E. M. – An explicit lower bound for a homogeneous rational linear forms in the logarithms of algebraic numbers. I and II. Izv. Akad. Nauk SSSR. Ser. Mat. **62** No. 4 (1998), 81–136; ibid. **64** No. 6 (2000), 125–180; Engl. transl.: Izv. Math. **62** No. 4 (1998), 723–772; ibid. **64** No. 6 (2000), 1217–1269.
21. Nesterenko, Y. V.; Waldschmidt, M. – On the approximation of the values of exponential function and logarithm by algebraic numbers. (Russian) *Diophantine approximations, Proceedings of papers dedicated to the memory of Prof. N. I. Fel'dman*, ed. Yu. V. Nesterenko, Centre for applied research under Mech.-Math. Faculty of MSU, Moscow (1996), 23–42.
`Number Theory xxx Math Archives : http://arXiv.org/abs/math/0002047`
22. Philippon, P. – Nouveaux lemmes de zéros dans les groupes algébriques commutatifs. *Symposium on Diophantine Problems* (Boulder, CO, 1994). Rocky Mountain J. Math. **26** No. 3 (1996), 1069–1088.
23. Philippon, P. – Une approche méthodique pour la transcendance et l'indépendance algébrique de valeurs de fonctions analytiques. J. Number Theory **64** No. 2 (1997), 291–338.
24. Philippon, P. and Waldschmidt, M. – Lower bounds for linear forms in logarithms. *New advances in transcendence theory* (Durham, 1986), 280–312, Cambridge Univ. Press, Cambridge-New York, 1988.
25. Philippon, P. and Waldschmidt, M. – Formes linéaires de logarithmes sur les groupes algébriques commutatifs. Illinois J. Math. **32** No. 2 (1988), 281–314.
26. Roy, D. – Interpolation formulas and auxiliary functions. J. Number Theory **94** No. 2 (2002), 248-285.
27. Schmidt, W. M.– *Diophantine approximation*. Lecture Notes in Mathematics, **785**. Springer, Berlin, 1980.
28. Sprindžuk, V. G. – *Classical Diophantine equations*. Translated from the 1982 Russian original (Nauka, Moscow). Translation edited by Ross Talent and Alf van der Poorten. Lecture Notes in Mathematics **1559**. Springer-Verlag, Berlin, 1993.
29. Tijdeman, R. – On the number of zeros of general exponential polynomials. Nederl. Akad. Wetensch. Proc. Ser. A **74** = Indag. Math. **33** (1971), 1–7.
30. Waldschmidt, M. – *Nombres transcendants*. Lecture Notes in Mathematics **402**. Springer-Verlag, Berlin-New York, 1974.
31. Waldschmidt, M. – A lower bound for linear forms in logarithms. Acta Arith. **37** (1980), 257–283.
32. Waldschmidt, M. – *Linear independence of logarithms of algebraic numbers*. The Institute of Mathematical Sciences, Madras, IMSc Report No. **116** (1992), 168 pp.
`http://www.math.jussieu.fr/~miw/articles/IMSc.Rpt.116.html`

33. Waldschmidt, M. – Integer valued functions on products. J. Ramanujan Math. Soc. **12** No. 1 (1997), 1–24.
34. Waldschmidt, M. – *Diophantine Approximation on Linear Algebraic Groups. Transcendence Properties of the Exponential Function in Several Variables.* Grundlehren der Mathematischen Wissenschaften **326**, Springer-Verlag, Berlin-Heidelberg, 2000.
35. Waldschmidt, M. – On a Problem of Mahler Concerning the Approximation of Exponentials and Logarithms. Publ. Math. Debrecen **56** No. 3-4 (2000), 713–738.
36. Wielonsky, F. – Hermite-Padé approximants to exponential functions and an inequality of Mahler. J. Number Theory **74** No. 2 (1999), 230–249.
37. Yu, Kun Rui – Linear forms in p-adic logarithms. Acta Arith. **53** (1989), No. 2, 107–186. II. Compositio Math. **74**, No. 1 (1990), 15–113 and **76** (1990), No. 1-2, 307. III. Compositio Math. **91**, No. 3 (1994), 241–276.
38. Yu, Kun Rui – p-adic logarithmic forms and Group Varieties I. J. reine angew. Math. **502** (1998), 29–92. II. Acta Arith. **89** No. 4 (1999), 337–378.

List of Participants

1. Ably Mohammed
2. Amoroso Francesco
3. Avanzi Roberto Maria
4. Basile Laura
5. Baxa Christoph
6. Bosser Vincent
7. Corticelli Davide
8. Corvaja Pietro
9. Dahari Shmuel
10. D'Ambros Paola
11. David Sinnou
12. Dekhissi Ahmed
13. De Santi Silvia
14. Dubickas Arturas
15. Dvornicich Roberto
16. Fischler Stéphane
17. Gasbarri Carlo
18. Gaudron Eric
19. Grinspan Pierre
20. Hirata-Kohno Noriko
21. Iskander Aliev
22. Locher Helmut
23. Marcovecchio Raffaele
24. Marmi Stefano
25. Masser David
26. Matveev Evgeny
27. Melfi Giuseppe
28. Pellarin Federico
29. Perelli Alberto
30. Rhin Georges
31. Schlickewei Hans Peter
32. Schmidt Wolfgang
33. Schulz Christoph
34. Summerer Leonhard
35. Surroca Andrea
36. Vasilyev Denis
37. Viola Carlo
38. Waldschmidt Michel
39. Welter Michael
40. Yu Kun Rui
41. Zaccagnini Alessandro
42. Zannier Umberto

LIST OF C.I.M.E. SEMINARS

1954	1. Analisi funzionale	C.I.M.E
	2. Quadratura delle superficie e questioni connesse	"
	3. Equazioni differenziali non lineari	"
1955	4. Teorema di Riemann-Roch e questioni connesse	"
	5. Teoria dei numeri	"
	6. Topologia	"
	7. Teorie non linearizzate in elasticità, idrodinamica, aerodinamic	"
	8. Geometria proiettivo-differenziale	"
1956	9. Equazioni alle derivate parziali a caratteristiche reali	"
	10. Propagazione delle onde elettromagnetiche	"
	11. Teoria della funzioni di più variabili complesse e delle funzioni automorfe	"
1957	12. Geometria aritmetica e algebrica (2 vol.)	"
	13. Integrali singolari e questioni connesse	"
	14. Teoria della turbolenza (2 vol.)	"
1958	15. Vedute e problemi attuali in relatività generale	"
	16. Problemi di geometria differenziale in grande	"
	17. Il principio di minimo e le sue applicazioni alle equazioni funzionali	"
1959	18. Induzione e statistica	"
	19. Teoria algebrica dei meccanismi automatici (2 vol.)	"
	20. Gruppi, anelli di Lie e teoria della coomologia	"
1960	21. Sistemi dinamici e teoremi ergodici	"
	22. Forme differenziali e loro integrali	"
1961	23. Geometria del calcolo delle variazioni (2 vol.)	"
	24. Teoria delle distribuzioni	"
	25. Onde superficiali	"
1962	26. Topologia differenziale	"
	27. Autovalori e autosoluzioni	"
	28. Magnetofluidodinamica	"
1963	29. Equazioni differenziali astratte	"
	30. Funzioni e varietà complesse	"
	31. Proprietà di media e teoremi di confronto in Fisica Matematica	"

1964	32. Relatività generale	C.I.M.E
	33. Dinamica dei gas rarefatti	"
	34. Alcune questioni di analisi numerica	"
	35. Equazioni differenziali non lineari	"
1965	36. Non-linear continuum theories	"
	37. Some aspects of ring theory	"
	38. Mathematical optimization in economics	"
1966	39. Calculus of variations	Ed. Cremonese, Firenze
	40. Economia matematica	"
	41. Classi caratteristiche e questioni connesse	"
	42. Some aspects of diffusion theory	"
1967	43. Modern questions of celestial mechanics	"
	44. Numerical analysis of partial differential equations	"
	45. Geometry of homogeneous bounded domains	"
1968	46. Controllability and observability	"
	47. Pseudo-differential operators	"
	48. Aspects of mathematical logic	"
1969	49. Potential theory	"
	50. Non-linear continuum theories in mechanics and physics and their applications	"
	51. Questions of algebraic varieties	"
1970	52. Relativistic fluid dynamics	"
	53. Theory of group representations and Fourier analysis	"
	54. Functional equations and inequalities	"
	55. Problems in non-linear analysis	"
1971	56. Stereodynamics	"
	57. Constructive aspects of functional analysis (2 vol.)	"
	58. Categories and commutative algebra	"
1972	59. Non-linear mechanics	"
	60. Finite geometric structures and their applications	"
	61. Geometric measure theory and minimal surfaces	"
1973	62. Complex analysis	"
	63. New variational techniques in mathematical physics	"
	64. Spectral analysis	"
1974	65. Stability problems	"
	66. Singularities of analytic spaces	"
	67. Eigenvalues of non linear problems	"
1975	68. Theoretical computer sciences	"
	69. Model theory and applications	"
	70. Differential operators and manifolds	"
1976	71. Statistical Mechanics	Ed. Liguori, Napoli
	72. Hyperbolicity	"
	73. Differential topology	"
1977	74. Materials with memory	"
	75. Pseudodifferential operators with applications	"
	76. Algebraic surfaces	"

1978	77. Stochastic differential equations		Ed. Liguori, Napoli & Birkhäuser
	78. Dynamical systems		"
1979	79. Recursion theory and computational complexity		"
	80. Mathematics of biology		"
1980	81. Wave propagation		"
	82. Harmonic analysis and group representations		"
	83. Matroid theory and its applications		"
1981	84. Kinetic Theories and the Boltzmann Equation	(LNM 1048)	Springer-Verlag
	85. Algebraic Threefolds	(LNM 947)	"
	86. Nonlinear Filtering and Stochastic Control	(LNM 972)	"
1982	87. Invariant Theory	(LNM 996)	"
	88. Thermodynamics and Constitutive Equations	(LN Physics 228)	"
	89. Fluid Dynamics	(LNM 1047)	"
1983	90. Complete Intersections	(LNM 1092)	"
	91. Bifurcation Theory and Applications	(LNM 1057)	"
	92. Numerical Methods in Fluid Dynamics	(LNM 1127)	"
1984	93. Harmonic Mappings and Minimal Immersions	(LNM 1161)	"
	94. Schrödinger Operators	(LNM 1159)	"
	95. Buildings and the Geometry of Diagrams	(LNM 1181)	"
1985	96. Probability and Analysis	(LNM 1206)	"
	97. Some Problems in Nonlinear Diffusion	(LNM 1224)	"
	98. Theory of Moduli	(LNM 1337)	"
1986	99. Inverse Problems	(LNM 1225)	"
	100. Mathematical Economics	(LNM 1330)	"
	101. Combinatorial Optimization	(LNM 1403)	"
1987	102. Relativistic Fluid Dynamics	(LNM 1385)	"
	103. Topics in Calculus of Variations	(LNM 1365)	"
1988	104. Logic and Computer Science	(LNM 1429)	"
	105. Global Geometry and Mathematical Physics	(LNM 1451)	"
1989	106. Methods of nonconvex analysis	(LNM 1446)	"
	107. Microlocal Analysis and Applications	(LNM 1495)	"
1990	108. Geometric Topology: Recent Developments	(LNM 1504)	"
	109. H_∞ Control Theory	(LNM 1496)	"
	110. Mathematical Modelling of Industrial Processes	(LNM 1521)	"
1991	111. Topological Methods for Ordinary Differential Equations	(LNM 1537)	"
	112. Arithmetic Algebraic Geometry	(LNM 1553)	"
	113. Transition to Chaos in Classical and Quantum Mechanics	(LNM 1589)	"
1992	114. Dirichlet Forms	(LNM 1563)	"
	115. D-Modules, Representation Theory, and Quantum Groups	(LNM 1565)	"
	116. Nonequilibrium Problems in Many-Particle Systems	(LNM 1551)	"

1993	117. Integrable Systems and Quantum Groups	(LNM 1620)	Springer-Verlag
	118. Algebraic Cycles and Hodge Theory	(LNM 1594)	"
	119. Phase Transitions and Hysteresis	(LNM 1584)	"
1994	120. Recent Mathematical Methods in Nonlinear Wave Propagation	(LNM 1640)	"
	121. Dynamical Systems	(LNM 1609)	"
	122. Transcendental Methods in Algebraic Geometry	(LNM 1646)	"
1995	123. Probabilistic Models for Nonlinear PDE's	(LNM 1627)	"
	124. Viscosity Solutions and Applications	(LNM 1660)	"
	125. Vector Bundles on Curves. New Directions	(LNM 1649)	"
1996	126. Integral Geometry, Radon Transforms and Complex Analysis	(LNM 1684)	"
	127. Calculus of Variations and Geometric Evolution Problems	(LNM 1713)	"
	128. Financial Mathematics	(LNM 1656)	"
1997	129. Mathematics Inspired by Biology	(LNM 1714)	"
	130. Advanced Numerical Approximation of Nonlinear Hyperbolic Equations	(LNM 1697)	"
	131. Arithmetic Theory of Elliptic Curves	(LNM 1716)	"
	132. Quantum Cohomology	(LNM 1776)	"
1998	133. Optimal Shape Design	(LNM 1740)	"
	134. Dynamical Systems and Small Divisors	(LNM 1784)	"
	135. Mathematical Problems in Semiconductor Physics	to appear	"
	136. Stochastic PDE's and Kolmogorov Equations in Infinite Dimension	(LNM 1715)	"
	137. Filtration in Porous Media and Industrial Applications	(LNM 1734)	"
1999	138. Computational Mathematics driven by Industrial Applications	(LNM 1739)	"
	139. Iwahori-Hecke Algebras and Representation Theory	(LNM 1804)	"
	140. Theory and Applications of Hamiltonian Dynamics	to appear	"
	141. Global Theory of Minimal Surfaces in Flat Spaces	(LNM 1775)	"
	142. Direct and Inverse Methods in Solving Nonlinear Evolution Equations	to appear	"
2000	143. Dynamical Systems	to appear	"
	144. Diophantine Approximation	(LNM 1819)	"
	145. Mathematical Aspects of Evolving Interfaces	(LNM 1812)	"
	146. Mathematical Methods for Protein Structure	to appear	"
	147. Noncommutative Geometry	to appear	"
2001	148. Topological Fluid Mechanics	to appear	"
	149. Spatial Stochastic Processes	(LNM 1802)	"
	150. Optimal Transportation and Applications	(LNM 1813)	"
	151. Multiscale Problems and Methods in Numerical Simulations	to appear	"
2002	152. Real methods in Complex and CR Geometry	to appear	"
	153. Analytic Number Theory	to appear	"
	154. Imaging	to appear	"

2003	155.	Stochastik Methods in Finance	announced
	156.	Hyperbolik Systems of Balance Laws	announced
	157.	Sympletic 4-Manifolds and Algebraic Surfaces	announced
	158.	Mathematical Foundation of Turbulent Viscous Flows	announced

Fondazione C.I.M.E.

Centro Internazionale Matematico Estivo
International Mathematical Summer Center
http://www.math.unifi.it/~cime
cime@math.unifi.it

2003 COURSES LIST

Stochastic Methods in Finance

July 6–13, Cusanus Akademie, Bressanone (Bolzano)
Joint course with European Mathematical Society

Course Directors:

Prof. Marco Frittelli (Univ. di Firenze), marco.frittelli@dmd.unifi.it
Prof. Wolfgang Runggaldier (Univ. di Padova), runggal@math.unipd.it

Hyperbolic Systems of Balance Laws

July 14–21, Cetraro (Cosenza)

Course Director:

Prof. Pierangelo Marcati (Univ. de L'Aquila), marcati@univaq.it

Symplectic 4-Manifolds and Algebraic Surfaces

September 2–10, Cetraro (Cosenza)

Course Directors:

Prof. Fabrizio Catanese (Bayreuth University)
Prof. Gang Tian (M.I.T. Boston)

Mathematical Foundation of Turbulent Viscous Flows

September 1–6, Martina Franca (Taranto)

Course Directors:

Prof. M. Cannone (Univ. de Marne-la-Vallée)
Prof.T. Miyakawa (Kobe University)

Printing and Binding: Strauss GmbH, Mörlenbach

Vol. 1726: V. Marić, Regular Variation and Differential Equations. X, 127 pages. 2000.
Vol. 1727: P. Kravanja M. Van Barel, Computing the Zeros of Analytic Functions. VII, 111 pages. 2000.
Vol. 1728: K. Gatermann Computer Algebra Methods for Equivariant Dynamical Systems. XV, 153 pages. 2000.
Vol. 1729: J. Azéma, M. Émery, M. Ledoux, M. Yor Séminaire de Probabilités XXXIV. VI, 431 pages. 2000.
Vol. 1730: S. Graf, H. Luschgy, Foundations of Quantization for Probability Distributions. X, 230 pages. 2000.
Vol. 1731: T. Hsu, Quilts: Central Extensions, Braid Actions, and Finite Groups. XII, 185 pages. 2000.
Vol. 1732: K. Keller, Invariant Factors, Julia Equivalences and the (Abstract) Mandelbrot Set. X, 206 pages. 2000.
Vol. 1733: K. Ritter, Average-Case Analysis of Numerical Problems. IX, 254 pages. 2000.
Vol. 1734: M. Espedal, A. Fasano, A. Mikelić, Filtration in Porous Media and Industrial Applications. Cetraro 1998. Editor: A. Fasano. 2000.
Vol. 1735: D. Yafaev, Scattering Theory: Some Old and New Problems. XVI, 169 pages. 2000.
Vol. 1736: B. O. Turesson, Nonlinear Potential Theory and Weighted Sobolev Spaces. XIV, 173 pages. 2000.
Vol. 1737: S. Wakabayashi, Classical Microlocal Analysis in the Space of Hyperfunctions. VIII, 367 pages. 2000.
Vol. 1738: M. Émery, A. Nemirovski, D. Voiculescu, Lectures on Probability Theory and Statistics. XI, 356 pages. 2000.
Vol. 1739: R. Burkard, P. Deuflhard, A. Jameson, J.-L. Lions, G. Strang, Computational Mathematics Driven by Industrial Problems. Martina Franca, 1999. Editors: V. Capasso, H. Engl, J. Periaux. VII, 418 pages. 2000.
Vol. 1740: B. Kawohl, O. Pironneau, L. Tartar, J.-P. Zolesio, Optimal Shape Design. Tróia, Portugal 1999. Editors: A. Cellina, A. Ornelas. IX, 388 pages. 2000.
Vol. 1741: E. Lombardi, Oscillatory Integrals and Phenomena Beyond all Algebraic Orders. XV, 413 pages. 2000.
Vol. 1742: A. Unterberger, Quantization and Non-holomorphic Modular Forms. VIII, 253 pages. 2000.
Vol. 1743: L. Habermann, Riemannian Metrics of Constant Mass and Moduli Spaces of Conformal Structures. XII, 116 pages. 2000.
Vol. 1744: M. Kunze, Non-Smooth Dynamical Systems. X, 228 pages. 2000.
Vol. 1745: V. D. Milman, G. Schechtman, Geometric Aspects of Functional Analysis. Israel Seminar 1999-2000. VIII, 289 pages. 2000.
Vol. 1746: A. Degtyarev, I. Itenberg, V. Kharlamov, Real Enriques Surfaces. XVI, 259 pages. 2000.
Vol. 1747: L. W. Christensen, Gorenstein Dimensions. VIII, 204 pages. 2000.
Vol. 1748: M. Ruzicka, Electrorheological Fluids: Modeling and Mathematical Theory. XV, 176 pages. 2001.
Vol. 1749: M. Fuchs, G. Seregin, Variational Methods for Problems from Plasticity Theory and for Generalized Newtonian Fluids. VI, 269 pages. 2001.
Vol. 1750: B. Conrad, Grothendieck Duality and Base Change. X, 296 pages. 2001.
Vol. 1751: N. J. Cutland, Loeb Measures in Practice: Recent Advances. XI, 111 pages. 2001.
Vol. 1752: Y. V. Nesterenko, P. Philippon, Introduction to Algebraic Independence Theory. XIII, 256 pages. 2001.
Vol. 1753: A. I. Bobenko, U. Eitner, Painlevé Equations in the Differential Geometry of Surfaces. VI, 120 pages. 2001.
Vol. 1754: W. Bertram, The Geometry of Jordan and Lie Structures. XVI, 269 pages. 2001.
Vol. 1755: J. Azéma, M. Émery, M. Ledoux, M. Yor, Séminaire de Probabilités XXXV. VI, 427 pages. 2001.
Vol. 1756: P. E. Zhidkov, Korteweg de Vries and Nonlinear Schrödinger Equations: Qualitative Theory. VII, 147 pages. 2001.
Vol. 1757: R. R. Phelps, Lectures on Choquet's Theorem. VII, 124 pages. 2001.
Vol. 1758: N. Monod, Continuous Bounded Cohomology of Locally Compact Groups. X, 214 pages. 2001.
Vol. 1759: Y. Abe, K. Kopfermann, Toroidal Groups. VIII, 133 pages. 2001.
Vol. 1760: D. Filipović, Consistency Problems for Heath-Jarrow-Morton Interest Rate Models. VIII, 134 pages. 2001.
Vol. 1761: C. Adelmann, The Decomposition of Primes in Torsion Point Fields. VI, 142 pages. 2001.
Vol. 1762: S. Cerrai, Second Order PDE's in Finite and Infinite Dimension. IX, 330 pages. 2001.
Vol. 1763: J.-L. Loday, A. Frabetti, F. Chapoton, F. Goichot, Dialgebras and Related Operads. IV, 132 pages. 2001.
Vol. 1764: A. Cannas da Silva, Lectures on Symplectic Geometry. XII, 217 pages. 2001.
Vol. 1765: T. Kerler, V. V. Lyubashenko, Non-Semisimple Topological Quantum Field Theories for 3-Manifolds with Corners. VI, 379 pages. 2001.
Vol. 1766: H. Hennion, L. Hervé, Limit Theorems for Markov Chains and Stochastic Properties of Dynamical Systems by Quasi-Compactness. VIII, 145 pages. 2001.
Vol. 1767: J. Xiao, Holomorphic Q Classes. VIII, 112 pages. 2001.
Vol. 1768: M.J. Pflaum, Analytic and Geometric Study of Stratified Spaces. VIII, 230 pages. 2001.
Vol. 1769: M. Alberich-Carramiñana, Geometry of the Plane Cremona Maps. XVI, 257 pages. 2002.
Vol. 1770: H. Gluesing-Luerssen, Linear Delay-Differential Systems with Commensurate Delays: An Algebraic Approach. VIII, 176 pages. 2002.
Vol. 1771: M. Émery, M. Yor, Séminaire de Probabilités 1967-1980. A Selection in Martingale Theory. IX, 553 pages. 2002.
Vol. 1772: F. Burstall, D. Ferus, K. Leschke, F. Pedit, U. Pinkall, Conformal Geometry of Surfaces in S^4. VII, 89 pages. 2002.
Vol. 1773: Z. Arad, M. Muzychuk, Standard Integral Table Algebras Generated by a Non-real Element of Small Degree. X, 126 pages. 2002.
Vol. 1774: V. Runde, Lectures on Amenability. XIV, 296 pages. 2002.
Vol. 1775: W. H. Meeks, A. Ros, H. Rosenberg, The Global Theory of Minimal Surfaces in Flat Spaces. Martina Franca 1999. Editor: G. P. Pirola. X, 117 pages. 2002.
Vol. 1776: K. Behrend, C. Gomez, V. Tarasov, G. Tian, Quantum Comohology. Cetraro 1997. Editors: P. de Bartolomeis, B. Dubrovin, C. Reina. VIII, 319 pages. 2002.
Vol. 1777: E. García-Río, D. N. Kupeli, R. Vázquez-Lorenzo, Osserman Manifolds in Semi-Riemannian Geometry. XII, 166 pages. 2002.
Vol. 1778: H. Kiechle, Theory of K-Loops. X, 186 pages. 2002.

Vol. 1779: I. Chueshov, Monotone Random Systems. VIII, 234 pages. 2002.

Vol. 1780: J. H. Bruinier, Borcherds Products on O(2,1) and Chern Classes of Heegner Divisors. VIII, 152 pages. 2002.

Vol. 1781: E. Bolthausen, E. Perkins, A. van der Vaart, Lectures on Probability Theory and Statistics. Ecole d' Eté de Probabilités de Saint-Flour XXIX-1999. Editor: P. Bernard. VIII, 466 pages. 2002.

Vol. 1782: C.-H. Chu, A. T.-M. Lau, Harmonic Functions on Groups and Fourier Algebras. VII, 100 pages. 2002.

Vol. 1783: L. Grüne, Asymptotic Behavior of Dynamical and Control Systems under Perturbation and Discretization. IX, 231 pages. 2002.

Vol. 1784: L.H. Eliasson, S. B. Kuksin, S. Marmi, J.-C. Yoccoz, Dynamical Systems and Small Divisors. Cetraro, Italy 1998. Editors: S. Marmi, J.-C. Yoccoz. VIII, 199 pages. 2002.

Vol. 1785: J. Arias de Reyna, Pointwise Convergence of Fourier Series. XVIII, 175 pages. 2002.

Vol. 1786: S. D. Cutkosky, Monomialization of Morphisms from 3-Folds to Surfaces. V, 235 pages. 2002.

Vol. 1787: S. Caenepeel, G. Militaru, S. Zhu, Frobenius and Separable Functors for Generalized Module Categories and Nonlinear Equations. XIV, 354 pages. 2002.

Vol. 1788: A. Vasil'ev, Moduli of Families of Curves for Conformal and Quasiconformal Mappings.IX, 211 pages. 2002.

Vol. 1789: Y. Sommerhäuser, Yetter-Drinfel'd Hopf algebras over groups of prime order. V, 157 pages. 2002.

Vol. 1790: X. Zhan, Matrix Inequalities. VII, 116 pages. 2002.

Vol. 1791: M. Knebusch, D. Zhang, Manis Valuations and Prüfer Extensions I: A new Chapter in Commutative Algebra. VI, 267 pages. 2002.

Vol. 1792: D. D. Ang, R. Gorenflo, V. K. Le, D. D. Trong, Moment Theory and Some Inverse Problems in Potential Theory and Heat Conduction. VIII, 183 pages. 2002.

Vol. 1793: J. Cortés Monforte, Geometric, Control and Numerical Aspects of Nonholonomic Systems. XV, 219 pages. 2002.

Vol. 1794: N. Pytheas Fogg, Substitution in Dynamics, Arithmetics and Combinatorics. Editors: V. Berthé, S. Ferenczi, C. Mauduit, A. Siegel. XVII, 402 pages. 2002.

Vol. 1795: H. Li, Filtered-Graded Transfer in Using Noncommutative Gröbner Bases. IX, 197 pages. 2002.

Vol. 1796: J.M. Melenk, hp-Finite Element Methods for Singular Perturbations. XIV, 318 pages. 2002.

Vol. 1797: B. Schmidt, Characters and Cyclotomic Fields in Finite Geometry. VIII, 100 pages. 2002.

Vol. 1798: W.M. Oliva, Geometric Mechanics. XI, 270 pages. 2002.

Vol. 1799: H. Pajot, Analytic Capacity, Rectifiability, Menger Curvature and the Cauchy Integral. XII,119 pages. 2002.

Vol. 1801: J. Azéma, M. Émery, M. Ledoux, M. Yor, Séminaire de Probabilités XXXVI. VIII, 499 pages. 2003.

Vol. 1802: V. Capasso, E. Merzbach, B.G. Ivanoff, M. Dozzi, R. Dalang, T. Mountford, Topics in Spatial Stochastic Processes. Martina Franca, Italy 2001. Editor: E. Merzbach. VIII, 253 pages. 2003.

Vol. 1803: G. Dolzmann, Variational Methods for Crystalline Microstructure - Analysis and Computation. VIII, 212 pages. 2003.

Vol. 1804: I. Cherednik, Ya. Markov, R. Howe, G. Lusztig, Iwahori-Hecke Algebras and their Representation Theory. Martina Franca, Italy 1999. Editors: V. Baldoni, D. Barbasch. X, 103 pages. 2003.

Vol. 1805: F. Cao, Geometric Curve Evolution and Image Processing. X, 187 pages. 2003.

Vol. 1806: H. Broer, I. Hoveijn. G. Lunther, G. Vegter, Bifurcations in Hamiltonian Systems. Computing Singularities by Gröbner Bases. XIV, 169 pages. 2003.

Vol. 1807: V. D. Milman, G. Schechtman, Geometric Aspects of Functional Analysis. Israel Seminar 2000-2002. VIII, 429 pages. 2003.

Vol. 1808: W. Schindler, Measures with Symmetry Properties.IX, 167 pages. 2003.

Vol. 1809: O. Steinbach, Stability Estimates for Hybrid Coupled Domain Decomposition Methods. VI, 120 pages. 2003.

Vol. 1810: J. Wengenroth, Derived Functors in Functional Analysis. VIII, 134 pages. 2003.

Vol. 1811: J. Stevens, Deformations of Singularities. VII, 157 pages. 2003.

Vol. 1812: L. Ambrosio, K. Deckelnick, G. Dziuk, M. Mimura, V. A. Solonnikov, H. M. Soner, Mathematical Aspects of Evolving Interfaces. Madeira, Funchal, Portugal 2000. Editors: P. Colli, J. F. Rodrigues. X, 237 pages. 2003.

Vol. 1813: L. Ambrosio, L. A. Caffarelli, Y. Brenier, G. Buttazzo, C. Villani, Optimal Transportation and its Applications. Martina Franca, Italy 2001. Editors: L. A. Caffarelli, S. Salsa. X, 164 pages. 2003.

Vol. 1815: A. M. Vershik, Asymptotic Combinatorics with Applications to Mathematical Physics. St. Petersbutg, Russia 2001. IX, 246 pages. 2003.

Vol. 1816: S. Albeverio, W. Schachermayer, M. Talagrand, Lectures on Probability Theory and Statistics. Ecole d' Eté de Probabilités de Saint-Flour XXX-2000. Editor: P. Bernard. VIII, 296 pages. 2003.

Vol. 1817: E. Koelink, Orthogonal Polynomials and Special Functions. Leuven 2002. X, 249 pages. 2003.

Vol. 1818: M. Bildhauer, Convex Variational Problems with Linear, nearly Linear and/or Anisotropic Growth Conditions. X, 217 pages. 2003.

Vol. 1819: D. Masser, Yu. V. Nesterenko, H. P. Schlickewei, W. M. Schmidt, M. Waldschmidt, Diophantine Approximation. Cetraro, Italy 2000. Editors: F. Amoroso, U. Zannier. XI,353 pages. 2003.

Recent Reprints and New Editions

Vol. 1200: V. D. Milman, G. Schechtman, Asymptotic Theory of Finite Dimensional Normed Spaces. 1986. – Corrected Second Printing. X, 156 pages. 2001.

Vol. 1618: G. Pisier, Similarity Problems and Completely Bounded Maps. 1995 – Second, Expanded Edition VII, 198 pages. 2001.

Vol. 1629: J. D. Moore, Lectures on Seiberg-Witten Invariants. 1997 – Second Edition. VIII, 121 pages. 2001.

Vol. 1638: P. Vanhaecke, Integrable Systems in the realm of Algebraic Geometry. 1996 – Second Edition. X, 256 pages. 2001.

Vol. 1702: J. Ma, J. Yong, Forward-Backward Stochastic Differential Equations and Their Applications. 1999. – Corrected Second Printing. XIII, 270 pages. 2000.